Monographs in Mathematics
Vol. 89

Herbert Amann

Linear and Quasilinear Parabolic Problems

Volume I
Abstract Linear Theory

1995

Birkhäuser Verlag
Basel · Boston · Berlin

Author:
Institut für Mathematik
Universität Zürich
Winterthurerstrasse 190
8057 Zürich
Switzerland

A CIP catalogue record for this book is available from the Library of Congress, Washington D.C., USA

Die Deutsche Bibliothek – CIP-Einheitsaufnahme
Amann, Herbert:
Linear and quasilinear parabolic problems / Herbert Amann. –
Basel ; Boston ; Berlin : Birkhäuser
Vol. 1. Abstract linear theory. – 1995
(Monographs in mathematics ; Vol. 89)
ISBN 3-7643-5114-4 (Basel...)
ISBN 0-8176-5114-4 (Boston)

Printed on acid-free paper produced of chlorine-free pulp
Printed in Germany
ISBN 3-7643-5114-4
ISBN 0-8176-5114-4

9 8 7 6 5 4 3 2 1

Aber — so fragen wir — wird es bei der Ausdehnung des mathematischen Wissens für den einzelnen Forscher nicht schließlich unmöglich, alle Teile dieses Wissens zu umfassen? Ich möchte als Antwort darauf hinweisen, wie sehr es im Wesen der mathematischen Wissenschaft liegt, daß jeder wirkliche Fortschritt stets Hand in Hand geht mit der Auffindung schärferer Hilfsmittel und einfacherer Methoden, die zugleich das Verständnis früherer Theorien erleichtern und umständliche ältere Entwicklungen beseitigen, und daß es daher dem einzelnen Forscher, indem er sich diese schärferen Hilfsmittel und einfacheren Methoden zu eigen macht, leichter gelingt, sich in den verschiedenen Wissenszweigen der Mathematik zu orientieren, als dies für irgend eine andere Wissenschaft der Fall ist.[1]

David Hilbert (1862–1943)

[1]But — so we ask — given the expansion of mathematical knowledge, will it eventually not be impossible for the individual researcher to encompass all parts of this knowledge? As an answer I want to point out how much it lies in the character of mathematical science that all real progress is intimately tied to the discovery of sharper tools and simpler methods that, at the same time, facilitate the comprehension of earlier theories and remove complicated older developments, and that therefore the researcher, adopting these sharper tools and simpler methods, succeeds in getting more easily acquainted with the diverse branches of mathematics than this is the case for any other field of science.

Preface

In this treatise we present the semigroup approach to quasilinear evolution equations of parabolic type that has been developed over the last ten years, approximately. It emphasizes the dynamic viewpoint and is sufficiently general and flexible to encompass a great variety of concrete systems of partial differential equations occurring in science, some of those being of rather 'nonstandard' type. In particular, to date it is the only general method that applies to noncoercive systems.

Although we are interested in nonlinear problems, our method is based on the theory of linear holomorphic semigroups. This distinguishes it from the theory of nonlinear contraction semigroups whose basis is a nonlinear version of the Hille-Yosida theorem: the Crandall-Liggett theorem. The latter theory is well-known and well-documented in the literature. Even though it is a powerful technique having found many applications, it is limited in its scope by the fact that, in concrete applications, it is closely tied to the maximum principle. Thus the theory of nonlinear contraction semigroups does not apply to systems, in general, since they do not allow for a maximum principle. For these reasons we do not include that theory.

Our approach is strongly motivated by the concept of weak solutions of differential equations. In fact, as one of the applications of our general results we eventually develop a theory of weak solutions for noncoercive quasilinear parabolic systems in divergence form in an L_p-setting. This is in contrast to the standard L_2-setting for coercive problems, that is, unfortunately, not suitable for noncoercive systems. Moreover, even in regular situations, where, in principle, we could work directly within the framework of strong solutions, the theory of weak solutions is of great importance. For instance, in connection with global existence it allows for a priori estimates in 'weak norms', which facilitates the establishing of those bounds considerably. For this reason we develop a general 'reflexive' theory taylored for applications in an L_p-setting. We treat evolution equations in spaces of continuous functions only marginally. An exposition of the latter theory, emphasizing fully nonlinear problems and strong solutions, can be found in the monograph by A. Lunardi [Lun95].

In order to obtain results that are sufficiently general and flexible to be applicable to a wide variety of concrete problems, we need a considerable amount of preparation. For this reason our treatise is divided in three volumes, carrying the respective titles:

Abstract Linear Theory,

Function Spaces and Linear Differential Operators,

Nonlinear Problems.

In the first volume we give a thorough discussion of linear parabolic evolution equations in general Banach spaces. This is the abstract basis for the nonlinear theory. The second volume is devoted to concrete realizations of linear parabolic evolution equations by general parabolic systems. There we discuss the various function spaces that are needed and useful, and the generation of analytic semigroups by general elliptic boundary value problems. The last volume contains the abstract nonlinear theory as well as various applications to concrete systems, illustrating the scope and the flexibility of the general results. Of course, each one of the three volumes contains much material of independent interest related to our main subject.

In writing this book I had help from many friends, collegues, and students. It is a pleasure to thank all of them, named or unnamed. I am particularly indebted to P. Quittner and G. Simonett, who critically and very carefully read, not only the whole manuscript of this first volume but also many earlier versions that were produced over the years and will never be published, and pointed out numerous mistakes and improvements. Large parts of the first volume, and of earlier versions as well, were also read and commented on by D. Daners, J. Escher, and P. Guidotti. Their constructive criticism, observations, and suggestions for improvements were enormously helpful. Of course, I am solely responsible for all remaining mistakes.

My son Andreas gave expert advice for taming the computer and kept open an emergency line. Finally, this book would never have appeared without the invaluable help of my 'comma sniffer', whose contributions are visible on every page. My heartfelt thanks go to both of them.

Through many years I obtained financial support by Schweizerischer Nationalfonds, that is gratefully acknowledged. It enabled me to maintain an active group in this field of research and to bring in visitors from outside. These contacts were enormously beneficial for my work. I also express my gratitude to Birkhäuser Verlag, in particular to Th. Hintermann, for the agreeable collaboration.

Zürich, December 1994 Herbert Amann

Contents

Preface vii

Introduction xiii

Notations and Conventions

1	Topological Spaces	1
2	Locally Convex Spaces	2
3	Complexifications	4
4	Unbounded Linear Operators	6
5	General Conventions	7

Chapter I Generators and Interpolation

1 Generators of Analytic Semigroups

1.1	Properties of Linear Operators	10
1.2	The Class $\mathcal{H}(E_1, E_0)$	11
1.3	Perturbation Theorems	14
1.4	Spectral Estimates	15
1.5	Compact Perturbations	20
1.6	Matrix Generators	21

2 Interpolation Functors

2.1	Definitions	24
2.2	Interpolation Inequalities	25
2.3	Retractions	26
2.4	Standard Interpolation Functors	28
2.5	Continuous Injections	30
2.6	Duality Properties	30
2.7	Compactness	31
2.8	Reiteration Theorems	31
2.9	Fractional Powers and Interpolation	32
2.10	Semigroups and Interpolation	33
2.11	Admissible Interpolation Functors	35

Chapter II Cauchy Problems and Evolution Operators

1 Linear Cauchy Problems

1.1 Hölder Spaces 40
1.2 Existence and Regularity Theorems 43

2 Parabolic Evolution Operators

2.1 Basic Properties 45
2.2 Determining Integral Equations 47

3 Linear Volterra Integral Equations

3.1 Weakly Singular Kernels 48
3.2 Resolvent Kernels 50
3.3 Singular Gronwall Inequalities 52

4 Existence of Evolution Operators

4.1 A Class of Parameter Integrals 53
4.2 Semigroup Estimates 55
4.3 Construction of Evolution Operators 57
4.4 The Main Result 63
4.5 Solvability of the Cauchy Problem 66

5 Stability Estimates

5.1 Estimates for Evolution Operators 68
5.2 Continuity Properties of Mild Solutions 71
5.3 Hölder Estimates 72
5.4 Boundedness of Mild Solutions 74

6 Invariance and Positivity

6.1 Yosida Approximations 75
6.2 Approximations of Evolution Operators 77
6.3 Invariance 80
6.4 Orderings and Positivity 84

Chapter III Maximal Regularity

1 General Principles

1.1 Sobolev Spaces 88
1.2 Absolutely Continuous Functions 89
1.3 Generalized Solutions 91
1.4 Trace Spaces 92
1.5 Pairs of Maximal Regularity 94
1.6 Stability 96

2 Maximal Hölder Regularity

2.1 Singular Hölder Spaces 98
2.2 Semigroup Estimates 102
2.3 Trace Spaces 106
2.4 Estimates for K_A 109
2.5 Maximal Regularity 113
2.6 Nonautonomous Problems 117

3 Maximal Continuous Regularity

3.1 Necessary Conditions 121
3.2 Higher Order Interpolation Spaces 123
3.3 Estimates for K_A 124
3.4 Maximal Regularity 126

4 Maximal Sobolev Regularity

4.1 Temperate Distributions 128
4.2 Fourier Transforms and Convolutions 130
4.3 The Hilbert Transform 135
4.4 UMD Spaces and Fourier Multipliers 141
4.5 Properties of UMD Spaces 144
4.6 Fractional Powers 147
4.7 Bounded Imaginary Powers 162
4.8 Perturbation Theorems 168
4.9 Sums of Closed Operators 173
4.10 Maximal Regularity 180

Chapter IV Variable Domains

1 Higher Regularity

1.1 Properties of Differentiable Functions 194
1.2 General Solvability Results for Cauchy Problems 195
1.3 Estimates for Evolution Operators 198
1.4 Evolution Operators on Interpolation Spaces 204
1.5 The Cauchy Problem 207

2 Constant Interpolation Spaces

2.1 Semigroup and Convergence Estimates 211
2.2 Assumptions and Consequences 214
2.3 Construction of Evolution Operators 218
2.4 Estimates for Evolution Operators 227
2.5 The Cauchy Problem 230
2.6 Abstract Boundary Value Problems 233

3 Maximal Regularity

3.1 Abstract Initial Boundary Value Problems 242
3.2 Isomorphism Theorems 245

Chapter V Scales of Banach Spaces

1 Banach Scales

1.1 General Concepts 250
1.2 Power Scales 255
1.3 Extrapolation Spaces 261
1.4 Dual Scales 267
1.5 Interpolation-Extrapolation Scales 275

2 Evolution Equations in Banach Scales

2.1 Semigroups in Interpolation-Extrapolation Scales 286
2.2 Parabolic Evolution Equations in Banach Scales 294
2.3 Duality 297
2.4 Approximation Theorems 300
2.5 Final Value Problems 304
2.6 Weak Solutions and Duality 307
2.7 Positivity 312
2.8 General Evolution Equations 314

Bibliography 321

List of Symbols 329

Index 333

Nichts setzt dem Fortgang der Wissenschaft mehr Hindernis entgegen als wenn man zu wissen glaubt, was man noch nicht weiß.[1]

Georg Christoph Lichtenberg (1742–1799)

Introduction

Partial differential equations of parabolic type are encountered in a variety of problems in mathematics, physics, chemistry, biology, and many other scientific subjects in which irreversible processes can be adequately described by mathematical models. For this reason parabolic equations have been thoroughly studied and there is a considerable mathematical literature in this field. However, most of the research has been concentrated on the study of a single second order parabolic equation for one scalar-valued unknown — at least as far as nonlinear equations are concerned — and on certain particular systems for a vector-valued unknown describing specific physical situations. The Navier-Stokes equations of hydrodynamics are among the most eminent representatives of the latter class.

During the last two or three decades, so-called reaction-diffusion equations have become a much favored object of study by application-oriented analysts. In contrast to the classical investigations in the theory of partial differential equations, that concentrate on questions of existence and uniqueness, there has been developed a more qualitative, dynamical-systems-type approach to reaction-diffusion equations. The basic idea of this method is to interprete the partial differential equation as an ordinary differential equation in an infinite-dimensional Banach space. This assigns a predominant rôle to the time variable and relegates the spacial dependence to the set-up, that is, to the correct choice of the underlying function spaces and to the properties of the operators representing the partial differential equations. Having found the right frame for this description one can try to mimic the finite-dimensional theory of ordinary differential equations to obtain information on the long-time behavior of solutions, their stability properties, bifurcation phenomena, etc., questions of paramount interest in applications.

The ordinary-differential-equations-approach to time-dependent partial differential equations has proven to be very powerful. It is by no means restricted to simple semilinear reaction-diffusion equations as they are studied in the literature most often. In fact, it is one of the main purposes of this treatise to extend this approach to general quasilinear parabolic systems encompassing a great variety of concrete equations from science. In addition, by our abstract approach we

[1]Nothing impedes the progress of science more than believing to know what one does not know yet.

are rewarded with a general flexible theory that is also applicable to many other problems not belonging to the class of parabolic systems in the narrow sense.

In the following, we describe our approach, as nontechnically as possible, and indicate the difficulties and problems that have to be resolved. By this way we are weaving a silver thread leading the reader through our treatise.

Semilinear Reaction-Diffusion Equations

Let X be a bounded open subset of $\mathbb{R}^n$ with smooth boundary ∂X, lying locally on one side of X. Most naturally, reaction-diffusion equations are derived from conservation laws of the form

$$\partial_t u + \operatorname{div} j = r \quad \text{in } X\ , \qquad t > 0\ , \tag{1}$$

by specifying the 'flux vector' j by means of phenomenological constitutive relations like

$$j := j(u) := -D \operatorname{grad} u - du\ . \tag{2}$$

Here r, the 'reaction rate', is a given smooth function of $(x,t) \in \overline{X} \times \mathbb{R}^+$ and the scalar-valued unknown u. The 'diffusion matrix' $D\colon \overline{X} \to \mathbb{R}^{n \times n}$ and the 'drift vector' $d\colon \overline{X} \to \mathbb{R}^n$ are also smooth, and $D(x)$ is symmetric and positive definite, uniformly with respect to $x \in \overline{X}$.

In concrete situations u may represent a concentration, a density, a temperature, or some other physical or mathematical quantity. Then (1) amounts to a mathematical formulation of the law of conservation of mass, if u is a concentration or a density, or of energy, if u is a temperature (and certain simplifications and constitutive assumptions for the entropy are imposed), etc. Moreover, in the very special case that D is a positive multiple of the identity matrix and $d = 0$, the constitutive relation (2) reduces to Fick's law or Darcy's law (depending on the model) if u is a concentration or a density, and to Fourier's law if u is the temperature, etc.

In addition to (1) and (2), the behavior of u on the boundary of X has to be specified. This can be done by prescribing the value of u on ∂X. By normalizing the boundary values, this condition can be formulated as a homogeneous Dirichlet condition:

$$u = 0 \quad \text{on } \partial X\ , \qquad t > 0\ . \tag{3}$$

Another possibility, being of chief importance in applications, consists of prescribing the flux through the boundary. Denoting by ν the outer unit-normal vector field on ∂X, the simplest situation occurs at an insulated boundary modeled by the 'no-flux' condition

$$\nu \cdot j(u) = 0 \quad \text{on } \partial X\ , \qquad t > 0\ . \tag{4}$$

Of course, there are situations where (3) occurs on a part $\partial_0 X$ of ∂X only and on the remaining part, $\partial_1 X := \partial X \backslash \partial_0 X$, the no-flux condition (4) is effective.

We always assume that $\partial_j X$ is open and closed in ∂X for $j = 0, 1$. This situation can also be described by fixing a continuous map $\delta : \partial X \to \{0, 1\}$, a 'boundary characterization map', and by letting

$$\partial_j X := \delta^{-1}(j) , \qquad j = 0, 1 .$$

Note that either one of the boundary parts $\partial_0 X$ and $\partial_1 X$ may be empty. Then we can formulate the more general boundary condition, thereby encompassing (3) and (4), as

$$-\delta\nu \cdot j(u) + (1 - \delta)u = 0 \quad \text{on } \partial X , \qquad t > 0 . \tag{5}$$

Lastly, in order to determine the time-evolution of u from (1), (2), and (5), that is, the functions $u(\cdot, t) : \overline{X} \to \mathbb{R}$ for $t > 0$, we have to specify its initial distribution:

$$u(\cdot, 0) = u^0 \qquad \text{on } \overline{X} . \tag{6}$$

By substituting (2) in (1) and (5), we can rewrite (1), (2), (5), and (6) as an initial-boundary value problem:

$$\left.\begin{aligned} \partial_t u + \mathcal{A}u &= r(\cdot, t, u) && \text{in } X \\ \mathcal{B}u &= 0 && \text{on } \partial X \end{aligned}\right\} \quad t > 0 , \qquad u(\cdot, 0) = u^0 \quad \text{on } \overline{X} . \tag{7}$$

Here we have put

$$\mathcal{A}u := -\operatorname{div}(D \operatorname{grad} u + du) \tag{8}$$

and

$$\mathcal{B}u := \delta\big(\nu \cdot D \operatorname{grad} u + (\nu \cdot d)u\big) + (1 - \delta)u . \tag{9}$$

Of course, the 'boundary operator' $\mathcal{B}$ has to be interpreted in the sense of traces. Note that

$$\nu \cdot D \operatorname{grad} u = D\nu \cdot \operatorname{grad} u = \partial_{D\nu} u$$

is the derivative with respect to the outer co-normal $D\nu$ on ∂X. Thus, in the very special case that D is the identity matrix, $d = 0$, and r is independent of t, system (7) reduces to an initial-boundary value problem for the autonomous semilinear heat equation:

$$\left.\begin{aligned} \partial_t u - \Delta u &= r(\cdot, u) && \text{in } X \\ u &= 0 && \text{on } \partial_0 X \\ \partial_\nu u &= 0 && \text{on } \partial_1 X \\ u(\cdot, 0) &= u^0 && \text{on } \overline{X} , \end{aligned}\right\} \quad t > 0 , \tag{10}$$

a problem having attracted much attention in the literature.

The Banach Space Formulation

In order to reformulate (7) as an ordinary differential equation we have to choose our basic Banach space E_0 in which we want to analyze the problem. Of course, E_0 will be a Banach space of distributions on X, that is, E_0 is a Banach space such that[2]

$$E_0 \hookrightarrow \mathcal{D}'(X) \ . \tag{11}$$

Next we define a linear operator A in E_0 by

$$\mathrm{dom}(A) := \{\, v \in E_0 \ ; \ \mathcal{A}v \in E_0 \text{ and } \mathcal{B}v = 0 \,\} \ , \qquad Av := \mathcal{A}v \ . \tag{12}$$

We also identify

$$u \colon X \times \mathbb{R}^+ \to \mathbb{R} \quad \text{and} \quad \big[t \mapsto u(\cdot,t)\big] : \mathbb{R}^+ \to \mathbb{R}^X$$

and denote by f the Nemytskii operator induced by r, that is, we put

$$f(t,u) := r\big(\cdot,t,u(\cdot)\big) \ , \qquad (t,u) \in \mathbb{R}^+ \times \mathbb{R}^X \ .$$

Then the initial-boundary value problem (7) can formally[3] be rewritten as an initial value problem for an ordinary differential equation in E_0:

$$\dot{u} + Au = f(t,u) \ , \quad t > 0 \ , \qquad u(0) = u^0 \ . \tag{13}$$

This follows from the fact that the boundary condition $\mathcal{B}u = 0$ has been incorporated in the domain of the linear operator A.

Of course, in order to render this procedure meaningful and to get a treatable abstract initial value problem (13) we have to impose certain minimal requirements. As for the linear operator A, we request that

$$\left.\begin{array}{l} A \text{ is closed and densely defined in } E_0, \\ \text{having a nonempty resolvent set.} \end{array}\right\} \tag{14}$$

Then, denoting by E_1 the domain of A, endowed with its graph norm, we see that

$$E_1 \overset{d}{\hookrightarrow} E_0 \ ,$$

that is, (E_0, E_1) is a 'densely injected Banach couple'. As for the nonlinearity f, we require that

$$\left.\begin{array}{l} f \in C(\mathbb{R}^+ \times E_1, E_0), \text{ and} \\ f(t,\cdot) : E_1 \to E_0 \text{ is (locally) Lipschitz continuous,} \\ \text{uniformly with respect to } t \text{ in bounded subintervals of } \mathbb{R}^+. \end{array}\right\} \tag{15}$$

---

[2]Cf. the sections 'Notations and Conventions' and 'List of Symbols' for the notations and definitions used without explanation in this introduction.

[3]Observe that our notation is inconsistent as far as we exhibit the time variable in the nonlinearity. Thus (13) is a formal relation only. In order to give it a precise meaning we have to define what is meant by a solution. Formal notations of this type are very suggestive and useful in the theory of differential equations and we use them throughout without fearing confusion.

These minimal conditions impose restrictions on the choice of the Banach space E_0. Observe from (12) that the distributions in $\operatorname{dom}(A)$, that is, in E_1, have to be regular enough to admit the traces $v \mapsto v|\partial_0 X$ and $v \mapsto \partial_{A\nu} v$ on $\partial_0 X$ and $\partial_1 X$, respectively. Hence the Banach space E_1 has to consist of sufficiently regular distributions. Since A is supposed to have a nonempty resolvent set, this requires, in turn, the Banach space E_0 to be not 'too large'. This stipulation is reinforced by the minimal requirements for f.

Except for the above somewhat implicit restrictions we are free in the choice of E_0. Of course, we have to keep in mind that, by fixing the space E_0, we may not rediscover all solutions of problem (7) as solutions of the abstract equation (13). This can be the case if we choose E_0, and thus E_1, to be 'too small', that is, if we require the elements of E_1 to be too regular. Of this danger one has to be aware, in particular, in the case where X is an unbounded domain, say $X = \mathbb{R}^n$ (a case not considered in this introduction), since the very definition of the Banach space E_0 often incorporates restrictions on the behavior of its elements 'near infinity'.

Using the relative freedom in the choice of E_0, we opt for simplicity. This means that we select spaces that are easy to describe and handle. At first sight the space $C := C(\overline{X})$ of continuous functions on $\overline{X}$ seems to be a good candidate. However, letting $E_0 := C$, there is no better description of $\operatorname{dom}(A)$ than the one of (12). In other words, although it is true that A is closed and has a nonempty resolvent set, the space E_1 does not coincide with any of the known function spaces. In particular, E_1 does not coincide with

$$C^2_{\mathcal{B}} := \left\{ v \in C^2(\overline{X}) \ ; \ \mathcal{B}v = 0 \right\} ,$$

but is a proper superspace thereof. Moreover, E_1 is not dense in E_0 if $\partial_0 X \neq \emptyset$. (The density condition is not indispensable for some parts of the general theory (cf. [Lun95]), but it is essential for others.) In addition, the domain of A depends on the diffusion matrix D in the sense that, in general, distinct (even constant) diffusion matrices D_1 and D_2 give rise to distinct domains $\operatorname{dom}(A_1) \neq \operatorname{dom}(A_2)$ of the corresponding operators induced by (8), (9), and (12) (cf. [Sob89]). Lastly, though the space C is rather simple from the analytical point of view, it is nonreflexive and thus lacks a very desirable and useful functional-analytical property.

The next class of simple spaces that comes to our mind is the class of Lebesgue spaces $L_p(X)$, $1 \le p \le \infty$. Since the spaces $L_1(X)$ and $L_\infty(X)$ show essentially the same 'deficiencies' as the space of continuous functions (cf. [Gui93]), we are naturally led to put

$$E_0 := L_p := L_p(X) \qquad \text{for some } p \in (1, \infty) \ . \tag{16}$$

In this case it turns out that the minimal requirement (14) is satisfied. Moreover, the space E_1 can be described explicitly by

$$E_1 \doteq W^2_{p,\mathcal{B}} := \left\{ v \in W^2_p(X) \ ; \ \mathcal{B}v = 0 \right\} . \tag{17}$$

Note that E_1 is independent of A if $\partial_0 X = \partial X$. But it does depend on A if $\partial_1 X \neq \emptyset$ (through the condition $\partial_{D\nu} u + (\nu \cdot d)u = 0$ on $\partial_1 X$).

As for the minimal requirement (15), we first recall Sobolev's embedding theorem:

$$W_p^s := W_p^s(X) \hookrightarrow \begin{cases} L_q & \text{if } 1/p \geq 1/q \geq 1/p - s/n \ , \\ C^\rho & \text{if } 0 \leq \rho \leq s - n/p \ \text{ and } \ s > n/p \ , \end{cases} \tag{18}$$

where $s, q \in \mathbb{R}^+$ and $\rho < s - n/p$ if $s - n/p \in \mathbb{N}$. Second, given a continuous function $g \colon \overline{X} \times \mathbb{R} \to \mathbb{R}$, it is known that the Nemytskii operator induced by g maps L_q into L_p iff it satisfies an estimate of the form

$$|g(x,\xi)| \leq c(1 + |\xi|^{q/p}) \ , \qquad (x,\xi) \in \overline{X} \times \mathbb{R} \ . \tag{19}$$

Moreover, to guarantee the Lipschitz continuity of this superposition operator one needs a polynomial growth restriction for $\partial_2 g$ in addition. Thus we see from $E_1 \hookrightarrow W_p^2$ and (18), (19) that, given $t \in \mathbb{R}^+$, the minimal requirement (15) implies a polynomial bound for the map $\xi \mapsto r(\cdot, t, \xi)$, if $p \leq n/2$. On the other hand, superposition operators have good properties in C. More precisely, if $g \in C^k(\overline{X} \times \mathbb{R}, \mathbb{R})$ for some $k \in \mathbb{N}$,

$$\big[u \mapsto g(\cdot, u(\cdot))\big] \in C^k\big(C(\overline{X}), C(\overline{X})\big) \ . \tag{20}$$

Thus, if we impose the condition $p > n/2$, it follows from $E_1 \hookrightarrow C \hookrightarrow E_0$ that the minimal requirement (15) is met.

Semilinear Parabolic Evolution Equations

The proof of the validity of (14) can be arranged to give more. Namely, it can be shown that the resolvent set of $-A$ contains a half-plane $[\operatorname{Re}\lambda \geq \lambda_0]$ for some $\lambda_0 \in \mathbb{R}$ and that there exists a constant κ such that

$$|\lambda| \, \|u\|_0 \leq \kappa \, \|(\lambda + A)u\|_0 \ , \qquad u \in E_1 \ , \quad \operatorname{Re}\lambda \geq \lambda_0 \ , \tag{21}$$

where $\|\cdot\|_0$ is the norm in E_0. This, together with (14), is equivalent to the assertion that $-A$ generates a strongly continuous analytic semigroup $\{\, e^{-tA} \ ; \ t \geq 0 \,\}$ on E_0. Thus

$$A \in \mathcal{H}(E_1, E_0) \ , \tag{22}$$

that is, $A \in \mathcal{L}(E_1, E_0)$ and $-A$, considered as an unbounded linear operator in E_0 with domain E_1, is the infinitesimal generator of a strongly continuous analytic semigroup on E_0.

Now suppose that J is a perfect compact subinterval of $\mathbb{R}^+$ containing 0 and that u is a solution of (13) on J. This means that, letting $\dot{J} := J\backslash\{0\}$,

$$u \in C(\dot{J}, E_1) \cap C^1(\dot{J}, E_0) \cap C(J, E_0) \tag{23}$$

and u satisfies (13) point-wise, that is, $\dot{u}(t) + Au(t) = f\big(t, u(t)\big)$ for $t \in \dot{J}$ and $u(0) = u^0$. Then — similarly as in the theory of ordinary differential equations — the variation-of-constants formula implies that

$$u(t) = e^{-tA}u^0 + \int_0^t e^{-(t-\tau)A} f\big(\tau, u(\tau)\big)\, d\tau\ , \qquad t \in J\ . \tag{24}$$

Note, however, that this similarity is a formal one only, since formula (24) does not make sense, given our assumptions. Indeed, it follows from (23) and (15) that

$$\big[\tau \mapsto f\big(\tau, u(\tau)\big)\big] \in C(\dot{J}, E_0)\ .$$

By combining this with the strong continuity of the semigroup $\{\, e^{-tA}\ ;\ t \geq 0\,\}$ on E_0, we see that

$$\big[\tau \mapsto e^{-(t-\tau)A} f\big(\tau, u(\tau)\big)\big] \in C\big((0,t], E_0\big)\ , \qquad t \in \dot{J}\ . \tag{25}$$

But this does not guarantee the existence of the integral in (24).

Clearly, there is an easy remedy. Namely, it suffices to strengthen the concept of a solution by replacing (23) by

$$u \in C(J, E_1) \cap C^1(J, E_0)\ . \tag{26}$$

This requires, of course, that we choose the initial value u^0 in E_1. From (26) we infer that (25) holds if $(0,t]$ is replaced by $[0,t]$. Then the right-hand side of (24) is well-defined and equation (24) holds in E_0.

Observe that (24) is a fixed point equation for u. More precisely, define

$$\Phi : C(J, E_1) \to C(J, E_0)$$

by

$$\Phi(u)(t) := e^{-tA}u^0 + \int_0^t e^{-(t-\tau)A} f\big(\tau, u(\tau)\big)\, d\tau\ , \qquad t \in J\ .$$

Then, since $E_1 \hookrightarrow E_0$ implies $C(J, E_1) \hookrightarrow C(J, E_0)$, we see from (24) that

$$u = \Phi(u) \tag{27}$$

in $C(J, E_0)$, provided u satisfies (26) and is a solution of (13) on J. Now, similarly as in the theory of ordinary differential equations, we would like to use this fixed point equation as a basis for an existence theory for the initial value problem (13). The advantage of such an approach is obvious: whereas (13) involves the unbounded linear operator A and the derivative $\dot{u}$ of u, in (27) there occur bounded linear operators only and no derivative at all. As a result, the nonlinear map Φ is rather easy to deal with. In fact, it is a simple consequence of (15) and (22) that Φ is a Lipschitz continuous map from $C(J, E_1)$ into $C(J, E_0)$. Furthermore, in a neighborhood of the constant map $(t \mapsto u^0) \in C(J, E_1)$, where Φ is uniformly Lipschitz continuous, the Lipschitz constant can be controlled and made smaller than 1 by making the interval J sufficiently small. Thus it is tempting to prove the existence of a fixed point of Φ by means of the method of successive iterations. However, this approach is bound to fail already at the first step since the image, u_1, of a function $u_0 \in C(J, E_1)$ under Φ is only known to lie in $C(J, E_0)$, so that u_1 is not necessarily in the domain of Φ, and $\Phi(u_1)$ is not defined, in general.

At this stage we recall that the semigroup generated by $-A$ is analytic, a property that has not been used so far. Analyticity implies that the image of e^{-tA} is contained in the domain of A whenever $t > 0$. More precisely, the map

$$(0, \infty) \to \mathcal{L}(E_0, E_1) \ , \quad t \mapsto e^{-tA} \tag{28}$$

is well-defined and analytic. This is an important smoothing property of analytic semigroups. It is the abstract counterpart to the well-known fact that parabolic differential equations regularize the initial data. This effect is seen most clearly in the case of the heat equation on $\mathbb{R}^n$:

$$\partial_t u - \Delta u = 0 \quad \text{in } \mathbb{R}^n \ , \qquad t > 0 \ , \qquad u(\cdot, 0) = u^0 \ .$$

Here the solution, being represented by the heat semigroup as $u(\cdot, t) = e^{t\Delta} u_0$ for $t \geq 0$, is analytic in $\mathbb{R}^n \times (0, \infty)$, even if u^0 belongs to $L_p(\mathbb{R}^n)$ only.

Of course, the map (28) cannot be continuous on $\mathbb{R}^+$. But it is known that there exist constants c and $\omega \in \mathbb{R}$ such that

$$\|e^{-tA}\|_{\mathcal{L}(E_0, E_1)} \leq c t^{-1} e^{\omega t} \ , \qquad t > 0 \ , \tag{29}$$

and this estimate is sharp, in general. At first sight this is not of much help: although it follows from (15) and (28) that

$$\left[\tau \mapsto e^{-(t-\tau)A} f\big(\tau, u(\tau)\big)\right] \in C\big((0, t], E_1\big) \ , \qquad t \in \dot{J} \ , \tag{30}$$

for $u \in C(J, E_1)$, thanks to (29) we cannot guarantee that (30) is integrable on J. Thus our minimal assumptions do not guarantee that Φ maps $C(J, E_1)$ into itself.

Suppose, however, there exist $\alpha \in (0, 1)$ and a Banach space E_α satisfying

$$E_1 \overset{d}{\hookrightarrow} E_\alpha \overset{d}{\hookrightarrow} E_0 \tag{31}$$

such that the estimate

$$\|e^{-tA}\|_{\mathcal{L}(E_0, E_\alpha)} \leq c t^{-\alpha} e^{\omega t} \ , \qquad t > 0 \ , \tag{32}$$

is valid and the semigroup $\{ e^{-tA} \ ; \ t \geq 0 \}$ on E_0 restricts to a strongly continuous semigroup on E_α. Also suppose that we strengthen the minimal requirement (15) — then calling it $(15)_\alpha$ — by replacing E_1 by E_α. Then it is easy to see that Φ is a well-defined Lipschitz continuous self-map of $C(J, E_\alpha)$ and that the fixed point approach indicated above can be carried through without difficulties. This amounts to a completely satisfactory theory of abstract semilinear parabolic evolution equations of the form (13). In particular, it can be shown that (13) generates a (local) semiflow $(t, u^0) \mapsto u(t, u^0)$ on the phase space E_α, where $u(\cdot, u^0)$ is the unique maximal solution of (13).

It is well-known that $E_\alpha := D\big((\omega + A)^\alpha\big)$, where $(\omega + A)^\alpha$ is the αth 'fractional power' of $\omega + A$ and ω is a sufficiently large real number, satisfies the above requirements. In general, the domains of $(\omega + A)^\alpha$ are not explicitly known. However, it can be shown that

$$D\big((\omega + A)^\alpha\big) \hookrightarrow W_p^s \ , \qquad 0 \leq s < 2\alpha \ . \tag{33}$$

Thus, choosing $p > n/2$ and $\alpha \in \big(n/(2p), 1\big)$, we infer from (18) that $E_\alpha \hookrightarrow C \hookrightarrow E_0$. Hence (20) implies the validity of the modified minimal requirement $(15)_\alpha$ in the case of problem (7) without any growth restriction for r. Furhermore, if we fix $p > n$, we deduce from (18) and (33) that $E_\alpha \hookrightarrow C^1 \hookrightarrow C \hookrightarrow E_0$, provided $\alpha \in (n/p, 1)$. In this case the reaction rate r can also be allowed to depend smoothly upon $\operatorname{grad} u$, that is,

$$r = r(x, t, u, \operatorname{grad} u) \ . \tag{34}$$

Indeed, again denoting by f the corresponding Nemytskii operator, that is,

$$f(t, u) := r\big(\cdot, t, u(\cdot), \operatorname{grad} u(\cdot)\big) \ , \qquad (t, u) \in \mathbb{R}^+ \times C^1(X) \ , \tag{35}$$

we infer from (20) that f satisfies the modified minimal requirement $(15)_\alpha$ in this case as well.

Quasilinear Reaction-Diffusion Systems

The above approach to semilinear parabolic initial-boundary value problems is well-known and well-documented in the literature, most notably in the rather influential monograph of Henry [Hen81]. It has been the basis for numerous investigations of reaction-diffusion equations, in particular of semilinear heat equations of the form (10), where u is sometimes admitted to be N-vector-valued.

However, most reaction-diffusion systems of interest in science are considerably more intricate than the simple laws (1), (2), and (5):

• First, as a rule, realistic models involve several species, say N, each of whom satisfies a conservation law (1). Thus we have to consider a system of equations:

$$\partial_t u_i + \operatorname{div} j_i = r_i \ , \qquad 1 \leq i \leq N \ . \tag{36}$$

• Second, the constitutive relations for the flux vectors j_i depend on all components of the vector

$$u := (u_1, \dots, u_N) \ .$$

Under rather general circumstances (e.g., [deGM84]) this dependence is of the form

$$j_i(u) := -\sum_{k=1}^{N} (D^{ik} \operatorname{grad} u_k + d^{ik} u_k) \ , \qquad 1 \le i \le N \ , \tag{37}$$

where D^{ik} and d^{ik} are $(n \times n)$-matrices and n-vectors, respectively.

• Third, for each species i there is a separate boundary characterization map δ_i. For simplicity, the boundary value of u_i on the 'Dirichlet boundary, $\delta_i^{-1}(0)$, of species i' is again normalized to zero. On the remaining part of ∂X, that is, on $\delta_i^{-1}(1)$, the no-flux condition is replaced by the more general 'prescribed flux condition'

$$-\nu \cdot j_i(u) = s_i$$

with a 'surface reaction rate' s_i. Thus there are the N boundary conditions:

$$-\delta_i \nu \cdot j_i(u) + (1 - \delta_i) u_i = \delta_i s_i \ , \qquad 1 \le i \le N \ . \tag{38}$$

• Fourth, the matrices D^{ik}, the vectors d^{ik}, and the functions r_i depend upon $(x,t) \in \overline{X} \times \mathbb{R}^+$, the functions s_i on $(y,t) \in \partial X \times \mathbb{R}^+$, and all of them on the solution vector u. Moreover, all these dependences are smooth, as a rule.

Of course, the initial condition (6) has to be replaced by N relations of that type:

$$u_i(\cdot, 0) = u_i^0 \ , \qquad 1 \le i \le N \ . \tag{39}$$

To rewrite system (36)–(39) as an initial-boundary value problem of a form similar to (7), we introduce the $(N \times N)$-matrices

$$a_{\alpha\beta}(t,u) := \left[D_{\alpha\beta}^{ik}(\cdot, t, u) \right]_{1 \le i,k \le N} \ , \quad a_\alpha(t,u) := \left[d_\alpha^{ik}(\cdot, t, u) \right]_{1 \le i,k \le N}$$

for $1 \le \alpha, \beta \le n$, where $D_{\alpha\beta}^{ik}$ and d_α^{ik} are the entries and components of the $(n \times n)$-matrices D^{ik} and n-vectors d^{ik}, respectively, and the N-vectors[4]

$$f(t,u) := \bigl(r_1(\cdot, t, u), \dots, r_N(\cdot, t, u) \bigr)$$

and

$$g(t,u) := \bigl(\delta_1 s_1(\cdot, t, u), \dots, \delta_N s_N(\cdot, t, u) \bigr) \ .$$

[4]Unless absolutely necessary, we do not distinguish between row and column vectors. Also we often use the same notation for a function and the Nemytskii operator induced by it, provided it is clear from the context which interpretation has to be chosen.

We also employ the summation convention for $\alpha, \beta \in \{1, \ldots, n\}$ and put

$$b(t,u) := \nu^\alpha a_\alpha(t,u) \ ,$$

where $\nu = (\nu^1, \ldots, \nu^n)$. Then, using euclidean coordinates, we define differential operators $\mathcal{A}(t,u)$ in X and boundary operators $\mathcal{B}(t,u)$ on ∂X by

$$\mathcal{A}(t,u)v := -\partial_\alpha \big(a_{\alpha\beta}(t,u)\partial_\beta v + a_\alpha(t,u)v\big) \tag{40}$$

and

$$\mathcal{B}(t,u)v := \delta\big(\nu^\alpha a_{\alpha\beta}(t,u)\partial_\beta v + b(t,u)u\big) + (1-\delta)u \ , \tag{41}$$

respectively, where δ is now the diagonal matrix with entries $\delta_1, \ldots, \delta_N$. Then the reaction-diffusion system (36)–(39) is rewritten as a quasilinear initial-boundary value problem:

$$\left.\begin{aligned} \partial_t u + \mathcal{A}(t,u)u &= f(t,u) && \text{in } X \\ \mathcal{B}(t,u)u &= g(t,u) && \text{on } \partial X \end{aligned}\right\} \quad t > 0 \ , \\ u(\cdot,0) = u^0 \qquad \text{on } \overline{X} \ , \tag{42}$$

where $u^0 := (u_1^0, \ldots, u_N^0)$.

Note that (42) is a *strongly coupled* system in the sense that the 'diffusion coefficients' $a_{\alpha\beta}$ are fully occupied $(N \times N)$-matrices, in general. Furthermore, these matrices depend nonlinearly on the solution vector u, that is, the system (42) is *quasilinear*, and it involves *nonlinear boundary conditions.* Clearly, the quasilinear initial-boundary value problem (42) for a vector-valued unknown is considerably more intricate than the semilinear problem (7) for a scalar unknown. In particular, (42) does not satisfy a maximum principle, in general, which means that we are lacking a tool that has proven to be very powerful in the study of a single parabolic differential equation. On the other hand, it is important to observe that (42) is of *divergence form*, a fact that turns out to be crucial for handling nonlinear boundary conditions.

Homogeneous Boundary Conditions

First we consider the case $g = 0$ and proceed as in the semilinear case. Thus we begin by interpreting (42) as an ordinary differential equation in the Banach space E_0. Of course, here and below it is always understood that all function spaces consist of N-vector-valued functions, e.g., $L_p = L_p(X, \mathbb{R}^n)$ etc., if we refer to systems, unless explicitly stated otherwise.

For the time being, by a solution of (42) on J we mean a function

$$u \in C(J, W_p^2) \cap C^1(J, L_p)$$

satisfying (42) in the obvious 'point-wise' sense. Given such a solution u, we put

$$A(t,u(t)) := \mathcal{A}(t,u(t)) | W^2_{p,\mathcal{B}(t,u(t))} ,$$

thereby using the notation of (17). Then u is, at least formally, a solution on J of the quasilinear Cauchy problem

$$\dot{u} + A(t,u)u = f(t,u) , \quad t \in \dot{J} , \qquad u(0) = u^0 . \tag{43}$$

Letting

$$A_u(t) := A(t,u(t)) , \qquad t \in J ,$$

u is also a solution of the nonautonomous semilinear Cauchy problem

$$\dot{v} + A_u(t)v = f(t,v) , \quad t \in \dot{J} , \qquad v(0) = u^0 . \tag{44}$$

Hence it can be expressed by the variation-of-constants formula

$$u(t) = U_u(t,0)u^0 + \int_0^t U_u(t,\tau) f(\tau, u(\tau))\, d\tau , \qquad t \in J , \tag{45}$$

where now the semigroup in (24) is replaced by the evolution operator U_u of the family $\{ A_u(t) \ ; \ t \in J \}$. Consequently, denoting by Φ the right-hand side of (45), we are again led to the fixed point equation (27) that we want to make the basis for our study of the quasilinear problem (43).

Now the situation is much more complex than in the semilinear case. First of all, to give a meaning to these formal considerations we have to guarantee the existence of the evolution operator U_u, given a function u possessing suitable regularity properties. Clearly, the existence and the properties of an evolution operator are intimately related to the solvability of the linear Cauchy problem

$$\dot{v} + A(t)v = f(t) , \quad t \in \dot{J} , \qquad v(0) = u^0 , \tag{46}$$

that is obtained from (44) by ignoring the u- and v-dependence of A and f, respectively, for the moment. This Cauchy problem corresponds to a Banach space formulation of the nonautonomous linear initial-boundary value problem

$$\left.\begin{aligned} \partial_t v + \mathcal{A}(t)v &= f(t) && \text{in } X \\ \mathcal{B}(t)v &= 0 && \text{on } \partial X \end{aligned}\right\} \quad t > 0 , \qquad v(\cdot,0) = u^0 \quad \text{on } \overline{X} , \tag{47}$$

where now the coefficients in (40) and (41) are independent of u.

In the simplest case of a scalar unknown, that is, for $N = 1$, we know already from (17) and (22) that

$$A(t) \in \mathcal{H}(E_1(t), E_0) , \tag{48}$$

where

$$E_1(t) \doteq W^2_{p,\mathcal{B}(t)} . \tag{49}$$

Note that (48) means that each $A(t)$ is the negative generator of an analytic semigroup on E_0, whose domain is t-dependent, in general.

In the case of systems, that is, for $N > 1$, as well as in the general abstract situation, condition (48) is a hypothesis. If $A(t)$ is induced by the linear initial-boundary value problem (47), it turns out that a necessary condition for (48) to be satisfied is that

$$\partial + \mathcal{A}(t) \text{ is Petrowskii parabolic for } t \in J\ . \tag{50}$$

This means that, given $t \in J$ and $\xi = (\xi^1, \ldots, \xi^n) \in \mathbb{R}^n \backslash \{0\}$,

$$\sigma\big(a_{\alpha\beta}(t)\xi^\alpha\xi^\beta\big) \subset [\operatorname{Re} z > 0]\ ,$$

or, in other words, that each eigenvalue of the $(N \times N)$-matrices $a_{\alpha\beta}(x,t)\xi^\alpha\xi^\beta$ has a positive real part, given any $x \in \overline{X}$ and $\xi \in \mathbb{R}^n \backslash \{0\}$.

In general, condition (50) is not sufficient, but there are additional requirements for the boundary symbols $\nu^\alpha a_{\alpha\beta}(t)\xi^\beta$ to be satisfied, that are too complicated to be given here. However, there are important special cases, that are of particular importance for applications, in which condition (50) already implies (48) and (49). For example, this is the case if $\mathcal{B}(t)$ equals the 'Dirichlet boundary operator', that is, $\mathcal{B}(t)u = u|\partial X$, for $t \in J$. To give a further example, suppose that, given $t \in J$,

$$a_\alpha(t) \in C^1\big(\overline{X}, \mathcal{L}(\mathbb{R}^N)\big)\ , \qquad 1 \le \alpha \le n\ , \tag{51}$$

and

$$a_{\alpha\beta}(t) = a(t)\delta_{\alpha\beta}\ , \quad 1 \le \alpha, \beta \le n\ , \qquad \text{where} \quad a(t) \in C^1\big(\overline{X}, \mathcal{L}(\mathbb{R}^N)\big)\ , \tag{52}$$

and that

$$\delta \in C\big(\partial X, \{0, 1_{\mathbb{R}^N}\}\big)\ . \tag{53}$$

Note that this implies the existence of a boundary decomposition $\partial X = \partial_0 X \cup \partial_1 X$ such that

$$\mathcal{B}(t)u = \begin{cases} u & \text{on } \partial_0 X\ , \\ a(t)\partial_\nu u + b(t)u & \text{on } \partial_1 X\ . \end{cases} \tag{54}$$

Then (48) and (49) are satisfied iff

$$\sigma\big(a(t)\big) \subset [\operatorname{Re} z > 0]\ , \tag{55}$$

that is, iff $\partial + \mathcal{A}(t)$ is Petrowskii parabolic for $t \in J$. Observe that (55) does not imply that the symmetric part of the matrix $a(t)$ is positive definite. Hence a Petrowskii parabolic system is not strongly parabolic, in general, that is, not coercive.

In the second volume of this treatise we discuss in detail the question under what conditions systems of partial differential operators (of arbitrary order), together with appropriate systems of boundary operators, generate analytic semigroups on Lebesgue and other spaces.

Constant Domains

Now let conditions (48) and (49) be satisfied. Then, to guarantee the existence of an evolution operator U for the linear Cauchy problem (46), one has to impose regularity assumptions on the function $t \mapsto A(t)$. It turns out that the required amount of regularity is proportional to the 'strength of variability' of the domains of $A(t)$, that is, the spaces $E_1(t)$, if t varies over J.

The simplest case occurs if the domains of $A(t)$ are independent of $t \in J$, that is, if $E_1(t) \doteq E_1(0) =: E_1$ for $t \in J$. Note that this is the case if $\mathcal{B}(t)$ equals the Dirichlet boundary operator for $t \in J$. It is also true if conditions (51)–(55) are satisfied and $b = 0$, since in this case the homogeneous boundary condition $\mathcal{B}(t)u = 0$ is equivalent to the constant boundary condition $\delta\partial_\nu u + (1-\delta)u = 0$ on ∂X.

If the operators $A(t)$ have a constant domain E_1, it suffices to require that

$$\bigl[t \mapsto A(t)\bigr] \in C^\rho\bigl(J, \mathcal{H}(E_1, E_0)\bigr) \tag{56}$$

for some $\rho \in (0,1)$. Then there exists a unique evolution operator U for the family $\{\, A(t) \ ; \ t \in J \,\}$ such that each solution of (46) can be represented by the variation-of-constants formula

$$u(t) = U(t,0)u^0 + \int_0^t U(t,\tau)f(\tau)\,d\tau \ , \qquad t \in J \ .$$

Furthermore, the evolution operator U possesses properties that are similar to the ones of the semigroup $\{\, e^{-tA} \ ; \ t \geq 0 \,\}$ of the autonomous problem. In particular, the following analogue of (29) is valid:

$$\|U(t,s)\|_{\mathcal{L}(E_0,E_1)} \leq c(t-s)^{-1} \ , \qquad t,s \in J \ , \quad s < t \ .$$

Recall that our interest belongs to the quasilinear equation (43), and thus to the fixed point equation (45). Since the dependence of A on u in (43) implies a dependence of U on u in (45), it is necessary to study carefully the dependence of U on A. The construction of the evolution operator, as well as the investigation of its dependence on A, is the content of the second chapter of this volume. There the general setting of an arbitrary densely injected Banach couple is considered, which greatly enhances the applicability of our results.

Of course, treating the quasilinear problem (43) by reducing it to the fixed point equation (45), we encounter the same difficulties we found when we had to give a meaning to the fixed point map Φ in the semilinear case (13). Thus we have to consider again an 'intermediate' space E_α satisfying (31) such that each one of the semigroups $\{\, e^{-tA(s)} \ ; \ t \geq 0 \,\}$, $s \in J$, restricts to a strongly continuous semigroup on E_α and (32) is replaced by

$$\|U(t,s)\|_{\mathcal{L}(E_0,E_\alpha)} \leq c(t-s)^{-\alpha} \ , \qquad t,s \in J \ , \quad 0 \leq s < t \ .$$

As in the semilinear case, $D\big(A^\alpha(0)\big)$ seems to be a natural candidate for E_α. However, this choice is not the best one, mainly since $D\big(A^\alpha(t)\big)$ is not known explicitly for $\alpha \in (0,1)$. Thus, for example, it is not clear if $D\big(A^\alpha(t)\big)$ is independent of $t \in J$, in general.

Although it is possible to develop a general theory of abstract quasilinear Cauchy problems by basing it on the theory of fractional powers — this has been done by P.E. Sobolevskii in [Sob66] (also cf. [Fri69] for an exposition of parts of that work) — , this approach does not give optimal results. Indeed, by Sobolevskii's approach it is not possible to prove that the solutions of (43) generate a semiflow, that is, a nonlinear local semigroup, since one has to require more smoothness for the initial datum u^0 than should be expected. In other words, the initial value u^0 has to belong to a slightly smaller space than the natural phase space of (43).

The semiflow property of the solutions of the abstract quasilinear Cauchy problem is both natural and fundamental for a qualitative study of the latter. Thus any good general theory has to produce this property. Therefore we are forced to look for other candidates for E_α. Fortunately, there are plenty of them: most of the interpolation spaces of exponent α between E_0 and E_1 turn out to lead to the desired results. Thus at this stage interpolation theory comes into play naturally.

Together with the theory of analytic semigroups, interpolation theory plays a predominant rôle in the abstract theory of linear and quasilinear Cauchy problems of 'parabolic type'. For this reason we collect in Chapter I the basic facts from these two fields, that are being used throughout our treatise. Furthermore, concrete characterizations of interpolation spaces are studied in detail in the second volume. In particular, the following is shown: let

$$W^s_{p,\mathcal{B}(t)} := \begin{cases} \{\, v \in W^s_p \ ;\ \mathcal{B}(t)v = 0 \,\}\ , & 1 + 1/p < s < \infty\ , \\ \{\, v \in W^s_p \ ;\ (1-\delta)v|\partial X = 0 \,\}\ , & 1/p < s < 1 + 1/p\ , \\ W^s_p\ , & 0 \le s < 1/p\ , \end{cases} \tag{57}$$

and put

$$W^s_{p,\mathcal{B}} := W^s_{p,\mathcal{B}(t)}\ , \qquad s \in [0, 1+1/p)\backslash\{1/p\}\ .$$

Then, given $\alpha \in (0,1)$ with $2\alpha \notin \{1/p, 1+1/p\}$, there exists an interpolation functor $(\cdot,\cdot)_\alpha$ of exponent α such that

$$E_\alpha(t) := \big(E_0, E_1(t)\big)_\alpha \doteq W^{2\alpha}_{p,\mathcal{B}(t)}\ , \tag{58}$$

where $E_0 := L_p$ and $E_1(t) := W^2_{p,\mathcal{B}(t)} = \mathrm{dom}\big(A(t)\big)$. Note that $E_\alpha := E_\alpha(t)$ is independent of $t \in J$ if $0 < 2\alpha < 1 + 1/p$ and $2\alpha \neq 1/p$.

The assumption of constant domains for $A\big(t, u(t)\big)$, which entails the independence of the interpolation spaces $E_\alpha(t)$ of $t \in J$, simplifies the approach to quasilinear parabolic problems considerably and results in a rather flexible theory.

This is essentially due to the fact that the existence of an evolution operator can be guaranteed by presupposing (56) with an arbitrarily small Hölder exponent ρ. As a rule, $A\big(t, u(t)\big)$ is well-defined only if $u(t)$ belongs to E_β for β sufficiently large. Indeed, to guarantee (49) we have to assume that

$$a_{\alpha\beta}\big(t, u(t)\big), a_\alpha\big(t, u(t)\big) \in C^1\big(\overline{X}, \mathcal{L}(\mathbb{R}^N)\big) \;, \qquad \alpha, \beta \in \{1, \ldots, n\} \;, \quad t \in J \;.$$

Thus we certainly have to require that $u \in C^1$. Consequently, we infer from (56) that we have to assume that

$$\big[t \mapsto u(t)\big] \in C^\rho(J, E_\beta) \;, \tag{59}$$

where $E_\beta \hookrightarrow C^1$. By (18), (57), and (58), for this embedding to be valid we have to impose the condition $(1 + n/p) < 2\beta < 2$, which entails $p > n$. Now we fix $\alpha \in (\beta, 1)$ and choose E_α as a phase space, that is, we study the fixed point equation (45) in $C(J, E_\alpha)$. Then Φ maps $C(J, E_\alpha)$ into $C(J, E_\alpha) \cap C^\rho(J, E_\beta)$, provided $\rho < \alpha - \beta$. Hence $u_1 := \Phi(u)$ satisfies (59) whenever $u \in C(J, E_\alpha)$. So the iteration process of the method of successive approximations can be carried out indefinitely, and it can indeed be shown that Φ has a unique fixed point in $C(J, E_\alpha)$, provided J is sufficiently short. Note that this imposes the restriction $2\rho < 1 - n/p$. It should also be remarked that, up to this point, the divergence structure of (40), (41) is not needed.

Extrapolation Theory

The approach outlined above uses in an essential way the fact that the domains of $A(t)$ are constant. More precisely, it is based on the fact that the interpolation spaces E_β and E_α are independent of t for $1 + n/p < 2\beta < 2\alpha < 2$. By (57) and (58), this is not the case if $\mathrm{dom}\big(A(t)\big) = W^2_{p,\mathcal{B}(t)}$ depends on $t \in J$.

On the other hand, thanks to its divergence structure, problem (42) has a natural weak formulation. For this put

$$\langle v, w\rangle := \int_X v \cdot w \, dx \;, \qquad (v, w) \in L_{p'} \times L_p \;.$$

Furthermore, given $t \in J$ and $u \in C$, let

$$\boldsymbol{a}(t, u)(v, w) := \big\langle \partial_\alpha v, a_{\alpha\beta}(t, u)\partial_\beta w + a_\alpha(t, u) w\big\rangle \;, \qquad (v, w) \in W^1_{p',\mathcal{B}} \times W^1_{p,\mathcal{B}} \;.$$

Then $\boldsymbol{a}(t, u)$ is a continuous bilinear form on $W^1_{p',\mathcal{B}} \times W^1_{p,\mathcal{B}}$, the 'Dirichlet form' belonging to $\big(\mathcal{A}(t, u), \mathcal{B}(t, u)\big)$. Thus, letting

$$W^{-s}_{p,\mathcal{B}} := (W^s_{p',\mathcal{B}})' \;, \qquad -2 + 1/p < s < 0 \;, \quad s \neq -1 + 1/p \;, \tag{60}$$

with respect to the duality pairing naturally induced by the L_p-duality pairing $\langle \cdot, \cdot \rangle$, it follows that $\boldsymbol{a}(t, u)$ induces

$$\mathbb{A}(t, u) \in \mathcal{L}(W^1_{p,\mathcal{B}}, W^{-1}_{p,\mathcal{B}}) \tag{61}$$

via

$$\langle v, \mathbb{A}(t,u)w\rangle := \boldsymbol{a}(t,u)(v,w) , \qquad (v,w) \in W^1_{p',\mathcal{B}} \times W^1_{p,\mathcal{B}} . \tag{62}$$

We also define a nonlinear map F by

$$\langle v, F(t,u)\rangle := \langle v, f(t,u)\rangle + \int_{\partial X} v \cdot g(t,u)\, d\sigma , \qquad v \in W^1_{p',\mathcal{B}} . \tag{63}$$

Then, thanks to the divergence structure of (40) and (41), the natural weak formulation of the initial-boundary value problem (42) takes the form:

find $u \in C(J, W^1_{p,\mathcal{B}})$ satisfying

$$\int_J \{-\langle \dot{v}, u\rangle + \boldsymbol{a}(t,u)(v,u)\}\, dt = \int_J \langle v, F(t,u)\rangle\, dt + \langle v(0), u^0\rangle \tag{64}$$

for each $v \in C^1(J, W^1_{p',\mathcal{B}})$ vanishing near the right endpoint of J.

It is easily seen that every solution of the initial-boundary value problem (42) is a weak solution in the sense of (64). Furthermore, (64) is equivalent to the abstract quasilinear Cauchy problem in $W^{-1}_{p,\mathcal{B}}$:

$$\dot{u} + \mathbb{A}(t,u)u = F(t,u) , \quad t \in \dot{J} , \qquad u(0) = u^0 . \tag{65}$$

Observe that, thanks to (61), the domains of the operators $\mathbb{A}(t,u)$ are all equal to $W^1_{p,\mathcal{B}}$. Thus (65) is amenable by the techniques described above for the constant domain problem (43), provided

$$\mathbb{A}(t,u) \in \mathcal{H}(W^1_{p,\mathcal{B}}, W^{-1}_{p,\mathcal{B}}) \tag{66}$$

with suitably regular dependence on (t,u).

It is a consequence of the divergence theorem that $\mathbb{A}(t,u)$ is an extension of

$$A(t,u) := \mathcal{A}(t,u) | W^2_{p,\mathcal{B}(t,u)} .$$

This fact can be used to prove (66). To outline this method, we first fix a negative generator B of a strongly continuous analytic semigroup on an arbitrary Banach space $F_0 := (F_0, \|\cdot\|_0)$ and assume, for simplicity, that B has a bounded inverse defined on F_0. Then

$$u \mapsto \|u\|_1 := \|Bu\|_0$$

defines a norm on $\mathrm{dom}(B)$ that is equivalent to the graph norm. This implies that $F_1 := \big(\mathrm{dom}(B), \|\cdot\|_1\big)$ is a Banach space such that (F_0, F_1) is a densely injected Banach couple, and $B \in \mathcal{H}(F_1, F_0)$.

The Banach space F_0 can be recovered from F_1 by observing that F_0 is a completion of F_1 in the norm $u \mapsto \|B_1^{-1}u\|_1 = \|u\|_0$, where B_1 is the F_1-realization of B. This suggests to introduce a superspace $F_{-1} := (F_{-1}, \|\cdot\|_{-1})$ of F_0 by choosing for

$$F_{-1} \text{ a completion of } F_0 \text{ in the norm } u \mapsto \|B^{-1}u\|_0 =: \|u\|_{-1}.$$

Then (F_{-1}, F_0) is a densely injected Banach couple as well, and it is not difficult to show that $B_0 := B$ extends continuously to an operator $B_{-1} \in \mathcal{H}(F_0, F_{-1})$. If B is not invertible, we simply replace B by $\omega + B$ for a sufficiently large $\omega \in \mathbb{R}^+$.

Now, given $\theta \in (0,1)$, we choose an interpolation functor $(\cdot,\cdot)_\theta$ of exponent θ with the property that G_1 is dense in $(G_0, G_1)_\theta$ whenever (G_0, G_1) is a densely injected Banach couple, an 'admissible interpolation functor'. Then we put $F_\theta := (F_0, F_1)_\theta$ and $F_{\theta-1} := (F_{-1}, F_0)_\theta$. This defines a 'scale of Banach spaces'

$$F_1 \overset{d}{\hookrightarrow} F_\alpha \overset{d}{\hookrightarrow} F_\beta \overset{d}{\hookrightarrow} F_{-1}\ , \qquad -1 < \beta < \alpha < 1\ .$$

Furthermore, denoting the $F_{\alpha-1}$-realization of B_{-1} by $B_{\alpha-1}$ for $\alpha \in [0,1]$, it follows that

$$B_{\alpha-1} \in \mathcal{H}(F_\alpha, F_{\alpha-1})\ , \qquad 0 \le \alpha \le 1\ .$$

These extensions are 'natural' in the sense that $e^{-tB_{\alpha-1}}$ is the restriction to $F_{\alpha-1}$ of $e^{-tB_{\beta-1}}$ for $0 \le \beta < \alpha \le 1$.

This 'interpolation-extrapolation' technique is rather flexible, has many applications, and is crucial for our approach to quasilinear parabolic evolution equations. It allows to measure very precisely regularity properties of operators and solutions. Scales of Banach spaces, interpolation-extrapolation techniques, and evolution equations in these scales are studied in detail in Chapter V.

Now we return to our quasilinear parabolic problem. In this case, given (t,u), we can construct an interpolation-extrapolation scale by starting with $A(t,u)$ on $E_0 = L_p$. Then, using suitable admissible interpolation functors, this scale, denoted by $E_\alpha(t,u)$ for $\alpha \in [-2,1]$, can be explicitly represented as

$$E_\alpha(t,u) \doteq W^{2\alpha}_{p,\mathcal{B}(t,u)}\ , \qquad 2\alpha \in (-2+1/p, 2] \backslash (\mathbb{Z} + 1/p)\ ,$$

in the range given. Thus we infer from (57) and (60) that

$$E_\alpha(t,u) \doteq E_\alpha := W^{2\alpha}_{p,\mathcal{B}}\ , \qquad 2\alpha \in (-2+1/p, 1+1/p) \backslash (\mathbb{Z} + 1/p)\ . \tag{67}$$

Therefore the interpolation-extrapolation spaces are independent of (t,u), that is, they are constant, if α belongs to the range given in (67). Moreover, denoting by $A_{\alpha-1}(t,u)$ the 'extrapolated operators', the general theory of Banach scales tells us that

$$A_{\alpha-1}(t,u) \in \mathcal{H}(E_\alpha, E_{\alpha-1})\ , \qquad 1/p < 2\alpha < 1 + 1/p\ . \tag{68}$$

Lastly, it can be shown that

$$\mathbb{A}(t,u) = A_{-1/2}(t,u)\ , \tag{69}$$

which, thanks to (67) and (68), proves (66).

Regularity

The interpolation-extrapolation technique outlined above enables us to solve the initial-boundary value problem (42) in its weak form (64), provided $\boldsymbol{a}$ and F are suitably regular. In fact, it suggests to replace the quasilinear evolution equation (65) by the family

$$\dot{u} + A_{\alpha-1}(t,u)u = F(t,u) \ , \quad t \in \dot{J} \ , \qquad u(0) = u^0 \tag{70}$$

for $1/p < 2\alpha < 1 + 1/p$. In particular, we can fix $2\alpha \in (1, 1 + 1/p)$ and can choose $E_{1/2} = W_{p,\mathcal{B}}^1$ as a phase space for the constant domain problem (70). Then, given $u^0 \in E_{1/2}$, problem (70) has a unique maximal solution

$$u_0 := u(\cdot, u^0) \in C(J_0, E_{1/2}) \cap C(\dot{J}_0, E_\alpha) \cap C^1(\dot{J}_0, E_{\alpha-1}) \ . \tag{71}$$

It follows from (69) and $A_{-1/2} \supset A_{\alpha-1}$ that u_0 is a weak solution of (42) in the sense of (64). But (71) shows that u_0 possesses better regularity properties.

This observation suggests a bootstrapping procedure. For this we observe that u_0 is a solution of the linear Cauchy problem

$$\dot{v} + A_{\alpha-1}\big(t, u_0(t)\big) = F\big(t, u_0(t)\big) \ , \quad t \in \dot{J}_0 \ , \qquad v(0) = u^0$$

in $E_{\alpha-1}$. Recall that $A_{\alpha-1}\big(t, u_0(t)\big) \supset A_{\beta-1}\big(t, u(t)\big)$ if $\beta > \alpha$. Thus, if the linear Cauchy problem

$$\dot{v} + A_{\beta-1}\big(t, u_0(t)\big)v = F\big(t, u_0(t)\big) \ , \quad t \in \dot{J}_0 \ , \qquad v(0) = u^0 \tag{72}$$

is well-posed in $E_{\beta-1}$ for some $\beta \in (\alpha, 1]$, its solution coincides with u_0. Note that a solution u of (72) satisfies $u(t) \in \mathrm{dom}\big(A_{\beta-1}\big(t, u_0(t)\big)\big)$ for $t \in \dot{J}_0$. Hence it follows that

$$u_0(t) \in \mathrm{dom}\big(A_{\beta-1}\big(t, u_0(t)\big)\big) \subset W_p^{2\beta-2} \tag{73}$$

for $t \in \dot{J}_0$, provided (72) is well-posed.

By these considerations we are led to study the linear Cauchy problem

$$\dot{v} + A_{\beta-1}(t)v = F(t) \ , \quad t \in \dot{J} \ , \qquad v(0) = u^0$$

in $E_{\beta-1}$, where we are particularly interested in the case $2\beta > 1 + 1/p$, so that $\mathrm{dom}\big(A_{\beta-1}(t)\big)$ is no longer constant. Since $A_{\alpha-1}(t) \supset A_{\beta-1}(t)$, it is to be expected that the restriction of the evolution operator $U_{\alpha-1}$ of the family $\big\{\, A_{\alpha-1}(t) \ ; \ t \in J \,\big\}$ to the space $E_{\beta-1} \hookrightarrow E_{\alpha-1}$ is an evolution operator for $\big\{\, A_{\beta-1}(t) \ ; \ t \in J \,\big\}$ on $E_{\beta-1}$. This is true in fact, provided

$$A_{\alpha-1} \in C^\rho\big(J, \mathcal{H}(E_\alpha, E_{\alpha-1})\big) \ , \qquad \rho > \beta - \alpha \ , \tag{74}$$

as follows from the general results derived in Chapter IV.

It turns out that $A_{\alpha-1}(t, u_0(t))$ can be characterized, similarly as $A_{-1/2}(t)$ in (69), through the bilinear form $\boldsymbol{a}(t, u_0(t))$, the latter now being continuous on $W^{2-2\alpha}_{p',\mathcal{B}} \times W^{2\alpha}_{p,\mathcal{B}}$. A careful study of point-wise multipliers on generalized Sobolev spaces, that is carried out in the second volume, reveals that

$$\big[(t,u) \mapsto A_{\alpha-1}(t,u)\big] \in C^{1\text{-}}\big(J \times C^{\sigma}, \mathcal{L}(E_\alpha, E_{\alpha-1})\big) \tag{75}$$

if $\sigma > 2\alpha - 1$. From (71) we infer, using suitable embedding theorems, that

$$u_0 \in C^{\rho}(\dot{J}_0, C^{\sigma}) , \tag{76}$$

provided $2\rho < 2\alpha - \sigma - n/p$. By combining (75) and (76) we see that

$$\big[t \mapsto A_{\alpha-1}(t, u_0(t))\big] \in C^{\rho}\big(\dot{J}_0, \mathcal{L}(E_\alpha, E_{\alpha-1})\big) \tag{77}$$

for $2\rho < 1 - 1/p$. Hence $A_{\alpha-1}(\cdot, u_0(\cdot))$ satisfies (74) for

$$2\beta < 2\alpha + 1 - n/p < 2 - (n-1)/p .$$

This way, and using the theory developed in Chapter IV, it follows that

$$u_0 \in C(\dot{J}_0, W^{2\beta}_p) \cap C^1(\dot{J}, W^{2\beta-2}_{p,\mathcal{B}}) \hookrightarrow C^{\rho}(\dot{J}, C^{\sigma}) ,$$

provided $2\beta < 2 - (n-1)/p$ and $\sigma + 2\rho < 2\beta - n/p < 2 - (2n-1)/p$. Thus, since we can choose an arbitrarily large value for p, we obtain

$$u_0 \in C^{\rho}(\dot{J}, C^{\sigma}) , \qquad \sigma + 2\rho < 2 . \tag{78}$$

At this stage we can invoke the classical Hölder theory for general parabolic systems, for example, to deduce that u_0 is in fact a classical solution, even a C^∞-solution, of the initial-boundary value problem (42).

Unfortunately, the situation is more complicated than just indicated. First, in the preceding outline we neglected the difficulties coming form $F(t, u_0(t))$. The above procedure requires $t \mapsto F(t, u_0(t))$ to belong to $C^{\sigma}(\dot{J}, E_{\beta-1})$ for β arbitrarily close to 1. This does not hold, in general, unless the boundary nonlinearity g satisfies appropriate compatibility conditions and the problem is suitably modified. For the latter modification one has to have a good theory of nonhomogeneous linear parabolic boundary value problems in W^s_p for $1 + 1/p < s \leq 2$ under mild regularity conditions on the coefficients. These questions are investigated in Volume Two.

Second, it is by no means clear that $A_{\alpha-1}(t, u_0(t)) \in \mathcal{H}(E_\alpha, E_{\alpha-1})$ if u_0 satisfies (76). Indeed, to derive (68) we started with $A(t,u) \in \mathcal{H}\big(E_1(t,u), E_0\big)$. In our concrete situation this can only be established if the coefficients of $\mathcal{A}$ and $\mathcal{B}$ are of class C^1. To guarantee this we would already have to know that $u_0(t) \in C^1$. Thus we have to prove directly that $-A_{\alpha-1}(t,u)$ generates a strongly continuous analytic semigroup on $E_{\alpha-1}$ if $\mathcal{A}$ and $\mathcal{B}$ have Hölder continuous coefficients only.

This problem is also studied in detail in the second volume by making use of the interpolation-extrapolation technique for smooth operators.

Having established the necessary ingredients, the above bootstrapping technique can be carried through, in principle. However, the technical details are somewhat delicate, and it is not possible, in general, to derive (78) in one step. Instead, a chain of bootstrapping arguments is needed. These arguments are given in detail in the third volume on the basis of the results of Volume Two.

Besides of existence and regularity theorems, in the last volume a qualitative theory of quasilinear parabolic evolution equations is developed as well. It is shown that they generate smooth local semiflows on appropriate natural phase spaces. Furthermore, questions of global existence, stability of critical points and periodic orbits, bifurcation phenomena, etc., are studied in the general abstract setting. In addition, those results are applied to a variety of concrete systems of partial differential equations stemming from diverse applications in physics, chemistry, etc.

Maximal Regularity

The theory of linear and quasilinear parabolic evolution equations, outlined so far, uses crucially the smoothing property of parabolic equations. This regularizing effect is the ultimate reason for the validity of global existence theorems based on a priori estimates in 'weak' norms, that is, norms involving little regularity of the solutions only.

On the other hand, it restricts the applicability of the general abstract theory to quasilinear reaction-diffusion systems whose 'diffusion matrices' depend on u only, at least in the divergence form case. If $\operatorname{grad} u$ also occurs in the diffusion matrices, this theory is not applicable since it uses the fact that the dependence of $A_{\alpha-1}(t,u) \in \mathcal{L}(E_\alpha, E_{\alpha-1})$ on (t,u) is smooth with respect to the topology of an interpolation space $(E_{\alpha-1}, E_\alpha)_\theta$ that is strictly weaker than the topology of E_α.

If the 'intermediate' smoothness requirement is not satisfied, e.g., if the diffusion matrices depend on $\operatorname{grad} u$, we have to invoke maximal regularity results. The maximal regularity approach is different in spirit from the evolution operator approach discussed so far. Whereas the latter is essentially a dynamical method, that is the infinite-dimensional analogue to the theory of ordinary differential equations, the former is basically a stationary technique. This means that the linear Cauchy problem (46) is interpreted as an 'algebraic' equation in a suitable Banach space $\mathbb{E}_0$ such that $\mathbb{E}_0 \hookrightarrow L_1(J, E_0)$. To be more precise, we consider the case of constant domains and suppose that $\mathbb{E}_1$ is a Banach space such that $\mathbb{E}_1 \hookrightarrow L_1(J, E_1) \cap W_1^1(J, E_0)$ and $\partial + A \in \mathcal{L}(\mathbb{E}_1, \mathbb{E}_0)$. Of course,

$$(\partial + A)u(t) := \dot{u}(t) + A(t)u(t) \;, \qquad u \in \mathbb{E}_0 \;, \quad \text{a.a. } t \in J \;.$$

Moreover, putting $\gamma u := u(0)$ for $u \in \mathbb{E}_1$, let $\gamma\mathbb{E}_1$ be the 'trace space' of γ such that $E_1 \hookrightarrow \gamma\mathbb{E}_1 \hookrightarrow E_0$ and γ is a continuous surjection from $\mathbb{E}_1$ onto $\gamma\mathbb{E}_1$. Then, given $(f, u^0) \in \mathbb{E}_0 \times \gamma\mathbb{E}_1$, we rewrite the linear Cauchy problem (46) in the form:

$$\text{find } u \in \mathbb{E}_1 \text{ such that } (\partial + A, \gamma)u = (f, u^0).$$

Finally, the pair of Banach spaces $(\mathbb{E}_0, \mathbb{E}_1)$ is said to be a pair of maximal regularity for $\{ A(t) \ ; \ t \in J \}$, provided $(\partial + A, \gamma)$ is a topological isomorphism from $\mathbb{E}_1$ onto $\mathbb{E}_0 \times \gamma\mathbb{E}_1$. If this is the case, it is to be expected that we can apply linearization techniques based on the implicit function theorem to solve the quasilinear Cauchy problem (43), that now takes the form of the nonlinear operator equation

$$\big(\partial + A(\cdot, u), \gamma\big)u = \big(f(\cdot, u), u^0\big) .$$

In fact, this way 'fully nonlinear problems' can be handled. However, in the quasilinear situation we can combine the maximal regularity approach with the interpolation-extrapolation method to deal with the case of nonconstant domains as well. Indeed, in the qualitative study of quasilinear evolution equations a combination of the dynamical evolution operator approach and the maximal regularity technique, used in the third volume, gives the best results.

It turns out that the requirement that $(\mathbb{E}_0, \mathbb{E}_1)$ be a pair of maximal regularity for $\{ A(t) \ ; \ t \in J \}$ imposes restrictions on the spaces and on the operators, in general. These questions are carefully discussed in Chapter III, where the most important and useful abstract maximal regularity results are derived. Concrete realizations necessitate rather deep information on various function spaces, Nikols'kii spaces, for example, and on linear elliptic differential operators. This is provided in the second volume.

The Scope of the Theory

By now the reader surely feels that our approach to the quasilinear initial-boundary value problem (42) is rather lenghty, highly complicated, and requires a lot of technical details. This being certainly true, one might wonder if it is worthwhile at all.

Of course, there are simpler methods for proving the existence of weak solution to problem (42). First, the theory of monotone operators can sometimes be applied. However, it requires rather stringent conditions for the operators and nonlinearties and applies to a small class of problems only, leaving aside most systems of interest.

Second, there is the Galerkin approximation approach, that is perhaps most often applied to produce weak solutions of problem (42). That method is basically a Hilbert space technique, being based on coercivity estimates. However, coercivity estimates do not hold for general Petrowskii parabolic systems. To guarantee coercivity estimates one has to restrict the consideration to certain 'strongly parabolic' systems, a class that is too narrow to cover many interesting problems occurring in science. Similar statements hold for other (space or time) discretization methods.

Even if the system is strongly parabolic, so that the Galerkin method is available, that approach gives weak W_2^1-solutions only, that is, solutions in the sense of (64) with $p = 2$. However, it is well-known (e.g., [Gia83]) that weak W_2^1-solutions of *systems* are not regular, in general. In fact, it is not even possible, as a rule, to prove that they are Hölder continuous. This lack of regularity renders it rather difficult — if not impossible — to develop a dynamical systems theory and to use techniques from nonlinear analysis to establish qualitative properties of the solution set.

In contrast, our approach applies to *noncoercive* problems as well and gives weak W_p^1-solutions, where $p > n$. Thus, thanks to (18), the solutions we find are already known to be Hölder continuous from the very beginning. As indicated above, this implies that all our solutions are classical if all data are sufficiently smooth. Of course, there may be singular solutions — in particular, if the coefficients of $\mathcal{A}$ and $\mathcal{B}$ lack regularity — that we 'do not see' in our approach. However, we obtain a completely satisfactory local theory in the class of regular functions. It tells us, roughly speaking, that the quasilinear Cauchy problem (43) has the same behavior — as far as local existence and continuous dependence of the solutions on the data are concerned — as an ordinary differential equation $\dot{y} = \varphi(t, y)$ in the finite-dimensional setting. In addition, since our solutions are weak W_2^1-solutions as well, they coincide on the common interval of existence with weak W_2^1-solutions obtained by any other method, provided there is a uniqueness theorem for weak W_2^1-solutions.

In addition, our approach is general. It is by no means restricted to second order reaction-diffusion systems, but applies to quasilinear systems of Petrowskii parabolic type of arbitrary order. Furthermore, the coefficients of $\mathcal{A}$ and $\mathcal{B}$, as well as the nonlinearities, can be nonlocal functions. This allows to deal rather easily with elliptic-parabolic systems, for example, by reducing them to abstract parabolic evolution equations. The general abstract theory can also be applied to produce nontrivial results for free boundary value problems, and to systems with 'dynamic boundary conditions'. It can handle 'highly degenerate' problems that are not even Petrowskii parabolic, as well as quasilinear parabolic problems involving pseudodifferential operators. Thus, at the end, we are rewarded by a rather flexible general theory that requires the verification of a few hypotheses only. Those can be met in many concrete situations occurring in applications, as is demonstrated by the examples given in the third volume.

> Wenn Du wissen willst, *was* bewiesen wurde, schau den Beweis an.[5]
>
> *Ludwig Wittgenstein*
> *(1889–1951)*

[5] If you want to know *what* has been proven, study the proof.

Notations and Conventions

In this section we collect some of the definitions and notations that we use throughout this treatise. The reader is expected to have a working knowledge of the basic theory of locally convex spaces and (unbounded) linear operators in Banach spaces. Thus we review here only briefly the most basic facts from linear functional analysis in order to fix the notations. We refer to the literature (in particular, to [Ber74], [DuS57], [Edw65], [Golb66], [HiP57], [Hor66], [Jar81], [Sch71], [Tre67], and [Yos65]) for proofs and many more details.

1 Topological Spaces

Let X and Y be nonempty sets. Then Y^X is the set of all maps $u\colon X \to Y$. If A is a subset of X, the characteristic function of A (in X) is denoted by χ_A, that is,

$$\chi_A(x) = 1 \text{ if } x \in A \quad \text{and} \quad \chi_A(x) = 0 \text{ if } x \in A^c := X \backslash A\ .$$

Occasionally, we put $\mathbf{1} := \mathbf{1}_X := \chi_X$.

We denote by

$$\mathbb{N} := \{0, 1, 2, \ldots\}$$

the set of all nonnegative integers, and $\bar{\mathbb{N}} := \mathbb{N} \cup \{\infty\}$. Similarly, $\bar{\mathbb{R}} := \mathbb{R} \cup \{\pm\infty\}$, etc. We put

$$x \wedge y := \min\{x, y\}\ , \quad x \vee y := \max\{x, y\}\ , \qquad x, y \in \mathbb{R}\ ,$$

and

$$x^+ := x \vee 0\ , \quad x^- := (-x) \vee 0$$

are the positive and negative parts, respectively, of $x \in \mathbb{R}$, more generally, of $x \in \mathbb{R}^X$.

The set of all $x \in X$ in which definitions and relations hold is often denoted by $[\ldots]$, where $\ldots$ stand for the definitions and relations. For example, if $u \in \mathbb{R}^X$ then

$$[u \geq 0] := \big\{\, x \in X\ ;\ u(x) \geq 0 \,\big\}$$

etc.

Let X and Y be topological spaces. Then $C(X, Y)$ is the set of all continuous maps in Y^X. We write

$$X \hookrightarrow Y \quad \text{or} \quad i\colon X \hookrightarrow Y\ ,$$

if X is **continuously injected** in Y, that is, $X \subset Y$ and the **natural injection**

$$i\colon X \to Y\ , \quad x \mapsto x$$

is continuous. If X is a dense subset of Y, we write $X \overset{d}{\subset} Y$. Thus $X \overset{d}{\hookrightarrow} Y$ means that X is densely and continuously injected in Y.

We often write 1_X or simply 1 for the identity mapping, $\mathrm{id}_X : X \to X,\ \ x \mapsto x$, if no confusion seems likely.

Let $X := (X, d)$ be a metric space and let M be a nonempty subset of X. Then

$$\mathbb{B}(M,\varepsilon) := \mathbb{B}_X(M,\varepsilon) := \bigl\{\, x \in X \ ;\ \operatorname{dist}(x,M) < \varepsilon \,\bigr\}$$

is the **open ε-neighborhood** of M in X and $\bar{\mathbb{B}}(M,\varepsilon) := \bigl\{\, x \in X \ ;\ \operatorname{dist}(x,M) \le \varepsilon \,\bigr\}$ is the corresponding **closed** ε-neighborhood. If $M = \{x\}$, we write $\mathbb{B}(x,\varepsilon)$ and $\bar{\mathbb{B}}(x,\varepsilon)$ for $\mathbb{B}\bigl(\{x\},\varepsilon\bigr)$ and $\bar{\mathbb{B}}\bigl(\{x\},\varepsilon\bigr)$, respectively, and call these sets open, resp. closed, **balls** with center at x and radius ε.

Let A and B be subsets of some topological space X. We use the symbols $\overline{A}$, $\operatorname{cl}(A)$, or $\operatorname{cl}_X(A)$ to denote the closure of A in X. Similarly, $\mathring{B}$, $\operatorname{int} B$, or $\operatorname{int}_X(B)$ stands for the interior of B in X. Then A is **compactly contained** in B, in symbols:

$$A \subset\subset B\ ,$$

if $\overline{A}$ is compact and contained in $\mathring{B}$.

2 Locally Convex Spaces

By a vector space, we always mean a vector space over $\mathbb{K}$ where $\mathbb{K} = \mathbb{R}$ or $\mathbb{K} = \mathbb{C}$. If M is a subset of a vector space, we put

$$\dot{M} := M \backslash \{0\}\ .$$

Let X be a nonempty set and let Y be a vector space. Then Y^X is given the canonical vector space structure defined by

$$(\alpha u + \beta v)(x) := \alpha u(x) + \beta v(x)\ , \qquad x \in X\ , \quad u, v \in Y^X\ , \quad \alpha, \beta \in \mathbb{K}\ .$$

If X is also a vector space (over the same field), $\operatorname{Hom}(X,Y)$ is the vector subspace of Y^X of all linear maps and $\operatorname{End}(X) := \operatorname{Hom}(X,X)$, the space of endomorphisms.

Let X and Y be topological vector spaces (*TVSs*). Then $\mathcal{L}(X,Y)$ is the vector subspace of $\operatorname{Hom}(X,Y)$ consisting of all continuous linear maps, and

$$\mathcal{L}(X) := \mathcal{L}(X,X)\ .$$

In this case $X \hookrightarrow Y$ means also that X is a vector subspace of Y, that is, $i \in \mathcal{L}(X,Y)$. Moreover,

$$\mathcal{L}\mathrm{is}(X,Y) := \bigl\{\, T \in \mathcal{L}(X,Y)\ ;\ T \text{ is bijective and } T^{-1} \in \mathcal{L}(Y,X) \,\bigr\}$$

is the set of all topological linear (**toplinear**) isomorphisms from X onto Y and

$$\mathcal{L}\mathrm{aut}(X) := \mathcal{L}\mathrm{is}(X,X)$$

is the group of all toplinear automorphisms of X, the general linear group, also denoted by $\mathcal{GL}(X)$.

By a **locally convex space** (*LCS*) we mean a Hausdorff locally convex *TVS*. If X is a *LCS*, there exists a separating family of continuous seminorms on X inducing the topology. Conversely, each separating family of seminorms $\mathcal{P}$ on a vector space X induces a coarsest locally convex Hausdorff topology on X such that each $p \in \mathcal{P}$ is continuous (cf. [Hor66, II.4]). Occasionally, we shall write $X := (X, \mathcal{P})$ if X is a *LCS* whose topology is induced by the family of seminorms $\mathcal{P}$.

Let $\mathcal{P}$ and $\mathcal{Q}$ be two families of seminorms on the same vector space X. Then $\mathcal{P}$ is said to be stronger than $\mathcal{Q}$ (and $\mathcal{Q}$ is weaker than $\mathcal{P}$) if

$$(X, \mathcal{P}) \hookrightarrow (X, \mathcal{Q}) .$$

If $\mathcal{P}$ is stronger and weaker than $\mathcal{Q}$, that is, if $\mathcal{P}$ and $\mathcal{Q}$ induce the same topology, then $\mathcal{P}$ and $\mathcal{Q}$ are equivalent. In this case we write

$$(X, \mathcal{P}) \doteq (X, \mathcal{Q}) .$$

Let X and $Y := (Y, \mathcal{Q})$ be *LCSs*. We write $\mathcal{L}_s(X, Y)$ if $\mathcal{L}(X, Y)$ is given the **simple convergence topology** induced by the family of seminorms

$$\big\{ T \mapsto q(Tx) \ ; \ x \in X, \ q \in \mathcal{Q} \big\} ,$$

whereas $\mathcal{L}(X, Y)$ *means that this vector space is always equipped with the* **bounded convergence topology** defined by the family of seminorms

$$\big\{ T \mapsto \sup_{x \in B} q(Tx) \ ; \ B \subset X \text{ bounded}, \ q \in \mathcal{Q} \big\} .$$

Observe that $\mathcal{L}_s(X, Y)$ and $\mathcal{L}(X, Y)$ are *LCSs* and

$$\mathcal{L}(X, Y) \hookrightarrow \mathcal{L}_s(X, Y) .$$

If X and Y are normed vector spaces, the topology of $\mathcal{L}_s(X, Y)$ is called **strong (operator) topology** and the one of $\mathcal{L}(X, Y)$ **uniform operator topology**. Observe that $\mathcal{L}(X, Y)$ is then a normed vector space with the **uniform operator norm** defined by

$$\|T\| := \sup\{ \|Tx\| \, / \, \|x\| \ ; \ x \in \dot{X} \}$$

(provided $\dot{X} \neq \emptyset$, of course). Moreover, $\mathcal{L}(X, Y)$ is a Banach space iff Y is one [Golb66, Theorem I.3.5 and Corollary I.5.8].

Let X and Y be *LCSs*. A linear map $T \colon X \to Y$ is **compact** if it maps bounded sets into compact sets. We put

$$\mathcal{K}(X, Y) := \big\{ T \in \mathcal{L}(X, Y) \ ; \ T \text{ is compact} \big\} ,$$

and $\mathcal{K}(X) := \mathcal{K}(X, X)$. Hence $T \in \mathcal{K}(X, Y)$ iff $T \in \mathcal{L}(X, Y)$ and $T(B)$ is relatively compact for each bounded subset B of X. Observe that the last condition implies the boundedness of T, hence its continuity, if X is metrizable. (It should be noted that our definition of a compact operator is not the one usually employed in the general theory of *TVSs* (e.g., [Jar81]).

We say that X is **compactly injected** in Y,

$$X \overset{d}{\hookrightarrow} Y ,$$

if $i \colon X \hookrightarrow Y$ and $i \in \mathcal{K}(X, Y)$.

If X is a normed vector space, we denote the open [resp. closed] unit ball in X by $\mathbb{B} := \mathbb{B}_X$ [resp. $\bar{\mathbb{B}}$]. Moreover, $x + \varepsilon\mathbb{B} := \mathbb{B}(x, \varepsilon)$ and $\mathbb{B}^n := \mathbb{B}_{\mathbb{R}^n}$, where $\mathbb{R}^n$ — more generally, $\mathbb{K}^n$ — is always given the euclidean norm, $|\cdot|$, unless explicitly stated otherwise.

Let X be a *TVS*. Then the **dual** (space) X' of X is given by $X' := \mathcal{L}(X, \mathbb{K})$. According to our conventions X' is always endowed with the bounded convergence topology that is called **strong topology of** X' in this context if X is a *LCS*. We often denote the value of $x' \in X'$ at $x \in X$ by $\langle x', x \rangle_X$, or simply by $\langle x', x \rangle$, and call the bilinear map

$$\langle \cdot, \cdot \rangle_X := \langle \cdot, \cdot \rangle : X' \times X \to \mathbb{K}$$

duality pairing between X' and X.

Let X be a *LCS*. We put $X'_{w^*} := \mathcal{L}_s(X, \mathbb{K})$ since the simple convergence topology is called w^***-topology** of X' in this context. The w^*-topology of X' is also denoted by $\sigma(X', X)$.

Due to the Hahn-Banach theorem [Hor66, III.1] the family

$$\mathcal{P}_{X'} := \{\, x \mapsto |\langle x', x \rangle| \ ; \ x' \in X' \,\}$$

is a separating family of seminorms on X. Hence $(X, \mathcal{P}_{X'})$ is a *LCS* and the topology induced by $\mathcal{P}_{X'}$ is called **weak topology** of X and denoted by $\sigma(X, X')$. We also write X_w for $\big(X, \sigma(X, X')\big)$. It is obvious that

$$X \hookrightarrow X_w \quad \text{and} \quad X' \hookrightarrow X'_{w^*} \ .$$

Sometimes we use $\rightharpoonup$ [resp. $\overset{w^*}{\rightharpoonup}$] to denote weak [resp. weak star] convergence (that is, convergence with respect to the weak [resp. w^*-] topology).

3 Complexifications

Let X be a real vector space. We denote by $X_{\mathbb{C}}$ the additive group $X \times X$ equipped with the multiplication by complex numbers $(\alpha, \beta) = \alpha + i\beta \in \mathbb{C}$ given by

$$(\alpha, \beta)(x, y) := (\alpha x - \beta y, \beta x + \alpha y) \ .$$

We identify $x \in X$ with $(x, 0) \in X \times X$ and find that $(0, 1)(x, 0) = (0, x)$. Thus each $z \in X \times X$ is uniquely represented as $z = x + iy$ with $x, y \in X$. Hence $X + iX := X_{\mathbb{C}}$ is a complex vector space, the **complexification of** X, where $X + iX$ is to be understood as the direct sum over $\mathbb{R}$ of the real vector spaces X and iX. In particular, $x + iy \mapsto x =: \operatorname{Re}(x + iy)$ and $x + iy \mapsto y =: \operatorname{Im}(x + iy)$ are $\mathbb{R}$-linear projections of $X_{\mathbb{C}}$ onto X. Observe that X is identified with an $\mathbb{R}$-linear subspace of $X_{\mathbb{C}}$, the **real subspace**. If X is a real *LCS* , $X_{\mathbb{C}}$ is a complex *LCS* and $X \hookrightarrow X_{\mathbb{C}}$ (as an $\mathbb{R}$-linear injection).

Let X, Y and Z be real vector spaces. Given $T \in \operatorname{Hom}(X, Y)$, define the **complexification**, $T_{\mathbb{C}}$, **of** T by

$$T_{\mathbb{C}}(x + iy) := Tx + iTy \ , \qquad x + iy \in X_{\mathbb{C}} \ .$$

Then $T_{\mathbb{C}} \in \operatorname{Hom}(X_{\mathbb{C}}, Y_{\mathbb{C}})$ and $T_{\mathbb{C}}(X) \subset Y$. If no confusion seems likely, we write T for $T_{\mathbb{C}}$, that is, we consider $\operatorname{Hom}(X, Y)$ as an $\mathbb{R}$-vector subspace of $\operatorname{Hom}(X_{\mathbb{C}}, Y_{\mathbb{C}})$, the real vector subspace of **real linear maps** in $\mathrm{Hom}(X_{\mathbb{C}},Y_{\mathbb{C}})$. The elements R of this space are characterized by $R(X)\subset Y$. Given $S\in\mathrm{Hom}(X_{\mathbb{C}},Y_{\mathbb{C}})$, there exist unique $S_1,S_2\in\mathrm{Hom}(X,Y)$ such that $Sx=S_1x+iS_2x$ for $x\in X$. Hence

$$\begin{aligned} Sz = Sx+iSy &= S_1x+iS_2x+i(S_1y+iS_2y)\\ &= S_1x+iS_1y+i(S_2x+iS_2y)\end{aligned} \tag{3.1}$$

for $z=x+iy\in X_{\mathbb{C}}$. Thus $Sz=S_1z+iS_2z$ for $z\in X_{\mathbb{C}}$, where we have used the identifications $S_j=(S_j)_{\mathbb{C}}$. Hence, letting $\mathrm{Re}\,S:=S_1$ and $\mathrm{Im}\,S:=S_2$,

$$S=\mathrm{Re}\,S+i\,\mathrm{Im}\,S\ ,\qquad S\in\mathrm{Hom}(X_{\mathbb{C}},Y_{\mathbb{C}})\ , \tag{3.2}$$

where $\mathrm{Re}\,S$ and $\mathrm{Im}\,S$ are real linear maps in $\mathrm{Hom}(X_{\mathbb{C}},Y_{\mathbb{C}})$. Observe that S is a real linear map in $\mathrm{Hom}(X_{\mathbb{C}},Y_{\mathbb{C}})$ iff $\mathrm{Im}\,S=0$. Also observe that

$$(ST)_{\mathbb{C}}=S_{\mathbb{C}}T_{\mathbb{C}}\ ,\qquad T\in\mathrm{Hom}(X,Y)\ ,\quad S\in\mathrm{Hom}(Y,Z)\ .$$

Given $S,T\in\mathrm{Hom}(X,Y)\subset\mathrm{Hom}(X_{\mathbb{C}},Y_{\mathbb{C}})$, it follows that

$$S+iT\in\big[\mathrm{Hom}(X,Y)\big]_{\mathbb{C}}\ .$$

Hence we deduce from (3.2) that, by identifying $\mathrm{Hom}(X,Y)$ with the vector subspace of real linear maps in $\mathrm{Hom}(X_{\mathbb{C}},Y_{\mathbb{C}})$, we can — and will — identify $\mathrm{Hom}(X_{\mathbb{C}},Y_{\mathbb{C}})$ with the complexification of $\mathrm{Hom}(X,Y)$, that is,

$$\mathrm{Hom}(X_{\mathbb{C}},Y_{\mathbb{C}})=\big[\mathrm{Hom}(X,Y)\big]_{\mathbb{C}}\ . \tag{3.3}$$

Now suppose that X and Y are real *LCSs*. Then

$$T\in\mathcal{L}(X,Y)\iff T_{\mathbb{C}}\in\mathcal{L}(X_{\mathbb{C}},Y_{\mathbb{C}})$$

and

$$T\in\mathcal{K}(X,Y)\iff T_{\mathbb{C}}\in\mathcal{K}(X_{\mathbb{C}},Y_{\mathbb{C}})\ .$$

Moreover, (3.3) easily implies

$$\mathcal{L}(X_{\mathbb{C}},Y_{\mathbb{C}})=\big[\mathcal{L}(X,Y)\big]_{\mathbb{C}}\ .$$

In particular, since $\mathbb{R}_{\mathbb{C}}=\mathbb{C}$ we see that

$$(X_{\mathbb{C}})'=(X')_{\mathbb{C}} \tag{3.4}$$

and it follows from (3.1) that the duality pairing $\langle\cdot,\cdot\rangle_{X_{\mathbb{C}}}:X_{\mathbb{C}}'\times X_{\mathbb{C}}\to\mathbb{C}$ is given by

$$\langle f+ig,x+iy\rangle_{X_{\mathbb{C}}}=\langle f,x\rangle-\langle g,y\rangle+i\big(\langle f,y\rangle+\langle g,x\rangle\big)\ .$$

Also observe that

$$(T_{\mathbb{C}})'=(T')_{\mathbb{C}}\ ,\qquad T\in\mathcal{L}(X,Y)\ .$$

Now suppose that $(X,\|\cdot\|)$ is a (complete) normed vector space over $\mathbb{R}$. Then $X_{\mathbb{C}}$ is a (complete) normable vector space over $\mathbb{C}$. It is easily verified that

$$\|z\|_{X_{\mathbb{C}}}:=\sup_{0\le\alpha\le2\pi}\|\mathrm{Re}(e^{i\alpha}z)\|=\sup_{0\le\alpha\le2\pi}\|x\cos\alpha+y\sin\alpha\|\ ,\qquad z=x+iy\in X_{\mathbb{C}}\ ,$$

defines a norm on $X_{\mathbb{C}}$ and that $X \hookrightarrow X_{\mathbb{C}}$ is a real linear isometry. If Y is a second real normed vector space, it is easily verified that

$$\mathcal{L}(X,Y) \hookrightarrow \mathcal{L}(X_{\mathbb{C}}, Y_{\mathbb{C}}) \quad \text{is a real linear isometry}$$

provided, of course, $\mathcal{L}(X_{\mathbb{C}}, Y_{\mathbb{C}})$ is given the uniform operator norm induced by $\|\cdot\|_{X_{\mathbb{C}}}$ and $\|\cdot\|_{Y_{\mathbb{C}}}$.

3.1 Remark Let $(F, \|\cdot\|_F)$ be a complex Banach space of complex-valued functions and let $(X, \|\cdot\|)$ be the real subspace of $\mathbb{R}$-valued functions. Then $X_{\mathbb{C}} \doteq F$ but $\|\cdot\|_{X_{\mathbb{C}}} \neq \|\cdot\|_F$, in general. For example, let $F := (\mathbb{C}^2, |\cdot|_1)$, where $|\cdot|_1$ stands for the ℓ_1-norm. Then $X = (\mathbb{R}^2, |\cdot|_1)$. Let $x, y \in X$ be given by $x := (2,0)$ and $y := (0,1)$. Then $x + iy = (2, i) \in \mathbb{C}^2$ and $|x + iy|_1 = 3$. On the other hand,

$$\begin{aligned} |x+iy|_{X_{\mathbb{C}}} &= \sup_{0\le\alpha\le 2\pi} |(2\cos\alpha, \sin\alpha)|_1 \\ &= \sup\{\, 2\,|\cos\alpha| + |\sin\alpha| \;;\; 0 \le \alpha \le 2\pi \,\} = \sqrt{5}\;. \end{aligned}$$ ■

Although (3.4) is true, it follows (from Remark 3.1, for example) that, in general, the dual norm of $(X_{\mathbb{C}})'$ is different from the norm $\|\cdot\|_{(X')_{\mathbb{C}}}$ if X is a normed vector space. Thus (3.4) has to be replaced by the more precise relation

$$(X_{\mathbb{C}})' \doteq (X')_{\mathbb{C}}\;,$$

if X is a real normed vector space.

Now suppose that $\big(X, (\cdot|\cdot)\big)$ is a real inner product space. Then $\big(X_{\mathbb{C}}, (\cdot|\cdot)_{X_{\mathbb{C}}}\big)$ is a complex inner product space where

$$(x+iy|u+iv)_{X_{\mathbb{C}}} := (x|u) + (y|v) + i\big[(y|u) - (x|v)\big]\;.$$

In this case the inner product norm $\|x + iy\| := (\|x\|^2 + \|y\|^2)^{1/2}$ is a norm on $X_{\mathbb{C}}$ such that $X \hookrightarrow X_{\mathbb{C}}$ is a real linear isometry.

4 Unbounded Linear Operators

Let X and Y be *LCSs*. If A is a linear operator with domain, $\mathrm{dom}(A)$, in X and range in Y, we write

$$A\colon \mathrm{dom}(A) \subset X \to Y\;.$$

Observe that $\mathrm{dom}(A)$ is a vector subspace of X and $A \in \mathrm{Hom}\big(\mathrm{dom}(A), Y\big)$. In general, the domain and the image of a map f are denoted by $\mathrm{dom}(f)$ and $\mathrm{im}(f)$, respectively.

We denote by $\mathcal{C}(X,Y)$ the set of all closed linear operators $A : \mathrm{dom}(A) \subset X \to Y$, and $\mathcal{C}(X) := \mathcal{C}(X,X)$. If X and Y are normed vector spaces,

$$D(A) := \big(\mathrm{dom}(A), \|\cdot\|_A\big)\;,$$

where

$$\|x\|_A := \|Ax\| + \|x\|\;, \qquad x \in \mathrm{dom}(A)\;,$$

is the **graph norm** of A. If X and Y are Banach spaces, $D(A)$ is a Banach space iff $A \in \mathcal{C}(X,Y)$.

Let X and Y be *LCSs* such that $X \hookrightarrow Y$, and let $A\colon \mathrm{dom}(A) \subset Y \to Y$ be linear in Y. Then the X-**realization**, A_X, (the part of A in X, or the maximal restriction of A to X) is the linear operator in X, defined by

$$\mathrm{dom}(A_X) := \{\, x \in X \cap \mathrm{dom}(A)\ ;\ Ax \in X \,\}\ , \qquad A_X x := Ax\ .$$

If no confusion seems likely, we often omit the index X. Observe that $A_X \in \mathcal{C}(X)$ if $A \in \mathcal{C}(Y)$.

5 General Conventions

Let X be a real *LCS*. If in a given formula there are implicit or explicit references to complex numbers, it is always understood that this formula is interpreted in its complexified version. Thus, for example, if X is a ($\mathbb{K}$-)Banach space and A is a linear operator in X, we mean by $\sigma(A)$ and $\rho(A)$ the spectrum and the resolvent set, respectively, of A if $\mathbb{K} = \mathbb{C}$ and of $A_{\mathbb{C}}$ if $\mathbb{K} = \mathbb{R}$.

We often use c to denote constants, that may differ from occurrence to occurrence, but are always independent of the specific free variables occurring at a given place. Thus we use c very much in the same way as the Landau symbol O. If the equations under consideration depend on additional parameters, say $\alpha, \beta, \ldots$, we sometimes write $c(\alpha, \beta, \ldots)$ to indicate this.

Chapter I

Generators and Interpolation

The abstract theory of linear parabolic problems, developed in the first part of this treatise, rests on two cornerstones: on the theory of analytic semigroups and on interpolation theory. In the first section of this chapter we discuss in some detail the classes of generators of analytic semigroups that are used throughout. In the second section we review the fundamentals of interpolation theory. We also introduce the class of 'admissible interpolation functors' that, on the one hand, is flexible enough to build an easy and general abstract theory of evolution equations on it, and, on the other hand, is general enough to cover all applications occurring in practice.

Throughout this chapter E, F, E_j, and F_j, $j = 0, 1, 2, \ldots$, denote Banach spaces.

1 Generators of Analytic Semigroups

In this section we study the properties of generators of analytic semigroups which are basic for the whole treatise. Although these results are well-known, in principle, our approach has some novel aspects. In particular, we develop quantitative versions of perturbation theorems and related results which show that much of the theory can be controlled by two numerical parameters, which — as we shall see in later chapters — are readily accessible in concrete applications.

We assume that the reader is familiar with the basic facts about strongly continuous semigroups on Banach spaces. We refer to [ClH+87], [Dav80], [Fat83], [Gols85], [HiP57], [Kat66], [KraZ76], and [Paz83] for these and many more details.

1.1 Properties of Linear Operators

For the reader's convenience we include an easy proof of the following useful perturbation theorem concerning bounded invertibility.

1.1.1 Proposition *Let X be a connected metric space, let*

$$B \in C\big(X, \mathcal{L}(E,F)\big) ,$$

and suppose that there exists $\beta > 0$ such that

$$\|B(x)e\| \geq \beta \,\|e\| , \qquad x \in X , \quad e \in E . \tag{1.1.1}$$

Then $B(X) \cap \mathcal{L}\mathrm{is}(E,F) \neq \emptyset$ implies $B(X) \subset \mathcal{L}\mathrm{is}(E,F)$.

Proof It follows from (1.1.1) that $B(x)$ is injective for $x \in X$. Thus, thanks to Banach's homomorphism theorem, it suffices to prove that $B(x)$ is surjective for each $x \in X$.

Let $X_0 := \big\{ x \in X \ ; \ B(x) \in \mathcal{L}\mathrm{is}(E,F) \big\}$. Then $X_0 \neq \emptyset$ by assumption. Since $\mathcal{L}\mathrm{is}(E,F)$ is open in $\mathcal{L}(E,F)$, the continuity of B implies that X_0 is open in X. Let (x_j) be a sequence in X_0 with $x_j \to x$ in X and let $f \in F$ be given. Letting $e_j := B^{-1}(x_j)f$, it follows that

$$B(x_j)(e_j - e_k) = f - B(x_j)e_k = \big[B(x_k) - B(x_j)\big]e_k .$$

Hence we deduce from (1.1.1) that

$$\beta \,\|e_j - e_k\| \leq \|B(x_j) - B(x_k)\| \,\|e_k\| \leq \beta^{-1} \,\|f\| \,\|B(x_j) - B(x_k)\| ,$$

which shows that (e_j) is a Cauchy sequence in E. Thus $e_j \to e$ for some $e \in E$ and

$$B(x)e - f = B(x)(e - e_j) + \big[B(x) - B(x_j)\big]e_j \to 0 .$$

This shows that X_0 is closed in X. Thus $X_0 = X$ by the connectedness of X. ■

Suppose that $E_1 \hookrightarrow E_0$. Given a linear map $A\colon E_1 \to E_0$, we can interpret it as a linear operator, A_0, *in* E_0 with domain E_1. If no confusion seems likely, we write again A for A_0.

1.1.2 Lemma *Let $A\colon E_1 \to E_0$ be linear. Then $A \in \mathcal{L}(E_1, E_0) \cap \mathcal{C}(E_0)$ iff the graph norm (of A_0) is an equivalent norm for E_1.*

Proof Let $A \in \mathcal{L}(E_1, E_0) \cap \mathcal{C}(E_0)$ and let $i\colon E_1 \hookrightarrow E_0$. Then

$$\|x\|_A = \|Ax\|_0 + \|x\|_0 \leq (\|A\| + \|i\|) \,\|x\|_1 , \qquad x \in E_1 ,$$

and, consequently, $E_1 \hookrightarrow D(A)$. Since A is closed, $D(A)$ is complete. Hence Banach's homomorphism theorem implies $D(A) \hookrightarrow E_1$.

Conversely, if the graph norm is equivalent to $\|\cdot\|_1$, there exists $\alpha > 0$ such that

$$\|Ax\|_0 \le \|x\|_A \le \alpha \, \|x\|_1 \ , \qquad x \in E_1 \ .$$

Hence $A \in \mathcal{L}(E_1, E_0)$. Since $D(A) \doteq E_1$, we see that $D(A)$ is complete. Consequently, $A \in \mathcal{C}(E_0)$. ∎

1.2 The Class $\mathcal{H}(E_1, E_0)$

By a **densely injected Banach couple** we mean a pair of Banach spaces, (E_0, E_1), such that $E_1 \overset{d}{\hookrightarrow} E_0$.

Throughout the remainder of Section 1 *we suppose that* (E_0, E_1) *is a densely injected Banach couple.*

We denote by

$$\mathcal{H}(E_1, E_0)$$

the set of all $A \in \mathcal{L}(E_1, E_0)$ such that $-A$, considered as a linear operator *in* E_0 with domain E_1, is the infinitesimal generator of a strongly continuous analytic semigroup $\{ e^{-tA} \ ; \ t \ge 0 \}$ on E_0, that is, in $\mathcal{L}(E_0)$.

In order to derive uniform estimates for these semigroups and related operators, it is important to possess quantitative descriptions of $\mathcal{H}(E_1, E_0)$. For this purpose, given $\kappa \ge 1$ and $\omega > 0$, we write

$$A \in \mathcal{H}(E_1, E_0, \kappa, \omega)$$

if $\omega + A \in \mathcal{L}\mathrm{is}(E_1, E_0)$ and

$$\kappa^{-1} \le \frac{\|(\lambda + A)x\|_0}{|\lambda| \, \|x\|_0 + \|x\|_1} \le \kappa \ , \qquad x \in \dot{E}_1 \ , \quad \mathrm{Re}\, \lambda \ge \omega \ , \tag{1.2.1}$$

where $\|\cdot\|_j$ is the norm in E_j.

1.2.1 Remarks (a) Inequality (1.2.1) has the advantage of being symmetric with respect to $\|x\|_1$ and $|\lambda| \, \|x\|_0$. However, it is easily verified that it suffices to require less in order that a linear map $A \colon E_1 \to E_0$ belongs to $\mathcal{H}(E_1, E_0, \kappa, \omega)$ for some $\kappa \ge 1$ and $\omega > 0$. In fact, suppose that there are $\kappa \ge 1$ and $\omega > 0$ such that $\omega + A \in \mathcal{L}\mathrm{is}(E_1, E_0)$ and

$$|\lambda| \, \|x\|_0 \le \kappa \, \|(\lambda + A)x\|_0 \ , \qquad x \in E_1 \ , \quad \mathrm{Re}\, \lambda \ge \omega \ . \tag{1.2.2}$$

Then, letting

$$\kappa_1 := \|A\|_{\mathcal{L}(E_1, E_0)} \vee \kappa\big(1 + 3 \, \|(\omega + A)^{-1}\|_{\mathcal{L}(E_0, E_1)}\big) \ ,$$

it follows that $A \in \mathcal{H}(E_1, E_0, \kappa_1, \omega)$. Similarly, if there are $\kappa \geq 1$ and $\omega > 0$ such that $\omega + A \in \mathcal{L}\mathrm{is}(E_1, E_0)$ and

$$\|x\|_1 \leq \kappa \, \|(\lambda + A)x\|_0 \ , \qquad x \in E_1 \ , \quad \operatorname{Re}\lambda \geq \omega \ , \tag{1.2.3}$$

then $A \in \mathcal{H}(E_1, E_0, \kappa_2, \omega)$, where $\kappa_2 := 1 + \kappa + \kappa \, \|A\|_{\mathcal{L}(E_1,E_0)}$. On the other hand, if $A \in \mathcal{H}(E_1, E_0, \kappa, \omega)$ for some $\kappa \geq 1$ and $\omega > 0$, estimates (1.2.2) and (1.2.3) are trivially true .

Proof Let (1.2.2) be satisfied. Then we infer from

$$x = (\omega + A)^{-1}\big[(\lambda + A)x + (\omega - \lambda)x\big] \ , \qquad x \in E_1 \ ,$$

and from $\omega \leq |\lambda|$ for $\operatorname{Re}\lambda \geq \omega$ that

$$\|x\|_1 \leq \|(\omega + A)^{-1}\|_{\mathcal{L}(E_0,E_1)} \, (1 + 2\kappa) \, \|(\lambda + A)x\|_0 \ , \qquad x \in E_1 \ , \quad \operatorname{Re}\lambda \geq \omega \ .$$

Hence

$$\begin{aligned} |\lambda| \, \|x\|_0 + \|x\|_1 &\leq \kappa\big(1 + 3 \, \|(\omega + A)^{-1}\|_{\mathcal{L}(E_0,E_1)}\big) \, \|(\lambda + A)x\|_0 \\ &\leq \kappa_1 \, \|(\lambda + A)x\|_0 \end{aligned}$$

for $x \in E_1$ and $\operatorname{Re}\lambda \geq \omega$. Moreover,

$$\begin{aligned} \|(\lambda + A)x\|_0 &\leq |\lambda| \, \|x\|_0 + \|A\|_{\mathcal{L}(E_1,E_0)} \, \|x\|_1 \\ &\leq \big(1 \vee \|A\|_{\mathcal{L}(E_1,E_0)}\big)\big(|\lambda| \, \|x\|_0 + \|x\|_1\big) \end{aligned} \tag{1.2.4}$$

for $x \in E_1$ and $\operatorname{Re}\lambda \geq \omega$. This implies the first assertion.

Similarly, if (1.2.3) is satisfied,

$$\begin{aligned} |\lambda| \, \|x\|_0 &\leq \|(\lambda + A)x\|_0 + \|A\|_{\mathcal{L}(E_1,E_0)} \, \|x\|_1 \\ &\leq \big(1 + \kappa \, \|A\|_{\mathcal{L}(E_1,E_0)}\big) \, \|(\lambda + A)x\|_0 \ . \end{aligned}$$

Hence

$$|\lambda| \, \|x\|_0 + \|x\|_1 \leq \kappa_2 \, \|(\lambda + A)x\|_0$$

for $x \in E_1$ and $\operatorname{Re}\lambda \geq \omega$. By taking (1.2.4) into consideration, the second assertion follows as well. ■

(b) Suppose that $E_1 = E_0 =: E$. Then $\mathcal{H}(E, E) = \mathcal{L}(E)$. Moreover,

$$A \in \mathcal{H}(E, E, 2, 1 + 2 \, \|A\|) \ , \quad A \in \mathcal{L}(E) \ .$$

Proof Since $\mathcal{H}(E, E) \subset \mathcal{C}(E)$, the closed graph theorem implies $\mathcal{H}(E, E) \subset \mathcal{L}(E)$. If $A \in \mathcal{L}(E)$ then, given $\lambda \in [\operatorname{Re} z \geq 1 + 2 \, \|A\|]$,

$$\begin{aligned} (1/2)(|\lambda| \, \|x\| + \|x\|) \leq (|\lambda| - \|A\|) \, \|x\| &\leq \|(\lambda + A)x\| \\ &\leq (|\lambda| + \|A\|) \, \|x\| \leq 2(|\lambda| \, \|x\| + \|x\|) \end{aligned}$$

for $x \in E$. Hence (1.2.1) is satisfied with $\kappa := 2$ and $\omega := 1 + 2 \, \|A\|$. Lastly, we infer from $\sigma(A) \subset [\,|z| \leq \|A\|\,]$ that $\omega + A \in \mathcal{L}\mathrm{aut}(E)$. ■

The basic result concerning the classes $\mathcal{H}(E_1, E_0, \kappa, \omega)$ is the following:

1.2.2 Theorem $\mathcal{H}(E_1, E_0) = \bigcup_{\substack{\kappa\geq 1\\ \omega>0}} \mathcal{H}(E_1, E_0, \kappa, \omega)$.

Proof Suppose that $A \in \mathcal{H}(E_1, E_0, \kappa, \omega)$. Lemma 1.1.2 implies $A \in \mathcal{C}(E_0)$. From (1.2.1) and Proposition 1.1.1 we easily deduce that

$$[\operatorname{Re} z \geq \omega] \subset \rho(-A)$$

and

$$|\lambda| \, \|(\lambda + A)^{-1}\|_{\mathcal{L}(E_0)} \leq \kappa \ , \qquad \operatorname{Re}\lambda \geq \omega \ . \tag{1.2.5}$$

This is well-known to imply $A \in \mathcal{H}(E_1, E_0)$ (e.g., [Fat83, Section 4.2]).

Suppose that $A \in \mathcal{H}(E_1, E_0)$. Then (cf. [Fat83], for example) $A \in \mathcal{C}(E_0)$ and there exist $\omega > 0$ and $\kappa \geq 1$ such that (1.2.5) is true. Now the assertion follows from Remark 1.2.1(a). ■

Since $\mathcal{H}(E_1, E_0) \subset \mathcal{C}(E_0)$, the dual $A' \in \mathcal{C}(E_0')$ is defined for $A \in \mathcal{H}(E_1, E_0)$. It is convenient to put $E_0^\sharp := E_0'$ and $A^\sharp := A'$. Then

$$E_1^\sharp := E_1^\sharp(A^\sharp) := D(A^\sharp)$$

is a Banach space such that $E_1^\sharp \hookrightarrow E_0^\sharp$. If E_0 is reflexive, $A_0^\sharp$ is densely defined so that $(E_0^\sharp, E_1^\sharp)$ is a densely injected Banach couple.

1.2.3 Proposition *Suppose that E_0 is reflexive and $A \in \mathcal{H}(E_1, E_0, \kappa, \omega)$. Then $A^\sharp \in \mathcal{H}(E_1^\sharp, E_0^\sharp, \kappa^\sharp, \omega)$, where $\kappa^\sharp := 10\kappa^2/(\omega \wedge 1)$.*

Proof Since $\rho(A) = \rho(A^\sharp)$ and $(\lambda + A^\sharp)^{-1} = \big[(\lambda + A)^{-1}\big]'$ for $\lambda \in \rho(-A^\sharp)$, it follows from (1.2.5) that $[\operatorname{Re} z \geq \omega] \subset \rho(-A^\sharp)$ and $|\lambda| \, \|x^\sharp\|_0 \leq \kappa \, \|(\lambda + A^\sharp)x^\sharp\|_1$ for $x^\sharp \in E_1^\sharp$ and $\operatorname{Re}\lambda \geq \omega$, where we denote the norm in $E_j^\sharp$ again by $\|\cdot\|_j$. Hence we infer from Remark 1.2.1(a) that $A^\sharp \in \mathcal{H}(E_1^\sharp, E_0^\sharp, \kappa_1, \omega)$, where

$$\kappa_1 := \kappa\big(1 + 3 \, \|(\omega + A^\sharp)^{-1}\|_{\mathcal{L}(E_0^\sharp, E_1^\sharp)}\big) \ ,$$

thanks to the fact that $\|A^\sharp\|_{\mathcal{L}(E_1^\sharp, E_0^\sharp)} \leq 1$. Note that

$$\|x^\sharp\|_1 \leq \|(\omega + A^\sharp)x^\sharp\|_0 + (1 + \omega) \, \|x^\sharp\|_0 \leq (1 + \kappa) \, \|(\omega + A^\sharp)x^\sharp\|_0 + \|x^\sharp\|_0$$

for $x^\sharp \in E_1^\sharp$. Hence

$$\begin{aligned}\|(\omega + A^\sharp)^{-1}x^\sharp\|_1 &\leq (1 + \kappa) \, \|x^\sharp\|_0 + \|(\omega + A^\sharp)^{-1}x^\sharp\|_0 \leq (1 + \kappa + \kappa/\omega) \, \|x^\sharp\|_0 \\ &\leq \big(3\kappa/(\omega \wedge 1)\big) \, \|x^\sharp\|_0\end{aligned}$$

for $x^\sharp \in E_0^\sharp$. Now the assertion follows. ■

The class $\mathcal{H}(E_1, E_0)$ has been introduced by the author in [Ama88a] and, independently, in [ClH$^+$87, Chapter 5], where it has been called $\mathrm{Hol}(E_1, E_0)$. The quantitative versions $\mathcal{H}(E_1, E_0, \kappa, \omega)$ are implicitly introduced in [Ama88a] since it is easily verified that a subset U of $\mathcal{H}(E_1, E_0)$ is regularly bounded (in the sense of [Ama88a]) iff there exist $\kappa \geq 1$ and $\omega > 0$ with $U \subset \mathcal{H}(E_1, E_0, \kappa, \omega)$.

1.3 Perturbation Theorems

Thanks to the quantitative information contained in Theorem 1.2.2 it is easy to give quantitative versions of important perturbation theorems. Some of them are collected in the following:

1.3.1 Theorem (i) *$\mathcal{H}(E_1, E_0)$ is open in $\mathcal{L}(E_1, E_0)$. In fact, given $r \in (0, 1/\kappa)$ and $A \in \mathcal{H}(E_1, E_0, \kappa, \omega)$,*

$$\bar{\mathbb{B}}_{\mathcal{L}(E_1,E_0)}(A, r) \subset \mathcal{H}\big(E_1, E_0, \kappa/(1-\kappa r), \omega\big) .$$

(ii) *Let $A \in \mathcal{H}(E_1, E_0, \kappa, \omega)$, $0 < r < 1/\kappa$, and $\beta \geq 0$. Then, given $B \in \mathcal{L}(E_1, E_0)$ satisfying*

$$\|Bx\|_0 \leq r\,\|x\|_1 + \beta\,\|x\|_0 , \qquad x \in E_1 ,$$

it follows that

$$A + B \in \mathcal{H}\big(E_1, E_0, \kappa/(1-\kappa r), \omega \vee \beta/r\big) .$$

Proof (i) Given $B \in \mathcal{L}(E_1, E_0)$ with $\|B\| \leq r$, we see that

$$\begin{aligned}(\kappa^{-1} - r)(|\lambda|\,\|x\|_0 + \|x\|_1) &\leq \kappa^{-1}(|\lambda|\,\|x\|_0 + \|x\|_1) - r\,\|x\|_1 \\ &\leq \|(\lambda + A)x\|_0 - \|Bx\|_0 \leq \|(\lambda + A + B)x\|_0 \leq \|(\lambda + A)x\|_0 + \|Bx\|_0 \\ &\leq \kappa(|\lambda|\,\|x\|_0 + \|x\|_1) + r\,\|x\|_1 \leq (\kappa + r)(|\lambda|\,\|x\|_0 + \|x\|_1)\end{aligned}$$

for $x \in E_1$ and $\mathrm{Re}\,\lambda \geq \omega$. Since $\kappa + r \leq \kappa/(1-\kappa r) =: \kappa_1$, it follows that $A + B$ satisfies (1.2.1) with κ replaced by κ_1. From this we deduce that

$$\kappa_1^{-1}\,\|x\|_1 \leq \|(\omega + A + tB)x\|_0 , \qquad x \in E_1 , \quad 0 \leq t \leq 1 .$$

Now Proposition 1.1.1 implies $\omega + A + B \in \mathcal{L}\mathrm{is}(E_1, E_0)$. Hence

$$A + B \in \mathcal{H}(E_1, E_0, \kappa_1, \omega) .$$

(ii) Observe that

$$|\lambda| \geq \beta/r \quad \text{iff} \quad \kappa^{-1} - \beta/|\lambda| \geq \kappa^{-1} - r \quad \text{iff} \quad \kappa + \beta/|\lambda| \leq \kappa + r .$$

Hence

$$\begin{aligned}(\kappa^{-1}-r)(|\lambda|\,\|x\|_0+\|x\|_1) &\le (\kappa^{-1}-\beta/|\lambda|)\,|\lambda|\,\|x\|_0+(\kappa^{-1}-r)\,\|x\|_1\\ &\le \|(\lambda+A)x\|_0-\|Bx\|_0\le\|(\lambda+A+B)x\|_0\le\|(\lambda+A)x\|_0+\|Bx\|_0\\ &\le(\kappa+\beta/|\lambda|)\,|\lambda|\,\|x\|_0+(\kappa+r)\,\|x\|_1\le(\kappa+r)(|\lambda|\,\|x\|_0+\|x\|_1)\end{aligned}$$

for $x \in E_1$ and $\operatorname{Re}\lambda \ge \omega \vee \beta/r$. Now the assertion follows from the last step of the proof of (i). ■

1.3.2 Corollary *Let $\mathcal{A}$ be a compact subset of $\mathcal{H}(E_1, E_0)$. Then there exist $\kappa \ge 1$ and $\omega > 0$ such that $\mathcal{H}(E_1, E_0, \kappa, \omega)$ is a closed neighborhood of $\mathcal{A}$ in $\mathcal{L}(E_1, E_0)$.*

Proof From Theorem 1.3.1(i) we know that each $A \in \mathcal{A}$ has a neighborhood of the form $\mathcal{H}(E_1, E_0, \kappa_A, \omega_A)$. Thus the compactness of $\mathcal{A}$ implies the existence of $\kappa_j \ge 1$ and $\omega_j > 0$, $0 \le j \le m$, with

$$\mathcal{A} \subset \bigcup \mathcal{H}(E_1, E_0, \kappa_j, \omega_j) \ .$$

Since $\kappa' \ge \kappa \ge 1$ and $\omega' \ge \omega > 0$ imply $\mathcal{H}(E_1, E_0, \kappa', \omega') \supset \mathcal{H}(E_1, E_0, \kappa, \omega)$, the assertion follows, since it is easily verified that each $\mathcal{H}(E_1, E_0, \kappa, \omega)$ is closed. ■

Theorem 1.3.1 is a quantitative formulation of well-known perturbation theorems for generators of analytic semigroups (e.g., [Paz83, Theorem 3.2.1]). The fact that $\mathcal{H}(E_1, E_0)$ is open in $\mathcal{L}(E_1, E_0)$ has also been observed in [ClH+87, Theorem 5.3].

1.4 Spectral Estimates

It is a consequence of Theorem 1.2.2 and known properties of generators of analytic semigroups that, given $A \in \mathcal{H}(E_1, E_0)$, there exist ω and a sector

$$\Sigma_\vartheta := [\,|\arg z| \le \vartheta + \pi/2] \cup \{0\} \ ,$$

where $0 < \vartheta < \pi/2$, such that $\omega + \Sigma_\vartheta$ belongs to the resolvent set of $-A$. The following proposition gives the sharper result that such a ϑ can be chosen uniformly for all $A \in \mathcal{H}(E_1, E_0, \kappa, \omega)$.

1.4.1 Proposition *Given $\kappa \ge 1$ and $\omega > 0$, there are $\omega_0 \in (0, \omega)$ and $\vartheta \in (0, \pi/2)$ such that $\omega_0 + \Sigma_\vartheta \subset \rho(-A)$ and*

$$(5\kappa)^{-1} \le \frac{\|(\lambda + A)x\|_0}{|\lambda|\,\|x\|_0 + \|x\|_1} \le 5\kappa \ , \qquad x \in \dot{E}_1 \ , \quad \lambda \in \omega_0 + \Sigma_\vartheta \ , \tag{1.4.1}$$

for $A \in \mathcal{H}(E_1, E_0, \kappa, \omega)$.

Proof Put

$$\theta := \arcsin\big[(2-\sqrt{2})/(2\kappa)\big] \wedge \arctan\big[(2-\sqrt{2})/4\big] \in (0,\pi/2)\ .$$

Suppose that $A \in \mathcal{H}(E_1, E_0, \kappa, \omega)$ and $\lambda \in 2\omega + \Sigma_\theta$ with $\operatorname{Re}\lambda \le \omega$. Then

$$|\arg(\lambda - 2\omega)| = \varphi + \pi/2$$

with $0 < \varphi \le \theta$ and $|\lambda| \ge |\operatorname{Im}\lambda| \ge \omega\cot\theta$. Let $\lambda_0 := 2\omega + i\eta$, where η is defined by $|\eta| := |\lambda - 2\omega| / \cos\varphi$ and $\operatorname{sign}(\eta) = \operatorname{sign}(\operatorname{Im}\lambda)$.

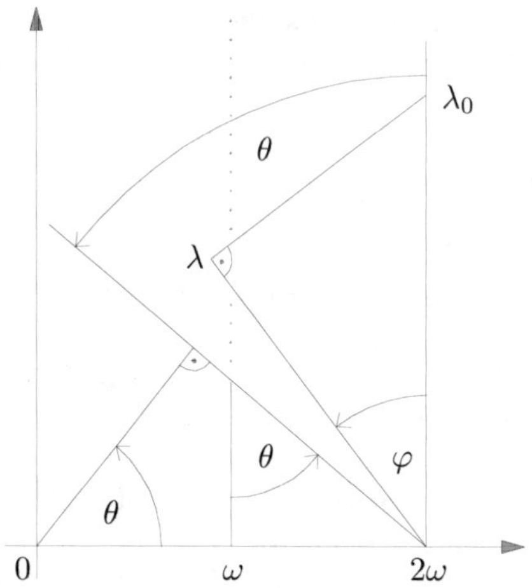

Then

$$\begin{aligned} \|(\lambda + A)x\|_0 &\ge \|(\lambda_0 + A)x\|_0 - |\lambda - \lambda_0|\,\|x\|_0 \\ &\ge (\kappa^{-1}|\lambda_0| - |\lambda - \lambda_0|)\,\|x\|_0 + \kappa^{-1}\|x\|_1\ . \end{aligned} \tag{1.4.2}$$

Since

$$|\eta| = |\lambda - \lambda_0| / \sin\varphi = |\lambda - 2\omega| / \cos\varphi \ge |\lambda - 2\omega|$$

and $|\lambda_0| \ge |\eta|$, it follows that

$$\begin{aligned} \kappa^{-1}|\lambda_0| - |\lambda - \lambda_0| &\ge |\eta|\,(\kappa^{-1} - \sin\varphi) \ge |\lambda - 2\omega|\,(\kappa^{-1} - \sin\theta) \\ &\ge |\lambda|\,(1 - 2\omega/|\lambda|)/(\sqrt{2}\kappa)\ . \end{aligned} \tag{1.4.3}$$

Since $\operatorname{Re}\lambda \le \omega$, we see that $2\omega/|\lambda| \le 2\tan\theta \le 1 - 1/\sqrt{2}$. By inserting the latter estimate in (1.4.3) we deduce from (1.4.2) that

$$\|(\lambda + A)x\|_0 \ge (2\kappa)^{-1}(|\lambda|\,\|x\|_0 + \|x\|_1)\ , \qquad x \in E_1\ .$$

On the other hand,

$$\begin{aligned} \|(\lambda + A)x\|_0 &\le \|(\omega + A)x\|_0 + |\lambda - \omega|\,\|x\|_0 \\ &\le \kappa(\omega\,\|x\|_0 + \|x\|_1) + (|\lambda| + \omega)\,\|x\|_0 \\ &\le 3\kappa(|\lambda|\,\|x\|_0 + \|x\|_1) \end{aligned} \tag{1.4.4}$$

for $\lambda \in 2\omega + \Sigma_\theta$ with $\operatorname{Re}\lambda \le \omega$, since $\kappa \ge 1$ and

$$|\lambda| \ge 2\omega \cos\theta = 2\omega\sqrt{1 - \left[(2-\sqrt{2})/(2\kappa)\right]^2} \ge \sqrt{2}\omega \ .$$

Hence estimate (1.4.1) is true if $\lambda \in 2\omega + \Sigma_\theta$ satisfies $\operatorname{Re}\lambda \le \omega$.

Put $\delta := 2\omega/(5\kappa)$ and suppose that $\omega - \delta \le \operatorname{Re}\lambda \le \omega$. Let $\lambda_0 := \omega + i \operatorname{Im}\lambda$ and observe that $\omega - \delta = \omega\bigl(1 - 2/(5\kappa)\bigr) \ge \omega/2$ implies

$$(|\lambda_0| - \kappa\,|\lambda - \lambda_0|)/|\lambda| \ge 1 - \kappa\delta/(\omega - \delta) \ge 1 - 2\kappa\delta/\omega = 1/5 \ .$$

Hence it follows from (1.4.2) that

$$\|(\lambda + A)x\|_0 \ge |\lambda|/(5\kappa)\,\|x\|_0 + \kappa^{-1}\,\|x\|_1 \ge (5\kappa)^{-1}(|\lambda|\,\|x\|_0 + \|x\|_1) \ .$$

On the other hand, $\omega - \delta \ge \omega/2$ implies $\omega \le 2\,|\lambda|$ for $\omega - \delta \le \operatorname{Re}\lambda \le \omega$. Hence we deduce from (1.4.4) that

$$\|(\lambda + A)x\|_0 \le (\kappa\omega + |\lambda| + \omega)\,\|x\|_0 + \kappa\,\|x\|_1 \le 5\kappa(|\lambda|\,\|x\|_0 + \|x\|_1) \ .$$

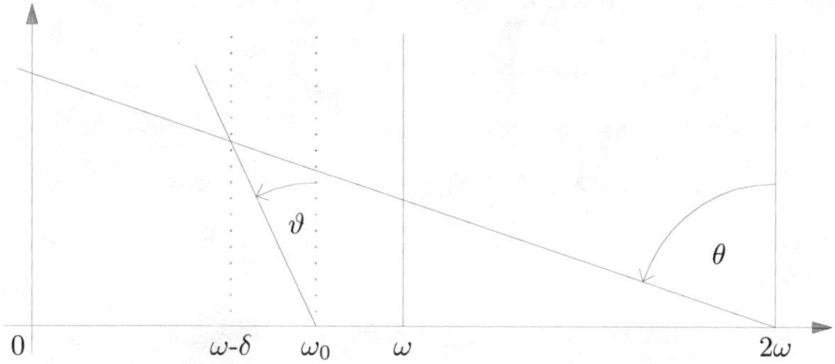

Consequently, estimate (1.4.1) is satisfied for $\lambda \in [\omega - \delta \le \operatorname{Re} z \le \omega]$. It is trivially satisfied if $\operatorname{Re}\lambda \ge \omega$. Put

$$\omega_0 := \omega - \delta/2 \quad \text{and} \quad \vartheta := \arctan\Bigl(\frac{\delta\tan\theta}{2(\omega + \delta)}\Bigr)$$

and observe that estimate (1.4.1) is satisfied for $\lambda \in \omega_0 + \Sigma_\vartheta$. Now the assertion is a consequence of Proposition 1.1.1. ∎

If A is a closed linear operator in E, we define its **spectral bound**, $s(A)$, by

$$s(A) := \sup\{\, \operatorname{Re}\lambda \ ; \ \lambda \in \sigma(A) \,\} \ ,$$

where $\sup\emptyset := -\infty$. Note that

$$s(\alpha + A) = \operatorname{Re}\alpha + s(A) \ , \qquad \alpha \in \mathbb{C} \ .$$

Also note that

$$s(-A) < \omega \ , \qquad A \in \mathcal{H}(E_1, E_0, \kappa, \omega) \ ,$$

where, of course, A is interpreted as a linear operator in E_0.

In many cases — in connection with stability problems for evolution equations, for example — it is important information if $s(-A) < 0$. Such information is not contained in the notation $\mathcal{H}(E_1, E_0, \kappa, \omega)$, since $\omega > 0$ by definition (which is a convenient assumption for an easy handling of the class $\mathcal{H}(E_1, E_0, \kappa, \omega)$). However, we have the following important inclusion:

1.4.2 Proposition *Given $A_0 \in \mathcal{H}(E_1, E_0, \kappa, \omega)$, any $\omega_1 > 0$, and an arbitrary $\sigma > s(-A_0)$, there exist $\kappa_1 \geq 1$ and a neighborhood $\mathcal{A}$ of A_0 in $\mathcal{L}(E_1, E_0)$ such that*

$$\sigma + \mathcal{A} \subset \mathcal{H}(E_1, E_0, \kappa_1, \omega_1) \ .$$

If $\sigma \geq \omega$ then $\mathcal{H}(E_1, E_0, \kappa, \omega) \subset \mathcal{A}$.

Proof By Proposition 1.4.1 we know that $\omega + \Sigma_\vartheta \subset \rho(-A_0)$ and

$$\|(\lambda + A_0)^{-1}\|_{\mathcal{L}(E_0,E_1)} \leq 5\kappa \ , \qquad \lambda \in \omega + \Sigma_\vartheta \ , \tag{1.4.5}$$

for some $\vartheta := \vartheta(\kappa, \omega) \in (0, \pi/2)$. Since

$$\Lambda := [\operatorname{Re}\lambda \geq \sigma] \cap (\omega + \mathring{\Sigma}_\vartheta)^c$$

is a compact subset of $\rho(-A_0)$ and $(\lambda \mapsto \lambda + A_0) \in C\big(\Lambda, \mathcal{L}\mathrm{is}(E_1, E_0)\big)$,

$$\max\big\{ \|(\lambda + A_0)^{-1}\|_{\mathcal{L}(E_0,E_1)} \ ; \ \lambda \in \Lambda \big\} < \infty \ , \tag{1.4.6}$$

where $\max \emptyset := -\infty$. From (1.4.5) and (1.4.6) we deduce the existence of a constant $M \geq 1$ such that

$$\|(\lambda + A_0)^{-1}\|_{\mathcal{L}(E_0,E_1)} \leq M \ , \qquad \operatorname{Re}\lambda \geq \sigma \ . \tag{1.4.7}$$

Put $r := 1/(2M)$ and observe that

$$\lambda + A_0 + B = \big[1 + B(\lambda + A_0)^{-1}\big](\lambda + A_0)$$

and $1 + B(\lambda + A_0)^{-1} \in \mathcal{L}\mathrm{aut}(E_0)$ with

$$\big\|\big[1 + B(\lambda + A_0)^{-1}\big]^{-1}\big\|_{\mathcal{L}(E_0)} \leq 2$$

for $\operatorname{Re}\lambda \geq \sigma$ and $B \in r\mathbb{B}_{\mathcal{L}(E_1,E_0)}$, thanks to (1.4.7). Hence

$$\lambda + A_0 + B \in \mathcal{L}\mathrm{is}(E_1, E_0)$$

and

$$\|(\lambda + A_0 + B)^{-1}\|_{\mathcal{L}(E_0,E_1)} \leq 2M \ , \qquad \operatorname{Re}\lambda \geq \sigma \ , \quad B \in r\mathbb{B}_{\mathcal{L}(E_1,E_0)} \ . \tag{1.4.8}$$

Consequently, we deduce from (1.4.8) that $[\operatorname{Re}\lambda \geq 0] \subset \rho\big(-(\sigma + A)\big)$ and

$$\|x\|_1 \leq 2M \, \big\|\big(\lambda + (\sigma + A)\big)x\big\|_0 \ , \qquad \operatorname{Re}\lambda \geq 0 \ , \quad A \in \mathbb{B}_{\mathcal{L}(E_1,E_0)}(A_0, r) \ .$$

Now the first part of the assertion follows easily from Remark 1.2.1(a).

If $\sigma \geq \omega$ then (1.2.1) implies

$$\|x\|_1 \leq \kappa \left\|\big((\lambda-\sigma)+\sigma+A\big)x\right\|_0 , \qquad \operatorname{Re}(\lambda-\sigma) \geq \omega-\sigma , \quad x \in E_1 ,$$

for each $A \in \mathcal{H}(E_1, E_0, \kappa, \omega)$. Thus the last part of the assertion again follows from Remark 1.2.1(a). ■

Let $\sigma > s(-A_0)$ be given. Fix $\sigma_1 \in \big(s(-A_0), \sigma\big)$ and put $\omega_1 := \sigma - \sigma_1$. Then Proposition 1.4.2 implies the existence of a neighborhood $\mathcal{A}$ of A_0 in $\mathcal{L}(E_1, E_0)$ such that

$$s(-A) = \sigma_1 + s\big(-(\sigma_1 + A)\big) < \sigma_1 + \omega_1 = \sigma , \qquad A \in \mathcal{A} , \tag{1.4.9}$$

Moreover, the following corollary is valid:

1.4.3 Corollary *Given $A_0 \in \mathcal{H}(E_1, E_0, \kappa, \omega)$ and $\sigma > s(-A_0)$, there exist $M > 0$, a neighborhood $\mathcal{A}$ of A_0 in $\mathcal{L}(E_1, E_0)$, and $\vartheta \in (0, \pi/2)$ such that*

$$\sigma + \Sigma_\vartheta \subset \rho(-A)$$

and

$$\|\sigma + A\|_{\mathcal{L}(E_1,E_0)} + (1+|\lambda|)^{1-j} \left\| \big[\lambda + (\sigma + A)\big]^{-1} \right\|_{\mathcal{L}(E_0,E_j)} \leq M$$

for $\lambda \in \Sigma_\vartheta$, $j = 0, 1$, and $A \in \mathcal{A}$. If $\sigma \geq \omega$ then $\mathcal{H}(E_1, E_0, \kappa, \omega) \subset \mathcal{A}$.

Proof Fix $\widetilde{\sigma} \in \big(s(-A_0), \sigma\big)$ and put $\varepsilon := \sigma - \widetilde{\sigma}$. By Proposition 1.4.2 there exist $\kappa_1 \geq 1$ and a neighborhood $\mathcal{A}$ of A_0 in $\mathcal{L}(E_1, E_0)$ such that

$$\widetilde{\sigma} + \mathcal{A} \subset \mathcal{H}(E_1, E_0, \kappa_1, \varepsilon) .$$

If $\sigma \geq \omega$ then $\mathcal{H}(E_1, E_0, \kappa, \omega) \subset \mathcal{A}$. Moreover, Proposition 1.4.1 implies the existence of $\vartheta \in (0, \pi/2)$ such that $\varepsilon + \Sigma_\vartheta \subset \rho(-A)$ and

$$(5\kappa_1)^{-1} \leq \frac{\|(\lambda + A)x\|_0}{|\lambda| \, \|x\|_0 + \|x\|_1} \leq 5\kappa_1 , \qquad x \in \dot{E}_1 , \quad \lambda \in \varepsilon + \Sigma_\vartheta ,$$

for $A \in \mathcal{H}(E_1, E_0, \kappa_1, \varepsilon)$. By replacing in this estimate A by $\widetilde{\sigma} + \widetilde{A}$ for $\widetilde{A} \in \mathcal{A}$, we see that

$$\|(\sigma + \widetilde{A})x\|_0 = \|(\varepsilon + \widetilde{\sigma} + \widetilde{A})x\|_0 \leq 5\kappa_1(\varepsilon \|i\| + 1) \|x\|_1 , \qquad x \in E_1 ,$$

where again $i \colon E_1 \hookrightarrow E_0$, and that

$$\|(\lambda + \sigma + \widetilde{A})x\|_0 = \left\|\big(\lambda + \varepsilon + (\widetilde{\sigma} + \widetilde{A})\big)x\right\|_0 \geq (5\kappa_1)^{-1}(|\lambda + \varepsilon| \, \|x\|_0 + \|x\|_1)$$

for $x \in E_1$, $\lambda \in \Sigma_\vartheta$, and $\widetilde{A} \in \mathcal{A}$. Now the assertion follows with

$$M := 5\kappa_1 \big[(\varepsilon \|i\| + 1) \vee \sup\nolimits_{\lambda \in \Sigma_\vartheta} (|\lambda| + 1)/\, |\lambda + \varepsilon|\big] ,$$

since $M \in \mathbb{R}^+$. ■

Proposition 1.4.2 and Corollary 1.4.3 will be important for getting uniform estimates for parabolic evolution operators for time-dependent linear evolution equations. However, a first application of Corollary 1.4.3 will lead to the perturbation result of Theorem 1.5.1.

1.5 Compact Perturbations

A family $\mathfrak{K}$ of linear operators from E to F is **collectively compact** if

$$\bigcup\{\, K(\mathbb{B}_E)\ ;\ K\in\mathfrak{K}\,\}\quad \text{is relatively compact in } F\ .$$

Observe that $\mathfrak{K}$ is a bounded subset of the Banach space $\mathcal{K}(E,F)$ of all compact linear maps from E to F. Using this definition we can formulate the following theorem about compact perturbations.

1.5.1 Theorem *Suppose that $\mathfrak{K}\subset\mathcal{L}(E_1,E_0)$ is collectively compact. Given $\kappa\geq 1$ and $\omega>0$, there exist $\kappa_1\geq\kappa$ and $\omega_1\geq\omega$ such that*

$$\mathcal{H}(E_1,E_0,\kappa,\omega)+\mathfrak{K}\subset\mathcal{H}(E_1,E_0,\kappa_1,\omega_1)\ .$$

Proof By Corollary 1.4.3 there exists M such that

$$\|x\|_1=\|(\omega+A)^{-1}(\omega+A)x\|_1\leq M\,\|(\omega+A)x\|_0\leq M^2\,\|x\|_1 \tag{1.5.1}$$

for $x\in E_1$, and

$$|\lambda|\,\|(\lambda+A)^{-1}\|_{\mathcal{L}(E_0)}\leq M\ ,\qquad \operatorname{Re}\lambda\geq\omega\ , \tag{1.5.2}$$

for $A\in\mathcal{H}(E_1,E_0,\kappa,\omega)$. From (1.5.1) and (1.5.2) we deduce that

$$\begin{aligned}\|\lambda(\lambda+A)^{-1}x\|_1&\leq M\,\|\lambda(\omega+A)(\lambda+A)^{-1}x\|_0\\&\leq M^2\,\|(\omega+A)x\|_0\leq M^3\,\|x\|_1\end{aligned} \tag{1.5.3}$$

for $x\in E_1$ and $\operatorname{Re}\lambda\geq\omega$.

Let $\mathcal{K}$ be the closure of $\bigcup\{\,K(\mathbb{B}_{E_1})\ ;\ K\in\mathfrak{K}\,\}$ in E_0. Since E_1 is dense in E_0, given $x_0\in\mathcal{K}$, there exists $x_1\in\mathbb{B}_{E_0}\big(x_0,1/(6\kappa)\big)\cap E_1$. Hence we deduce from (1.5.3) and Remark 1.2.1(a) that

$$\begin{aligned}\|(\lambda+A)^{-1}x\|_1&=\big\|(\lambda+A)^{-1}\big[x_1+(x_0-x_1)+(x-x_0)\big]\big\|_1\\&\leq|\lambda|^{-1}\,M^3\,\|x_1\|_1+1/6+\kappa\,\|x-x_0\|_0\end{aligned}$$

for $x\in E_0$, $\operatorname{Re}\lambda\geq\omega$, and $A\in\mathcal{H}(E_1,E_0,\kappa,\omega)$. Consequently, there exists $\omega(x_0)\geq\omega$ such that

$$\|(\lambda+A)^{-1}x\|_1\leq 1/2\ ,\qquad \operatorname{Re}\lambda\geq\omega(x_0)\ ,\quad x\in\mathbb{B}_{E_0}\big(x_0,1/(6\kappa)\big)$$

for $A \in \mathcal{H}(E_1, E_0, \kappa, \omega)$. By the compactness of $\mathcal{K}$ we find $x_0, \ldots, x_m \in \mathcal{K}$ such that $\{ \mathbb{B}_{E_0}(x_j, 1/(6\kappa)) \; ; \; j = 0, \ldots, m \}$ is a covering of $\mathcal{K}$. Thus, letting

$$\omega_1 := \max\{ \omega(x_j) \; ; \; 0 \leq j \leq m \} ,$$

we see that

$$\|(\lambda + A)^{-1} x\|_1 \leq 1/2 , \qquad \operatorname{Re} \lambda \geq \omega_1 , \qquad x \in \mathcal{K} ,$$

which implies

$$\|(\lambda + A)^{-1} K\|_{\mathcal{L}(E_1)} \leq 1/2 , \qquad \operatorname{Re} \lambda \geq \omega_1 , \quad K \in \mathfrak{K} . \tag{1.5.4}$$

Since

$$\lambda + A + K = (\lambda + A)\big(1 + (\lambda + A)^{-1} K\big) ,$$

we obtain from (1.5.4) that $\lambda \in \rho\big(-(A + K)\big)$ and

$$\|(\lambda + A + K)^{-1}\|_{\mathcal{L}(E_0, E_1)} \leq 2\kappa$$

for $A \in \mathcal{H}(E_1, E_0, \kappa, \omega)$, $K \in \mathfrak{K}$, and $\operatorname{Re} \lambda \geq \omega_1$. Now the assertion follows from Remark 1.2.1(a). ■

Theorem 1.5.1 — in the case that $\mathfrak{K}$ consists of one operator only — is proven in [ClH+87, Theorem 5.6] and, independently, by Desch and Schappacher [DeS88], although it was around before as a 'folk theorem'. Our extension to a collectively compact set of linear operators is trivial of course but useful.

1.6 Matrix Generators

Let (F_0, F_1) be a densely injected Banach couple. Then $(E_0 \times F_0, E_1 \times F_1)$ is a densely injected Banach couple as well. Suppose that

$$A = \begin{bmatrix} A_{11} & A_{12} \\ A_{21} & A_{22} \end{bmatrix} \in \mathcal{L}(E_1 \times F_1, E_0 \times F_0) ,$$

where we use obvious matrix notation. For definiteness, in the following theorems we use the ℓ_1-norm on the product Banach spaces.

1.6.1 Theorem *Suppose that*

$$A_{11} \in \mathcal{H}(E_1, E_0, \kappa_1, \omega_1) \quad \textit{and} \quad A_{22} \in \mathcal{H}(F_1, F_0, \kappa_2, \omega_2) .$$

Put

$$\kappa := \kappa_1(1 + \kappa_2 \, \|A_{21}\|_{\mathcal{L}(E_1, F_0)}) \vee \kappa_2 \quad \textit{and} \quad \omega := \omega_1 \vee \omega_2$$

and suppose that there are $r \in (0, 1/\kappa)$ and $\beta \geq 0$ such that

$$\|A_{12} y\|_{E_0} \leq r \, \|y\|_{F_1} + \beta \, \|y\|_{F_0} , \qquad y \in F_1 . \tag{1.6.1}$$

Then $A \in \mathcal{H}\big(E_1 \times F_1, E_0 \times F_0, \kappa/(1 - \kappa r), \omega \vee \beta/r\big)$.

Proof Put

$$C := \begin{bmatrix} A_{11} & 0 \\ A_{21} & A_{22} \end{bmatrix} . \tag{1.6.2}$$

Given $(\xi,\eta) \in E_0 \times F_0$ and $\lambda \in [\operatorname{Re} z \geq \omega]$, the equation $(\lambda + C)(x,y) = (\xi,\eta)$ has the unique solution

$$x = (\lambda + A_{11})^{-1}\xi \ , \quad y = (\lambda + A_{22})^{-1}\bigl(\eta - A_{21}(\lambda + A_{11})^{-1}\xi\bigr) \ .$$

Hence it follows from (1.2.1) that

$$|\lambda| \, \|x\|_{E_0} + \|x\|_{E_1} \leq \kappa_1 \, \|\xi\|_{E_0} \leq \kappa_1 \, \|(\xi,\eta)\|_{E_0 \times F_0}$$

and

$$\begin{aligned} |\lambda| \, \|y\|_{F_0} + \|y\|_{F_1} &\leq \kappa_2 \, \|\eta - A_{21}(\lambda + A_{11})^{-1}\xi\|_{F_0} \\ &\leq \kappa_2(\|\eta\|_{F_0} + \|A_{21}\| \, \kappa_1 \, \|\xi\|_{E_0}) \\ &\leq (\kappa_2 \vee \kappa_1\kappa_2 \, \|A_{21}\|) \, \|(\xi,\eta)\|_{E_0 \times F_0} \ . \end{aligned}$$

On the other hand,

$$\begin{aligned} \|(\lambda + C)(x,y)\|_{E_0 \times F_0} &= \|(\lambda + A_{11})x\|_{E_0} + \|A_{21}x + (\lambda + A_{22})y\|_{F_0} \\ &\leq \kappa_1(|\lambda| \, \|x\|_{E_0} + \|x\|_{E_1}) + \|A_{21}\| \, \|x\|_{E_1} + \kappa_2(|\lambda| \, \|y\|_{F_0} + \|y\|_{F_1}) \\ &\leq \bigl[(\kappa_1 + \|A_{21}\|) \vee \kappa_2\bigr] \bigl(|\lambda| \, \|(x,y)\|_{E_0 \times F_0} + \|(x,y)\|_{E_1 \times F_1}\bigr) \ . \end{aligned}$$

From these estimates we deduce that

$$C \in \mathcal{H}(E_1 \times F_1, E_0 \times F_0, \kappa, \omega) \ . \tag{1.6.3}$$

Put

$$B := \begin{bmatrix} 0 & A_{12} \\ 0 & 0 \end{bmatrix} \tag{1.6.4}$$

and note that $A = C + B$ and, thanks to (1.6.1),

$$\|B(x,y)\|_{E_0 \times F_0} = \|A_{12}y\|_{E_0} \leq r \, \|(x,y)\|_{E_1 \times F_1} + \beta \, \|(x,y)\|_{E_0 \times F_0}$$

for $(x,y) \in E_1 \times F_1$. Now (1.6.3) and Theorem 1.3.1(ii) imply the assertion. ■

1.6.2 Remarks **(a)** Suppose that A_{11} and A_{22} satisfy the hypotheses of Theorem 1.6.1 and replace κ by

$$\widetilde{\kappa} := \kappa_2(1 + \kappa_1 \, \|A_{12}\|_{\mathcal{L}(F_1,E_0)}) \vee \kappa_1 \ .$$

Then $A \in \mathcal{H}\bigl(E_1 \times F_1, E_0 \times F_0, \widetilde{\kappa}/(1 - \widetilde{\kappa}r), \omega \vee \beta\bigr)$, provided

$$\|A_{21}x\|_{F_0} \leq r \, \|x\|_{E_1} + \beta \, \|x\|_{E_0} \ , \qquad x \in E_1 \ .$$

Proof This follows from Theorem 1.6.1 by replacing the densely injected Banach couple $(E_0 \times F_0, E_1 \times F_1)$ by $(F_0 \times E_0, F_1 \times E_1)$. ■

(b) The hypotheses that $A_{11} \in \mathcal{H}(E_1, E_0)$ and $A_{22} \in \mathcal{H}(F_1, F_0)$ are necessary for the validity of Theorem 1.6.1 if $A_{12} = 0$.

Proof Suppose that (1.6.3) is true. Given $(\xi, \eta) \in E_0 \times F_0$ and λ with $\operatorname{Re}\lambda \geq \omega$, there exists a unique $(x, y) \in E_1 \times F_1$ satisfying

$$(\lambda + A_{11})x = \xi \ , \quad A_{21}x + (\lambda + A_{22})y = \eta \ .$$

Hence $\lambda + A_{11} \in \mathcal{L}\mathrm{is}(E_1, E_0)$ by Banach's theorem. Letting $\xi = 0$, the second equation reduces to

$$(\lambda + A_{22})y = \eta \ .$$

Thus $\lambda + A_{22} \in \mathcal{L}\mathrm{is}(F_1, F_0)$. Moreover, it follows from (1.6.3) that, letting $\eta := 0$,

$$\|x\|_{E_1} \leq \|(x,y)\|_{E_1 \times F_1} \leq \kappa \left\|(\lambda + A_{11})x\right\|_{E_0} \ , \qquad x \in E_1 \ ,$$

and, letting $\xi := 0$,

$$\|y\|_{F_1} \leq \|(x,y)\|_{E_1 \times F_1} \leq \kappa \left\|(\lambda + A_{22})y\right\|_{E_0} \ , \qquad y \in F_1 \ ,$$

Now the assertion follows from Remark 1.2.1(a). ■

1.6.3 Corollary *Suppose that $F_1 = F_0 =: F$. Then*

$$A \in \mathcal{H}(E_1 \times F, E_0 \times F) \quad \textit{iff} \quad A_{11} \in \mathcal{H}(E_1, E_0) \ .$$

More precisely, if $A_{11} \in \mathcal{H}(E_1, E_0, \kappa_1, \omega_1)$ then

$$A \in \mathcal{H}(E_1 \times F, E_0 \times F, \kappa, \omega) \ ,$$

where $\kappa := 2\kappa_1(1 + 2\,\|A_{21}\|) \vee 4$ and

$$\omega := \omega_1 \vee (1 + 2\,\|A_{22}\|) \vee (\|A_{12}\|\,\kappa) \ .$$

Proof Suppose that $A_{11} \in \mathcal{H}(E_1, E_0, \kappa_1, \omega_1)$. Thanks to Remark 1.2.1(b) we know that $A_{22} \in \mathcal{H}(F_1, F_0, 2, 1 + 2\,\|A_{22}\|)$. Let

$$\widetilde{\kappa} := \kappa_1(1 + 2\,\|A_{21}\|) \vee 2 \quad \text{and} \quad \widetilde{\omega} := \omega_1 \vee (1 + 2\,\|A_{22}\|)$$

and note that (1.6.1) is satisfied with $r := 1/(2\widetilde{\kappa})$ and $\beta := \|A_{12}\|$. Now the second part of the assertion follows from Theorem 1.6.1.

Let B and C be defined by (1.6.4) and (1.6.2), resp. Since $B \in \mathcal{L}(E_0 \times F)$ and $A = B + C$, it follows from Theorem 1.3.1(ii) that $A \in \mathcal{L}(E_1 \times F, E_0 \times F)$ iff $C \in \mathcal{L}(E_1 \times F, E_0 \times F)$. Now Remark 1.6.2(ii) shows that $A_{11} \in \mathcal{H}(E_1, E_0)$ is implied by $A \in \mathcal{H}(E_1 \times F, E_0 \times F)$. ■

It should be remarked that throughout this whole section we have not tried to find the best choices for κ and ω. In fact, we have tried to choose simple estimates for these numbers. It is the only purpose of the explicit formulas for κ and ω in Theorem 1.3.1, and in the theorems of this subsection, to exhibit the dependence of these constants on the various quantities involved. This information is useful if one has to consider subsets of $\mathcal{H}$ and uniform estimates are needed.

If $A_{12} = 0$, a qualitative version of Theorem 1.6.1 has been proven — by a different method — in [Nag89, Corollary 3.3]. A quantitative version of Corollary 1.6.3 appears in [Ang90, Lemma 2.6] and a qualitative form of it in [Ama91].

For the sake of simplicity and for the importance in applications we restrict ourselves to the case of densely injected Banach couples (E_0, E_1). Everything proven — with the notable exception of Theorem 1.5.1 — remains valid without the density assumption, provided we define $\mathcal{H}(E_1, E_0)$ by the formula of Theorem 1.2.2. On the other hand, since $\mathcal{H}(E_1, E_0) \subset \mathcal{C}(E_0)$, a result of Kato [Kat59] implies that E_1 is dense in E_0 if $\mathcal{H}(E_1, E_0) \neq \emptyset$ and either E_1 or E_0 is reflexive. (Note that E_1 is reflexive iff E_0 is reflexive in the case that $\mathcal{H}(E_1, E_0) \neq \emptyset$ since then $\mathcal{L}\mathrm{is}(E_1, E_0) \neq \emptyset$.) We refer to the monograph of Lunardi [Lun95] for a detailed exposition of the theory of analytic semigroups with nondense domains of the generators.

2 Interpolation Functors

In this section we collect the basic facts from the theory of interpolation spaces that we use freely in the remainder of this work. For proofs and many more details we refer to [BerL76], [BenS88], [ButB67], [KrPS82], and [Tri78].

2.1 Definitions

The pair (E_0, E_1) is said to be an **interpolation couple** if there exists a *LCS* X such that $E_j \hookrightarrow X$, $j = 0, 1$. In this case $E_0 \cap E_1$ and $E_0 + E_1$ are well-defined Banach spaces. Observe that $E_0 \cap E_1 \doteq E_1$ and $E_0 + E_1 \doteq E_0$ if $E_1 \hookrightarrow E_0$ so that we can choose X to be E_0. If (E_0, E_1) is an interpolation couple and

$$E_0 \cap E_1 \hookrightarrow E \hookrightarrow E_0 + E_1$$

then E is said to be an **intermediate space** with respect to (E_0, E_1).

Let $\mathcal{B}$ be the category of ($\mathbb{K}$-)Banach spaces. Thus the objects of $\mathcal{B}$ are the $\mathbb{K}$-Banach spaces, the morphisms of $\mathcal{B}$ are the bounded linear operators, and the composition is the usual composition of maps. We denote by $\mathcal{B}_1$ the category of interpolation couples, that is, the objects of $\mathcal{B}_1$ are the interpolation couples, the morphisms of $\mathcal{B}_1$ are the elements A of $\mathcal{L}(E_0+E_1, F_0+F_1)$ satisfying $A \in \mathcal{L}(E_j, F_j)$,

$j = 0, 1$, where (E_0, E_1) and (F_0, F_1) are interpolation couples, and the composition is the natural composition of maps. We write $A\colon (E_0, E_1) \to (F_0, F_1)$ if (E_0, E_1) and (F_0, F_1) are interpolation couples and A is a morphism of $\mathcal{B}_1$.

Let (E_0, E_1) and (F_0, F_1) be interpolation couples. Then E and F are said to be **interpolation spaces** with respect to (E_0, E_1) and (F_0, F_1) if E and F are intermediate spaces with respect to (E_0, E_1) and (F_0, F_1), respectively, and $A \in \mathcal{L}(E, F)$ whenever $A\colon (E_0, E_1) \to (F_0, F_1)$. Moreover, E and F are said to be **interpolation spaces of exponent** θ, where $0 < \theta < 1$, with respect to (E_0, E_1) and (F_0, F_1) if there exists $c(\theta) > 0$ such that

$$\|A\|_{\mathcal{L}(E,F)} \le c(\theta)\, \|A\|^{1-\theta}_{\mathcal{L}(E_0,F_0)}\, \|A\|^{\theta}_{\mathcal{L}(E_1,F_1)} \tag{2.1.1}$$

for $A\colon (E_0, E_1) \to (F_0, F_1)$. If $c(\theta) = 1$ then E and F are **exact** interpolation spaces of exponent θ with respect to (E_0, E_1) and (F_0, F_1).

Lastly, a covariant functor $\mathfrak{F}$ from $\mathcal{B}_1$ into $\mathcal{B}$ is said to be an [**exact**] **interpolation functor** [**of exponent** θ] if, given interpolation couples (E_0, E_1) and (F_0, F_1), it follows that $\mathfrak{F}(E_0, E_1)$ and $\mathfrak{F}(F_0, F_1)$ are [exact] interpolation spaces [of exponent θ] with respect to (E_0, E_1) and (F_0, F_1) and if

$$\mathfrak{F}(A) = A \in \mathcal{L}\big(\mathfrak{F}(E_0, E_1), \mathfrak{F}(F_0, F_1)\big)\ , \qquad A\colon (E_0, E_1) \to (F_0, F_1)\ .$$

2.2 Interpolation Inequalities

The following lemma shows that the norms of intermediate spaces 'generated by interpolation methods' possess an important convexity property.

2.2.1 Proposition *Let $\mathfrak{F}_\theta$ be an interpolation functor of exponent θ. Given an interpolation couple (E_0, E_1), put $E_\theta := \mathfrak{F}_\theta(E_0, E_1)$. Then*

$$\|x\|_{E_\theta} \le c_\theta\, \|x\|^{1-\theta}_{E_0}\, \|x\|^{\theta}_{E_1}\ , \qquad x \in E_0 \cap E_1\ . \tag{2.2.1}$$

Proof Since $(\mathbb{K}, \mathbb{K})$ is an interpolation couple, $\mathbb{K}_\theta \doteq \mathbb{K}$. Given $x \in E_0 \cap E_1$, define $A\colon (\mathbb{K}, \mathbb{K}) \to (E_0, E_1)$ by $A\lambda := \lambda x$. Observe that

$$\|A\|_{\mathcal{L}(\mathbb{K},E_j)} = \|x\|_{E_j}\ , \qquad j = 0, 1\ .$$

Hence we deduce from (2.1.1) the existence of a constant $c(\theta) > 0$ such that

$$c_1(\theta)\, \|x\|_{E_\theta} \le \sup_{|\lambda|_{\mathbb{K}_\theta}=1} \|\lambda x\|_{E_\theta} = \|A\|_{\mathcal{L}(\mathbb{K}_\theta,E_\theta)} \le c(\theta)\, \|x\|^{1-\theta}_{E_0}\, \|x\|^{\theta}_{E_1}\ ,$$

which implies the assertion. ■

Estimate (2.2.1) is also said to be an **interpolation inequality**. It follows from (2.2.1) and Young's inequality, that is, from

$$0 \le x^\theta y^{1-\theta} \le \theta x + (1-\theta) y \ , \qquad x, y \in \mathbb{R}^+ \ , \quad 0 < \theta < 1 \ ,$$

that

$$\|x\|_{E_\theta} \le \varepsilon \|x\|_1 + c(\theta) \varepsilon^{\theta/(\theta-1)} \|x\|_0 \ , \qquad x \in E_0 \cap E_1 \ , \quad \varepsilon > 0 \ , \tag{2.2.2}$$

where $c(\theta) := (1-\theta)\theta^{\theta/(1-\theta)} c_\theta^{1/(1-\theta)}$. Note that $c_\theta = 1$ in (2.2.1) if $\mathfrak{F}_\theta$ is an exact interpolation functor such that $\mathfrak{F}_\theta(\mathbb{K}, \mathbb{K}) = \mathbb{K}$.

2.3 Retractions

An object, Y, in a given category is a **retract** of the object X if there are morphisms, $r \colon X \to Y$, a **retraction** onto Y, and $r^c \colon Y \to X$, a **coretraction** for r, such that the diagram

$$\begin{array}{ccc} Y & \xrightarrow{\ \text{id}\ } & Y \\ & {\scriptstyle r^c} \searrow \quad \nearrow {\scriptstyle r} & \\ & X & \end{array} \tag{2.3.1}$$

is commutative. In other words: the morphism $r \colon X \to Y$ is a retraction of X onto Y if there exists a right inverse for r in the given category.

2.3.1 Lemma *Let $r \in \mathcal{L}(X, Y)$ be a retraction in the category of LCSs and let $r^c \in \mathcal{L}(Y, X)$ be a coretraction for r. Then $p := r^c r \in \mathcal{L}(X)$ is a projection and X is a topological direct sum,*

$$X = X_p \oplus X_{1-p} \ , \quad X_p := \text{im}(p) \ , \quad X_{1-p} := \text{im}(1-p) \ .$$

Furthermore, $r^c \in \mathcal{L}\text{is}(Y, X_p)$ and $X_{1-p} = \ker(p) = \ker(r)$.

Proof Observe that $p^2 = (r^c r)(r^c r) = r^c r = p$. Moreover,

$$\text{im}(1-p) = \ker(p) \supset \ker(r) \ .$$

Since $p(x) = 0$ implies $0 = rp(x) = r(x)$, it follows that $\ker(p) = \ker(r)$. Now the assertion is obvious. ■

2.3.2 Proposition *Let $r \colon (E_0, E_1) \to (F_0, F_1)$ be a retraction in the category $\mathcal{B}_1$ of interpolation couples, and let $r^c \colon (F_0, F_1) \to (E_0, E_1)$ be a coretraction for r. Suppose $\mathfrak{F}$ is an arbitrary interpolation functor from $\mathcal{B}_1$ into $\mathcal{B}$. Then*

$$r \in \mathcal{L}\big(\mathfrak{F}(E_0, E_1), \mathfrak{F}(F_0, F_1)\big)$$

is a retraction and r^c is a coretraction for $r = \mathfrak{F}(r)$.

Proof This is an immediate consequence of (2.3.1) and the functorial properties of $\mathfrak{F}$. ■

As an application of the considerations we now show — by a proof which we owe essentially to D. Daners (also cf. [DaK92, Proposition 3.4]) — that, in case we are dealing with product spaces, we can often interpolate 'component-wise'.

2.3.3 Proposition *Let (E_0, E_1) and (F_0, F_1) be interpolation couples such that $E_1 \hookrightarrow E_0$ and $F_1 \hookrightarrow F_0$. Let $\mathfrak{F}$ be an arbitrary interpolation functor from $\mathcal{B}_1$ into $\mathcal{B}$. Then $(E_0 \times F_0, E_1 \times F_1)$ is also an interpolation couple, and*

$$E_1 \times F_1 \hookrightarrow \mathfrak{F}(E_0, E_1) \times \mathfrak{F}(F_0, F_1) \doteq \mathfrak{F}(E_0 \times F_0, E_1 \times F_1) \hookrightarrow E_0 \times F_0 \ .$$

Proof It is obvious that $(E_0 \times F_0, E_1 \times F_1)$ is an interpolation couple with

$$E_1 \times F_1 \hookrightarrow E_0 \times F_0 \ .$$

Let $r_E : E_0 \times F_0 \to E_0$ be the natural projection and $i_E : E_0 \to E_0 \times F_0$ the natural injection. Observe that

$$r_E : (E_0 \times F_0, E_1 \times F_1) \to (E_0, E_1)$$

and

$$i_E : (E_0, E_1) \to (E_0 \times F_0, E_1 \times F_1) \ .$$

Moreover, r_E is a retraction and i_E is a coretraction for r_E in the category $\mathcal{B}_1$. Thus it follows from Lemma 2.3.1 and Proposition 2.3.2 that

$$\mathfrak{F}(E_0 \times F_0, E_1 \times F_1) = \mathrm{im}(p) \oplus \ker(p) \ ,$$

where

$$p := i_E r_E \in \mathcal{L}\bigl(\mathfrak{F}(E_0 \times F_0, E_1 \times F_1)\bigr) \ ,$$

and that

$$i_E \in \mathcal{L}\mathrm{is}\bigl(\mathfrak{F}(E_0, E_1), \mathrm{im}(p)\bigr) \ .$$

Let $r_F : E_0 \times F_0 \to F_0$ be the natural projection and $i_F : F_0 \to E_0 \times F_0$ the natural injection. Of course, r_F and i_F enjoy properties that are completely analoguous to the ones of r_E and i_E, respectively. Moreover, $i_F r_F = 1 - p$ and, consequently, since $\ker(p) = \mathrm{im}(1-p)$,

$$i_F \in \mathcal{L}\mathrm{is}\bigl(\mathfrak{F}(F_0, F_1), \ker(p)\bigr) \ .$$

From this we deduce that

$$\bigl[(e, f) \mapsto i_E(e) + i_F(f)\bigr] \in \mathcal{L}\mathrm{is}\bigl(\mathfrak{F}(E_0, E_1) \times \mathfrak{F}(F_0, F_1), \mathfrak{F}(E_0 \times F_0, E_1 \times F_1)\bigr) \ .$$

Now the assertion is obvious. ■

2.4 Standard Interpolation Functors

In this subsection we introduce briefly the 'classical' interpolation functors which are of utmost importance in concrete applications.

2.4.1 Example: Real Interpolation Functors Let (E_0, E_1) be an interpolation couple. Given $x \in E_0 + E_1$, define the K-**functional** by

$$K(t,x) := K(t,x,E_0,E_1) := \inf\{\, \|x_0\|_{E_0} + t\,\|x_1\|_{E_1} \ ;\ x = x_0 + x_1 \,\}$$

for $t > 0$. Put

$$\|x\|_{\theta,q} := \|t^{-\theta}K(t,x)\|_{L_q(\dot{\mathbb{R}}^+,dt/t)} , \qquad 0 < \theta < 1 , \quad 1 \le q \le \infty ,$$

and

$$(E_0, E_1)_{\theta,q} := \big(\{\, x \in E_0 + E_1 \ ;\ \|x\|_{\theta,q} < \infty \,\}, \|\cdot\|_{\theta,q}\big)$$

for $0 < \theta < 1,\ 1 \le q \le \infty$. Let

$$\mathfrak{F}_{\theta,q}(E_0, E_1) := (E_0, E_1)_{\theta,q} \quad \text{and} \quad \mathfrak{F}_{\theta,q}(A) := A$$

for $A\colon (E_0, E_1) \to (F_0, F_1)$. Then, given any $q \in [1,\infty]$ and $\theta \in (0,1)$, it follows that $\mathfrak{F}_{\theta,q}$ is an exact interpolation functor of exponent θ. Henceforth we denote it by

$$(\cdot,\cdot)_{\theta,q}$$

and call it **the real interpolation functor** of exponent θ and parameter q. ■

2.4.2 Example: Complex Interpolation Functors Suppose that $\mathbb{K} = \mathbb{C}$ and let (E_0, E_1) be an interpolation couple. Let $\mathcal{F}(E_0, E_1)$ be the set of all bounded and continuous functions f from $\overline{S}$ into $E_0 + E_1$, where $S := [0 < \operatorname{Re} z < 1]$, such that $f|S$ is holomorphic and

$$f|S_j \in C_0(S_j, E_j) , \qquad j = 0, 1 ,$$

where $S_j := [\operatorname{Re} z = j]$ and C_0 is the space of continuous functions vanishing at infinity. Then $\mathcal{F}(E_0, E_1)$ is a Banach space with the norm

$$\|f\|_{\mathcal{F}(E_0,E_1)} := \sup_{t\in\mathbb{R}} \|f(it)\|_{E_0} \vee \sup_{t\in\mathbb{R}} \|f(1+it)\|_{E_1} .$$

Given $\theta \in (0,1)$, put

$$[E_0, E_1]_\theta := \big(\{\, x \in E_0 + E_1 \ ;\ f(\theta) = x \text{ for some } f \in \mathcal{F}(E_0, E_1) \,\}, \|\cdot\|_\theta\big) ,$$

where

$$\|x\|_\theta := \inf\{\, \|f\|_{\mathcal{F}(E_0,E_1)} \ ;\ f(\theta) = x \,\} .$$

Letting $\mathfrak{F}_\theta(E_0, E_1) := [E_0, E_1]_\theta$ and $\mathfrak{F}_\theta(A) := A$ for $A\colon (E_0, E_1) \to (F_0, F_1)$, it follows that $\mathfrak{F}_\theta$ is an exact interpolation functor of exponent θ.

Now suppose that $\mathbb{K} = \mathbb{R}$. Given an interpolation couple (E_0, E_1), we put

$$\mathfrak{F}_\theta(E_0, E_1) := \left(\left[(E_0)_\mathbb{C}, (E_1)_\mathbb{C}\right]_\theta \cap (E_0 + E_1), \|\cdot\|_\theta\right) .$$

Thus, in order to determine $\mathfrak{F}_\theta(E_0, E_1)$ we complexify E_0 and E_1, apply the interpolation functor $\mathfrak{F}_\theta$ for complex spaces, and 'decomplexify' afterwards, that is, restrict ourselves to the real subspace of the interpolation space $[(E_0)_\mathbb{C}, (E_1)_\mathbb{C}]_\theta$. Defining $\mathfrak{F}_\theta(A) := A$ for $A\colon (E_0, E_1) \to (F_0, F_1)$, it follows that $\mathfrak{F}_\theta$ is an exact interpolation functor of exponent θ in this case as well.

In the general case, that is, if $\mathbb{K} = \mathbb{R}$ or $\mathbb{C}$, we denote the interpolation functor introduced by

$$[\cdot, \cdot]_\theta$$

and call it **the complex interpolation functor.** ■

2.4.3 Remark Let $(E, \|\cdot\|_E)$ be a complex Banach space of complex-valued functions and let $(F, \|\cdot\|)$ be the real subspace of real-valued functions. Then we know from Remark 3.1 that $F_\mathbb{C} \doteq E$ but $\|\cdot\|_{F_\mathbb{C}} \neq \|\cdot\|_E$, in general. This fact has to be kept in mind when employing the complex interpolation functor in the real case. ■

2.4.4 Example: Continuous Interpolation Functors We denote by $\mathcal{B}_2$ the subcategory of $\mathcal{B}_1$ of all densely injected Banach couples, that is, $\mathcal{B}_2$ is the subcategory of $\mathcal{B}_1$ whose objects are the densely injected Banach couples. Given two such couples (E_0, E_1) and (F_0, F_1), it follows from the closed graph theorem that $A\colon (E_0, E_1) \to (F_0, F_1)$ iff $A \in \mathcal{L}(E_0, F_0)$ and $A(E_1) \subset F_1$. On the category $\mathcal{B}_2$ we now define another important interpolation functor.

Let (E_0, E_1) be a densely injected Banach couple. Given $\theta \in (0, 1)$, put

$$(E_0, E_1)^0_{\theta,\infty} := \text{closure of } E_1 \text{ in } (E_0, E_1)_{\theta,\infty} .$$

Let (F_0, F_1) be a second object of $\mathcal{B}_2$ and suppose that $A\colon (E_0, E_1) \to (F_0, F_1)$. Since $A(E_1) \subset F_1$, it follows that

$$A \in \mathcal{L}\left((E_0, E_1)^0_{\theta,\infty}, (F_0, F_1)^0_{\theta,\infty}\right) .$$

Hence, putting $\mathfrak{F}_\theta(E_0, E_1) := (E_0, E_1)^0_{\theta,\infty}$ and $\mathfrak{F}_\theta(A) := A$, we obtain an exact interpolation functor of exponent θ. Henceforth we denote it by

$$(\cdot, \cdot)^0_{\theta,\infty}$$

and call it **the continuous interpolation functor**. It is known that

$$(E_0, E_1)^0_{\theta,\infty} = \left\{ x \in E_0 \ ; \ \lim_{t \to 0} t^{-\theta} K(t, x) = 0 \right\} . \tag{2.4.1}$$

It is also known (e.g., [DorF87]) that $(\cdot, \cdot)^0_{\theta,\infty}$ coincides with the 'continuous interpolation method' of Da Prato and Grisvard [DaPG79]. ■

2.5 Continuous Injections

Suppose that (E_0, E_1) is an interpolation couple. Given $\theta \in (0,1)$, we define an intermediate space with respect to (E_0, E_1) by

$$(E_0, E_1)^0_{\theta,\infty} := \text{closure of } E_0 \cap E_1 \text{ in } (E_0, E_1)_{\theta,\infty} \,. \tag{2.5.1}$$

Then the following important injections are valid:

$$\begin{aligned} E_0 \cap E_1 &\overset{d}{\hookrightarrow} (E_0, E_1)_{\xi,q} \overset{d}{\hookrightarrow} (E_0, E_1)_{\eta,1} \overset{d}{\hookrightarrow} [E_0, E_1]_\eta \\ &\overset{d}{\hookrightarrow} (E_0, E_1)^0_{\eta,\infty} \hookrightarrow (E_0, E_1)_{\eta,\infty} \overset{d}{\hookrightarrow} (E_0, E_1)_{\zeta,q} \hookrightarrow E_0 + E_1 \end{aligned} \tag{2.5.2}$$

for $1 \le q < \infty$ and $0 < \zeta < \eta < \xi < 1$. It is also true that

$$(E_0, E_1)_{\theta,q} \overset{d}{\hookrightarrow} (E_0, E_1)_{\theta,r} \overset{d}{\hookrightarrow} (E_0, E_1)^0_{\theta,\infty} \tag{2.5.3}$$

for $1 \le q < r < \infty,\ \ 0 < \theta < 1$. Moreover,

$$(E_0, E_1)_{\theta,q} = (E_1, E_0)_{1-\theta,q} \,, \quad [E_0, E_1]_\theta = [E_1, E_0]_{1-\theta} \tag{2.5.4}$$

for $0 < \theta < 1$ and $1 \le q \le \infty$.

2.6 Duality Properties

Given $q \in [1,\infty]$, we define the 'dual exponent' $q' \in [1,\infty]$ by

$$\frac{1}{q} + \frac{1}{q'} = 1 \,.$$

Then we have the following duality properties:

$$(E_0, E_1)'_{\theta,q} \doteq (E_0', E_1')_{\theta,q'} \,, \qquad 0 < \theta < 1 \,, \quad 1 \le q < \infty \,, \tag{2.6.1}$$

and

$$\big[(E_0, E_1)^0_{\theta,\infty}\big]' \doteq (E_0', E_1')_{\theta,1} \tag{2.6.2}$$

with respect to the duality pairing naturally induced by $\langle\cdot,\cdot\rangle_{E_0\cap E_1}$, provided $E_0 \cap E_1 \overset{d}{\hookrightarrow} E_j$ for $j = 0, 1$.

If either E_0 or E_1 is reflexive and $E_0 \cap E_1 \overset{d}{\hookrightarrow} E_j,\ \ j = 0, 1$, then

$$[E_0, E_1]'_\theta \doteq [E_0', E_1']_\theta \,, \qquad 0 < \theta < 1 \,, \tag{2.6.3}$$

with respect to the duality pairing naturally induced by $\langle\cdot,\cdot\rangle_{E_0\cap E_1}$. Moreover, $[E_0, E_1]_\theta$ is reflexive for $0 < \theta < 1$.

2.7 Compactness

In the following, we put

$$(E_0, E_1)_{j,q} := E_j \ , \qquad j = 0,1 \ , \quad 1 \le q \le \infty \ . \tag{2.7.1}$$

Suppose that $E_1 \overset{d}{\hookrightarrow} E_0$. Then

$$(E_0, E_1)_{\theta_1,q_1} \overset{c}{\hookrightarrow} (E_0, E_1)_{\theta_0,q_0} \tag{2.7.2}$$

for $0 \le \theta_0 < \theta_1 \le 1$, $1 \le q_0, q_1 \le \infty$.

2.8 Reiteration Theorems

As for the stability of the interpolation methods, we have the following reiteration theorem for the real methods: Suppose that

$$(E_0, E_1)_{\theta_j,1} \hookrightarrow F_j \hookrightarrow (E_0, E_1)_{\theta_j,\infty} \ , \qquad j = 0,1 \ , \tag{2.8.1}$$

where $0 \le \theta_j \le 1$ and $\theta_0 \ne \theta_1$. Then

$$(F_0, F_1)_{\eta,q} \doteq (E_0, E_1)_{(1-\eta)\theta_0+\eta\theta_1,q} \ , \qquad 0 < \eta < 1 \ , \quad 1 \le q \le \infty \ . \tag{2.8.2}$$

Similarly, as in (2.7.1), we put

$$[E_0, E_1]_j := E_j \ , \qquad j = 0,1 \ . \tag{2.8.3}$$

Then assuming, for simplicity, that (E_0, E_1) is a densely injected Banach couple, the reiteration theorem for the complex method says that

$$\big[[E_0, E_1]_{\theta_0}, [E_0, E_1]_{\theta_1}\big]_\eta = [E_0, E_1]_{(1-\eta)\theta_0+\eta\theta_1} \ , \qquad \theta_0, \theta_1, \eta \in [0,1] \ . \tag{2.8.4}$$

Observe that (2.5.2) and (2.8.1), (2.8.2) imply

$$\big([E_0, E_1]_{\theta_0}, [E_0, E_1]_{\theta_1}\big)_{\eta,q} \doteq (E_0, E_1)_{(1-\eta)\theta_0+\eta\theta_1,q} \tag{2.8.5}$$

for $\theta_0, \theta_1, \eta \in (0,1)$ with $\theta_0 \ne \theta_1$ and $1 \le q \le \infty$. On the other hand, it can be shown that

$$\big[(E_0, E_1)_{\theta_0,q_0}, (E_0, E_1)_{\theta_1,q_1}\big]_\eta \doteq (E_0, E_1)_{(1-\eta)\theta_0+\eta\theta_1,q} \tag{2.8.6}$$

for $\theta_0, \theta_1, \eta \in (0,1)$ with $\theta_0 \ne \theta_1$ and $1 \le q_j \le \infty$, where

$$1/q := (1-\eta)/q_0 + \eta/q_1$$

and where the case $q_0 = q_1 = \infty$ is excluded. (We refer to [Gri66] and [MS90] for further commutation properties for interpolation functors.)

Let (2.8.1) be satisfied. Then (2.8.2) implies that $(F_0, F_1)^0_{\eta,\infty}$ equals the closure of $F_0 \cap F_1$ in $(E_0, E_1)_{(1-\eta)\theta_0+\eta\theta_1,\infty}$. From (2.5.2), (2.5.3), and (2.8.2) we deduce that

$$F_0 \cap F_1 \overset{d}{\hookrightarrow} (F_0, F_1)_{\eta,1} \doteq (E_0, E_1)_{(1-\eta)\theta_0+\eta\theta_1,1} \overset{d}{\hookrightarrow} (E_0, E_1)^0_{(1-\eta)\theta_0+\eta\theta_1,\infty} .$$

Hence

$$(F_0, F_1)^0_{\eta,\infty} \doteq (E_0, E_1)^0_{(1-\eta)\theta_0+\eta\theta_1,\infty} , \qquad 0 < \eta < 1 . \tag{2.8.7}$$

2.9 Fractional Powers and Interpolation

Let $A \in \mathcal{C}(E)$ be such that $\mathbb{R}^+ \subset \rho(-A)$ and there exists a constant M with

$$(1+\lambda) \, \|(\lambda + A)^{-1}\|_{\mathcal{L}(E)} \leq M , \qquad \lambda \in \mathbb{R}^+ . \tag{2.9.1}$$

Assumption (2.9.1) is easily seen to imply the existence of constants $c := c(M)$ and $\varphi := \varphi(M) \in (0, \pi)$ such that

$$R_\varphi := [\,|\arg z| \leq \varphi] \cup [\,|z| \leq \varphi] \subset \rho(-A) \tag{2.9.2}$$

and

$$|\lambda| \, \|(\lambda + A)^{-1}\| \leq c , \qquad \lambda \in R_\varphi . \tag{2.9.3}$$

Let Γ be any piece-wise smooth simple curve in R_φ running from $\infty e^{-i\varphi}$ to $\infty e^{i\varphi}$ and avoiding $\mathbb{R}^+$. Then, given any $z \in \mathbb{C}$ with $\operatorname{Re} z < 0$, put

$$A^z := \frac{1}{2\pi i} \int_\Gamma (-\lambda)^z (\lambda + A)^{-1} \, d\lambda . \tag{2.9.4}$$

Observe that A^z is a well-defined bounded linear operator in E, thanks to estimate (2.9.3).

If A is densely defined, it is well-known and will be proven in Section III.4.6 that the definition of the 'fractional power' A^z of A can be extended to all $z \in \mathbb{C}$ in such a way that $A^z \in \mathcal{C}(E)$ with dense domain, $A^{z_1} A^{z_2} x = A^{z_1+z_2} x$ for any $x \in \operatorname{dom}(A^{2m})$, where $m \in \mathbb{N}$ with $m \geq 2$ and $\operatorname{Re} z_1 \vee \operatorname{Re} z_2 < m$, and such that A^z is the 'ordinary power' if $z \in \mathbb{Z}$. It follows that

$$D(A^{z_1}) \overset{d}{\hookrightarrow} D(A^{z_2}) , \qquad 0 < \operatorname{Re} z_2 < \operatorname{Re} z_1 , \tag{2.9.5}$$

and that $z \mapsto A^z x$ is continuous at $z_0 \in \mathbb{C}$ if $x \in \operatorname{dom}(A^{z_0+1})$, for example. Furthermore (e.g., [Tri78, Theorem 1.15.2]),

$$\big(E, D(A^m)\big)_{\operatorname{Re} z/m,1} \overset{d}{\hookrightarrow} D(A^z) \overset{d}{\hookrightarrow} \big(E, D(A^m)\big)^0_{\operatorname{Re} z/m,\infty} \tag{2.9.6}$$

for $0 < \operatorname{Re} z < m$ and $m \in \dot{\mathbb{N}}$.

If the purely imaginary powers of A are locally uniformly bounded, that is, $A^{it} \in \mathcal{L}(E)$ for $t \in [-1,1]$ and there exists a constant K such that

$$\|A^{it}\|_{\mathcal{L}(E)} \le K \;, \qquad |t| \le 1 \;, \tag{2.9.7}$$

then

$$\big[D(A^\alpha), D(A^\beta)\big]_\theta \doteq D(A^{(1-\theta)\alpha+\theta\beta}) \tag{2.9.8}$$

for $0 \le \operatorname{Re}\alpha < \operatorname{Re}\beta,\ 0 < \theta < 1$. More precisely, let

$$E_A^\alpha := \big(\operatorname{dom}(A^\alpha), \|A^\alpha \cdot\|\big) \;, \qquad \operatorname{Re}\alpha \ge 0 \;.$$

Then, given $\theta \in (0,1)$, there exists a constant $c := c(K) \ge 1$ such that

$$c^{-1}\,\|\cdot\|_{E_A^{(1-\theta)\alpha+\theta\beta}} \le \|\cdot\|_{[E_A^\alpha, E_A^\beta]_\theta} \le c\,\|\cdot\|_{E_A^{(1-\theta)\alpha+\theta\beta}} \tag{2.9.9}$$

for $0 \le \operatorname{Re}\alpha < \operatorname{Re}\beta$ and all densely defined $A \in \mathcal{C}(E)$ satisfying (2.9.7). This follows by an obvious modification of the proof of [See71, Theorem 3].

Suppose that H is a Hilbert space and A is a self-adjoint linear operator in H. Denoting by $\{\, E_\lambda \;;\; \lambda \in \mathbb{R} \,\}$ the spectral resolution of A, we can define A^z by

$$A^z := \int_{\sigma(A)} \lambda^z \, dE_\lambda \;, \qquad z \in \mathbb{C} \;, \tag{2.9.10}$$

provided $\sigma(A) \subset \mathbb{R}^+$, that is, $A \ge 0$. If A is positive definite, i.e., $A \ge \alpha$ for some $\alpha > 0$, it follows that $\{\, A^z \;;\; z \in \mathbb{C} \,\}$ are the fractional powers of A in the sense of the definition. If $A \ge 0$, we have the following improvement over (2.9.8):

$$\big[D(A^\alpha), D(A^\beta)\big]_\theta \doteq \big(D(A^\alpha), D(A^\beta)\big)_{\theta,2} \doteq D(A^{(1-\theta)\alpha+\theta\beta}) \tag{2.9.11}$$

for $0 \le \operatorname{Re}\alpha < \operatorname{Re}\beta$ and $0 < \theta < 1$ (e.g., [Tri78, Theorem 1.18.10]).

Lastly, we recall that, given $m \in \dot{\mathbb{N}}$,

$$x \mapsto \big\| t^{\theta m} \big[A(t+A)^{-1}\big]^m x \big\|_{L_q(\dot{\mathbb{R}}^+, dt/t; E)} \tag{2.9.12}$$

is an equivalent norm on $\big(E, D(A^m)\big)_{\theta,q}$ for $0 < \theta < 1$ and $1 \le q \le \infty$, uniformly with respect to every densely defined $A \in \mathcal{C}(E)$ satisfying (2.9.1). This is a consequence of (the proof of) [Tri78, Theorem 1.14.3].

2.10 Semigroups and Interpolation

Next we give some estimates for the K-functional in terms of semigroups. For this subsection we refer, in particular, to [Tri78] and [Lun95].

Given $M \geq 1$ and $\sigma \in \mathbb{R}$, we denote by

$$\mathcal{G}(E, M, \sigma)$$

the set of all $A \in \mathcal{C}(E)$ such that $-A$ is the infinitesimal generator of a strongly continuous semigroup $\{\, e^{-tA} \ ; \ t \geq 0, \}$ on E satisfying

$$\|e^{-tA}\| \leq Me^{\sigma t} , \qquad t \geq 0 .$$

Moreover,

$$\mathcal{G}(E) := \bigcup_{\substack{\sigma \in \mathbb{R} \\ M \geq 1}} \mathcal{G}(E, M, \sigma)$$

is the set of all negative generators of strongly continuous semigroups on E. Given $A \in \mathcal{G}(E)$,

$$\operatorname{type}(-A) := \inf\bigl\{\, \sigma \in \mathbb{R} \ ; \ \text{there exists } M \geq 1 \text{ with } A \in \mathcal{G}(E, M, \sigma) \,\bigr\}$$

is the **type** of $-A$ (or the 'exponential growth bound' of $-A$).

Suppose that $A_1, \ldots, A_n \in \mathcal{G}(E, M, \sigma)$ for some $\sigma \leq 0$ and that the semigroups $\{\, e^{-tA_j} \ ; \ t \geq 0 \,\}$, $j = 1, \ldots, n$, are pair-wise commuting. Put

$$\mathbb{E} := \bigcap_{j=1}^{n} D(A_j) \tag{2.10.1}$$

and

$$e^{-s\mathbb{A}} := e^{-s_1 A_1} \cdots e^{\ s_n A_n} , \qquad s := (s_1, \ldots, s_n) \in (\mathbb{R}^+)^n . \tag{2.10.2}$$

Then $(E, \mathbb{E})$ is a densely injected Banach couple and $\{\, e^{-s\mathbb{A}} \ ; \ s \in (\mathbb{R}^+)^n \,\}$ is a strongly continuous n-parameter semigroup on E that has $\mathbb{E}$ as an invariant subspace. Put

$$\omega(t, x) := \sup_{s \in (0,t)^n} \|(1 - e^{-s\mathbb{A}})x\|_E , \qquad t > 0 , \quad x \in E . \tag{2.10.3}$$

Then there exists a constant $c := c(M) \geq 1$ such that

$$c^{-1} K(t, x, E, \mathbb{E}) \leq (1 \wedge t)\, \|x\| + \omega(t, x) \leq cK(t, x, E, \mathbb{E}) \tag{2.10.4}$$

for $t > 0$ and $x \in E$. If $\sigma < 0$, there exists a constant $c := c(M, \sigma) \geq 1$ such that

$$c^{-1} K(t, x, E, \mathbb{E}) \leq \omega(t, x) \leq cK(t, x, E, \mathbb{E}) \tag{2.10.5}$$

for $t > 0$ and $x \in E$. Using these estimates it is obvious how to get equivalent characterizations and equivalent norms for the real interpolation spaces $(E, \mathbb{E})_{\theta,q}$ as well as for the continuous interpolation spaces $(E, \mathbb{E})^0_{\theta,\infty}$.

Lastly, let (E_0, E_1) be a densely injected Banach couple and $A \in \mathcal{H}(E_1, E_0)$. Also suppose that there exists $M \geq 1$ such that

$$\|(tA)^j e^{-tA}\|_{\mathcal{L}(E_0)} \leq M , \qquad t > 0 , \quad j = 0, 1 . \tag{2.10.6}$$

Given $\theta \in (0,1)$ and $1 \leq q \leq \infty$, put

$$\|x\|_{\theta,q,A} := \left\| t^{-\theta} \, \|tAe^{-tA}x\|_0 \right\|_{L_q(\dot{\mathbb{R}}^+, dt/t)} , \qquad x \in E_0 . \tag{2.10.7}$$

Then

$$(E_0, E_1)_{\theta,q} = \{\, x \in E_0 \ ; \ \|x\|_{\theta,q,A} < \infty \,\}$$

and there exists a constant $c := c(M) \geq 1$ such that

$$c^{-1} \|x\|_{\theta,q} \leq \|x\|_0 + \|x\|_{\theta,q,A} \leq c \, \|x\|_{\theta,q} , \qquad x \in (E_0, E_1)_{\theta,q} , \tag{2.10.8}$$

whenever $A \in \mathcal{H}(E_1, E_0)$ satisfies (2.10.6).

Given $\sigma < 0$, there exists a constant $c := c(M, \sigma) \geq 1$ such that

$$c^{-1} \|x\|_{\theta,q} \leq \|x\|_{\theta,q,A} \leq c \, \|x\|_{\theta,q} , \qquad x \in (E_0, E_1)_{\theta,q} , \tag{2.10.9}$$

for all $A \in \mathcal{H}(E_1, E_0)$ satisfying (2.10.6) with $\operatorname{type}(-A) \leq \sigma$. Finally,

$$(E_0, E_1)^0_{\theta,\infty} = \{\, x \in E_0 \ ; \ \lim_{t \to 0} t^{-\theta} \, \|tAe^{-tA}x\|_0 = 0 \,\} \tag{2.10.10}$$

if $A \in \mathcal{H}(E_1, E_0)$ satisfies (2.10.6).

Estimate (2.10.9) has the following generalization. Suppose $m \in \dot{\mathbb{N}}$ and $\sigma < 0$ and, given $A \in \mathcal{H}(E_1, E_0)$ with $\operatorname{type}(-A) \leq \sigma$, put

$$E^m_A := \big(\operatorname{dom}(A^m), \|A^m \cdot\|_0\big) .$$

Then, given $M \geq 1$, there exists a constant $c := c(M) \geq 1$ such that

$$\begin{aligned} c^{-1} \|\cdot\|_{(E_0, E^m_A)_{\theta,q}} &\leq \left\| t^{-m\theta} \, \|t^m A^m e^{-tA} \cdot\|_0 \right\|_{L_q(\dot{\mathbb{R}}^+, dt/t)} \\ &\leq c \, \|\cdot\|_{(E_0, E^m_A)_{\theta,q}} \end{aligned} \tag{2.10.11}$$

for all $A \in \mathcal{H}(E_1, E_0)$ satisfying $\operatorname{type}(-A) \leq \sigma$ and (2.10.6) for $j = 0, 1, \ldots, m$. This follows from the proof of [Tri78, Theorem 1.14.5].

2.11 Admissible Interpolation Functors

Suppose that $0 < \theta < 1$. By an **admissible interpolation functor**, denoted by $(\cdot,\cdot)_\theta$, we mean an interpolation functor of exponent θ for the category of densely injected Banach couples such that

$$E_1 \text{ is dense in } (E_0, E_1)_\theta ,$$

whenever (E_0, E_1) is such a couple. Observe that the real interpolation functors $(\cdot,\cdot)_{\theta,p}$, $1 \le p < \infty$, the complex interpolation functor $[\cdot,\cdot]_\theta$, and the 'continuous' interpolation functor $(\cdot,\cdot)^0_{\theta,\infty}$ are admissible. If no confusion seems possible, we put

$$E_\theta := (E_0, E_1)_\theta$$

and denote the norm in E_θ by $\|\cdot\|_\theta$. In the following, we fix for each $\theta \in (0,1)$ an admissible interpolation functor arbitrarily.

2.11.1 Theorem *If (E_0, E_1) is a densely injected Banach couple,*

$$E_1 \overset{d}{\hookrightarrow} E_\alpha \overset{d}{\hookrightarrow} E_\beta \overset{d}{\hookrightarrow} E_0 \ , \qquad 0 < \beta < \alpha < 1 \ . \tag{2.11.1}$$

Given $\alpha, \beta, \gamma \in [0,1]$ with $0 \le \gamma < \beta < \alpha \le 1$, there is $c := c(\alpha,\beta,\gamma)$ such that

$$\|x\|_\beta \le c\, \|x\|_\gamma^{(\alpha-\beta)/(\alpha-\gamma)} \, \|x\|_\alpha^{(\beta-\gamma)/(\alpha-\gamma)} \ , \qquad x \in E_\alpha \ . \tag{2.11.2}$$

If $E_1 \overset{c}{\hookrightarrow} E_0$ then

$$E_\alpha \overset{c}{\hookrightarrow} E_\beta \ , \qquad 0 \le \beta < \alpha \le 1 \ . \tag{2.11.3}$$

Conversely, $E_1 \overset{c}{\hookrightarrow} E_0$ if $E_\alpha \overset{c}{\hookrightarrow} E_\beta$ for some $\alpha, \beta \in [0,1]$ with $\beta < \alpha$.

Proof It follows from the extremal property of the real method (e.g., [BerL76, Theorem 3.9.1]) that

$$(E_0, E_1)_{\theta,1} \overset{d}{\hookrightarrow} E_\theta \overset{d}{\hookrightarrow} (E_0, E_1)^0_{\theta,\infty} \ , \qquad 0 < \theta < 1 \ . \tag{2.11.4}$$

Hence (2.11.1) is an immediate consequence of (2.5.2). Thus (E_β, E_α) is a densely injected Banach couple whenever $0 \le \beta < \alpha \le 1$. Letting

$$\eta := (\beta - \gamma)/(\alpha - \gamma) \ ,$$

we infer from (2.11.4) and the reiteration property (2.8.2) of the real method that

$$(E_\gamma, E_\alpha)_{\eta,1} \hookrightarrow (E_0, E_1)_{\beta,1} \hookrightarrow E_\beta \ .$$

Now (2.11.2) follows from Proposition 2.2.1 and Example 2.4.1.

Assertion (2.11.3) is a consequence of (2.7.2), (2.11.4), and (2.5.2). The last part of the statement follows from $E_1 \hookrightarrow E_\alpha \overset{c}{\hookrightarrow} E_\beta \hookrightarrow E_0$ and the fact that compositions of linear operators are compact provided at least one of them is compact. ■

2.11.2 Remarks **(a)** Let (E_0, E_1) be a densely injected Banach couple. Then the scale of Banach spaces $\{\, E_\alpha \ ; \ 0 \le \alpha \le 1 \,\}$ possesses the following **almost reiteration property**:

$$\text{if} \quad 0 \le \alpha < \beta \le 1 \quad \text{and} \quad 0 < \eta_- < \eta < \eta_+ < 1 \quad \text{then}$$
$$(E_\alpha, E_\beta)_{\eta_+} \hookrightarrow E_{(1-\eta)\alpha+\eta\beta} \hookrightarrow (E_\alpha, E_\beta)_{\eta_-} \ .$$

Proof It follows from (2.5.2), (2.11.4), and (2.8.2) that

$$(E_\alpha, E_\beta)_\eta \hookrightarrow (E_\alpha, E_\beta)_{\eta,\infty} \hookrightarrow (E_0, E_1)_{(1-\eta)\alpha+\eta\beta,\infty}$$

and

$$(E_0, E_1)_{(1-\alpha)\eta+\eta\beta,1} \doteq \big((E_0, E_1)_{\alpha,1}, (E_0, E_1)_{\beta,1}\big)_{\eta,1} \hookrightarrow (E_\alpha, E_\beta)_\eta$$

for $0 < \eta < 1$. This implies the assertion. ■

(b) Of course, it is a consequence of the reiteration theorems for the real and complex methods that

$$(E_\alpha, E_\beta)_\eta \doteq E_{(1-\eta)\alpha+\eta\beta} , \qquad 0 \le \alpha < \beta \le 1 , \quad 0 < \eta < 1 , \tag{2.11.5}$$

if $(\cdot,\cdot)_\theta$ is one of the real interpolation functors or $(\cdot,\cdot)_\theta = (\cdot,\cdot)^0_{\theta,\infty}$ *for each* $\theta \in (0,1)$, or if $(\cdot,\cdot)_\theta = [\cdot,\cdot]_\theta$ *for each* $\theta \in (0,1)$. Moreover, (2.8.5) and (2.8.6) imply the validity of (2.11.5) for appropriate other choices of $(\cdot,\cdot)_\theta$ as well.

(c) Let (E_0, E_1) be a densely injected Banach couple such that $\mathcal{H}(E_1, E_0) \ne \emptyset$. Then it is easily verified that $E_1 \overset{c}{\hookrightarrow} E_0$ iff there exists $A \in \mathcal{H}(E_1, E_0)$ possessing a compact resolvent. This is the case iff each $A \in \mathcal{H}(E_1, E_0)$ has a compact resolvent (considered as 'unbounded' linear operators in E_0, of course). ■

Chapter II

Cauchy Problems and Evolution Operators

Any theory of abstract quasilinear parabolic problems requires, of course, a good understanding of the theory of linear parabolic evolution equations. In this chapter we develop that part of the linear theory which is based upon the concept of evolution operators. The latter correspond to the fundamental matrices in the theory of ordinary differential equations. This 'classical' theory is particularly well-suited for the study of quasilinear parabolic problems exhibiting smoothing phenomena since it exploits the fact that the solution of a linear parabolic evolution equation has in general better regularity properties than its initial value.

Besides its importance for the study of quasilinear parabolic problems, the concept of evolution operators is of intrinsic interest for many questions concerning the behavior of solutions to nonautonomous linear problems — in particular, under periodicity assumptions.

In the first section of this chapter we essentially describe the basic results about linear parabolic Cauchy problems. In Section 2 we introduce the concept of a parabolic evolution operator. Section 3 contains some technical results about operator-valued Volterra integral equations. These are needed for the construction of parabolic evolution operators in Section 4. Section 5 is devoted to the proof of some stability estimates that are basic for the study of abstract quasilinear parabolic problems. In Section 6 we investigate invariance properties of evolution operators. In particular, we derive sufficient conditions for the positivity of these operators in the case of ordered Banach spaces. These results can be considered to be abstract versions of the maximum principle for second order elliptic and parabolic differential equations.

Throughout this chapter E, F, E_j, and F_j, $j = 0, 1, \ldots$, are Banach spaces.

1 Linear Cauchy Problems

In this section we collect some of the basic existence and regularity results for linear parabolic evolution equations whose proofs are given in later sections. For a concise formulation we first introduce some classes of function spaces.

1.1 Hölder Spaces

Let X be a nonempty set. We write

$$B(X,E)$$

for the Banach space of all bounded maps from X to E, equipped with the supremum norm $\|\cdot\|_\infty$. If X is a topological space,

$$BC(X,E)$$

is the closed linear subspace of $B(X,E)$ consisting of all bounded and continuous functions.

Let $X = (X,d)$ be a metric space. Then

$$BUC(X,E)$$

is the closed linear subspace of $BC(X,E)$ consisting of all bounded and uniformly continuous functions. Given $\rho \in (0,1]$, we put

$$[u]_\rho := [u]_{\rho,X} := [u]_{C^\rho(X,E)} := \sup_{\substack{x,y\in X \\ x\neq y}} \frac{\|u(x)-u(y)\|}{[d(x,y)]^\rho} \tag{1.1.1}$$

for $u\colon X \to E$, and

$$\|\cdot\|_{C^\rho} := \|\cdot\|_\infty + [\cdot]_\rho \ .$$

Then

$$BUC^\rho(X,E) := \bigl(\{\, u \in B(X,E)\ ;\ [u]_\rho < \infty \,\},\ \|\cdot\|_{C^\rho}\bigr) \tag{1.1.2$_\rho$}$$

is the Banach space of all bounded and uniformly ρ-Hölder continuous E-valued functions on X. If $\rho = 1$, we write $BUC^{1\text{-}}(X,E)$ for $(1.1.2)_1$ in order to avoid confusion with spaces of differentiable functions. Of course, the elements of $BUC^{1\text{-}}(X,E)$ are the bounded and uniformly Lipschitz continuous functions from X to E. In general, we use 1- to denote Lipschitz continuity. However, in numerical calculations we identify 1- with the number 1. In particular, $s < 1\text{-}$ for each $s \in \mathbb{R}$ with $s < 1$.

Given $\rho \in (0,1) \cup \{1\text{-}\}$, we denote by

$$C^\rho(X,E)$$

the set of all $u \in C(X,E)$ such that each point in X has a neighborhood Y such that $u|Y \in BUC^\rho(Y,E)$, that is, if u is (locally) **ρ-Hölder continuous** (Lipschitz continuous if $\rho = 1\text{-}$).

1.1.1 Lemma *Given $u \in C^\rho(X,E)$ and $K \subset\subset X$, there exists a neighborhood Y of K in X such that $u|Y \in BUC^\rho(Y,E)$. Thus Hölder (resp. Lipschitz) continuous functions are uniformly Hölder (resp. Lipschitz) continuous on compact sets.*

Proof This follows from an obvious modification of the proof of [Ama90b, Proposition 6.4]. ■

Lemma 1.1.1 implies that $\{\, \|\cdot\|_{C^\rho(K,E)}\ ;\ K \subset\subset X\,\}$ is a separating family of seminorms on $C^\rho(X,E)$ inducing a locally convex Hausdorff topology. Thus $C^\rho(X,E)$ is a *LCS*. It is well-known and not difficult to prove that

$$BUC^\rho(X,E) \hookrightarrow BUC^\sigma(X,E) \hookrightarrow C^\sigma(X,E) \hookrightarrow C^\tau(X,E) \qquad (1.1.3)$$

for $\rho,\sigma,\tau \in [0,1) \cup \{1\text{-}\}$ with $\tau \le \sigma \le \rho$, where $BUC^0 := BUC$ and $C^0 := C$.

Now let X be an open subset of F or a perfect subset of $\mathbb{K}$. Then, given $\rho \in [0,1) \cup \{1\text{-}\}$, we denote by

$$BUC^{1+\rho}(X,E)$$

the Banach space of all continuously differentiable functions u in $BUC(X,E)$ such that $\partial u \in BUC\big(X, \mathcal{L}(F,E)\big)$, and $[\partial u]_{\rho,X} < \infty$ if $\rho > 0$, equipped with the norm

$$u \mapsto \|u\|_{BUC^{1+\rho}(X,E)} := \|u\|_{C^{1+\rho}} := \max_{j=0,1} \|\partial^j u\|_{\infty,X} + [\partial u]_{\rho,X}\ ,$$

where ∂ denotes the Fréchet derivative and $\partial^0 := \mathrm{id}$, and where the last term is omitted if $\rho = 0$. Similarly, $u \in C^{1+\rho}(X,E)$ if each point in X has a neighborhood Y such that $u|Y \in BUC^{1+\rho}(X,E)$. This space is a *LCS* with respect to the obvious topology. Lastly, we put $2\text{-} := 1 + 1\text{-}$.

Of course,

$$C^\sigma(X,Y) := \big\{\, u \in C^\sigma(X,E)\ ;\ u(X) \subset Y \,\big\}$$

for $\sigma \in [0,2) \cup \{1\text{-}, 2\text{-}\}$, etc., if Y is a nonempty subset of E.

Hölder spaces occur naturally by interpolation as is witnessed by the following 'embedding' theorem.

1.1.2 Proposition *Let (E_0,E_1) be a densely injected Banach couple and let $(\cdot,\cdot)_\theta$ be an admissible interpolation functor. Let X be open in F or a convex perfect subset of $\mathbb{K}$. Then*

$$C(X,E_1) \cap C^1(X,E_0) \hookrightarrow C^{1-\theta}(X,E_\theta)\ ,$$

where $E_\theta := (E_0,E_1)_\theta$.

Proof Given $x_0 \in X$, there exists a convex neighborhood Y of x_0 in X such that $u|Y \in BC(X, E_1)$ and $\partial u|Y \in BC\big(Y, \mathcal{L}(F, E_0)\big)$. Hence the mean-value theorem implies

$$\|u(x) - u(y)\|_0 \le \|\partial u\|_{C(Y,\mathcal{L}(F,E_0))} \, \|x - y\| \ , \qquad x, y \in Y \ .$$

Now we deduce from the interpolation inequality (I.2.11.2) that

$$\begin{aligned}\|u(x) - u(y)\|_\theta &\le c\, \|u(x) - u(y)\|_0^{1-\theta} \, \|u(x) - u(y)\|_1^\theta \\ &\le 2^\theta c\, \|\partial u\|_{C(Y,\mathcal{L}(F,E_0))}^{1-\theta} \, \|u\|_{C(Y,E_1)}^\theta \, \|x - y\|^{1-\theta} \\ &\le c(\theta)\, \|u\|_{C(Y,E_1)\cap C^1(Y,E_0)} \, \|x - y\|^{1-\theta}\end{aligned}$$

for $x, y \in Y$. Now the assertion is obvious. ■

A bilinear and continuous map of norm at most one,

$$E_1 \times E_2 \to E \ , \quad (x_1, x_2) \mapsto x_1 \bullet x_2 \ , \tag{1.1.4}$$

is said to be a **multiplication**. If S is a nonempty set, we define **point-wise multiplication**

$$E_1^S \times E_2^S \to E^S \ , \quad (u_1, u_2) \mapsto u_1 \bullet u_2$$

induced by (1.1.4) by

$$u_1 \bullet u_2(s) := u_1(s) \bullet u_2(s) \ , \qquad s \in S \ .$$

It is obvious how to extend this definition to more than two factors.

The following proposition shows that Hölder spaces are 'stable' under point-wise multiplication.

1.1.3 Proposition *Let X be a metric space and let $\rho \in [0,1) \cup \{1-\}$. Then the point-wise multiplication induced by (1.1.4) is a bilinear and continuous map:*
(i) $BUC^\rho(X, E_1) \times BUC^\rho(X, E_2) \to BUC^\rho(X, E)$;
(ii) $C^\rho(X, E_1) \times C^\rho(X, E_2) \to C^\rho(X, E)$.

Proof Observe that, given a nonempty subset Y of X,

$$\|u_1 \bullet u_2\|_{\infty,Y} \le \|u_1\|_{\infty,Y} \, \|u_2\|_{\infty,Y} \tag{1.1.5}$$

and, if $\rho > 0$,

$$[u_1 \bullet u_2]_{\rho,Y} \le \|u_1\|_{\infty,Y} \, [u_2]_{\rho,Y} + [u_1]_{\rho,Y} \, \|u_2\|_{\infty,Y} \tag{1.1.6}$$

for $u_j \in BUC^\rho(Y, E_j)$. Now the assertion is obvious. ■

Observe that (1.1.5) and (1.1.6) imply that the map (i) has norm at most one. In particular, if $E_1 = E_2 = E = \mathbb{K}$, it follows that

$$BUC^\rho(X) := BUC^\rho(X, \mathbb{K})$$

is a Banach algebra (with unit $\mathbf{1}$).

1.2 Existence and Regularity Theorems

Let $J \subset \mathbb{R}$ be a perfect interval containing its left endpoint $s := \min J$. For each $t \in J$ let $A(t)$ be a linear operator in E, and let $f : J \to E$. By a **solution** of the **linear evolution equation**

$$\dot{u} + A(t)u = f(t) \ , \qquad t \in J \backslash \{s\} \ , \tag{1.2.1}$$

where $\dot{u} := \partial_t u$, we mean a function $u \in C^1\big(J\backslash\{s\}, E\big)$ such that $u(t) \in \operatorname{dom}\big(A(t)\big)$ and $\dot{u}(t) + A(t)u(t) = f(t)$ for $t \in J\backslash\{s\}$. If, in addition, $u \in C(J, E)$ and $u(s) = x$ then u is a solution of the **linear Cauchy problem** (in E)

$$\dot{u} + A(t)u = f(t) \ , \quad t \in J\backslash\{s\} \ , \qquad u(s) = x \ . \qquad (1.2.2)_{(s,x,A,f)}$$

Lastly, u is a **strict solution** of (1.2.1) if $u \in C^1(J, E)$ and $u(s) \in \operatorname{dom}\big(A(s)\big)$.

The linear evolution equation (1.2.1) and the corresponding linear Cauchy problem $(1.2.2)_{(s,x,A,f)}$ are said to be **parabolic** if $-A(t)$ is the infinitesimal generator of a strongly continuous analytic semigroup on E for each $t \in J$.

As for the solvability of parabolic linear Cauchy problems, we have the following fundamental existence and regularity result:

1.2.1 Theorem *Suppose that (E_0, E_1) is a densely injected Banach couple and*

$$(x, A, f) = \big(x, (A, f)\big) \in E_0 \times C^\rho\big(J, \mathcal{H}(E_1, E_0) \times E_0\big)$$

for some $\rho \in (0,1)$. Then the Cauchy problem $(1.2.2)_{(s,x,A,f)}$ possesses a unique solution $u := u(\cdot, s, x, A, f)$, and

$$u \in C^\rho\big(J\backslash\{s\}, E_1\big) \cap C^{1+\rho}\big(J\backslash\{s\}, E_0\big) \ . \tag{1.2.3}$$

If $x \in E_1$ then u is a strict solution.

The existence of a unique solution, which is strict if $x \in E_1$, has been proven by Sobolevskii [Sob66] and, independently, by Tanabe [Tan60]. A different proof has more recently been given by Acquistapace and Terreni [AcT85]. The regularity result (1.2.3) is contained in [Sob66] and [AcT85]. A complete proof of Theorem 1.2.1 is given in Subsections 4 and III.2.6.

It is an important property of parabolic evolution equations that they 'regularize'. In fact, (1.2.3) shows that the solution is more regular than its initial value in general. The following 'automatic regularity theorem' implies that the solution of $(1.2.2)_{(s,x,A,f)}$ has even better regularity properties if f is more regular.

1.2.2 Theorem *Suppose that* (E_0, E_1) *is a densely injected Banach couple. Given* $\theta \in (0,1)$, *let* $(\cdot,\cdot)_\theta$ *be an admissible interpolation functor and put*

$$E_\theta := (E_0, E_1)_\theta \ .$$

Suppose that $0 < \gamma < \rho < 1$ *and* $0 < \varepsilon < 1-\gamma$, *and that*

$$(x, A) \in E_0 \times C^\rho\big(J, \mathcal{H}(E_1, E_0)\big)$$

and

$$f \in C^\varepsilon(J, E_\gamma) + C(J, E_{\gamma+\varepsilon}) \ .$$

Then the Cauchy problem $(1.2.2)_{(s,x,A,f)}$ *has a unique solution* u, *and*

$$u \in C(J, E_0) \cap C\big(J\backslash\{s\}, E_1\big) \cap C^1\big(J\backslash\{s\}, E_\gamma\big) \ .$$

It is also a solution of the parabolic equation

$$\dot{v} + A_\gamma(t)v = f(t) \ , \qquad t \in J\backslash\{s\} \ , \tag{1.2.4}$$

in E_γ, *where* A_γ *is the* E_γ*-realization of* A. *If* $x \in E_\gamma$ *then* $u \in C(J, E_\gamma)$, *and if* $x \in \operatorname{dom}\big(A_\gamma(s)\big)$, *it is a strict solution of* (1.2.4).

This theorem is due to the author (cf. [Ama88c, Theorem 8.2]). Its proof is given in Subsection IV.1.5.

It should be noted that the domains, $\operatorname{dom}\big(A_\gamma(t)\big)$, of the linear operators $A_\gamma(t)$ depend on $t \in J$, in general, whereas the operators $A(t)$, $t \in J$, have 'constant domains'. Also observe that the assertion that u is a solution of (1.2.4) implies $u(t) \in \operatorname{dom}\big(A_\gamma(t)\big)$ for $t \in J\backslash\{s\}$. This means that $u(t)$ has, in general, better 'spacial' regularity. This will be clear from concrete applications of Theorem 1.2.2 where it is often possible to characterize $\operatorname{dom}\big(A_\gamma(t)\big)$ explicitly.

As we shall see, parabolic evolution equations with 'constant domains' are particularly well-suited for studying quasilinear problems. This is not so for problems with variable domains. In later sections show how — by extrapolation techniques — we often can associate with a given evolution equation with 'nonconstant domains' a generalized problem possessing 'constant domains'. Then we can apply the general theory for evolution equations with 'constant domains', developed below, to the generalized problem. Afterwards Theorem 1.2.2 can be used to carry out an abstract 'bootstrapping' argument to prove that the generalized solutions are, in fact, already solutions of the original problem.

2 Parabolic Evolution Operators

Let J be a perfect subinterval of $\mathbb{R}$. Then

$$J_\Delta := \{\, (t,s) \in J \times J \; ; \; s \le t \,\}$$

and we put

$$J_\Delta^* := \{\, (t,s) \in J_\Delta \; ; \; s < t \,\} \, .$$

If φ is an operator-valued function on J_Δ^* and ψ is such a function on J, we write

$$\psi\varphi(t,s) := \psi(t)\varphi(t,s) \; , \quad \varphi\psi(t,s) := \varphi(t,s)\psi(s)$$

if these compositions are meaningful and no confusion seems likely.

2.1 Basic Properties

Suppose that $F \hookrightarrow E$ and that

$$A \colon J \to \mathcal{C}(E) \qquad \text{with} \quad \operatorname{dom}\bigl(A(t)\bigr) \subset F \; , \quad t \in J \, . \tag{2.1.1}$$

Then a map $U : J_\Delta \to \mathcal{L}(E)$ is said to be a **parabolic evolution operator** (propagator, fundamental solution) for A with **regularity subspace** F if

$$U \in C\bigl(J_\Delta, \mathcal{L}_s(E)\bigr) \cap C\bigl(J_\Delta^*, \mathcal{L}(E,F)\bigr) \tag{2.1.2}$$

and

$$U(t,t) = 1 \; , \quad U(t,s) = U(t,\tau)U(\tau,s) \; , \qquad s \le \tau \le t \; , \quad (t,s) \in J_\Delta \; , \tag{2.1.3}$$

if

$$\operatorname{im}\bigl(U(t,s)\bigr) \subset \operatorname{dom}\bigl(A(t)\bigr) \subset F \; , \qquad (t,s) \in J_\Delta^* \; , \tag{2.1.4}$$

and

$$AU \in C\bigl(J_\Delta^*, \mathcal{L}(E)\bigr)$$

with

$$\sup_{(t,s)\in I_\Delta^*} (t-s) \, \|AU(t,s)\|_{\mathcal{L}(E)} < \infty \; , \qquad I \subset\subset J \; , \tag{2.1.5}$$

if

$$U(\cdot,s) \in C^1\bigl(J \cap (s,\infty), \mathcal{L}(E)\bigr) \; , \quad s \in J \; , \tag{2.1.6}$$

and

$$\partial_1 U = -AU \; , \tag{2.1.7}$$

and if

$$U(t,\cdot) \in C^1\bigl(J \cap (-\infty,t), \mathcal{L}_s(F,E)\bigr) \; , \qquad t \in J \; , \tag{2.1.8}$$

and

$$\partial_2 U \supset UA \, . \tag{2.1.9}$$

2.1.1 Example: Analytic Semigroups Suppose that (E_0, E_1) is a densely injected Banach couple and $A \in \mathcal{H}(E_1, E_0)$. Put $J := \mathbb{R}^+$ and

$$U_A(t,s) := e^{-(t-s)A} \ , \qquad (t,s) \in J_\Delta \ .$$

Then U_A is a parabolic evolution operator for the constant map $J \to \mathcal{C}(E),\ t \mapsto A$ with regularity subspace E_1. ■

2.1.2 Remarks **(a)** Suppose that $f \in C\big(J \cap [s,\infty), E\big)$ and u is a solution of the Cauchy problem

$$\dot{u} + A(t)u = f(t) \ , \qquad t \in J \cap (s,\infty) \ , \quad u(s) = x \ . \qquad (2.1.10)_{(s,x,A,f)}$$

Also suppose that $V \in C\big(J_\Delta, \mathcal{L}_s(E)\big)$ satisfies $V(t,t) = 1$ and

$$V(t,\cdot) \in C^1\big(J \cap (s,t), \mathcal{L}_s(F,E)\big)$$

with $\partial_2 V \supset VA$ for $t \in J \cap (s,\infty)$. Then

$$u(t) = V(t,s)u(s) + \int_s^t V(t,\tau)f(\tau)\,d\tau \ , \qquad t \in J \cap [s,\infty) \ . \qquad (2.1.11)$$

Proof It is easily verified that, given $t \in J \cap (s,\infty)$,

$$V(t,\cdot)u \in C^1\big((s,t),E\big) \cap C\big([s,t],E\big)$$

and that $\partial_2(Vu)(t,\tau) = Vf(t,\tau)$ for $s < \tau < t$. Hence

$$V(t,t')u(t') - V(t,s')u(s') = \int_{s'}^{t'} V(t,\tau)f(\tau)\,d\tau \ , \qquad s < s' < t' < t \ ,$$

and the assertion follows by letting $t' \to t$ and $s' \to s$. ■

(b) Supppose that there exists a parabolic evolution operator U for A. Then, given $(s,x) \in J \times E$ and $f \in C\big(J \cap [s,\infty), E\big)$, the Cauchy problem $(2.1.10)_{(s,x,A,f)}$ has at most one solution. It is given by the **variation of constants formula** (2.1.11) with $V := U$.

Proof This is an immediate consequence of (a). ■

(c) There exists at most one parabolic evolution operator for A. More precisely: if U_j is a parabolic evolution operator for A with regularity subspace F_j for $j = 1,2$ then $U_1 = U_2 =: U$ and $F_1 \cap F_2$ is a regularity subspace for U.

Proof Given $(s,x) \in J \times E$, it follows from (2.1.2), (2.1.3), (2.1.6), and (2.1.7) that $U_j(\cdot,s)x$ is a solution of $(2.1.10)_{(s,x,A,0)}$. Now the assertion is an easy consequence of (b). ■

(d) Let U_A be a parabolic evolution operator for A with regularity subspace F. Given $\alpha \in \mathbb{K}$, put

$$U_{\alpha+A}(t,s) := e^{-\alpha t} U_A(t,s) e^{\alpha s} , \qquad (t,s) \in J_\Delta .$$

Then $U_{\alpha+A}$ is a parabolic evolution operator for $\alpha + A$ with regularity subspace F.

(e) Suppose there exists a (necessarily unique) parabolic evolution operator U_A for A. Given

$$(x,f) \in E \times L_{1,\mathrm{loc}}(J,E) ,$$

where $J \subset \mathbb{R}^+$ with $0 \in J$, put

$$u(t,x,A,f) := U_A(t,0)x + \int_0^t U_A(t,\tau) f(\tau)\, d\tau , \qquad t \in J .$$

Then $u(\cdot,x,f,A)$ is a **mild solution** of the linear Cauchy problem $(2.1.10)_{(0,x,A,f)}$. Observe that

$$u(\cdot,x,A,f) \in C(J,E) .$$

Theorems 1.2.1 and 1.2.2 contain sufficient conditions for f to guarantee that the mild solution is in fact a solution. Note that, thanks to (d),

$$u(\cdot,x,A,f) = e^{\alpha t} u(\cdot,x,\alpha+A,e^{-\alpha t} f) , \qquad \alpha \in \mathbb{R} ,$$

where $e^{\alpha t} v := \big[t \mapsto e^{\alpha t} v(t)\big]$. ■

2.2 Determining Integral Equations

Let (E_0,E_1) be a densely injected Banach couple and suppose that J is a perfect subinterval of $\mathbb{R}^+$ containing 0. Also suppose that

$$A \in C^\rho\big(J,\mathcal{H}(E_1,E_0)\big) \tag{2.2.1}$$

for some $\rho \in (0,1)$. Lastly, suppose that U_A is a parabolic evolution operator for A with regularity subspace E_1. Then it follows from (2.1.2), (2.1.6)–(2.1.9), and Example 2.1.1 that

$$\big[\tau \mapsto U_A(t,\tau) e^{-(\tau-s)A(s)}\big] \in C\big([s,t],\mathcal{L}_s(E_0)\big) \cap C^1\big((s,t),\mathcal{L}_s(E_0)\big)$$

and that

$$\partial_\tau \big[U_A(t,\tau) e^{-(\tau-s)A(s)}\big] = U_A(t,\tau)\big[A(\tau)-A(s)\big] e^{-(\tau-s)A(s)}$$

for $(t,s) \in J_\Delta^*$ and $s < \tau < t$. By integrating we find that U_A solves the integral equation

$$U_A(t,s) = e^{-(t-s)A(s)} - \int_s^t U_A(t,\tau)\big[A(\tau)-A(s)\big] e^{-(\tau-s)A(s)}\, d\tau \tag{2.2.2}$$

for $(t,s) \in J_\Delta$. Similarly,

$$\partial_\tau \big[e^{-(t-\tau)A(t)} U_A(\tau,s) \big] = e^{-(t-\tau)A(t)} \big[A(t) - A(\tau) \big] U_A(\tau,s)$$

in $\mathcal{L}_s(E_0)$ for $0 \le s < \tau < t$ with $t \in J$, and, consequently,

$$U_A(t,s) = e^{-(t-s)A(t)} + \int_s^t e^{-(t-\tau)A(t)} \big[A(t) - A(\tau) \big] U_A(\tau,s)\, d\tau \qquad (2.2.3)$$

in $\mathcal{L}_s(E_0)$ for $(t,s) \in J_\Delta$.

The integral equations (2.2.2) and (2.2.3) are the starting point for the proof of Theorems 1.2.1 and 1.2.2. Namely, we shall show that these integral equations are uniquely solvable, that their solutions coincide, and that the so-defined function is a parabolic evolution operator for A. Afterwards, we discuss the solvability of the Cauchy problem $(2.1.10)_{(0,x,A,f)}$.

3 Linear Volterra Integral Equations

Motivated by the considerations at the end of the preceding section, we prove some simple facts about linear Volterra integral equations in spaces of bounded linear operators.

3.1 Weakly Singular Kernels

Suppose that J is a perfect subinterval of $\mathbb{R}^+$ containing 0 and put

$$J_T := J \cap [0,T] \ , \qquad T \in \mathbb{R}^+ \ .$$

Given $\alpha \in \mathbb{R}$, denote by

$$\mathfrak{K}(E,F,\alpha)$$

the Fréchet space of all $k \in C\big(J_\Delta^*, \mathcal{L}(E,F)\big)$ satisfying

$$\|k\|_{(\alpha),T} := \|k\|_{(\alpha),T,\mathcal{L}(E,F)} := \sup_{0 \le s < t \le T} (t-s)^\alpha \, \|k(t,s)\|_{\mathcal{L}(E,F)} < \infty \ , \qquad T \in \dot{J} \ ,$$

equipped with the topology induced by the seminorms $\{ \|\cdot\|_{(\alpha),T} \ ; \ T \in \dot{J} \}$. Moreover, $\mathfrak{K}(E,\alpha) := \mathfrak{K}(E,E,\alpha)$. Observe that

$$\|\cdot\|_{(\alpha),T} \le T^{\alpha-\beta} \, \|\cdot\|_{(\beta),T} \ , \qquad \alpha > \beta \ , \quad T \in \dot{J} \ , \qquad (3.1.1)$$

so that

$$\mathfrak{K}(E,F,\beta) \hookrightarrow \mathfrak{K}(E,F,\alpha) \ , \qquad \alpha > \beta \ . \qquad (3.1.2)$$

We put

$$\|k\|_{(\alpha)} := \sup_{(t,s)\in J_\Delta^*} (t-s)^\alpha \, \|k(t,s)\|_{\mathcal{L}(E,F)}$$

and denote by

$$\mathfrak{K}_\infty(E,F,\alpha)$$

the Banach space consisting of all $k \in \mathfrak{K}(E,F,\alpha)$ satisfying $\|k\|_{(\alpha)} < \infty$, equipped with the norm $\|\cdot\|_{(\alpha)}$. Observe that

$$\mathfrak{K}_\infty(E,F,\alpha) \hookrightarrow \mathfrak{K}(E,F,\alpha) \tag{3.1.3}$$

and

$$\mathfrak{K}_\infty(E,F,0) = BC\big(J_\Delta^*, \mathcal{L}(E,F)\big) . \tag{3.1.4}$$

If $\alpha < 0$, each $k \in \mathfrak{K}(E,F,\alpha)$ can be continuously extended over J_Δ by putting $k(t,t) = 0$ for $t \in J$ so that

$$\mathfrak{K}(E,F,\alpha) \hookrightarrow C\big(J_\Delta, \mathcal{L}(E,F)\big) , \qquad \alpha < 0 . \tag{3.1.5}$$

If $E = \mathbb{K}$, we identify $\mathcal{L}(\mathbb{K},F)$ naturally with F via

$$\mathcal{L}(\mathbb{K},F) \ni B \quad \longleftrightarrow \quad B \cdot 1 \in F .$$

Then $k \in \mathfrak{K}(\mathbb{K},F,\alpha)$ iff $k \in C(J_\Delta^*, F)$ and

$$\sup_{0\le s<t\le T} (t-s)^\alpha \, \|k(t,s)\|_F < \infty , \qquad T \in \dot{J} .$$

In particular,

$$BC(\dot{J},F) \hookrightarrow \mathfrak{K}_\infty(\mathbb{K},F,0) = BC(J_\Delta^*,F) \tag{3.1.6}$$

by the identification

$$C(\dot{J},F) \ni u \quad \longleftrightarrow \quad \big[(t,s) \mapsto u(t)\big] \in C(J_\Delta^*,F) . \tag{3.1.7}$$

Let G be a Banach space. Suppose that $k \in \mathfrak{K}(E,F,\alpha)$ and $h \in \mathfrak{K}(F,G,\beta)$ with $\alpha, \beta \in (-\infty, 1)$, and put

$$h \star k(t,s) := \int_s^t h(t,\tau)k(\tau,s)\, d\tau , \qquad (t,s) \in J_\Delta .$$

It is easily verified that

$$h \star k \in \mathfrak{K}(E,G,\alpha+\beta-1) \tag{3.1.8}$$

and

$$\|h \star k\|_{(\alpha+\beta-1),T} \le \mathsf{B}(1-\alpha, 1-\beta)\, \|h\|_{(\alpha),T} \, \|k\|_{(\beta),T} , \qquad T \in \dot{J} , \tag{3.1.9}$$

where B is the Euler beta function. It is a consequence of Fubini's theorem that the operation $\star$ is associative.

3.2 Resolvent Kernels

Suppose that $k \in \mathfrak{K}(E,\alpha)$ for some $\alpha \in [0,1)$. By an easy induction argument we see that

$$\| \underbrace{k \star k \star \cdots \star k}_{n}(t,s)\|_{\mathcal{L}(E)} \le \frac{\left[\Gamma(1-\alpha)\,\|k\|_{(\alpha),T}\right]^n}{\Gamma\big(n(1-\alpha)\big)}(t-s)^{n(1-\alpha)-1} \tag{3.2.1}$$

for $n \in \dot{\mathbb{N}}$ and $0 \le s < t \le T$. Put

$$w := \sum_{j=1}^{\infty} \underbrace{k \star \cdots \star k}_{j} . \tag{3.2.2}$$

Then we have the following:

3.2.1 Lemma *$w \in \mathfrak{K}(E,\alpha)$ and, given $\varepsilon > 0$,*

$$(t-s)^{\alpha}\,\|w(t,s)\|_{\mathcal{L}(E)} \le c(\alpha,\varepsilon) m e^{(1+\varepsilon)m^{1/(1-\alpha)}(t-s)}$$

for $0 \le s < t \le T$ and $T \in \dot{J}$, where $m := \Gamma(1-\alpha)\,\|k\|_{(\alpha),T}$.

Proof Let $\beta := 1-\alpha \in (0,1]$. Thanks to (3.2.1) it suffices to prove

$$\sum_{j=1}^{\infty} \frac{x^{j-1}}{\Gamma(\beta j)} \le c(\beta,\varepsilon) e^{(1+\varepsilon)x^{1/\beta}} , \qquad x > 0 . \tag{3.2.3}$$

Stirling's formula implies the existence of $\theta(t) \in (0,1)$ such that

$$\Gamma(t) = \sqrt{2\pi}\, t^{t-1/2} e^{-t+\theta(t)/(12t)} , \qquad t > 0 .$$

From this we deduce for $j \in \dot{\mathbb{N}}$ that

$$\frac{\Gamma(j+1)^{\beta}}{\Gamma(\beta j)} = \frac{[j\Gamma(j)]^{\beta}}{\Gamma(\beta j)} \le (2\pi)^{\frac{\beta-1}{2}} e^{\frac{\beta}{12}} \beta^{1/2} \frac{j^{(1+\beta)/2}}{\beta^{\beta j}} .$$

Hence, by Hölder's inequality,

$$\begin{aligned}\sum_{j=1}^{\infty} \frac{x^j}{\Gamma(\beta j)} &= \sum_{j=1}^{\infty} \frac{x^j}{(j!)^{\beta}} \frac{\Gamma(j+1)^{\beta}}{\Gamma(\beta j)} \\ &\le c(\beta) \Big[\sum_{j=1}^{\infty} \frac{(\eta x^{1/\beta})^j}{j!}\Big]^{\beta} \Big[\sum_{j=1}^{\infty} \frac{j^{(1+\beta)/(2(1-\beta))}}{(\eta\beta)^{j\beta/(1-\beta)}}\Big]^{1-\beta}\end{aligned}$$

where $\eta > 0$ is arbitrary. Since the last series converges for $\eta > 1/\beta$,

$$\sum_{j=1}^{\infty} \frac{x^{j-1}}{\Gamma(\beta j)} \le c(\beta,\eta)\Big(\frac{e^{\eta x^{1/\beta}} - 1}{\eta x^{1/\beta}}\Big)^{\beta} \le c(\beta,\eta) e^{\beta\eta x^{1/\beta}}$$

for $x > 0$ and $\eta > 1/\beta$. Now the assertion follows by setting $\eta := (1+\varepsilon)/\beta$. ■

Now it is easy to prove the following existence and uniqueness theorem for abstract linear Volterra equations.

3.2.2 Theorem *Suppose that* $\alpha, \beta \in [0,1)$ *and* $k \in \mathfrak{K}(E,\alpha)$*. Then the* **linear Volterra equations**

$$u = a + u \star k \ , \quad v = b + k \star v \tag{3.2.4}$$

possess for each $a \in \mathfrak{K}(E,F,\beta)$ *and* $b \in \mathfrak{K}(F,E,\beta)$ *unique solutions*

$$u \in \mathfrak{K}(E,F,\beta) \quad \text{and} \quad v \in \mathfrak{K}(F,E,\beta) \ ,$$

respectively. They are given by

$$u = a + a \star w \ , \quad v = b + w \star b \ , \tag{3.2.5}$$

respectively, where w*, the* **resolvent kernel** *of* (3.2.4)*, belongs to* $\mathfrak{K}(E,\alpha)$ *and is given by* (3.2.2)*.*

Proof We consider the first equation in (3.2.4). The second one can be treated analogously.

Define w by (3.2.2) and u by (3.2.5), resp., and observe that $w \in \mathfrak{K}(E,\alpha)$ and $u \in \mathfrak{K}(E,F,\beta)$ by Lemma 3.2.1 and by (3.1.2) and (3.1.8), respectively. It is obvious that u solves (3.2.4).

Let $T \in \dot{J}$ be fixed. By replacing J by J_T, it follows from (3.1.1), (3.1.2), (3.1.8), (3.1.9), and (3.2.1) that $\star\, k \in \mathcal{L}(\mathfrak{K}_\infty(E,F,\beta))$ and that the spectral radius of this operator equals zero. Hence (3.2.4) has at most one solution 'on J_T' for each $T \in \dot{J}$. This proves the assertion. ■

3.2.3 Remark In the definition of $\mathfrak{K}(E,F,\alpha)$ we can replace the assumption that $k \in C(J_\Delta^*, \mathcal{L}(E,F))$ by $k \in L_{\infty,\mathrm{loc}}(J_\Delta^*, \mathcal{L}(E,F))$. Then everything remains true provided:

(i) $\sup_{0 \le s < t \le T}$ is replaced by $\text{ess-sup}_{0 \le s < t \le T}$ everywhere.

(ii) (3.1.5) is replaced by

$$\mathfrak{K}(E,F,\alpha) \cap C(J_\Delta^*, \mathcal{L}(E,F)) \hookrightarrow C(J_\Delta, \mathcal{L}(E,F)) \ , \qquad \alpha < 0 \ .$$

(iii) $C(J_\Delta^*, F)$ is replaced by $L_{\infty,\mathrm{loc}}(J_\Delta^*, F)$ in the interpretation of $\mathfrak{K}(\mathbb{K},F,\alpha)$.

(iv) BC is replaced by L_∞ in (3.1.4) and (3.1.6).

Observe that with this new definition of $\mathfrak{K}(E,F,\alpha)$, and by using obvious notation,

$$\mathfrak{K}(F,G,\beta) \star \mathfrak{K}(E,F,\alpha) \hookrightarrow \mathfrak{K}(E,G,\alpha+\beta-1) \cap C(J_\Delta, \mathcal{L}(E,G))$$

if $\alpha + \beta < 1$.

These generalizations are occasionally useful — in Theorem 3.3.1, for example. ■

3.3 Singular Gronwall Inequalities

As a simple application of Theorem 3.2.2 we prove the following **generalized Gronwall inequality**.

3.3.1 Theorem *Given $\alpha, \beta \in [0,1)$ and $\varepsilon > 0$, there exists a positive constant $c := c(\alpha, \beta, \varepsilon)$ such that the following is true:*

If $u\colon J \to \mathbb{R}$ satisfies

$$\big[t \mapsto t^{\beta} u(t)\big] \in L_{\infty,\mathrm{loc}}(J, \mathbb{R}) \tag{3.3.1}$$

and

$$u(t) \le A t^{-\beta} + B \int_0^t (t-\tau)^{-\alpha} u(\tau)\, d\tau , \qquad a.a.\ t \in \dot{J} , \tag{3.3.2}$$

where A and B are positive constants, then

$$u(t) \le A t^{-\beta} \big(1 + cBt^{1-\alpha} e^{(1+\varepsilon)\mu(\alpha,B)t}\big) , \qquad a.a.\ t \in \dot{J} , \tag{3.3.3}$$

where $\mu(\alpha, B) := \big(\Gamma(1-\alpha)B\big)^{1/(1-\alpha)}$.

Proof Let $E := F := \mathbb{R}$ and put $k(t,s) := B(t-s)^{-\alpha}$ for $(t,s) \in J_{\Delta}^{*}$. Then we see that $k \in \mathfrak{K}(E, \alpha)$ and $\|k\|_{(\alpha),T} = B$ for $T \in \dot{J}$. Let $a(t) := At^{-\beta}$ and observe that (3.1.7) implies $a \in \mathfrak{K}(E, \beta)$ and $\|a\|_{(\beta),T} = A$ for $T \in \dot{J}$. Since $u \in \mathfrak{K}(E, \beta)$ by (3.3.1) and Remark 3.2.3, it follows from (3.1.8) and (3.1.2) that

$$b := a + k \star u - u \in \mathfrak{K}(E, \beta) .$$

Hence $u = a - b + k \star u$, and Theorem 3.2.2 implies that

$$u = (a-b) + w \star (a-b) .$$

Observe that $b \ge 0$ by (3.3.2) and that $k \ge 0$ implies $w \ge 0$. Thus $u \le a + w \star a$, that is,

$$u(t) \le A t^{-\beta} + A \int_0^t w(t-\tau) \tau^{-\beta}\, d\tau , \qquad \text{a.a. } t \in \dot{J} .$$

By Lemma 3.2.1,

$$w(t) \le c(\alpha, \varepsilon) B t^{-\alpha} e^{(1+\varepsilon)\mu(\alpha,B)t} , \qquad t > 0 , \quad \varepsilon > 0 .$$

Since

$$\begin{aligned}\int_0^t e^{\nu(t-\tau)} (t-\tau)^{-\alpha} \tau^{-\beta}\, d\tau &\le e^{\nu t} \int_0^t (t-\tau)^{-\alpha} \tau^{-\beta}\, d\tau \\ &= \mathsf{B}(1-\alpha, 1-\beta) t^{1-\alpha-\beta} e^{\nu t}\end{aligned}$$

for $t > 0$ and $\nu \ge 0$, the assertion follows. ■

3.3.2 Corollary *Let (3.3.1) and (3.3.2) be satisfied. Then, given $\varepsilon > 0$, there exists a constant $c := c(\varepsilon, \alpha, \beta, B)$ such that*

$$u(t) \leq Act^{-\beta} e^{(1+\varepsilon)\mu(\alpha,B)t} , \qquad a.a.\ t \in \dot{J} .$$

3.3.3 Remarks **(a)** It should be noted that, in general, the constant $c(\alpha, \varepsilon)$ in the estimate of Lemma 3.2.1 — and, consequently, the constants c in Theorem 3.3.1 and Corollary 3.3.2 as well — go to infinity if $\varepsilon \to 0$. Of course, if $\alpha = 0$ then $\varepsilon = 0$ is possible and $c(0,0) = 1$. In this case the constant c of Theorem 3.3.1 equals $1/(1-\beta)$ and (3.3.3) is then a consequence of the classical Gronwall inequality (e.g., [Ama90b, Corollary (6.2)]).

(b) Of course, the factor $e^{(1+\varepsilon)\mu(t-s)}$ in Lemma 3.2.1 and in Theorem 3.3.1 and its corollary — where $\mu := m^{1/(1-\alpha)}$ in Lemma 3.2.1 — can be replaced by $e^{(\mu+\varepsilon)(t-s)}$. ■

It is well-known that the generalized Gronwall inequality is a very useful tool in the theory of semilinear parabolic evolution equations. It has been proven — in a form somewhat less precise than the one of Theorem 3.3.1 — by the author in [Ama78, Lemma 2.3] and, independently, by means of Laplace transform techniques, by Henry [Hen81, Lemma 7.1.1]. The trick, used in the proof of Lemma 3.2.1 for estimating the majorant of the series (3.2.2) by means of Stirling's formula, is taken from [vW85, p. 6].

4 Existence of Evolution Operators

In this section we carry out the construction of a parabolic evolution operator, given the classical Sobolevskii-Tanabe assumptions. It should be observed that we pay particular attention to clarifying the dependence of our estimates on a few explicitly specified parameters.

4.1 A Class of Parameter Integrals

In order to prove some useful uniform estimates for the semigroups generated by the elements of $\mathcal{H}(E_1, E_0)$ we use the following technical lemma. Recall that

$$\dot{\Sigma}_\vartheta := \Sigma_\vartheta \backslash \{0\} = [\,|\arg z| \leq \vartheta + \pi/2]$$

for $\vartheta \in [0, \pi/2]$.

4.1.1 Lemma *Given $\vartheta \in (0, \pi/2)$, let Γ be an arbitrary piece-wise smooth simple curve in $\dot{\Sigma}_\vartheta$ running from $\infty e^{-i(\vartheta+\pi/2)}$ to $\infty e^{i(\vartheta+\pi/2)}$, and let X be a metric space. Suppose that the map $f : \dot{\Sigma}_\vartheta \times X \times \dot{\mathbb{R}}^+ \to E$ has the following properties:*

(i) *$f(\cdot, x, t) : \dot{\Sigma}_\vartheta \to E$ is holomorphic for $(x, t) \in X \times \dot{\mathbb{R}}^+$.*

(ii) *$f(z, \cdot, \cdot) \in C(X \times \dot{\mathbb{R}}^+, E)$ for $z \in \dot{\Sigma}_\vartheta$.*

(iii) *There are constants $\alpha \in \mathbb{R}$ and $M > 0$ such that*

$$\|f(z, x, t)\| \le M\, |z|^{\alpha-1}\, e^{t\,\mathrm{Re}\, z} , \qquad (z, x, t) \in \dot{\Sigma}_\vartheta \times X \times \dot{\mathbb{R}}^+ .$$

Then

$$\Big[(x, t) \mapsto \int_\Gamma f(z, x, t)\, dz\Big] \in C(X \times \dot{\mathbb{R}}^+, E)$$

and

$$\Big\| \int_\Gamma f(z, x, t)\, dz \Big\| \le cMt^{-\alpha} , \qquad (x, t) \in X \times \dot{\mathbb{R}}^+ .$$

Proof Let $(x, t) \in X \times (0, \infty)$ be fixed and put $\Gamma = \Gamma_t^- + S_t + \Gamma_t^+$, where

$$\Gamma_t^\pm := \{\, re^{\pm i(\vartheta+\pi/2)} \ ;\ t^{-1} \le r < \infty \,\}$$

and

$$S_t := \{\, t^{-1} e^{i\varphi} \ ;\ |\varphi| \le \vartheta + \pi/2 \,\} .$$

Then

$$\begin{aligned}\int_\Gamma f(z, x, t)\, dz &= \Big(\int_{\Gamma_t^-} + \int_{S_t} + \int_{\Gamma_t^+}\Big) f(z, x, t)\, dz \\ &= \int_{t^{-1}}^\infty \Big\{ f(re^{-i(\vartheta+\pi/2)}, x, t) e^{-i\vartheta} + f(re^{i(\vartheta+\pi/2)}, x, t) e^{i\vartheta} \Big\} i\, dr \\ &\quad + \int_{-\vartheta-\pi/2}^{\vartheta+\pi/2} f(t^{-1} e^{i\varphi}, x, t) i t^{-1} e^{i\varphi}\, d\varphi .\end{aligned}$$

By substituting $s := tr \sin\vartheta$ in the first integral after the second equality sign, it transforms into

$$\int_{\sin\vartheta}^\infty \Big\{ f\Big(\frac{-ise^{-i\vartheta}}{t\sin\vartheta}, x, t\Big) e^{-i\vartheta} + f\Big(\frac{ise^{i\vartheta}}{t\sin\vartheta}, x, t\Big) e^{i\vartheta} \Big\} \frac{ds}{t\sin\vartheta} .$$

Since $\mathrm{Re}(\pm ise^{\pm i\vartheta}/t\sin\vartheta) = -s/t$, it follows from (iii) that

$$\begin{aligned}\Big\| \int_\Gamma f(z, x, t)\, dz \Big\| &\le t^{-\alpha} \Big\{ 2M(\sin\vartheta)^{-\alpha} \int_{\sin\vartheta}^\infty s^{\alpha-1} e^{-s}\, ds + M \int_{-\vartheta-\pi/2}^{\vartheta+\pi/2} e^{\cos\varphi}\, d\varphi \Big\} \\ &= Mc(\alpha, \vartheta) t^{-\alpha} .\end{aligned}$$

Now the assertion is an easy consequence of the theorem on the continuity of parameter integrals and of Cauchy's theorem. ■

It should be remarked that estimates of the type given in this lemma are standard in the theory of analytic semigroups (e.g., [Tan79]).

4.2 Semigroup Estimates

Now let (E_0, E_1) be a densely injected Banach couple, J a perfect subinterval of $\mathbb{R}^+$ containing 0, and $\rho \in (0,1)$. Suppose that

$$\left.\begin{array}{l} \mathcal{A} \subset C^{\rho}\big(J, \mathcal{H}(E_1, E_0)\big) \text{ and there are constants } M, \eta \in \mathbb{R}^+ \text{ and } \\ \vartheta \in (0, \pi/2) \text{ such that} \\ \qquad [A]_{\rho,J} \le \eta\ , \qquad A \in \mathcal{A}\ , \\ \text{that } \Sigma_\vartheta \subset \rho\big(-A(s)\big), \text{ and that} \\ \quad \|A(s)\|_{\mathcal{L}(E_1,E_0)} + (1+|\lambda|)^{1-j}\, \big\|\big(\lambda + A(s)\big)^{-1}\big\|_{\mathcal{L}(E_0,E_j)} \le M \\ \text{for } (s,\lambda,A) \in J \times \Sigma_\vartheta \times \mathcal{A} \text{ and } j = 0,1. \end{array}\right\} \tag{4.2.1}$$

Observe that (4.2.1) implies

$$M^{-1}\,\|x\|_1 \le \|A(s)x\|_0 \le M\,\|x\|_1\ , \qquad (s,x,A) \in J \times E_1 \times \mathcal{A}\ . \tag{4.2.2}$$

Assumption (4.2.1) also implies uniform estimates for the semigroups generated by $-A(s)$, as well as their continuous dependence on $s \in J$, as is seen from the following result:

4.2.1 Lemma *The estimate*

$$\big\|\big[tA(s)\big]^k e^{-tA(s)}\big\|_{\mathcal{L}(E_j)} + t\,\big\|\big[tA(s)\big]^k e^{-tA(s)}\big\|_{\mathcal{L}(E_0,E_1)} \le c(k) \tag{4.2.3}$$

is valid for $k \in \mathbb{N}$, $(t,s,A) \in \mathbb{R}^+ \times J \times \mathcal{A}$, and $j = 0,1$. Moreover,

$$\big[(t,s) \mapsto e^{-tA(s)}\big] \in C\big(\mathbb{R}^+ \times J, \mathcal{L}_s(E_k, E_j)\big) \cap C\big(\dot{\mathbb{R}}^+ \times J, \mathcal{L}(E_j, E_k)\big)$$

for $j,k \in \{0,1\}$ with $j \le k$.

Proof It is well-known from the theory of analytic semigroups that

$$A^k(s)e^{-tA(s)} = \frac{(-1)^k}{2\pi i}\int_\Gamma \lambda^k e^{t\lambda}\big(\lambda + A(s)\big)^{-1}\,d\lambda\ , \qquad t > 0\ , \quad k \in \mathbb{N}\ ,$$

Γ being a piece-wise smooth curve running in $\dot{\Sigma}_\vartheta$ from $\infty e^{-i(\vartheta+\pi/2)}$ to $\infty e^{i(\vartheta+\pi/2)}$. From the resolvent estimates of (4.2.1) and from Lemma 4.1.1 we deduce that

$$\big\|\big[tA(s)\big]^k e^{-tA(s)}\big\|_{\mathcal{L}(E_0)} \le c(k)\ , \qquad (t,s,A) \in \dot{\mathbb{R}}^+ \times J \times \mathcal{A}\ , \quad k \in \mathbb{N}\ . \tag{4.2.4}$$

Now (4.2.3) is an easy consequence of (4.2.2) and the fact that a semigroup commutes with its generator. Since $A \in C^\rho\big(J, \mathcal{L}(E_1, E_0)\big)$ and the inversion map $B \mapsto B^{-1}$ is smooth (in fact: analytic), it follows that

$$\big[s \mapsto \big(\lambda + A(s)\big)^{-1}\big] \in C\big(J, \mathcal{L}(E_j, E_k)\big) , \qquad j, k \in \{0, 1\} , \quad j \leq k ,$$

where we use again (4.2.2) and the fact that a closed linear operator commutes with its resolvent. Now Lemma 4.1.1 implies

$$\big[(t, s) \mapsto e^{-tA(s)}\big] \in C\big(\dot{\mathbb{R}}^+ \times J, \mathcal{L}(E_j, E_k)\big) \tag{4.2.5}$$

for $j, k \in \{0, 1\}$ with $j \leq k$.

Given $x \in E_1$, we deduce from

$$e^{-tA(s)}x - x = \int_0^t \partial_\tau e^{-\tau A(s)} x \, d\tau = -\int_0^t e^{-\tau A(s)} A(s) x \, d\tau$$

and from (4.2.2) and (4.2.3) that

$$\|e^{-tA(s)}x - x\|_0 \leq ct \, \|x\|_1 , \qquad t \in \mathbb{R}^+ , \quad s \in J .$$

Moreover,

$$\begin{aligned} e^{-tA(s)} - e^{-tA(s')} &= \frac{1}{2\pi i} \int_\Gamma e^{t\lambda} \big[\big(\lambda + A(s)\big)^{-1} - \big(\lambda + A(s')\big)^{-1}\big] \, d\lambda \\ &= \frac{1}{2\pi i} \int_\Gamma e^{t\lambda} \big(\lambda + A(s)\big)^{-1} \big[A(s') - A(s)\big] \big(\lambda + A(s')\big)^{-1} \, d\lambda , \end{aligned}$$

(4.2.1), (4.2.2), and Lemma 4.1.1 imply

$$\|e^{-tA(s)} - e^{-tA(s')}\|_{\mathcal{L}(E_1, E_0)} \leq ct \, |s - s'|^\rho , \qquad t \in \mathbb{R}^+ , \quad s, s' \in J .$$

Thus we see from

$$\begin{aligned} &e^{-tA(s)}x - e^{-t'A(s')}x \\ &\quad = (e^{-tA(s)} - e^{-tA(s')})x + (e^{-tA(s')}x - x) - (e^{-t'A(s')}x - x) \end{aligned} \tag{4.2.6}$$

and (4.2.5) that

$$\big[(t, s) \mapsto e^{-tA(s)}\big] \in C\big(\mathbb{R}^+ \times J, \mathcal{L}_s(E_1, E_0)\big) . \tag{4.2.7}$$

Now we obtain

$$\big[(t, s) \mapsto e^{-tA(s)}\big] \in C\big(\mathbb{R}^+ \times J, \mathcal{L}_s(E_0)\big) \tag{4.2.8}$$

from (4.2.7), from (4.2.3) with $j = k = 0$, and from the density of E_1 in E_0. Lastly, note that (4.2.2) and (4.2.8) imply

$$\|e^{-tA(s)}x - x\|_1 \leq M \, \|e^{-tA(s)} A(s)x - A(s)x\|_0 \to 0$$

for $t \to 0$ and $x \in E_1$. Thus we infer from (4.2.5) and (4.2.6) the validity of (4.2.8) with E_0 replaced by E_1. ∎

4.3 Construction of Evolution Operators

We put

$$a_A(t,s) := e^{-(t-s)A(s)}\ , \quad k_A(t,s) := -\big[A(t)-A(s)\big]a_A(t,s)$$

for $(t,s)\in J^*_\Delta$ and $A\in\mathcal{A}$. Then we deduce from (4.2.2) and Lemma 4.2.1 that

$$a_A \in \mathfrak{K}_\infty(E_0,0)\cap\mathfrak{K}_\infty(E_1,0)\cap\mathfrak{K}_\infty(E_0,E_1,1) \tag{4.3.1}$$

and that

$$\|a_A\| \le c_0\ , \qquad A\in\mathcal{A}\ , \tag{4.3.2}$$

where $\|\cdot\|$ is the norm of the space appearing in (4.3.1). Note that (4.2.2) and Lemma 4.2.1 imply

$$k_A \in \mathfrak{K}(E_0,1-\rho)\cap\mathfrak{K}(E_1,E_0,-\rho) \tag{4.3.3}$$

and

$$(t-s)^{1-j-\rho}\,\|k_A(t,s)\|_{\mathcal{L}(E_j,E_0)} \le c \tag{4.3.4}$$

for $(t,s)\in J^*_\Delta,\ \ A\in\mathcal{A}$, and $j=0,1$.

Lastly, we put

$$w_A := \sum_{n=1}^{\infty}\underbrace{k_A\star\cdots\star k_A}_{n} \tag{4.3.5}$$

and

$$\mu := (\Gamma(\rho)c_0\eta)^{1/\rho}\ . \tag{4.3.6}$$

In the following computations we often omit the index A if no confusion seems likely.

4.3.1 Lemma *The function w_A is well-defined and satisfies*

$$w_A \in \mathfrak{K}(E_0,1-\rho)\cap\mathfrak{K}(E_1,E_0,-\rho)$$

with

$$(t-s)^{1-j-\rho}\,\|w_A(t,s)\|_{\mathcal{L}(E_j,E_0)} \le c(\varepsilon)e^{(\mu+\varepsilon)(t-s)} \tag{4.3.7$_j$}$$

for $\varepsilon>0,\ \ (t,s)\in J^*_\Delta,\ \ A\in\mathcal{A}$, and $j=0,1$.

Proof It follows from (4.3.4), Lemma 3.2.1, and Remark 3.3.3(b) that

$$w_A \in \mathfrak{K}(E_0,1-\rho)$$

and that $(4.3.7)_0$ is true. It is an obvious consequence of (4.3.5) that

$$w = k + k\star w = k + w\star k\ . \tag{4.3.8}$$

Thus we deduce from (4.3.4), $(4.3.7)_0$, (3.1.1), and (3.1.8) that

$$w \in \mathfrak{K}(E_1,E_0,-\rho) \tag{4.3.9}$$

and that $(4.3.7)_1$ is valid. ■

Note that (4.3.1), (4.3.3), and Theorem 3.2.2 imply that the Volterra integral equation

$$u = a_A + u \star k_A \tag{4.3.10}$$

possesses a unique solution

$$U_A \in \mathfrak{K}(E_0, 0) \tag{4.3.11}$$

and that it is given by

$$U_A = a_A + a_A \star w_A \ . \tag{4.3.12}$$

Observe that the integral equation (4.3.10) coincides with (2.2.2). Thus U_A of (4.3.12) is our candidate for the parabolic evolution operator for A. In the remainder of this section we show that U_A has the desired properties. In addition, we derive important uniform estimates.

4.3.2 Lemma *Put*

$$e_A(t,s) := A(t)e^{-(t-s)A(t)} - A(s)e^{-(t-s)A(s)} \ , \qquad (t,s) \in J_\Delta^* \ .$$

Then $e_A \in \mathfrak{K}(E_0, 1-\rho)$ *and*

$$(t-s)^{1-\rho} \|e_A(t,s)\|_{\mathcal{L}(E_0)} \le c \ , \qquad (t,s) \in J_\Delta^* \ , \quad A \in \mathcal{A} \ .$$

Proof It is a consequence of (4.2.1) that

$$\begin{aligned}
&\big\|\big(\lambda + A(t)\big)^{-1} - \big(\lambda + A(s)\big)^{-1}\big\|_{\mathcal{L}(E_0)} \\
&\qquad \le \big\|\big(\lambda + A(t)\big)^{-1}\big\|_{\mathcal{L}(E_0)} \|A(t) - A(s)\|_{\mathcal{L}(E_1,E_0)} \big\|\big(\lambda + A(s)\big)^{-1}\big\|_{\mathcal{L}(E_0,E_1)} \\
&\qquad \le c(t-s)^\rho \, |\lambda|^{-1}
\end{aligned}$$

for $\lambda \in \Sigma_\vartheta$, $0 \le s < t \le T$, and $A \in \mathcal{A}$. From this and from Lemma 4.1.1 we deduce that

$$\begin{aligned}
&\|A(t)e^{-(t-s)A(t)} - A(s)e^{-(t-s)A(s)}\|_{\mathcal{L}(E_0)} \\
&\qquad = \Big\|\frac{1}{2\pi i}\int_\Gamma \lambda e^{\lambda(t-s)} \big[\big(\lambda + A(t)\big)^{-1} - \big(\lambda + A(s)\big)^{-1}\big] \, d\lambda\Big\|_{\mathcal{L}(E_0)} \le c(t-s)^{\rho-1}
\end{aligned}$$

for $(t,s) \in J_\Delta^*$ and $A \in \mathcal{A}$. Now the assertion follows from Lemma 4.2.1. ■

The next technical lemma gives estimates for $w_A(t,s) - w_A(\tau,s)$ as a function of t, τ, s with $s < \tau < t$. Here and in the following, it is always understood that $(t,s) \in J_\Delta^*$.

4.3.3 Lemma *Suppose that* $0 < \beta < \rho$. *Then*

$$\begin{aligned}&\|w_A(t,s) - w_A(\tau,s)\|_{\mathcal{L}(E_j,E_0)} \\ &\qquad \le c(\varepsilon)\big\{\delta_{1,j}(t-\tau)^\rho + (t-\tau)^\beta(\tau-s)^{j+\rho-\beta-1}\big\}e^{(\mu+\varepsilon)(t-s)}\end{aligned} \tag{4.3.13}$$

for $\varepsilon > 0,\ \ s < \tau < t,\ \ A \in \mathcal{A}$, *and* $j \in \{0,1\}$.

Proof First we derive a bound for

$$k(t,s) - k(\tau,s) = \big[A(\tau) - A(t)\big]a(t,s) - \big[A(\tau) - A(s)\big]\big[a(t,s) - a(\tau,s)\big]\ , \tag{4.3.14}$$

where $0 \le s < \tau < t$. From (4.3.2) it follows that

$$(t-s)^{1-j}\,\big\|\big[A(t) - A(\tau)\big]a(t,s)\big\|_{\mathcal{L}(E_j,E_0)} \le c(t-\tau)^\rho \tag{4.3.15}$$

for $j \in \{0,1\}$. Since

$$a(t,s) - a(\tau,s) = -\int_\tau^t A(s)e^{-(\sigma-s)A(s)}\,d\sigma\ ,$$

we deduce from Lemma 4.2.1 that

$$\|a(t,s) - a(\tau,s)\|_{\mathcal{L}(E_j,E_1)} \le M\int_\tau^t (\sigma-s)^{j-2}\,d\sigma \le M(t-\tau)(\tau-s)^{j-2}$$

for $j \in \{0,1\}$. Hence

$$\big\|\big[A(\tau) - A(s)\big]\big[a(t,s) - a(\tau,s)\big]\big\|_{\mathcal{L}(E_j,E_0)} \le c(t-\tau)(\tau-s)^{\rho+j-2} \tag{4.3.16}$$

for $j \in \{0,1\}$. On the other hand,

$$\begin{aligned}&\big\|\big[A(\tau) - A(s)\big]\big[a(t,s) - a(\tau,s)\big]\big\|_{\mathcal{L}(E_j,E_0)} \\ &\qquad \le c(\tau-s)^\rho\Big\{\|a(t,s)\|_{\mathcal{L}(E_j,E_1)} + \|a(\tau,s)\|_{\mathcal{L}(E_j,E_1)}\Big\} \\ &\qquad \le c\big((t-s)^{j-1}(\tau-s)^\rho + (\tau-s)^{\rho-1+j}\big) \le c(\tau-s)^{\rho-1+j}\ .\end{aligned} \tag{4.3.17}$$

By combining (4.3.16) and (4.3.17) we see that

$$\begin{aligned}&\big\|\big[A(\tau) - A(s)\big]\big[a(t,s) - a(\tau,s)\big]\big\|_{\mathcal{L}(E_j,E_0)} \\ &\qquad \le c(t-\tau)^\rho(\tau-s)^{\rho(\rho-2+j)}(\tau-s)^{(\rho-1+j)(1-\rho)} \\ &\qquad = c(t-\tau)^\rho(\tau-s)^{j-1}\ .\end{aligned} \tag{4.3.18}$$

Thus we obtain from (4.3.14), (4.3.15), and (4.3.18) the estimate

$$(\tau-s)^{1-j}\,\|k(t,s) - k(\tau,s)\|_{\mathcal{L}(E_j,E_0)} \le c(t-\tau)^\rho\ , \qquad j = 0,1\ . \tag{4.3.19}$$

On the other hand,

$$\|k(t,s)-k(\tau,s)\|_{\mathcal{L}(E_0)} \le c\big[(t-s)^{\rho-1}+(\tau-s)^{\rho-1}\big] \le c(\tau-s)^{\rho-1} \tag{4.3.20}$$

by (4.3.4). Hence it follows from (4.3.19) and (4.3.20) that

$$\begin{aligned}\|k(t,s)-k(\tau,s)\|_{\mathcal{L}(E_0)} &\le c\big[(t-\tau)^{\rho}(\tau-s)^{-1}\big]^{\beta/\rho}(\tau-s)^{(\rho-1)(1-\beta/\rho)} \\ &= c(t-\tau)^{\beta}(\tau-s)^{\rho-\beta-1}\ .\end{aligned} \tag{4.3.21}$$

Observe that (4.3.8) implies

$$\begin{aligned} w(t,s)-w(\tau,s) = k(t,s)-k(\tau,s) & \\ +\int_s^{\tau}\big[k(t,\sigma)-k(\tau,\sigma)\big]w(\sigma,s)\,d\sigma & \\ +\int_{\tau}^{t}k(t,\sigma)w(\sigma,s)\,d\sigma\ . & \end{aligned} \tag{4.3.22}$$

Thus it is an easy consequence of (4.3.21), (4.3.4), (3.1.9), and Lemma 4.3.1 that

$$\begin{aligned} &\|w(t,s)-w(\tau,s)\|_{\mathcal{L}(E_0)} \\ &\quad\le c(\varepsilon)\Big\{(t-\tau)^{\beta}(\tau-s)^{\rho-\beta-1}+\Big[\int_s^{\tau}(t-\tau)^{\beta}(\tau-\sigma)^{\rho-\beta-1}(\sigma-s)^{\rho-1}\,d\sigma \\ &\qquad\qquad +\int_{\tau}^{t}(t-\sigma)^{\rho-1}(\sigma-s)^{\rho-1}\,d\sigma\Big]e^{(\mu+\varepsilon)(t-s)}\Big\} \\ &\quad\le c(\varepsilon)\Big[(t-\tau)^{\beta}(\tau-s)^{\rho-\beta-1}\big(1+(t-s)^{\rho}\big)+(t-\tau)^{\rho}(\tau-s)^{\rho-1}\Big]e^{(\mu+\varepsilon)(t-s)}\end{aligned}$$

for $\varepsilon>0$. From this we conclude the validity of the assertion for $j=0$. (Straightforward, we obtain in the last inequality the exponent $(\mu+2\varepsilon)(t-s)$, for example. Since this is true for every $\varepsilon>0$, we can replace 2ε by ε by changing the constant $c(\varepsilon)$, of course. This fact is frequently used in the following without further mention.)

Similarly, thanks to (4.3.19) and (4.3.22),

$$\begin{aligned} &\|w(t,s)-w(\tau,s)\|_{\mathcal{L}(E_1,E_0)} \\ &\qquad\le c(\varepsilon)\big\{(t-\tau)^{\rho}+(t-\tau)^{\beta}(\tau-s)^{2\rho-\beta}+(t-\tau)^{\rho}(t-s)^{\rho}\big\}e^{(\mu+\varepsilon)(t-s)} \\ &\qquad\le c(\varepsilon)\big\{(t-\tau)^{\rho}+(t-\tau)^{\beta}(\tau-s)^{\rho-\beta}\big\}e^{(\mu+\varepsilon)(t-s)}\end{aligned}$$

for $\varepsilon>0$, which proves the assertion for $j=1$. ■

Put

$$d_{\varepsilon}(t,s) := \int_s^{t-\varepsilon}a(t,\tau)w(\tau,s)\,d\tau\ , \qquad 0<\varepsilon<t-s\ ,$$

and $d_\varepsilon(t,s) := 0$ for $s \le t \le s+\varepsilon$, and note that

$$d_\varepsilon(t,s) \to a \star w(t,s) \quad \text{in } \mathcal{L}(E_0) \qquad \text{as} \quad \varepsilon \to 0 \ . \tag{4.3.23}$$

Moreover,

$$\partial_1 d_\varepsilon(t,s) = e^{-\varepsilon A(t-\varepsilon)} w(t-\varepsilon,s) - \int_s^{t-\varepsilon} A(\tau)a(t,\tau)w(\tau,s)\,d\tau \tag{4.3.24}$$

for $t > s+\varepsilon$ in $\mathcal{L}(E_0)$. Since $A(t)e^{-(t-\tau)A(t)} = \partial_\tau e^{-(t-\tau)A(t)}$, we see that

$$\begin{aligned}\partial_1 d_\varepsilon(t,s) = {} & e^{-\varepsilon A(t-\varepsilon)} w(t-\varepsilon,s) - [e^{-\varepsilon A(t)} - e^{-(t-s)A(t)}]w(t,s)\\ & + \int_s^{t-\varepsilon} e(t,\tau)w(\tau,s)\,d\tau\\ & - \int_s^{t-\varepsilon} A(t)e^{-(t-\tau)A(t)}\big[w(\tau,s)-w(t,s)\big]\,d\tau\end{aligned} \tag{4.3.25}$$

for $t > s+\varepsilon$. Put

$$\begin{aligned}\dot{d}_A(t,s) := {} & e^{-(t-s)A(t)} w(t,s)\\ & + \int_s^t e(t,\tau)w(\tau,s)\,d\tau\\ & + \int_s^t A(t)e^{-(t-\tau)A(t)}\big[w(t,s)-w(\tau,s)\big]\,d\tau \ .\end{aligned} \tag{4.3.26}$$

It is an easy consequence of Lemma 4.2.1 and of Lemmas 4.3.1 to 4.3.3, letting $\beta := \rho/2$ in Lemma 4.3.3, that

$$\dot{d}_A \in \mathfrak{K}(E_j, E_0, 1-j-\rho) \tag{4.3.27}$$

and

$$(t-s)^{1-j-\rho}\, \|\dot{d}_A(t,s)\|_{\mathcal{L}(E_j,E_0)} \le c(\varepsilon)e^{(\mu+\varepsilon)(t-s)} \tag{4.3.28}$$

for $\varepsilon > 0$, $(t,s) \in J_\Delta^*$, $A \in \mathcal{A}$, and $j = 0,1$, and that

$$\partial_1 d_\varepsilon \to \dot{d}_A \quad \text{in } C\big(J_\Delta^*, \mathcal{L}_s(E_0)\big) \qquad \text{as} \quad \varepsilon \to 0 \ . \tag{4.3.29}$$

Thus we deduce from (4.3.23) that

$$a_A \star w_A(\cdot,s) \in C^1\big(J \cap (s,\infty), \mathcal{L}_s(E_0)\big) \ , \qquad s \in J \ ,$$

and

$$\partial_1(a_A \star w_A) = \dot{d}_A \ . \tag{4.3.30}$$

Consequently,

$$a_A \star w(t,s) - a_A \star w(t',s) = \int_{t'}^t \dot{d}_A(\tau,s)\,d\tau \ , \qquad s < t' < t \ , \tag{4.3.31}$$

in $\mathcal{L}_s(E_0)$. By employing (4.3.27) and (4.3.28) we infer that (4.3.31) even holds in $\mathcal{L}(E_0)$, which, in turn, implies

$$a_A \star w_A(\cdot, s) \in C^1\big(J \cap (s,\infty), \mathcal{L}(E_0)\big) , \qquad s \in J . \tag{4.3.32}$$

Now it is not difficult to prove the following technical assertions:

4.3.4 Lemma *Given $A \in \mathcal{A}$,*

$$U_A(\cdot, s) = a_A(\cdot, s) + a_A \star w_A(\cdot, s) \in C^1\big(J \cap (s,\infty), E_0\big) , \qquad s \in J ,$$

and

$$\partial_1 U_A = -AU_A . \tag{4.3.33}$$

Moreover,

$$a_A \star w_A \in \mathfrak{K}(E_j, E_k, k-j-\rho) \tag{4.3.34}$$

and

$$(t-s)^{k-j-\rho} \, \|a_A \star w_A(t,s)\|_{\mathcal{L}(E_j,E_k)} \leq c(\varepsilon) e^{(\mu+\varepsilon)(t-s)} \tag{4.3.35}$$

for $\varepsilon > 0$, $(t,s) \in J^*_\Delta$, $A \in \mathcal{A}$, and $j,k \in \{0,1\}$.

Proof Put

$$U_A^\varepsilon(t,s) := e^{-(t-s)A(s)} + d_\varepsilon(t,s) , \qquad 0 < \varepsilon < t-s .$$

Then it is easily verified that

$$\begin{aligned} &\partial_1 U_A^\varepsilon(t,s) + A(t) U_A^\varepsilon(t,s) \\ &\quad = -k_A(t,s) + e^{-\varepsilon A(t-\varepsilon)} w_A(t-\varepsilon, s) - \int_s^{t-\varepsilon} k_A(t,\tau) w_A(\tau,s)\, d\tau \end{aligned}$$

for $0 < \varepsilon < t-s$. Letting $\varepsilon \to 0$, we infer from (4.3.23), (4.3.29), (4.3.30), and the closedness of $A(t)$ that

$$\partial_1 U_A + AU_A = w_A - k_A - k_A \star w_A$$

on J^*_Δ. Since $w_A - k_A = k_A \star w_A$ by (3.2.2), the first assertion is true.

If $k = 0$, the second assertion is an easy consequence of (4.3.1), (4.3.2), of Lemma 4.3.1, and of (3.1.8) and (3.1.9). Observe that, thanks to (4.3.33),

$$\begin{aligned} A(a_A \star w) &= AU_A - Aa_A = -\partial_1 U_A - Aa_A \\ &= -\partial_1 a_A - Aa_A - \partial_1(a_A \star w_A) = k_A - \partial_1(a_A \star w_A) . \end{aligned}$$

Hence we deduce from (4.2.2) that

$$\|a_A \star w_A(t,s)\|_{\mathcal{L}(E_j,E_1)} \leq M\big(\|k_A(t,s)\|_{\mathcal{L}(E_j,E_0)} + \|\partial_1(a_A \star w_A)(t,s)\|_{\mathcal{L}(E_j,E_0)}\big)$$

for $(t,s) \in J^*_\Delta$, $A \in \mathcal{A}$, and $j \in \{0,1\}$. Now (4.3.34) and (4.3.35) with $k = 1$ are consequences of (4.3.3), (4.3.4), (4.3.27), (4.3.28), and (4.3.30). ■

4.4 The Main Result

After these preparations we can prove the first main result of this section, namely the following existence and regularity result:

4.4.1 Theorem *Let assumption* (4.2.1) *be satisfied. For each* $A \in \mathcal{A}$ *there exists a unique parabolic evolution operator* U_A *possessing* E_1 *as a regularity subspace and satisfying*

$$U_A \in C\big(J_\Delta, \mathcal{L}_s(E_1)\big) \cap \mathfrak{K}(E_0, 0) \cap \mathfrak{K}(E_1, 0) \cap \mathfrak{K}(E_0, E_1, 1) \ . \tag{4.4.1}$$

There exists a constant $c(\rho) > 0$, *which is independent of* η, *such that, putting*

$$\mu := \mu(\eta) := c(\rho)\eta^{1/\rho} \ , \tag{4.4.2}$$

the estimate

$$\|U_A(t,s)\|_{\mathcal{L}(E_j)} + (t-s)\,\|U_A(t,s)\|_{\mathcal{L}(E_0,E_1)} \le c(\varepsilon)e^{(\mu+\varepsilon)(t-s)} \tag{4.4.3}$$

is valid for $\varepsilon > 0$, $(t,s) \in J_\Delta^*$, $A \in \mathcal{A}$, *and* $j = 0, 1$.

Proof Define U_A by (4.3.12). Then it follows from Lemmas 4.2.1 and 4.3.4 and from (3.1.5) that U_A satisfies (2.1.2), (2.1.4)–(2.1.7) with $E := E_0$ and $F := E_1$, as well as (4.4.1) and (4.4.3). Of course,

$$U_A(t,t) = 1 \ , \qquad t \in J \ . \tag{4.4.4}$$

Now let $A \in \mathcal{A}$ be fixed, suppose that $A \in C^1\big(J, \mathcal{L}(E_1, E_0)\big)$, and that J is compact. Then

$$\big[\lambda + A(\cdot)\big]^{-1} \in C^1\big(J, \mathcal{L}(E_0)\big) \tag{4.4.5}$$

and $\partial\big[\lambda + A(\cdot)\big]^{-1} = -(\lambda + A)^{-1}\partial A(\lambda + A)^{-1}$ for $\lambda \in \Sigma_\vartheta$. Thus (4.2.1) implies

$$\big\|\partial_s\big[\lambda + A(s)\big]^{-1}\big\|_{\mathcal{L}(E_0)} \le c\,|\lambda|^{-1} \ , \qquad (\lambda, s) \in \Sigma_\vartheta \times J \ . \tag{4.4.6}$$

Now it is an easy consequence of Lemma 4.1.1 that

$$\partial_2 a \in C\big(J_\Delta^*, \mathcal{L}(E_0)\big)$$

and

$$\partial_2 a(t,s) = A(s)e^{-(t-s)A(s)} + m(t,s) \ , \tag{4.4.7}$$

where

$$m(t,s) := \frac{1}{2\pi i}\int_\Gamma e^{\lambda(t-s)}\partial_s\big(\lambda + A(s)\big)^{-1}\,d\lambda \ .$$

Observe that (4.4.5), (4.4.6), and Lemma 4.1.1 imply

$$m \in \mathfrak{K}_\infty(E_0, 0) \ .$$

Hence we infer from Theorem 3.2.2 the existence of a unique solution

$$v \in \mathfrak{K}(E_0, 0) \tag{4.4.8}$$

of the integral equation

$$v = m + v \star m \; .$$

Put

$$V := a + v \star a$$

and note that, thanks to Lemma 4.2.1, (4.4.8), and (3.1.5),

$$V \in C\big(J_\Delta, \mathcal{L}_s(E_0)\big) \cap \mathfrak{K}(E_0, 0) \tag{4.4.9}$$

and

$$V(t,t) = 1 \; , \qquad t \in J \; . \tag{4.4.10}$$

Given $t \in \dot{J}$, it is easily seen that

$$V(t,\cdot) \in C^1\big((0,t), \mathcal{L}_s(E_1, E_0)\big)$$

and that

$$\begin{aligned} \partial_2 V &= \partial_2 a - v + (v \star a)A + v \star m \\ &= (a + v \star a)A + m + v \star m - v = VA \end{aligned} \tag{4.4.11}$$

in $\mathcal{L}_s(E_1, E_0)$. Observe that $U(\cdot, s)x$ is for each $(s,x) \in J \times E_0$ a solution of the Cauchy problem

$$\dot{u} + A(t)u = 0 \; , \quad t \in J \cap (s, \infty) \; , \qquad u(s) = x \; . \qquad (4.4.12)_{(s,x)}$$

Hence we obtain from (4.4.9)–(4.4.11) and Remark 2.1.2(a) that $V = U$ and that $(4.4.12)_{(s,x)}$ has a unique solution. This shows that (2.1.8) and (2.1.9) are satisfied. Now (2.1.3) follows easily from the unique solvability of $(4.4.12)_{(s,x)}$ for each $(s,x) \in J \times E_0$. This proves the theorem, provided each $A \in \mathcal{A}$ is continuously differentiable.

To remove this latter assumption we can again assume that J is compact. Then we extend an arbitrarily fixed $A \in \mathcal{A}$ trivially, that is, by zero, over $\mathbb{R}$ and put

$$A_\varepsilon := (\varphi_\varepsilon * A) | J \; ,$$

where $\{\, \varphi_\varepsilon \; ; \; \varepsilon > 0 \,\}$ is a mollifier on $\mathbb{R}$ (cf. Section III.4.2, where we recall the basic facts about convolutions). Then $A_\varepsilon \in C^\infty\big(J, \mathcal{L}(E_1, E_0)\big)$ and

$$\|A_\varepsilon\|_{C(J,\mathcal{L}(E_1,E_0))} \le \|A\|_{C(J,\mathcal{L}(E_1,E_0))} \; , \qquad \varepsilon > 0 \; , \tag{4.4.13}$$

by Young's inequality. Moreover,

$$A_\varepsilon(s) - A_\varepsilon(t) = \int \varphi_1(\tau)\big[A(s - \varepsilon\tau) - A(t - \varepsilon\tau)\big]\, d\tau$$

implies

$$[A_\varepsilon]_{\rho,J} \le \eta\ , \qquad \varepsilon > 0\ . \tag{4.4.14}$$

Since, as is well-known and easily seen,

$$A_\varepsilon \to A \quad \text{in } C\big(J, \mathcal{L}(E_1, E_0)\big) \qquad \text{as} \quad \varepsilon \to 0\ , \tag{4.4.15}$$

it follows from (4.4.14) and Theorem I.1.3.1(i) that there exists $\varepsilon_0 > 0$ such that

$$A_\varepsilon \in C^\rho\big(J, \mathcal{H}(E_1, E_0)\big)\ , \qquad 0 < \varepsilon < \varepsilon_0\ .$$

Note that the spectral bound $s(-A)$ of $-A$ is negative. It is an easy consequence of (4.4.15) and of Corollary I.1.4.3 that there exist $M' > 0$, $\vartheta' \in (0, \pi/2)$, and $\varepsilon' > 0$ such that the family

$$\mathcal{A}' := \{\, A_\varepsilon\ ;\ 0 < \varepsilon < \varepsilon' \,\}$$

satisfies assumption (4.2.1) (with M and ϑ being replaced by M' and ϑ', respectively.)

Thus, by what has already been shown, we know that there exists a unique parabolic evolution operator U_ε for A_ε, $0 < \varepsilon < \varepsilon'$. Given $(t,s) \in J_\Delta^*$,

$$\partial_\tau \big[U_\varepsilon(t,\tau) U(\tau,s)\big] = U_\varepsilon(t,\tau)\big[A_\varepsilon(\tau) - A(\tau)\big]U(\tau,s)\ .$$

Upon integrating this identity,

$$U(t,s) - U_\varepsilon(t,s) = \int_s^t U_\varepsilon(t,\tau)\big[A_\varepsilon(\tau) - A(\tau)\big]U(\tau,s)\, d\tau\ .$$

From this we deduce that

$$\|U_\varepsilon - U\|_{C(J_\Delta, \mathcal{L}(E_1,E_0))} \le c\, \|A_\varepsilon - A\|_{C(J,\mathcal{L}(E_1,E_0))}\ , \qquad 0 < \varepsilon < \varepsilon'\ .$$

Now (4.4.15), the fact that U_ε is uniformly bounded in $\mathcal{L}(E_0)$ for $0 < \varepsilon < \varepsilon'$, and the density of E_1 in E_0 imply

$$U_\varepsilon \to U \quad \text{in } C\big(J_\Delta, \mathcal{L}_s(E_0)\big) \qquad \text{as} \quad \varepsilon \to 0\ . \tag{4.4.16}$$

Since

$$U_\varepsilon(t,s)x - U_\varepsilon(t,s')x = \int_{s'}^s \partial_\tau U_\varepsilon(t,\tau)x\, d\tau = \int_{s'}^s U_\varepsilon(t,\tau)A_\varepsilon(\tau)x\, d\tau$$

for $s' < s < t$, we obtain from (4.4.15) and (4.4.16) that

$$U(t,s) - U(t,s') = \int_{s'}^s U(t,\tau)A(\tau)\, d\tau\ , \qquad 0 \le s' < s < t\ ,$$

in $\mathcal{L}_s(E_1, E_0)$. From this it follows that U satisfies (2.1.8) and (2.1.9). Consequently, (2.1.3) is again a consequence of Remark 2.1.2(a). ■

4.4.2 Corollary *Suppose that*

$$A \in C^{\rho}\big(J, \mathcal{H}(E_1, E_0)\big)$$

for some $\rho \in (0,1)$. *Then there exists a unique parabolic evolution operator* U_A *for* A *possessing* E_1 *as a regularity subspace.*

Proof First let J be compact. Then Theorem I.1.2.2 and Corollary I.1.4.3 imply the existence of $\sigma \in \mathbb{R}$ such that $\mathcal{A} := \{\sigma + A\}$ satisfies assumption (4.2.1). Hence the assertion follows in this case from Theorem 4.4.1 and Remark 2.1.2(d). If J is not compact, we obtain the assertion by applying this result to every compact subinterval of J. ■

4.5 Solvability of the Cauchy Problem

Now it is easy to give a proof of the first part of the basic existence and regularity result of this section:

Proof of Theorem 1.2.1: Part 1 In this part of the proof we show that the Cauchy problem $(1.2.2)_{(s,x,A,f)}$ has a unique solution that is strict if $x \in E_1$. The proof of the asserted Hölder regularity is postponed to Section III.2.6.

First suppose that $J \subset \mathbb{R}^+$ with $0 \in J$, that $s = 0$, and that $\mathcal{A} := \{A\}$ satisfies condition (4.2.1). Thanks to Remarks 2.1.2(a) and (b) it suffices to show that the mild solution

$$u := (Ux + U \star f)(\cdot, 0)$$

is a solution of $(1.2.2)_{(0,x,A,f)}$, where $U := U_A$. Since $U(\cdot,0)x$ is a solution of $(1.2.2)_{(0,x,A,0)}$ by (2.1.2), (2.1.6) and (2.1.7), in order to prove that u is a solution it remains to show that $U \star f := U \star f(\cdot, 0)$ is a solution of $(1.2.2)_{(0,0,A,f)}$.

Recall that $U = a + a \star w$. It is an easy consequence of (4.3.28), (4.3.30), and the continuity of $a \star w$ on J_Δ that

$$\partial_t \int_0^t (a \star w)(t,\tau) f(\tau)\, d\tau = \int_0^t \partial_1 (a \star w)(t,\tau) f(\tau)\, d\tau \ , \tag{4.5.1}$$

which shows, thanks to (4.3.28) and (4.3.30), that

$$(a \star w) \star f(\cdot, 0) \in C^1(J, E_0) \ . \tag{4.5.2}$$

Put

$$v_\varepsilon(t) := \int_0^{t-\varepsilon} e^{-(t-\tau)A(\tau)} f(\tau)\, d\tau \ , \qquad \varepsilon > 0 \ , \quad t \in J \ , \quad t \geq \varepsilon \ ,$$

with $v_\varepsilon(t) := 0$ for $0 \leq t < \varepsilon$ and note that

$$v_\varepsilon(t) \to v(t) := a \star f(t,0) \ \text{ in } E_0 \qquad \text{as} \quad \varepsilon \to 0 \ . \tag{4.5.3}$$

Also observe that

$$\partial v_\varepsilon(t) = e^{-\varepsilon A(t-\varepsilon)} f(t-\varepsilon) - \int_0^{t-\varepsilon} A(\tau) e^{-(t-\tau)A(\tau)} f(\tau)\, d\tau$$

and that

$$\begin{aligned}\int_0^{t-\varepsilon} & A(\tau) e^{-(t-\tau)A(\tau)} f(\tau)\, d\tau \\ &= -\int_0^{t-\varepsilon} e(t,\tau) f(\tau)\, d\tau + \int_0^{t-\varepsilon} A(t) e^{-(t-\tau)A(t)} f(\tau)\, d\tau \ .\end{aligned}$$

Since the last integral can be rewritten as

$$\int_0^{t-\varepsilon} A(t) e^{-(t-\tau)A(t)} \big(f(\tau) - f(t)\big)\, d\tau - [e^{-\varepsilon A(t)} - e^{-tA(t)}] f(t) \ ,$$

we deduce from the Hölder continuity of f and from Lemmas 4.2.1 and 4.3.2 that there exists $w \in C(J, E_0)$ such that

$$\partial v_\varepsilon \to w \quad \text{in } C(J, E_0) \qquad \text{as} \quad \varepsilon \to 0 \ .$$

Thus, thanks to (4.5.3), we see that $v \in C^1(J, E_0)$ which, together with (4.5.2), implies

$$U \star f(\cdot, 0) \in C^1(J, E_0) \ . \tag{4.5.4}$$

Note that the closedness of $A(t)$ and (2.1.7) imply

$$\partial_t \int_0^{t-\varepsilon} U(t,\tau) f(\tau)\, d\tau + A(t) \int_0^{t-\varepsilon} U(t,\tau) f(\tau)\, d\tau = U(t, t-\varepsilon) f(t-\varepsilon) \ .$$

From this, the closedness of $A(t)$, and the strong continuity of U on J_Δ we infer, by letting $\varepsilon \to 0$, that $U \star f(\cdot, 0)$ is a solution of $(1.2.2)_{(0,0,A,f)}$.

Suppose that $x \in E_1$. Then from (4.4.1), that is, from $U \in C\big(J_\Delta, \mathcal{L}_s(E_1)\big)$, we infer $U(\cdot, 0)x \in C(J, E_1)$. Thus $AU(\cdot, 0)x \in C(J, E_0)$ since $A \in C\big(J, \mathcal{L}(E_1, E_0)\big)$. Now we see from

$$\partial\big[U(\cdot, 0)x\big] = -AU(\cdot, 0)x$$

that $U(\cdot, 0)x$ is a strict solution of $(1.2.2)_{(0,x,A,0)}$. By combining this with (4.5.4) we see that u is a strict solution of $(1.2.2)_{(0,x,A,f)}$. This proves the part under consideration of the theorem in this case. Now the general case follows by an obvious translation argument and by Remark 2.1.2(d). ■

The proof of Theorem 4.4.1 and the proof of part 1 of Theorem 1.2.1 follow essentially Tanabe [Tan60] and [Tan79]. It should be remarked, however, that the regularity statement (4.4.1) and the uniform estimates (4.4.3), which will be crucial in the following, are contained in these references implicitly at most.

5 Stability Estimates

Let (E_0, E_1) be a densely injected Banach couple, let J be a perfect subinterval of $\mathbb{R}^+$ containing 0, and let $\rho \in (0,1)$. Throughout this section we assume that

$$\left.\begin{aligned}&\mathcal{A} \subset C^\rho\big(J, \mathcal{L}(E_1, E_0)\big) \text{ and there exist constants } \eta \geq 0,\\ &\omega > 0,\ \ \kappa \geq 1, \text{ and } \sigma \in \mathbb{R} \text{ such that}\\ &\qquad\qquad [A]_{\rho,J} \leq \eta\ , \qquad A \in \mathcal{A}\ ,\\ &\text{and}\\ &\qquad\qquad \sigma + \mathcal{A} \subset C\big(J, \mathcal{H}(E_1, E_0, \kappa, \omega)\big)\ .\end{aligned}\right\} \tag{5.0.1}$$

In the following, we prove some continuity and decay estimates for parabolic evolution equations and for mild solutions of linear Cauchy problems. These results are of basic importance for our study of quasilinear problems.

5.1 Estimates for Evolution Operators

As an easy consequence of the results of Sections I.1 and 4 we obtain the following fundamental estimates:

5.1.1 Theorem *There exists a constant $c_0(\rho) > 0$, which is independent of η, such that, letting*

$$\nu := c_0(\rho)\eta^{1/\rho} + \sigma + \omega\ , \tag{5.1.1}$$

the following is true: for each $A \in \mathcal{A}$ there exists a unique parabolic evolution operator U_A for A possessing E_1 as a regularity subspace, and

$$\|U_A(t,s)\|_{\mathcal{L}(E_j)} + (t-s)\,\|U_A(t,s)\|_{\mathcal{L}(E_0,E_1)} \leq ce^{\nu(t-s)}$$

for $(t,s) \in J^*_\Delta$, $A \in \mathcal{A}$, and $j = 0, 1$.

Proof It follows from (5.0.1), Proposition I.1.4.1, and Corollary I.1.4.3 that there exist $\widetilde{\omega} < \omega$, $\vartheta \in (0, \pi/2)$, and $M > 0$ such that

$$\sigma + \widetilde{\omega} + \Sigma_\vartheta \subset \rho\big(-A(s)\big)$$

and

$$\|\sigma + \widetilde{\omega} + A(s)\|_{\mathcal{L}(E_1,E_0)} + (1+|\lambda|)^{1-j}\,\big\|\big[\lambda + \big(\sigma + \widetilde{\omega} + A(s)\big)\big]^{-1}\big\|_{\mathcal{L}(E_0,E_j)} \leq M$$

for $(s, \lambda, A) \in J \times \Sigma_\vartheta \times \mathcal{A}$ and $j = 0, 1$. Thus Theorem 4.4.1 implies the existence of a constant $c_0(\rho) > 0$, being independent of η, such that for each $A \in \mathcal{A}$ there exists a unique parabolic evolution operator $U_{\sigma+\widetilde{\omega}+A}$ for $\sigma+\widetilde{\omega}+A$ possessing E_1 as a regularity subspace and satisfying

$$U_{\sigma+\widetilde{\omega}+A} \in C\big(J_\Delta, \mathcal{L}_s(E_1)\big) \tag{5.1.2}$$

such that

$$\|U_{\sigma+\widetilde{\omega}+A}(t,s)\|_{\mathcal{L}(E_j)} + (t-s)\left\|U_{\sigma+\widetilde{\omega}+A}(t,s)\right\|_{\mathcal{L}(E_0,E_1)} \le ce^{(\mu+\omega-\widetilde{\omega})(t-s)}$$

for $(t,s) \in J_\Delta^*$, $A \in \mathcal{A}$, and $j=0,1$, where $\mu := c_0(\rho)\eta^{1/\rho}$. Since, by Remark 2.1.2(d),

$$U_A(t,s) := e^{(\sigma+\widetilde{\omega})(t-s)}U_{\sigma+\widetilde{\omega}+A}(t,s)\ , \qquad (t,s) \in J_\Delta\ , \tag{5.1.3}$$

is a parabolic evolution operator for A with regularity subspace E_1, the assertion follows. ■

5.1.2 Remark Suppose that $A \in \mathcal{H}(E_1, E_0)$ and $\nu > s(-A)$. Then

$$\|e^{-tA}\|_{\mathcal{L}(E_j)} + t\,\|e^{-tA}\|_{\mathcal{L}(E_0,E_1)} \le ce^{\nu t} \tag{5.1.4}$$

for $t > 0$ and $j = 0, 1$. Hence

$$s(-A) = \text{type}(-A)\ , \qquad A \in \mathcal{H}(E_1, E_0)\ , \tag{5.1.5}$$

that is, the spectral bound and the exponential type of a generator of a strongly continuous analytic semigroup coincide.

Proof Fix $\sigma \in \big(s(-A), \nu\big)$ and put $\omega := \nu - \sigma$. Then Proposition I.1.4.2 implies the existence of $\kappa \ge 1$ such that $\sigma + A \in \mathcal{H}(E_1, E_0, \kappa, \omega)$. Hence $\mathcal{A} := \{A\}$ satisfies (5.0.1) with $\eta = 0$. Consequently, (5.1.4) follows from Theorem 5.1.1 and Example 2.1.1. Now it is obvious that $\text{type}(-A) \le s(-A)$. On the other hand, the Hille-Yosida theorem implies the converse estimate. ■

Throughout the remainder of Section 5 we denote by U_A the parabolic fundamental solution for $A \in \mathcal{A}$, and ν is the number defined in (5.1.1). Now we study the behavior of U_A in the interpolation spaces $E_\alpha,\ 0 < \alpha < 1$.

5.1.3 Lemma *Suppose that $0 \le \beta_- \le \beta \le \alpha \le 1$. Also assume that $\beta_- < \beta$ iff $0 < \beta < \alpha < 1$ and there does not exist an interpolation functor $\mathfrak{F}$ of exponent β/α such that $\mathfrak{F}(E_0, E_\alpha) \doteq E_\beta$. Then*

$$U_A \in C\big(J_\Delta, \mathcal{L}_s(E_\alpha)\big) \cap C\big(J_\Delta^*, \mathcal{L}(E_\beta, E_\alpha)\big)$$

and

$$\|U_A(t,s)\|_{\mathcal{L}(E_\alpha)} + (t-s)^{\alpha-\beta_-}\,\|U_A(t,s)\|_{\mathcal{L}(E_\beta,E_\alpha)} \le ce^{\nu(t-s)}$$

for $(t,s) \in J_\Delta^*$, and $A \in \mathcal{A}$.

Proof If $\alpha, \beta \in \{0,1\}$, this follows from Theorem 5.1.1 and from (5.1.2) and (5.1.3). From this we easily obtain the assertion by interpolation if either $\beta = 0$ and $0 \le \alpha \le 1$, or $0 \le \beta \le 1$ and $\alpha = 1$, or if $\beta = \alpha$. (For the proof of the strong continuity one uses in an essential way the fact that E_1 is dense in E_α).

Suppose that $0 < \alpha < 1$. By interpolating the morphisms

$$U_A(t,s) : (E_0, E_\alpha) \to (E_\alpha, E_\alpha)$$

by means of an interpolation functor $\mathfrak{F}$ of exponent ξ it follows that

$$U_A \in C\big(J_\Delta^*, \mathcal{L}(\mathfrak{F}(E_0, E_\alpha), E_\alpha)\big)$$

and

$$(t-s)^{(1-\xi)\alpha} \, \|U_A(t,s)\|_{\mathcal{L}\left(\mathfrak{F}(E_0,E_\alpha),E_\alpha\right)} \le ce^{\nu(t-s)}$$

for $(t,s) \in J_\Delta^*$, $A \in \mathcal{A}$, and $0 < \xi < 1$. If there does exist an interpolation functor $\mathfrak{F}$ of exponent β/α such that $\mathfrak{F}(E_0, E_\alpha) \doteq E_\beta$, the assertion follows by letting $\xi := \beta/\alpha$. Otherwise we put $\mathfrak{F} := (\cdot,\cdot)_\xi$ with $\xi := \beta_-/\alpha$ and obtain the desired result from the almost reiteration property of Remark I.2.11.2(a). ■

The following lemma gives continuity estimates for U_A as a function of $A \in \mathcal{A}$.

5.1.4 Lemma *Suppose that $0 \le \beta < 1$ and $0 < \alpha \le 1$. Then*

$$\begin{aligned}(t-s)^{\beta-\alpha} \, &\|(U_A - U_B)(t,s)\|_{\mathcal{L}(E_\alpha, E_\beta)} \\ &\le ce^{\nu(t-s)} \max_{s \le \tau \le t} \|A(\tau) - B(\tau)\|_{\mathcal{L}(E_1, E_0)}\end{aligned}$$

for $(t,s) \in J_\Delta^*$, and $A, B \in \mathcal{A}$.

Proof By replacing in the argument leading to the integral equation (2.2.2) the function $e^{-(\tau-s)A(s)}$ by $U_B(\tau,s)$, it follows that

$$U_A(t,s) - U_B(t,s) = -\int_s^t U_A(t,\tau)\big[A(\tau) - B(\tau)\big]U_B(\tau,s)\,d\tau\ , \qquad (t,s) \in J_\Delta\ .$$

Now we deduce from Lemma 5.1.3 that

$$\begin{aligned}\|(U_A - U_B)(t,s)\|_{\mathcal{L}(E_\alpha,E_\beta)} &\le ce^{\nu(t-s)} \int_s^t (t-\tau)^{-\beta}(\tau-s)^{\alpha-1}\,d\tau\, |||A-B||| \\ &= c\mathsf{B}(\alpha, 1-\beta)\, |||A-B|||\, (t-s)^{\alpha-\beta} e^{\nu(t-s)}\ ,\end{aligned}$$

where

$$|||A-B||| := \max_{s \le \tau \le t} \|(A-B)(\tau)\|_{\mathcal{L}(E_1,E_0)}$$

and B is the Euler beta function. ■

5.2 Continuity Properties of Mild Solutions

Given

$$(x, A, f) \in E_0 \times \mathcal{A} \times L_{1,\mathrm{loc}}(J, E_0) ,$$

we consider the linear Cauchy problem

$$\dot{u} + A(t)u = f(t) , \qquad t \in \dot{J} , \quad u(0) = x . \qquad (5.2.1)_{(x,A,f)}$$

Recall from Remark 2.1.2(e) that the mild solution of $(5.2.1)_{(x,A,f)}$ is given by

$$u(\cdot, x, A, f) := (U_A x + U_A \star f)(\cdot, 0) . \qquad (5.2.2)$$

In the following, we prove some stability estimates that are of importance for the study of quasilinear problems. The next theorem implies, in particular, that the mild solutions depend continuously upon all data, provided f is locally bounded.

5.2.1 Theorem *Suppose that $0 \le \beta \le \alpha \le 1$ with $\alpha > 0$ and $\beta < 1$ and $0 < \gamma \le 1$. Then*

$$\begin{aligned} &\|u(t, x_0, A_0, f_0) - u(t, x_1, A_1, f_1)\|_\beta \\ &\quad \le c\Big\{ t^{\alpha-\beta} \|A_0 - A_1\|_{C([0,t],\mathcal{L}(E_1,E_0))} \left[\|x_0\|_\alpha + t^{1-\alpha+\gamma} \|f_0\|_{L_\infty((0,t),E_\gamma)} \right] \\ &\qquad + \|x_0 - x_1\|_\beta + t^{1-\beta} \|f_0 - f_1\|_{L_\infty((0,t),E_0)} \Big\} e^{\nu t} \end{aligned}$$

for $t \in J$ and $(x_j, A_j, f_j) \in E_\alpha \times \mathcal{A} \times L_{\infty,\mathrm{loc}}(J, E_\gamma)$.

Proof Put $u_j := u(\cdot, x_j, A_j, f_j)$ and $U_j := U_{A_j}$ and observe that

$$u_0 - u_1 = (U_0 - U_1)x_0 + (U_0 - U_1) \star f_0 + U_1(x_0 - x_1) + U_1 \star (f_0 - f_1) .$$

Now the assertion is an easy consequence of Lemmas 5.1.3 and 5.1.4 and the continuous embedding of E_α in E_β. ■

5.2.2 Remark Suppose that $0 \le \beta \le \alpha \le 1$ with $\alpha > 0$ and $\beta < 1$ and $0 < \gamma \le 1$. Let X_ξ, $\xi \in \{\alpha, \beta, \gamma\}$, be Banach spaces satisfying

$$E_1 \hookrightarrow X_\alpha \hookrightarrow E_\alpha \hookrightarrow E_\beta \hookrightarrow X_\beta \hookrightarrow E_0$$

and

$$E_1 \hookrightarrow X_\gamma \hookrightarrow E_\gamma ,$$

respectively. Observe that the X-spaces are not supposed to be interpolation spaces between E_0 and E_1. Also suppose that $A \in \mathcal{A}$.

It follows from Lemma 5.1.3 and the factorization

$$X_\xi \hookrightarrow E_0 \xrightarrow{U_A} E_1 \hookrightarrow X_\xi$$

that

$$U_A \in C\big(J_\Delta^*, \mathcal{L}(X_\xi)\big) , \qquad \xi \in \{\alpha, \beta, \gamma\} .$$

Similarly,

$$X_\gamma \hookrightarrow E_\gamma \xrightarrow{U_A} E_\beta \hookrightarrow X_\beta$$

and Lemma 5.1.2 imply

$$U_A \in C\big(J_\Delta^*, \mathcal{L}(X_\gamma, X_\beta)\big)$$

and

$$(t-s)^\beta \, \|U_A(t,s)\|_{\mathcal{L}(X_\gamma, X_\beta)} \le c e^{\nu(t-s)}$$

for $(t,s) \in J_\Delta^*$. Hence we deduce from the representation (5.2.2) that

$$u(\cdot, x, A, f) \in C(\dot{J}, X_\beta)$$

for $x \in X_\beta$ and $f \in L_{\infty,\mathrm{loc}}(J, X_\beta)$.

Suppose, in addition, that

$$\|U_A(t,s)\|_{\mathcal{L}(X_\beta)} \le c e^{\nu(t-s)} , \qquad (t,s) \in J_\Delta^* , \quad A \in \mathcal{A} .$$

Then it is obvious from the proof of Theorem 5.2.1 and factorization arguments of the type used that

$$\begin{aligned} &\|u(t, x_0, A_0, f_0) - u(t, x_1, A_1, f_1)\|_{X_\beta} \\ &\quad \le c\Big\{ t^{\alpha-\beta} \, \|A_0 - A_1\|_{C([0,t], \mathcal{L}(E_1, E_0))} \, \big[\, \|x_0\|_{X_\alpha} + t^{1-\alpha+\gamma} \, \|f_0\|_{L_\infty((0,t), X_\gamma)} \big] \\ &\qquad + \|x_0 - x_1\|_{X_\beta} + t^{1-\beta} \, \|f_0 - f_1\|_{L_\infty((0,t), E_0)} \Big\} e^{\nu t} \end{aligned}$$

for $t \in \dot{J}$ and $(x_j, A_j, f_j) \in X_\alpha \times \mathcal{A} \times L_{\infty,\mathrm{loc}}(J, X_\gamma)$. This simple generalization of Theorem 5.2.1 is useful in some applications. ∎

5.3 Hölder Estimates

Our next theorem provides uniform Hölder estimates for mild solutions.

5.3.1 Theorem *Suppose that $0 \le \beta \le \alpha < 1$.*
If $x \in E_\alpha$ and $f \in L_{\infty,\mathrm{loc}}(J, E_0)$ then

$$u := u(\cdot, x, A, f) \in C^{\alpha-\beta}(J, E_\beta) . \tag{5.3.1}$$

More precisely,

$$\|u(t) - u(s)\|_\beta \le c(t-s)^{\alpha-\beta} e^{\nu t} \big(\|x\|_\alpha + \|f\|_{L_\infty((0,t), E_0)} \big) \tag{5.3.2}$$

for $(t,s) \in J_\Delta$ and $(x, A, f) \in E_\alpha \times \mathcal{A} \times L_{\infty,\mathrm{loc}}(J, E_0)$.

Proof Observe that

$$\|A(t)U_A(t,s)\|_{\mathcal{L}(E_\alpha,E_0)} \le \|A(t)\|_{\mathcal{L}(E_1,E_0)} \|U_A(t,s)\|_{\mathcal{L}(E_\alpha,E_1)} \,. \tag{5.3.3}$$

Hence it follows from the proof of Theorem 5.1.1 and from Lemma 5.1.3 that there exists $\varepsilon > 0$ such that

$$\|U_A(t,s)\|_{\mathcal{L}(E_\alpha)} + (t-s)^{1-\alpha} \|A(t)U_A(t,s)\|_{\mathcal{L}(E_\alpha,E_0)} \le ce^{(\nu-\varepsilon)(t-s)} \tag{5.3.4}$$

for $(t,s) \in J_\Delta^*$ and $A \in \mathcal{A}$. Using the first half of this estimate, Lemma 5.1.3 with $\beta = 0$, and the strong continuity of U_A, the assertion is obvious if $\alpha = 0$. Hence we can assume from now on that $\alpha > 0$.

Put $v(t,s) := U_A(t,s)x$ for $(t,s) \in J_\Delta$ and $x \in E_\alpha$ and observe that

$$v(\cdot,s) \in C\big(J \cap [s,\infty), E_\alpha\big) \cap C^1\big(J \cap (s,\infty), E_0\big) \tag{5.3.5}$$

with

$$\partial_1 v(t,s) = -A(t)U_A(t,s)x \,.$$

Hence it follows from (5.3.4) that

$$v(t,s) - v(r,s) = -\int_r^t (\tau - s)^{1-\alpha} A(\tau)U_A(\tau,s)x(\tau-s)^{\alpha-1}\, d\tau$$

and

$$\|v(t,s) - v(r,s)\|_0 \le c(t-r)^\alpha e^{(\nu-\varepsilon)(t-s)} \|x\|_\alpha \tag{5.3.6}$$

for $s \le r \le t$ and $(t,s) \in J_\Delta$. Note that (5.3.4) also implies

$$\|v(t,s) - v(r,s)\|_\alpha \le \|v(t,s)\|_\alpha + \|v(r,s)\|_\alpha \le ce^{(\nu-\varepsilon)(t-s)} \|x\|_\alpha \,. \tag{5.3.7}$$

Thus we deduce from (5.3.6), (5.3.7), the interpolation inequality (I.2.11.2), and the definition of v that

$$\|U_A(t,s) - U_A(r,s)\|_{\mathcal{L}(E_\alpha,E_\beta)} \le c(t-r)^{\alpha-\beta} e^{(\nu-\varepsilon)(t-s)} \tag{5.3.8}$$

for $s \le r \le t$, $(t,s) \in J_\Delta$, and $A \in \mathcal{A}$.

Put $w := U_A \star f(\cdot,0)$ and observe that

$$w(t) - w(s) = \int_s^t U_A(t,\tau)f(\tau)\, d\tau + \int_0^s \big[U_A(t,\tau) - U_A(s,\tau)\big] f(\tau)\, d\tau \tag{5.3.9}$$

for $(t,s) \in J_\Delta$. Thanks to Lemma 5.1.3 the first integral is estimated in the norm of E_β by

$$ce^{(\nu-\varepsilon)t} \int_s^t (t-\tau)^{-\beta}\, d\tau \, \|f\|_{L_\infty((0,t),E_0)} \,.$$

Since $(t-s)^{1-\beta} \le t^{1-\alpha}(t-s)^{\alpha-\beta}$, we see that the first integral in (5.3.9) is estimated in the norm of E_β by

$$c(t-s)^{\alpha-\beta} e^{\nu t} \|f\|_{L_\infty((0,t),E_0)} \ . \tag{5.3.10}$$

From Lemma 5.1.3 and (5.3.8) we deduce that

$$\begin{aligned}
&\|U_A(t,\tau) - U_A(s,\tau)\|_{\mathcal{L}(E_0,E_\beta)} \\
&\quad \le \|U_A(t,s) - U_A(s,s)\|_{\mathcal{L}(E_\alpha,E_\beta)} \|U(s,\tau)\|_{\mathcal{L}(E_0,E_\alpha)} \\
&\quad \le c(t-s)^{\alpha-\beta}(s-\tau)^{-\alpha} e^{(\nu-\varepsilon)(t-\tau)}
\end{aligned}$$

for $\tau \le s \le t$. This estimate shows that the second integral in (5.3.9) is bounded in the E_β-norm by

$$c(t-s)^{\alpha-\beta} e^{\nu t} \|f\|_{L_\infty((0,t),E_0)} \tag{5.3.11}$$

for $(t,s) \in J_\Delta$, $A \in \mathcal{A}$, and $f \in L_{\infty,\mathrm{loc}}(J,E_0)$. Now estimate (5.3.2) is a consequence of (5.3.8)–(5.3.11). Since it is easily verified that $w \in C(J,E_\beta)$ (if $\alpha=\beta$), it follows from (5.3.5) that $u \in C(J,E_\beta)$. This proves the theorem. ∎

5.4 Boundedness of Mild Solutions

Lastly, we prove a useful boundedness result that again illustrates the smoothing effects of parabolic evolution equations. For this we assume without loss of generality that $\nu \ne 0$.

5.4.1 Theorem *Suppose that* $0 \le \beta_- \le \beta \le \alpha \le 1$. *Also assume that* $\beta_- < \beta$ *iff* $0 < \beta < \alpha < 1$ *and there does not exist an interpolation functor* $\mathfrak{F}$ *of exponent* β/α *such that* $\mathfrak{F}(E_0,E_\alpha) \doteq E_\beta$. *Also suppose that* $\gamma = 0$ *if* $\alpha < 1$ *and* $0 < \gamma < 1$ *if* $\alpha = 1$. *Letting*

$$b_{\alpha-\gamma}(t,r) := |r|^{\alpha-\gamma-1} \int_0^{|r|t} \xi^{\gamma-\alpha} e^{\xi \operatorname{sign}(r)}\, d\xi$$

for $t \in \mathbb{R}^+$ *and* $r \in \dot{\mathbb{R}}$, *it follows that*

$$\|u(t,x,A,f)\|_\alpha \le c\big\{t^{\beta_- -\alpha} e^{\nu t}\|x\|_\beta + b_{\alpha-\gamma}(t,\nu)\|f\|_{L_\infty((0,t),E_\gamma)}\big\}$$

for $t \in J$ *and* $(x,A,f) \in E_\beta \times \mathcal{A} \times L_{\infty,\mathrm{loc}}(J,E_\gamma)$.

Proof This is an easy consequence of Lemma 5.1.3. ∎

5.4.2 Corollary *Let the assumptions of Theorem* 5.4.1 *be satisfied and suppose* $\nu < 0$. *Then*

$$\|u(t,x,A,f)\|_\alpha \le c\big\{t^{\beta_- -\alpha} e^{\nu t}\|x\|_\beta + \|f\|_{L_\infty((0,t),E_\gamma)}\big\} \tag{5.4.1}$$

for $t \in J$ *and* $(x,A,f) \in E_\beta \times \mathcal{A} \times L_{\infty,\mathrm{loc}}(J,E_\gamma)$.

6 Invariance and Positivity

It is the main purpose of this section to show that the evolution operator $U := U_A$ of $A \in C^\rho(J, \mathcal{H}(E_1, E_0))$ enjoys essentially the same invariance properties as each one of the semigroups $\{ e^{-tA(s)} ; t \geq 0 \}$ for $s \in J$. In particular, if E_0 is an ordered Banach space, we derive a useful criterion for A guaranteeing the positivity of U_A.

6.1 Yosida Approximations

Suppose that

$$A \in \mathcal{C}(E) \quad \text{and} \quad (\omega, \infty) \subset \rho(-A) \tag{6.1.1}$$

for some $\omega \geq 0$. Then

$$A_\varepsilon := A(1 + \varepsilon A)^{-1} \in \mathcal{L}(E) \tag{6.1.2}$$

for $0 < \varepsilon < 1/\omega$, and

$$\{ A_\varepsilon ; 0 < \varepsilon < 1/\omega \}$$

is the **Yosida approximation** of A.

6.1.1 Lemma *Suppose that*

$$\|\lambda(\lambda + A)^{-1}\| \leq \kappa , \qquad \lambda > \omega . \tag{6.1.3}$$

Then

$$\lim_{\varepsilon \to 0} (1 + \varepsilon A)^{-1} x = x , \qquad x \in \overline{\operatorname{dom}(A)} , \tag{6.1.4}$$

and

$$\lim_{\varepsilon \to 0} A_\varepsilon x = Ax , \qquad x \in \operatorname{dom}(A) . \tag{6.1.5}$$

Proof Note that (6.1.3) implies

$$\|(1 + \varepsilon A)^{-1}\| \leq \kappa , \qquad 0 < \varepsilon < 1/\omega . \tag{6.1.6}$$

Hence it follows from

$$(1 + \varepsilon A)^{-1} x = x - \varepsilon (1 + \varepsilon A)^{-1} Ax , \qquad x \in \operatorname{dom}(A) ,$$

that

$$\|(1 + \varepsilon A)^{-1} x - x\| \leq \varepsilon \kappa \, \|Ax\| , \qquad x \in \operatorname{dom}(A) .$$

Thus $(1 + \varepsilon A)^{-1} x \to x$ for $x \in \operatorname{dom}(A)$ as $\varepsilon \to 0$. Now (6.1.4) is an easy consequence of (6.1.6). Since $A_\varepsilon x = (1 + \varepsilon A)^{-1} Ax$ for $x \in \operatorname{dom}(A)$, the last assertion follows from (6.1.4). ■

Let (E_0, E_1) be a densely injected Banach couple and suppose that

$$A \in \mathcal{H}(E_1, E_0, \kappa, \omega) \tag{6.1.7}$$

for some $\kappa \geq 1$ and $\omega > 0$. Then (6.1.1) is satisfied (with $E := E_0$). Thus $A_\varepsilon \in \mathcal{L}(E_0)$ is well-defined for $0 < \varepsilon < 1/\omega$. Since

$$\|(1+\varepsilon A)^{-1}\|_{\mathcal{L}(E_0,E_1)} \leq \kappa/\varepsilon\ , \qquad 0 < \varepsilon < 1/\omega\ , \tag{6.1.8}$$

by Remark I.1.2.1(a), it follows that

$$\|A_\varepsilon\|_{\mathcal{L}(E_0)} \leq \varepsilon^{-1}\kappa\,\|A\|_{\mathcal{L}(E_1,E_0)}\ , \qquad 0 < \varepsilon < 1/\omega\ . \tag{6.1.9}$$

Hence

$$A_\varepsilon \in \mathcal{H}\big(E_0, E_0, 2, 1 + 2\varepsilon^{-1}\kappa\,\|A\|_{\mathcal{L}(E_1,E_0)}\big)$$

by Remark I.1.2.1(b). The following lemma implies, however, that $\lambda(\lambda + A_\varepsilon)^{-1}$ exists in $\big[\operatorname{Re} z \geq \omega/(1-\varepsilon\omega)\big]$ and can be estimated independently of ε.

6.1.2 Lemma *Suppose that $A \in \mathcal{C}(E)$ and there exist $\kappa, \omega \in \mathbb{R}^+$ such that $[\operatorname{Re}\lambda \geq \omega] \subset \rho(-A)$ and*

$$\|\lambda(\lambda + A)^{-1}\| \leq \kappa\ , \qquad \operatorname{Re}\lambda \geq \omega\ . \tag{6.1.10}$$

Then

$$\big[\operatorname{Re} z \geq \omega/(1-\varepsilon\omega)\big] \subset \rho(-A_\varepsilon)$$

and

$$\|\lambda(\lambda + A_\varepsilon)^{-1}\| \leq \kappa + 1\ , \qquad \operatorname{Re}\lambda \geq \omega/(1-\varepsilon\omega)\ , \tag{6.1.11}$$

for $0 < \varepsilon < 1/\omega$.

Proof Note that

$$\lambda + A_\varepsilon = \lambda + A(1+\varepsilon A)^{-1} = (1+\varepsilon\lambda)\Big(\frac{\lambda}{1+\varepsilon\lambda} + A\Big)(1+\varepsilon A)^{-1}$$

for $\lambda \in \mathbb{C}$. It is easily verified that

$$\operatorname{Re}\lambda \geq \omega/(1-\varepsilon\omega) \quad \text{implies} \quad \operatorname{Re}\big(\lambda/(1+\varepsilon\lambda)\big) \geq \omega\ .$$

Hence

$$\begin{aligned}(\lambda + A_\varepsilon)^{-1} &= \frac{1}{1+\varepsilon\lambda}(1+\varepsilon A)\Big(\frac{\lambda}{1+\varepsilon\lambda} + A\Big)^{-1} \\ &= \frac{1}{1+\varepsilon\lambda}\Big[\varepsilon + \frac{1}{1+\varepsilon\lambda}\Big(\frac{\lambda}{1+\varepsilon\lambda} + A\Big)^{-1}\Big]\end{aligned} \tag{6.1.12}$$

for $\operatorname{Re}\lambda \geq \omega/(1-\varepsilon\omega)$ and $0 < \varepsilon < 1/\omega$. Consequently, thanks to (6.1.10),

$$\|(\lambda + A_\varepsilon)^{-1}\| \leq \frac{1}{|\lambda|}\frac{\varepsilon\,|\lambda| + \kappa}{|1+\varepsilon\lambda|}\ , \qquad \operatorname{Re}\lambda \geq \omega/(1-\varepsilon\omega)\ . \tag{6.1.13}$$

If $\varepsilon|\lambda| \geq 1$ and $\operatorname{Re}\lambda \geq 0$ then

$$\frac{\varepsilon|\lambda|+\kappa}{|1+\varepsilon\lambda|} \leq \frac{\varepsilon|\lambda|+\kappa}{\varepsilon|\lambda|} \leq \kappa+1 . \tag{6.1.14}$$

Now the assertion follows since $|1+\varepsilon\lambda| \geq 1$ for $\operatorname{Re}\lambda \geq 0$. ■

6.1.3 Corollary *Suppose that $A \in \mathcal{C}(E)$, $[\operatorname{Re} z \geq 0] \subset \rho(-A)$, and there exists $\kappa > 0$ such that*

$$\|\lambda(\lambda+A)^{-1}\| \leq \kappa , \qquad \operatorname{Re}\lambda \geq 0 .$$

Then there are $\vartheta := \vartheta(\kappa) \in (0,\pi/2)$ and $c := c(\kappa) > 0$ such that $\Sigma_\vartheta \subset \rho(-A_\varepsilon)$ and

$$\|\lambda(\lambda+A_\varepsilon)^{-1}\| \leq c , \qquad \lambda \in \Sigma_\vartheta ,$$

for $\varepsilon > 0$.

Proof This is an easy consequence of the first part of the proof of Proposition I.1.4.1 by observing that for the analogue of (I.1.4.2) (where the term containing $\|x\|_1$ is omitted) estimate (6.1.11) suffices. ■

6.2 Approximations of Evolution Operators

Let (E_0, E_1) be a densely injected Banach couple, let J be a perfect subinterval of $\mathbb{R}^+$ containing 0, and suppose that $\rho \in (0,1)$, $\kappa \geq 1$, and $\omega > 0$. Also suppose that

$$A \in C^\rho\big(J, \mathcal{H}(E_1, E_0, \kappa, \omega)\big) . \tag{6.2.1}$$

Then the Yosida approximations $\{\, A_\varepsilon(s) \in \mathcal{L}(E_0) \ ; \ 0 < \varepsilon < 1/\omega \,\}$ are well-defined for $s \in J$. Since the maps

$$\mathcal{H}(E_1, E_0, \kappa, \omega) \to \mathcal{L}(E_0, E_1) , \quad B \mapsto (1+\varepsilon B)^{-1}$$

are analytic for $0 < \varepsilon < 1/\omega$, it follows from (6.2.1) that $A_\varepsilon \in C^\rho\big(J, \mathcal{L}(E_0)\big)$ for $0 < \varepsilon < 1/\omega$. Hence, thanks to Remark I.1.2.1(b) and Corollary 4.4.2, there exists a unique parabolic evolution operator U_{A_ε} for A_ε possessing E_0 as a regularity subspace. It is the purpose of the following considerations to show that, given $(t,s) \in J_\Delta$,

$$U_{(\omega+A)_\varepsilon-\omega}(t,s) \to U_A(t,s) \qquad \text{in } \mathcal{L}_s(E_0)$$

as $\varepsilon \to 0$.

Let

$$B := \omega + A$$

and note that (6.1.2) implies

$$[\operatorname{Re} z \geq 0] \subset \rho\big(-B(s)\big) , \qquad s \in J , \tag{6.2.2}$$

and

$$\left\|\left(\lambda+B(s)\right)^{-1}\right\|_{\mathcal{L}(E_0)}\leq\frac{\kappa}{|\lambda+\omega|}\leq\frac{\kappa}{|\lambda|}\ ,\qquad \operatorname{Re}\lambda\geq 0\ ,\quad s\in J\ . \tag{6.2.3}$$

Hence the Yosida approximations

$$\left\{\,B_\varepsilon(s):=\left(B(s)\right)_\varepsilon\ ;\ \varepsilon>0\,\right\}$$

are well-defined for $s\in J$. The following lemma gives uniform estimates for the semigroups generated by $-B_\varepsilon(s)$.

6.2.1 Lemma *There exists $c>0$, depending on κ and ω only, such that*

$$\left\|\left[B_\varepsilon(s)\right]^k e^{-tB_\varepsilon(s)}\right\|_{\mathcal{L}(E_0)}\leq ct^{-k}$$

for $t>0,\ \ s\in J,\ \ \varepsilon>0$, and $k\in\{0,1\}$.

Proof Corollary 6.1.3 and (6.2.3) imply the existence of $\vartheta\in(0,\pi/2)$ and of a constant c with

$$\left\|\lambda\left(\lambda+B_\varepsilon(s)\right)^{-1}\right\|_{\mathcal{L}(E_0)}\leq c\ ,\qquad \lambda\in\Sigma_\vartheta\ ,\quad s\in J\ ,\quad \varepsilon>0\ .$$

Since

$$\left[B_\varepsilon(s)\right]^k e^{-tB_\varepsilon(s)}=\frac{(-1)^k}{2\pi i}\int_\Gamma \lambda^k e^{t\lambda}\left(\lambda+B_\varepsilon(s)\right)^{-1}d\lambda\ ,$$

Γ being a piece-wise smooth curve running in $\dot{\Sigma}_\vartheta$ from $\infty e^{-i(\vartheta+\pi/2)}$ to $\infty e^{i(\vartheta+\pi/2)}$, the assertion follows from Lemma 4.1.1. ■

By (6.2.1) and Remark I.1.2.1(a),

$$\omega\left\|\left(\omega+A(s)\right)^{-1}\right\|_{\mathcal{L}(E_0)}=\omega\left\|B^{-1}(s)\right\|_{\mathcal{L}(E_0)}\leq\kappa\ ,\qquad s\in J\ .$$

Hence $0\in\rho\left(B_\varepsilon(s)\right)$ and

$$B_\varepsilon^{-1}(s)=\left(1+\varepsilon B(s)\right)B^{-1}(s)\ ,\qquad s\in J\ ,\quad \varepsilon>0\ ,$$

so that $\left[B_\varepsilon(s)-B_\varepsilon(t)\right]B_\varepsilon^{-1}(s)$ is well-defined for $s,t\in J$ and $\varepsilon>0$.

6.2.2 Lemma *Given $s,t\in J$,*

$$\left\|\left[B_\varepsilon(s)-B_\varepsilon(t)\right]B_\varepsilon^{-1}(s)\right\|_{\mathcal{L}(E_0)}\leq\kappa^2[A]_{\rho,J}\,|s-t|^\rho\ ,\qquad \varepsilon>0\ .$$

Proof Observe that $\varepsilon B_\varepsilon = 1 - (1+\varepsilon B)^{-1}$ for $\varepsilon > 0$. Hence

$$\begin{aligned} B_\varepsilon(s) - B_\varepsilon(t) &= \varepsilon^{-1}\big[\big(1+\varepsilon B(t)\big)^{-1} - \big(1+\varepsilon B(s)\big)^{-1}\big] \\ &= \big(1+\varepsilon B(t)\big)^{-1}\big(B(s)-B(t)\big)\big(1+\varepsilon B(s)\big)^{-1} . \end{aligned}$$

Consequently,

$$\big[B_\varepsilon(s) - B_\varepsilon(t)\big]B_\varepsilon^{-1}(s) = \big(1+\varepsilon B(t)\big)^{-1}\big[A(s)-A(t)\big]B^{-1}(s) .$$

Since (6.2.3) and Remark I.1.2.1(a) imply

$$\big\|\big(1+\varepsilon B(t)\big)^{-1}\big\|_{\mathcal{L}(E_0)} \vee \|B^{-1}(t)\|_{\mathcal{L}(E_0,E_1)} \le \kappa ,$$

the assertion follows. ∎

After these preparations we can prove the following uniform estimate for the evolution operator U_{B_ε}.

6.2.3 Lemma *Suppose that $[A]_{\rho,J} \le \eta$. Then there exist positive constants $c := c(\kappa,\omega,\eta)$ and $\mu := \mu(\kappa,\omega,\eta,\rho)$ such that*

$$\|U_{B_\varepsilon}(t,s)\|_{\mathcal{L}(E_0)} \le ce^{\mu(t-s)} , \qquad (t,s)\in J_\Delta , \quad \varepsilon > 0 .$$

Proof Using the notations of Section 4.3, it follows from Lemma 6.2.1 that there exists $c_0 := c_0(\kappa,\omega)$ such that $a_{B_\varepsilon} \in \mathfrak{K}(E_0,0)$ and

$$\|a_{B_\varepsilon}(t,s)\|_{\mathcal{L}(E_0)} \le c_0 , \qquad (t,s)\in J_\Delta , \quad \varepsilon > 0 . \tag{6.2.4}$$

Hence, putting $\mu_0 := \big(\Gamma(\rho)c_0\eta\big)^{1/\rho}$, we infer from

$$k_{B_\varepsilon}(t,s) = -\big[B_\varepsilon(t) - B_\varepsilon(s)\big]B_\varepsilon^{-1}(s)B_\varepsilon(s)e^{-(t-s)B_\varepsilon(s)}$$

and from Lemmas 6.2.1 and 6.2.3 that $k_{B_\varepsilon} \in \mathfrak{K}(E_0, 1-\rho)$ and

$$(t-s)^{1-\rho}\,\|k_{B_\varepsilon}(t,s)\|_{\mathcal{L}(E_0)} \le c , \qquad (t,s)\in J_\Delta^* , \quad \varepsilon > 0 .$$

Thus (cf. the proof of Lemma 4.3.1) $w_{B_\varepsilon} \in \mathfrak{K}(E_0,1-\rho)$ and

$$(t-s)^{1-\rho}\,\|w_{B_\varepsilon}(t,s)\|_{\mathcal{L}(E_0)} \le ce^{\mu(t-s)} , \qquad (t,s)\in J_\Delta^* , \quad \varepsilon > 0 , \tag{6.2.5}$$

where $\mu := \mu_0 + 1$. Now we infer from (6.2.4), (6.2.5), (3.1.8), and (3.1.9) that $a_{B_\varepsilon} \star w_{B_\varepsilon} \in \mathfrak{K}(E_0,-\rho)$ and

$$(t-s)^{-\rho}\,\|a_{B_\varepsilon}\star w_{B_\varepsilon}(t,s)\|_{\mathcal{L}(E_0)} \le ce^{\mu(t-s)} , \qquad (t,s)\in J_\Delta , \quad \varepsilon > 0 . \tag{6.2.6}$$

Finally, the assertion follows from (6.2.4) and (6.2.6) since $U_{B_\varepsilon} = a_{B_\varepsilon} + a_{B_\varepsilon}\star w_{B_\varepsilon}$ by the proof of Theorem 4.4.1. ∎

Now we are ready for the proof of the main result of this section, namely the approximation assertion:

6.2.4 Theorem *Given* $(t,s) \in J_\Delta$,

$$U_{(\omega+A)_\varepsilon - \omega}(t,s) \to U_A(t,s) \qquad \textit{in } \mathcal{L}_s(E_0)$$

as $\varepsilon \to 0$.

Proof Thanks to Remark 2.1.2(d) it suffices to show that $U_{B_\varepsilon}(t,s) \to U_B(t,s)$ in $\mathcal{L}_s(E_0)$ as $\varepsilon \to 0$. Moreover, we can assume that $(t,s) \in J_\Delta^*$ and, by replacing J by $[0,t]$, that J is compact.

By integrating the identity

$$\partial_\tau \big[U_{B_\varepsilon}(t,\tau)U_B(\tau,s)\big] = U_{B_\varepsilon}(t,\tau)\big[B_\varepsilon(\tau) - B(\tau)\big]U_B(\tau,s)\ , \qquad s < \tau < t\ ,$$

it follows that

$$U_{B_\varepsilon}(t,s)x - U_B(t,s)x = -\int_s^t U_{B_\varepsilon}(t,\tau)\big[B_\varepsilon(\tau) - B(\tau)\big]U_B(\tau,s)x\, d\tau \tag{6.2.7}$$

for $x \in E_0$. Note that $B_\varepsilon(\tau) - B(\tau) = \big[\big(1+\varepsilon B(\tau)\big)^{-1} - 1\big]B(\tau)$ and (4.4.3) imply

$$\big\|\big[B_\varepsilon(\tau) - B(\tau)\big]U_B(\tau,s)\big\|_{\mathcal{L}(E_1,E_0)} \le c(1+\kappa)\,\|B\|_{C(J,\mathcal{L}(E_1,E_0))}\ , \qquad s \le \tau \le t\ .$$

Hence we deduce from Lemma 6.2.3 that the integrand in (6.2.7) is bounded in the norm of E_0, uniformly for $\varepsilon \in (0,1/\omega)$, provided $x \in E_1$. Consequently, given $x \in E_1$,

$$U_{B_\varepsilon}(t,s)x \to U_B(t,s)x \qquad \text{in } E_0$$

as $\varepsilon \to 0$, thanks to (6.1.5) and the dominated convergence theorem. Now the assertion follows from Lemma 6.2.3 and the density of E_1 in E_0. ■

It should be remarked that the 'Yosida approximation technique' was first used by Kato [Kat61] to construct parabolic evolution operators (cf. Subsection IV.2.3). Moreover, Theorem 6.2.4 can be improved to show that the convergence is locally uniform with respect to $(t,s) \in J_\Delta$ (cf. [DaK92]).

6.3 Invariance

In this section we are interested in conditions guaranteeing that a closed convex subset X of E_0 is mapped into itself by the evolution operator U_A, that is,

$$U_A(t,s)(X) \subset X\ , \qquad (t,s) \in J_\Delta\ .$$

For this we first prove the following simple technical result:

6.3.1 Lemma *Let Ω be a nonempty set and let μ be a probability measure on (a σ-algebra of subsets of) Ω. Let X be a nonempty closed convex subset of some Banach space E. If $f \in \mathcal{L}_1(\Omega, E)$ and $f(\omega) \in X$ for μ-a.a. $\omega \in \Omega$ then $\int_\Omega f\,d\mu \in X$.*

Proof Suppose that $e' \in E'$ and $\alpha \in \mathbb{R}$ are such that $X \subset [\operatorname{Re}\langle e', x\rangle \le \alpha]$. Then

$$\operatorname{Re}\Big\langle e', \int_\Omega f\,d\mu\Big\rangle = \int_\Omega \operatorname{Re}\langle e', f\rangle\,d\mu \le \alpha ,$$

that is, $\int_\Omega f\,d\mu \in [\operatorname{Re}\langle e', x\rangle \le \alpha]$. Now the assertion follows since, by the Hahn-Banach theorem, X is the intersection of all closed real half-spaces $[\operatorname{Re}\langle e', x\rangle \le \alpha]$ containing X. ■

For the reader's convenience we include the following well-known invariance theorem from semigroup theory (e.g., [HiP57, Section 11.7]):

6.3.2 Theorem *Suppose that $A \in \mathcal{G}(E, M, \sigma)$. Then*

$$e^{-tA}(X) \subset X , \qquad t \ge 0 , \tag{6.3.1}$$

iff

$$(1 + \varepsilon A)^{-1}(X) \subset X , \qquad 0 < \varepsilon < 1/\sigma^+ . \tag{6.3.2}$$

Proof Recall from semigroup theory that

$$(1 + \varepsilon A)^{-1}x = \int_0^\infty e^{-t}e^{-\varepsilon At}x\,dt , \qquad x \in E , \quad 0 < \varepsilon < 1/\sigma^+ . \tag{6.3.3}$$

Thus, letting $\mu := e^{-t}dt$ on $\Omega := \mathbb{R}^+$, Lemma 6.3.1 and (6.3.1) imply (6.3.2). On the other hand,

$$e^{-tA}x = \lim_{k\to\infty}\Big(1 + \frac{t}{k}A\Big)^{-k}x , \qquad t \ge 0 , \quad x \in E . \tag{6.3.4}$$

Hence (6.3.2) implies (6.3.1). ■

Throughout the remainder of this section we assume that

$$\left.\begin{array}{l}(E_0, E_1) \text{ is a densely injected Banach couple,}\\ J \text{ is a perfect subinterval of } \mathbb{R}^+ \text{ containing zero,}\\ \text{and } X \text{ is a nonempty closed convex subset of } E_0.\end{array}\right\}$$

The following invariance theorem is a partial extension of Theorem 6.3.2 to the case of parabolic evolution operators. The main idea of its proof is essentially due to Daners and Koch (cf. [DaK92, Proposition 10.9]). Recall that $s(-A)$ denotes the spectral bound of $A \in \mathcal{C}(E_0)$.

6.3.3 Theorem *Suppose that $A \in C^\rho\big(J, \mathcal{H}(E_1, E_0)\big)$ for some $\rho \in (0,1)$ and that*

$$s\big(-A(t)\big) \leq -\sigma < 0\ , \qquad t \in J\ . \tag{6.3.5}$$

Then

$$\big(1 + \varepsilon A(t)\big)^{-1}(X) \subset X\ , \qquad t \in J\ , \quad \varepsilon > 0\ , \tag{6.3.6}$$

implies

$$U_A(t,s)(X) \subset X\ , \qquad (t,s) \in J_\Delta\ . \tag{6.3.7}$$

Proof Let $(t_1, s) \in J_\Delta^*$ be fixed and replace J by $[0, t_1]$, again denoted by J. Then $A(J)$ is compact in $\mathcal{H}(E_1, E_0)$. Thus, thanks to Corollary I.1.3.2, there exist $\kappa \geq 1$ and $\omega_0 > 0$ such that $A \in C^\rho\big(J, \mathcal{H}(E_1, E_0, \kappa, \omega_0)\big)$.

Put $\omega := \sigma/2$. Then, given $t \in J$, Proposition I.1.4.2 guarantees the existence of $\kappa_t \geq 1$ and of an open neighborbood $\mathcal{A}_t$ of $A(t)$ in $\mathcal{L}(E_1, E_0)$ such that

$$-\omega + \mathcal{A}_t \subset \mathcal{H}(E_1, E_0, \kappa_t, \omega)\ .$$

By the compactness of $A(J)$ we can find $t_0, \ldots, t_n \in J$ such that $A(J) \subset \bigcup \mathcal{A}_{t_j}$. Thus, letting $\kappa := \max \kappa_{t_j}$, it follows that

$$\widetilde{A}(t) := -\omega + A(t) \in \mathcal{H}(E_1, E_0, \kappa, \omega)\ , \qquad t \in J\ .$$

Theorem 6.2.4 and Remark 2.1.2(d) imply that

$$U_{A_\varepsilon}(t,s) = U_{(\omega + \widetilde{A})_\varepsilon}(t,s) \to U_{\omega + \widetilde{A}}(t,s) = U_A(t,s)$$

in $\mathcal{L}_s(E_0)$ as $\varepsilon \to 0$. Thus it suffices to prove that (6.3.6) implies

$$U_{A_\varepsilon}(t,s)(X) \subset X\ , \qquad (t,s) \in J_\Delta\ , \quad \varepsilon > 0\ .$$

Recall that $u(\cdot, x) := U_{A_\varepsilon}(\cdot, s)x$ is, for each $s \in [0, t_1)$ and $x \in E_0$, the unique solution of the initial value problem

$$\dot{u} + A_\varepsilon(t)u = 0\ , \quad s < t \leq t_1\ , \qquad u(s) = x\ .$$

Since $A_\varepsilon = \varepsilon^{-1}\big(1 - (1 + \varepsilon A)^{-1}\big)$, we see that $u(\cdot, x)$ solves

$$\dot{u} + \varepsilon^{-1}u = \varepsilon^{-1}\big(1 + \varepsilon A(t)\big)^{-1}u\ , \quad s \leq t < t_1\ , \qquad u(s) = x\ .$$

Hence $u(\cdot, x)$ is given by the variation-of-constants formula with respect to the semigroup $\{\, e^{-t/\varepsilon}1_{E_0}\ ;\ t \geq 0 \,\}$, that is,

$$u(t,x) = e^{-(t-s)/\varepsilon}x + \varepsilon^{-1}\int_s^t e^{-(t-\tau)/\varepsilon}\big(1 + \varepsilon A(\tau)\big)^{-1}u(\tau, x)\, d\tau\ , \qquad s \leq t \leq t_1\ .$$

Let $v(t) := e^{-(t-s)/\varepsilon}x$ for $t \in I := [s, t_1]$ and denote by K the Volterra integral operator in $C(I, E_0)$ with kernel

$$k(t,\tau) := \varepsilon^{-1} e^{-(t-\tau)/\varepsilon} \big(1 + \varepsilon A(\tau)\big)^{-1} , \qquad (t,\tau) \in I_\Delta .$$

Since $\|k(t,\tau)\|_{\mathcal{L}(E_0)} \le \kappa/\varepsilon$ for $(t,\tau) \in I_\Delta$, it follows from (3.2.1) that K is quasi-nilpotent. Hence $(1-K)^{-1}$ exists and is given by the Neumann series. Thus

$$u(\cdot, x) = (1-K)^{-1} v = \sum_{k=0}^{\infty} K^n v .$$

From this we deduce, letting

$$f(w) := v + Kw , \qquad w \in C(I, E_0) ,$$

that $u(\cdot, x) = \lim u_k$ in $C(I, E_0)$, where

$$u_0 := v , \quad u_{k+1} = f(u_k) , \quad k \in \mathbb{N} .$$

Note that

$$f(w) = \alpha x + (1-\alpha) M w ,$$

where $\alpha(t) := \alpha(t,s) := e^{-(t-s)/\varepsilon}$ and

$$Mw(t) := \int_s^t \varphi(t,\tau) \big(1 + \varepsilon A(\tau)\big)^{-1} w(\tau)\, d\tau$$

with

$$\varphi(t,\tau) := \big[1 - \alpha(t)\big]^{-1} \varepsilon^{-1} e^{-(t-\tau)/\varepsilon} , \qquad s \le \tau \le t .$$

Hence, letting $\mu_t := \varphi(t, \cdot)\, d\tau$ on $\Omega := [s,t]$ for $t \in I$, it follows from (6.3.6) and Lemma 6.3.1 that $Mw(t) \in X$ for $t \in I$, provided $w \in C(I, X)$. Thus

$$f\big(C(I,X)\big) \subset C(I,X) \qquad \text{if} \quad x \in X ,$$

and, consequently, $u(t,x) = \lim u_k(t) \in X$ for $t \in I$ if $x \in X$. ∎

A nonempty subset C of a vector space is a **cone** if $\mathbb{R}^+ C \subset C$. Thus every cone contains 0. A cone is **proper** if $C \cap (-C) = \{0\}$.

The following theorem shows that we can omit condition (6.3.5) if X is a cone.

6.3.4 Theorem *Suppose that $A \in C^\rho\big(J, \mathcal{H}(E_1, E_0)\big)$ for some $\rho \in (0,1)$. If X is a cone,*

$$e^{-tA(s)}(X) \subset X , \qquad t \ge 0 , \quad s \in J , \tag{6.3.8}$$

implies

$$U_A(t,s)(X) \subset X , \qquad (t,s) \in J_\Delta .$$

Proof Having fixed $(t_1, s) \in J_\Delta$, we can assume that J is compact. Hence, by invoking Corollary I.1.3.2, we can assume that $A \in C^\rho\big(J, \mathcal{H}(E_1, E_0, \kappa, \omega)\big)$ for some $\kappa \geq 1$ and $\omega > 0$. Since $s(-A) < \omega$ we see that

$$s\big(-(2\omega + A(t))\big) = -2\omega + s\big(-A(t)\big) \leq -\omega \ , \qquad t \in J \ .$$

Let $B := 2\omega + A$. Since $e^{-tB(s)} = e^{-2\omega t} e^{-tA(s)}$ we infer from (6.3.8) and the fact that X is a cone that $e^{-tB(s)}(X) \subset X$ for $t \geq 0$ and $s \in J$. Thus, thanks to Theorem 6.3.2,

$$\big(1 + \varepsilon B(t)\big)^{-1}(X) \subset X \ , \qquad t \in J \ , \quad \varepsilon > 0 \ ,$$

and Theorem 6.3.3 guarantees that

$$U_B(t, s)(X) \subset X \ , \qquad (t, s) \in J_\Delta \ .$$

Now the assertion follows from Remark 2.1.2(d) and from $\mathbb{R}^+ X \subset X$. ■

Of course, thanks to Theorem 6.3.2, condition (6.3.6) can also be replaced by the equivalent requirement (6.3.8).

6.4 Orderings and Positivity

Let E be a real vector space and let P be a convex cone in E. Then we define a reflexive and transitive binary relation in E, a **preorder** $\leq$, by putting

$$x \leq y \qquad \text{iff} \qquad y - x \in P \ .$$

It is obvious that this preorder is **linear**, that is, $x \leq y$ implies $x + z \leq y + z$ and $\alpha x \leq \alpha y$ for $z \in E$ and $\alpha \in \mathbb{R}^+$. Conversely, given a linear preorder $\leq$ in E, we put $P := \{\, x \in E \ ; \ x \geq 0 \,\}$. Then P is a convex cone in E, the **positive cone** of the preordered vector space E. The preorder is an **order**, that is, $x \leq y$ and $y \leq x$ imply $x = y$ iff its positive cone is proper. By an **ordered vector space** $(E, \leq)$ we mean a real vector space E together with a linear order $\leq$. Instead of $(E, \leq)$ we often write (E, P), where P is the positive proper cone. Of course, $x \geq y$ means that $y \leq x$, and we write $x < y$ iff $x \leq y$ and $x \neq y$.

If $E := (E, P)$ is an ordered vector space which, in addition, is a *TVS* then E is said to be an **ordered** *TVS* (*OTVS*) if its positive cone is closed. Of course, if E is an *OTVS* such that it is a *LCS* (resp. Fréchet space, resp. Banach space, etc.) with its topology then it is said to be an **ordered locally convex space** (*OLCS*) (resp. **ordered Fréchet space**, resp. **ordered Banach space** (*OBS*), etc.). Note that $\mathbb{R}$ is an *OBS* with its **natural order** induced by the positive cone $\mathbb{R}^+$.

Let (E, P) and (F, Q) be preordered vector spaces. Then a linear map T from E to F is **positive**, in symbols, $T \geq 0$, if $T(P) \subset Q$, that is, $Tx \geq 0$ whenever

$x \geq 0$ (where we usually use the same symbol for the preorder in either of the spaces, if no confusion seems likely). Now let (E, P) and (F, Q) be *OTVSs*. Then

$$\mathcal{L}^+(E, F) := \{ T \in \mathcal{L}(E, F) \ ; \ T \geq 0 \}$$

is easily seen to be a closed cone in $\mathcal{L}(E, F)$ which is said to induce the **natural preorder** in $\mathcal{L}(E, F)$. It is a proper cone iff P is total, that is, $\overline{P - P} = E$ (e.g., [Sch71, Proposition V.5.1]). Of course, $\mathcal{L}^+(E) := \mathcal{L}^+(E, E)$.

Let E be an *OLCS*. Then $P' := \mathcal{L}^+(E, \mathbb{R})$ is the **dual cone** of E'. Thus (E', P') is an *OLCS* iff P is total.

Now suppose that E is a Banach space, preordered by a closed convex cone, and $A \in \mathcal{G}(E)$. Then the **semigroup** $\{ e^{-tA} \ ; \ t \geq 0 \}$ is **positive** if $e^{-tA} \geq 0$ for $t \geq 0$. A linear operator A in E is **resolvent positive** if $(\omega, \infty) \subset \rho(-A)$ for some $\omega \geq 0$ and

$$(1 + \varepsilon A)^{-1} \geq 0 \ , \qquad 0 < \varepsilon < 1/\omega \ .$$

If $A \in \mathcal{G}(E)$, these two concepts are equivalent, as is seen from the next assertion:

6.4.1 Theorem *Let E be a Banach space, preordered by a closed convex cone, and suppose that $A \in \mathcal{G}(E)$. Then the semigroup $\{ e^{-tA} \ ; \ t \geq 0 \}$ is positive iff A is resolvent positive.*

Proof This is an immediate consequence of Theorem 6.3.2. ■

Now suppose that J is a perfect interval and $A \colon J \to \mathcal{C}(E)$. Also suppose that U is a parabolic evolution operator for A (with some regularity subspace). Then U is said to be **positive** if $U(t, s) \geq 0$ for $(t, s) \in J_\Delta$.

6.4.2 Theorem *Let (E_0, E_1) be a densely injected Banach couple, let J be a perfect subinterval of $\mathbb{R}^+$ containing 0, and suppose that*

$$A \in C^\rho\big(J, \mathcal{H}(E_1, E_0)\big)$$

for some $\rho \in (0, 1)$. Also suppose that E_0 is preordered by a closed convex cone. Then the unique parabolic evolution operator U_A for A is positive, provided each one of the semigroups $\{ e^{-tA(s)} \ ; \ t \geq 0 \}$, $s \in J$, is positive.

Proof This follows from Theorem 6.3.4. ■

Chapter III

Maximal Regularity

In Chapter II we studied the solvability of the Cauchy problem for linear parabolic evolution equations by taking into consideration, in particular, the smoothing effects of analytic semigroups. The regularizing properties deduced from these effects have important implications for the qualitative theory of quasilinear parabolic equations. However, they have the disadvantage that, in general, the derivative of a solution is less regular than the right-hand side of the corresponding parabolic evolution equation. This 'loss of regularity' leads to some difficulties in the treatment of nonlinear evolution equations. For this reason we investigate in this chapter situations where such a loss of regularity does not occur, that is, cases of 'maximal regularity'.

In Section 1 we generalize the concept of solutions of linear parabolic evolution equations to admit distributional solutions belonging to suitable Sobolev spaces. In addition, we introduce the general concept of maximal regularity. The remaining sections are devoted to the study of concrete cases of maximal regularity. Namely, in Section 2 we show that we can always have maximal regularity if we require all functions to be Hölder continuous with a prescribed singularity of the Hölder seminorm at zero. If we wish to consider continuous functions only, the class of admissible Banach spaces has to be restricted considerably — in dependence of the generator of the analytic semigroup under consideration. This is shown in Section 3. Finally, in Section 4 we prove that we can have maximal regularity in the Sobolev space setting provided we consider a restricted class of Banach spaces and, in addition, restrict the class of generators as well.

1 General Principles

In this section we give a general discussion of the concept of maximal regularity for linear evolution equations. Concrete realizations are discussed in Sections 2–4. For the reader's convenience we begin by recalling some known facts about Sobolev spaces and absolutely continuous functions.

1.1 Sobolev Spaces

Let X be an open subset of $\mathbb{R}^n$ and E a Banach space. Then $\mathcal{D}(X,E)$ is the space of E-valued **test functions** on X, that is, of all E-valued C^∞-functions on X with compact supports. We equip $\mathcal{D}(X,E)$ with the usual inductive limit topology, and $\mathcal{D}(X) := \mathcal{D}(X,\mathbb{K})$. Then

$$\mathcal{D}'(X,E) := \mathcal{L}\big(\mathcal{D}(X),E\big)$$

is the space of E-valued **distributions** on X, endowed with the topology of uniform convergence on bounded subsets of $\mathcal{D}(X)$. Thus $\mathcal{D}'(X,E)$ is a *LCS* and

$$\mathcal{D}'(X) := \mathcal{D}'(X,\mathbb{K}) = \mathcal{D}(X)' .$$

Recall that $\mathcal{D}(X)$ and $\mathcal{D}'(X)$ are Montel spaces, that is, they are barreled and bounded subsets are relatively compact. Thus $\mathcal{D}(X)$ (consequently, $\mathcal{D}'(X)$ as well) is reflexive.

As usual, given a multiindex $\alpha \in \mathbb{N}^n$, the distributional derivative $\partial^\alpha T$ of $T \in \mathcal{D}'(X,E)$ is defined by

$$(\partial^\alpha T)(\varphi) := (-1)^{|\alpha|}T(\partial^\alpha \varphi) , \qquad \varphi \in \mathcal{D}(X) .$$

It is easily verified that

$$\partial^\alpha \in \mathcal{L}\big(\mathcal{D}'(X,E)\big) .$$

Given $u \in L_{1,\mathrm{loc}}(X,E)$, we put

$$T_u(\varphi) := \int_X \varphi u \, dx , \qquad \varphi \in \mathcal{D}(X) .$$

Since $T_u : \mathcal{D}(X) \to E$ is obviously linear, and since

$$\|T_u(\varphi)\| \le \|u\|_{1,K} \, \|\varphi\|_{\infty,K} , \qquad K \subset\subset X , \quad \varphi \in \mathcal{D}(X) , \quad \mathrm{supp}(\varphi) \subset K ,$$

where, given any measurable subset Y of X,

$$\|\cdot\|_p := \|\cdot\|_{p,Y} , \qquad 1 \le p \le \infty ,$$

is the norm in $L_p(Y,E)$, it follows that $T_u \in \mathcal{D}'(X,E)$. As in the classical case, it is not difficult to see that the map

$$L_{1,\mathrm{loc}}(X,E) \to \mathcal{D}'(X,E) , \quad u \mapsto T_u \tag{1.1.1}$$

is linear and injective, Thus we identify $L_{1,\mathrm{loc}}(X,E)$ with its image in $\mathcal{D}'(X,E)$ under the map (1.1.1) so that

$$L_{1,\mathrm{loc}}(X,E) \hookrightarrow \mathcal{D}'(X,E) .$$

The elements of $L_{1,\mathrm{loc}}(X,E)$ are the **regular** E-valued distributions on X.

Given $k \in \mathbb{N}$ and $p \in [1, \infty]$, we define the **Sobolev space**

$$W_p^k(X, E) := \big(W_p^k(X, E); \|\cdot\|_{k,p}\big)$$

to be the subspace of $L_p(X, E)$ consisting of all u for which $\partial^\alpha u \in L_p(X, E)$ for $|\alpha| \leq k$, and

$$\|\cdot\|_{k,p} := \|\cdot\|_{k,p,X} := \begin{cases} \Big(\sum\limits_{|\alpha|\leq k} \|\partial^\alpha \cdot\|_p^p\Big)^{1/p}, & 1 \leq p < \infty , \\ \max\limits_{|\alpha|\leq k} \|\partial^\alpha \cdot\|_\infty , & p = \infty . \end{cases}$$

Since $L_p(X, E) \hookrightarrow L_{1,\mathrm{loc}}(X, E)$, this definition is meaningful and it is easily verified that $W_p^k(X, E)$ is a Banach space. Note that $W_p^0 = L_p$.

We also define the Fréchet spaces

$$W_{p,\mathrm{loc}}^k(X, E) , \qquad 1 \leq p \leq \infty , \quad k \in \mathbb{N} ,$$

to consist of all $u \in L_{p,\mathrm{loc}}(X, E)$ such that $\varphi u \in W_p^k(X, E)$ for $\varphi \in \mathcal{D}(X)$. Then $\{\, u \mapsto \|\varphi u\|_{k,p} \ ; \ \varphi \in \mathcal{D}(X) \,\}$ is a generating family of seminorms for the topology of $W_{p,\mathrm{loc}}^k(X, E)$.

Now let J be a perfect subinterval of $\mathbb{R}$. Then we put

$$W_p^k(J, E) := W_p^k(\mathring{J}, E) , \qquad 1 \leq p \leq \infty , \quad k \in \mathbb{N} .$$

Moreover, given $k \in \mathbb{N}$ and $p \in [1, \infty]$, we write $u \in W_{p,\mathrm{loc}}^k(J, E)$ iff $u \in W_{p,\mathrm{loc}}^k(\mathring{J}, E)$ and $\partial^j u \in L_{p,\mathrm{loc}}(J, E)$ for $0 \leq j \leq k$. Note that this means that $\partial^j u \in L_p(I, E)$ for each compact subinterval I of J (and not only of $\mathring{J}$).

Of course, if $E = \mathbb{K}$, we omit this symbol everywhere, if no confusion seems likely, and denote by

$$\langle \cdot, \cdot \rangle := \langle \cdot, \cdot \rangle_{\mathcal{D}}$$

the duality pairing between $\mathcal{D}'(X)$ and $\mathcal{D}(X)$.

1.2 Absolutely Continuous Functions

Let E be a Banach space. A function $u \colon J \to E$ is **locally absolutely continuous** (on J) if there exists $v \in \mathcal{L}_{1,\mathrm{loc}}(J, E)$ such that, given $a \in J$,

$$u(t) = u(a) + \int_a^t v(\tau)\, d\tau , \qquad t \in J , \tag{1.2.1}$$

(Here and in the following, $\mathcal{L}_p$ and $\mathcal{L}_{p,\mathrm{loc}}$ are the vector spaces of all measurable functions $u \colon J \to E$ such that $t \mapsto |u(t)|^p$ is integrable, respectively locally integrable, whereas we use L_p and $L_{p,\mathrm{loc}}$ to denote the respective Banach and Fréchet spaces of equivalence classes of these functions.) The function u is **absolutely continuous** (on J) if there exists $v \in \mathcal{L}_1(J,E)$ such that (1.2.1) is true. If u is locally absolutely continuous, it is continuous, possesses a.e. a derivative $\dot{u}$, and $\dot{u} = v$ a.e., thanks to Lebesgue's differentiation theorem. Moreover, given $a, b \in J$ with $a < b$ and a sequence (J_k) of pair-wise disjoint open subintervals, $J_k := (a_k, b_k)$, of (a,b), it follows from (1.2.1) that

$$\sum_k |u(b_k) - u(a_k)| \le \int_{\bigcup_k J_k} |v| \, d\tau \ .$$

Thus Lebesgue's dominated convergence theorem implies that, given $\varepsilon > 0$, there exists $\delta > 0$ such that

$$\sum_k |u(b_k) - u(a_k)| < \varepsilon \tag{1.2.2}$$

for each pair $a, b \in J$ with $a < b$ and each sequence of pair-wise disjoint subintervals $J_k := (a_k, b_k)$ of (a,b) satisfying

$$\sum_k (b_k - a_k) < \delta \ . \tag{1.2.3}$$

If E is reflexive, $u \colon J \to E$ is locally absolutely continuous iff (1.2.2), (1.2.3) are true, thanks to a theorem of Kōmura [Kōm67].

1.2.1 Remark The above definition of absolute continuity differs from the classical one of the finite-dimensional case which is based on property (1.2.2), (1.2.3). In general Banach spaces the two definitions are not equivalent (e.g., [Yos65, Section V.5]). Since the integral characterization (1.2.1) is the more useful one, we have decided to take it as definition of an absolutely continuous function. ■

In the following, we adopt the usual convention not to distinguish (in notation) between a measurable function $J \to E$ and its equivalence class modulo functions vanishing almost everywhere.

1.2.2 Theorem *Suppose that* $1 \le p \le \infty$. *Then*

$$u \in W_p^1(J,E) \ [\textit{resp. } W_{p,\mathrm{loc}}^1(J,E)]$$

iff u *is locally absolutely continuous and*

$$u, \dot{u} \in L_p(J,E) \quad [\textit{resp. } L_{p,\mathrm{loc}}(J,E)] \ .$$

If this is the case, $\partial u = \dot{u}$.

Proof See [Bar76, Section I.2.2], for example. ■

1.3 Generalized Solutions

Let (E_0, E_1) be a densely injected Banach couple. Suppose that

$$A \in \mathcal{L}_{\infty,\mathrm{loc}}\big(J, \mathcal{L}(E_1, E_0)\big) \quad \text{and} \quad (x, f) \in E_0 \times L_{1,\mathrm{loc}}(J, E_0) .$$

Then, given $p \in [1, \infty]$, we say u is a $W^1_{p,\mathrm{loc}}$**-solution** (a **generalized solution**) of the linear Cauchy problem

$$\dot{u} + A(t)u = f(t) , \quad t \in J \cap (s, \infty) , \qquad u(s) = x \tag{1.3.1$_{(s,x,A,f)}$}$$

if

$$u \in W^1_{p,\mathrm{loc}}(J^s, E_0) \cap L_{p,\mathrm{loc}}(J^s, E_1) \tag{1.3.2}$$

and

$$(\partial + A)u = f , \quad u(s) = x , \tag{1.3.3}$$

where $(Au)(t) := A(t)u(t)$ for $t \in J^s := J \cap [s, \infty)$. Note that $u(s) \in E_0$ is well-defined by Theorem 1.2.2. Of course, u is a W^1_p**-solution** of $(1.3.1)_{(s,x,A,f)}$ if

$$u \in W^1_p(J^s, E_0) \cap L_p(J^s, E_1) \tag{1.3.4}$$

and (1.3.3) is true.

Suppose that there exists a parabolic evolution operator U_A for A. The following proposition shows that every $W^1_{p,\mathrm{loc}}$-solution is then a mild solution of $(1.3.1)_{(s,x,A,f)}$. Thus, in this case, $W^1_{p,\mathrm{loc}}$-solutions are unique, if they exist at all.

1.3.1 Proposition *Suppose $1 \leq p \leq \infty$, there exists a parabolic evolution operator for A, and u is a $W^1_{p,\mathrm{loc}}$-solution of the Cauchy problem $(1.3.1)_{(s,x,A,f)}$. Then*

$$u = u(\cdot, s, x, A, f) = (U_A x + U_A \star f)(\cdot, s) .$$

Proof Given $t \in J \cap (s, \infty)$, put $v(\tau) := U_A(t, \tau)u(\tau)$ for $s < \tau < t$. Then it is an easy consequence of (II.2.1.2), (II.2.1.8), (II.2.1.9), and Theorem 1.2.2 that

$$v \in W^1_{p,\mathrm{loc}}\big((s,t), E_0\big) \cap L_{p,\mathrm{loc}}\big((s,t), E_1\big)$$

and that

$$\dot{v}(\tau) = U_A(t, \tau)\big[A(\tau)u(\tau) + \dot{u}(\tau)\big] = U_A(t, \tau)f(\tau) , \qquad \text{a.e. } \tau \in (s, t) .$$

Now the assertion follows from (1.2.1). ∎

1.4 Trace Spaces

Let J be a perfect subinterval of $\mathbb{R}^+$ containing 0 and let E and $\mathbb{E}(J)$ be Banach spaces such that

$$\mathbb{E}(J) \hookrightarrow W^1_{1,\mathrm{loc}}(J, E) .$$

It follows from Theorem 1.2.2 that the **trace map**

$$\gamma := \gamma_0 : \mathbb{E}(J) \to E , \quad u \mapsto u(0)$$

is a well-defined continuous linear operator. We put

$$\gamma\mathbb{E}(J) := \mathrm{im}(\gamma) ,$$

equipped with the quotient topology of the Banach space $\mathbb{E}(J)/\ker(\gamma)$. Thus, $\gamma\mathbb{E}(J)$ is a Banach space with the norm

$$u \mapsto \|u\|_{\gamma\mathbb{E}(J)} = \inf\{\, \|v\|_{\mathbb{E}(J)} \;;\; v \in \mathbb{E}(J),\ \gamma v = u \,\} ,$$

the **trace space** of $\mathbb{E}(J)$.

Let J_1 be another subinterval of $\mathbb{R}^+$ with $J_1 \supset J$, and let $\mathbb{E}(J_1)$ be a Banach space such that $\mathbb{E}(J_1) \hookrightarrow W^1_{1,\mathrm{loc}}(J_1, E)$. Then the trace space $\gamma\mathbb{E}(J_1)$ is also well-defined. Suppose that $\mathbb{E}(J_1)$ is a **regular extension** of $\mathbb{E}(J)$, that is, there exist a continuous retraction r from $\mathbb{E}(J_1)$ onto $\mathbb{E}(J)$ and a corresponding coretraction r^c such that r is naturally induced by the 'point-wise restriction' $u \mapsto u|J$. Observe that this implies

$$ru_1(0) = u_1(0) , \qquad u_1 \in \mathbb{E}(J_1) ,$$

and

$$r^c u(0) = u(0) , \qquad u \in \mathbb{E}(J) .$$

Thus it is reasonable to expect that the trace spaces $\gamma\mathbb{E}(J)$ and $\gamma\mathbb{E}(J_1)$ are the same since, at least formally, the trace γu depends on the values of u near zero only. This is made precise in the following:

1.4.1 Proposition *If $\mathbb{E}(J_1)$ is a regular extension of $\mathbb{E}(J)$ then $\gamma\mathbb{E}(J_1) \doteq \gamma\mathbb{E}(J)$.*

Proof Let $x \in \gamma\mathbb{E}(J_1)$ be given. Then there exists $u_1 \in \mathbb{E}(J_1)$ with $\gamma u_1 = x$. Consequently, $ru_1 \in \mathbb{E}(J)$ and $\gamma r u_1 = x$. This shows that $\gamma\mathbb{E}(J_1) \subset \gamma\mathbb{E}(J)$ and

$$\|x\|_{\gamma\mathbb{E}(J)} \le \|ru_1\|_{\mathbb{E}(J)} \le \|r\|\, \|u_1\|_{\mathbb{E}(J_1)} .$$

Since this is true for every $u_1 \in \mathbb{E}(J_1)$ with $\gamma u_1 = x$,

$$\|x\|_{\gamma\mathbb{E}(J)} \le \|r\|\, \|x\|_{\gamma\mathbb{E}(J_1)} ,$$

that is, $\gamma\mathbb{E}(J_1) \hookrightarrow \gamma\mathbb{E}(J)$. A similar argument implies

$$\|x\|_{\gamma\mathbb{E}(J_1)} \le \|r^c\|\, \|x\|_{\gamma\mathbb{E}(J)} ,$$

that is, $\gamma\mathbb{E}(J) \hookrightarrow \gamma\mathbb{E}(J_1)$. ∎

We denote by $\{\lambda_s \,;\, s \geq 0\}$ the semigroup of **left translations** on $E^{\mathbb{R}^+}$, that is,

$$\lambda_s u(t) = u(s+t) \ , \qquad u \in E^{\mathbb{R}^+} \ , \quad s,t \in \mathbb{R}^+ \ .$$

Then $\mathbb{E}(\mathbb{R}^+)$ is **translation invariant** if $\{\lambda_s \,;\, s \geq 0\}$ induces naturally (that is, by restriction) a uniformly bounded semigroup on $\mathbb{E}(\mathbb{R}^+)$. The space $\mathbb{E}(\mathbb{R}^+)$ is **continuously translation invariant** if $\{\lambda_s \,;\, s \geq 0\}$ is a strongly continuous uniformly bounded semigroup on $\mathbb{E}(\mathbb{R}^+)$.

Using these concepts it is easy to prove the following important embedding result.

1.4.2 Proposition *Let $\mathbb{E}(\mathbb{R}^+)$ be a translation invariant regular extension of $\mathbb{E}(J)$. Then*

$$\mathbb{E}(J) \hookrightarrow B\big(J, \gamma\mathbb{E}(J)\big) \ . \tag{1.4.1}$$

If, in addition, $\mathbb{E}(\mathbb{R}^+)$ is continuously translation invariant,

$$\mathbb{E}(J) \hookrightarrow BUC\big(J, \gamma\mathbb{E}(J)\big) \ . \tag{1.4.2}$$

Proof Note that $\gamma\lambda_s u = u(s)$. Hence

$$\|u(s)\|_{\gamma\mathbb{E}(\mathbb{R}^+)} \leq \|\lambda_s u\|_{\mathbb{E}(\mathbb{R}^+)} \leq c\,\|u\|_{\mathbb{E}(\mathbb{R}^+)} \ , \qquad s \in \mathbb{R}^+ \ ,$$

implies $\mathbb{E}(\mathbb{R}^+) \hookrightarrow B\big(\mathbb{R}^+, \gamma\mathbb{E}(\mathbb{R}^+)\big)$. If $\mathbb{E}(\mathbb{R}^+)$ is continuously translation invariant, we see from

$$\|u(r) - u(s)\|_{\gamma\mathbb{E}(\mathbb{R}^+)} \leq \|\lambda_s(\lambda_{r-s} - 1)u\|_{\mathbb{E}(\mathbb{R}^+)} \leq c\,\|(\lambda_{r-s} - 1)u\|_{\mathbb{E}(\mathbb{R}^+)}$$

for $0 \leq s < r < \infty$ and $u \in \mathbb{E}(\mathbb{R}^+)$ that $\mathbb{E}(\mathbb{R}^+) \hookrightarrow BUC\big(\mathbb{R}^+, \gamma\mathbb{E}(\mathbb{R}^+)\big)$. Observe that the following diagram

$$\begin{array}{ccc} \mathbb{E}(J) & \xrightarrow{\ r^c\ } & \mathbb{E}(\mathbb{R}^+) \\ \big\downarrow & & \big\downarrow \\ B\big(J, \gamma\mathbb{E}(\mathbb{R}^+)\big) & \xleftarrow{\ r\ } & B\big(\mathbb{R}^+, \gamma\mathbb{E}(\mathbb{R}^+)\big) \end{array}$$

is commutative, where the left vertical arrow represents the natural injection and where B has to be replaced by BUC if $\mathbb{E}(\mathbb{R}^+)$ is continuously translation invariant. Now the assertion follows from Proposition 1.4.1. ■

1.4.3 Remark Suppose that $\mathbb{E}(\mathbb{R}^+)$ is a regular extension of $\mathbb{E}(J)$ for each $J \subset \mathbb{R}^+$ containing $[0,T]$ for some $T > 0$ such that the corresponding retractions and coretractions are bounded independently of J. Then it is obvious that the injections (1.4.1) and (1.4.2), respectively, are uniformly bounded with respect to J. If this is the case, we say $\mathbb{E}(\mathbb{R}^+)$ is a regular extension of $\mathbb{E}(J)$ **uniformly with respect to** $J \supset [0,T]$. ■

1.5 Pairs of Maximal Regularity

Let J be a perfect subinterval of $\mathbb{R}^+$ containing 0 and let (E_0, E_1) be a densely injected Banach couple. Suppose that

$$(\mathbb{E}_0, \mathbb{E}_1) := \big(\mathbb{E}_0(J), \mathbb{E}_1(J)\big)$$

is a pair of Banach spaces such that

$$\mathbb{E}_0 \hookrightarrow L_{1,\mathrm{loc}}(J, E_0) \tag{1.5.1}$$

and

$$\mathbb{E}_1 \hookrightarrow L_{1,\mathrm{loc}}(J, E_1) \cap W^1_{1,\mathrm{loc}}(J, E_0) \ , \tag{1.5.2}$$

respectively. Lastly, let

$$\mathcal{A} \subset \mathcal{L}_{\infty,\mathrm{loc}}\big(J, \mathcal{L}(E_1, E_0)\big) \ .$$

Then $(\mathbb{E}_0, \mathbb{E}_1)$ is said to possess the **property of maximal regularity** with respect to $\mathcal{A}$, or to be a **pair of maximal regularity** for $\mathcal{A}$, if

$$(\partial + A, \gamma) \in \mathcal{L}\mathrm{is}(\mathbb{E}_1, \mathbb{E}_0 \times \gamma\mathbb{E}_1) \ , \qquad A \in \mathcal{A} \ . \tag{1.5.3}$$

Of course, by a pair of maximal regularity for A we mean one for $\mathcal{A} := \{A\}$. Moreover, each $A \in \mathcal{L}(E_1, E_0)$ is canonically identified with the constant map

$$(t \mapsto A) \in \mathcal{L}_{\infty,\mathrm{loc}}\big(J, \mathcal{L}(E_1, E_0)\big) \ ,$$

again denoted by A.

1.5.1 Lemma *Suppose that $\mathcal{A} \subset \mathcal{H}(E_1, E_0)$. Then the pair $(\mathbb{E}_0, \mathbb{E}_1)$ has the property of maximal regularity with respect to $\mathcal{A}$ iff*

$$\partial + A \in \mathcal{L}(\mathbb{E}_1, \mathbb{E}_0) \ , \qquad A \in \mathcal{A} \ , \tag{1.5.4}$$

and, given $(x, A, f) \in \gamma\mathbb{E}_1 \times \mathcal{A} \times \mathbb{E}_0$, the Cauchy problem

$$\dot u + Au = f(t) \ , \quad t \in \dot J \ , \qquad u(0) = x \tag*{$(1.5.5)_{(x,A,f)}$}$$

has a $W^1_{1,\mathrm{loc}}$-solution u with $u \in \mathbb{E}_1$.

Proof Since, trivially, $\gamma \in \mathcal{L}(\mathbb{E}_1, \gamma\mathbb{E}_1)$, we see that (1.5.4) is equivalent to

$$(\partial + A, \gamma) \in \mathcal{L}(\mathbb{E}_1, \mathbb{E}_0 \times \gamma\mathbb{E}_1) , \qquad A \in \mathcal{A} .$$

If $(\mathbb{E}_0, \mathbb{E}_1)$ is a pair of maximal regularity for $\mathcal{A}$ then, given

$$(x, A, f) \in \gamma\mathbb{E}_1 \times \mathcal{A} \times \mathbb{E}_0 ,$$

it follows from (1.5.2) that $u := (\partial + A, \gamma)^{-1}(f, x)$ is a $W^1_{1,\mathrm{loc}}$-solution of the Cauchy problem $(1.5.5)_{(x,A,f)}$. Conversely, the assumptions imply that $(\partial + A, \gamma)$ is surjective. Since it is also injective, thanks to Proposition 1.3.1 and Example II.2.1.1, the assertion follows from the open mapping theorem. ■

Given $(x, A, f) \in E_0 \times \mathcal{H}(E_1, E_0) \times L_{1,\mathrm{loc}}(J, E_0)$, put

$$R_A x \colon J \to E_0 , \quad t \mapsto e^{-tA} x \tag{1.5.6}$$

and

$$K_A f \colon J \to E_0 , \quad t \mapsto \int_0^t e^{-(t-\tau)A} f(\tau)\, d\tau . \tag{1.5.7}$$

Note that

$$R_A \in \mathcal{L}\bigl(E_0, C(J, E_0)\bigr) , \quad K_A \in \mathcal{L}\bigl(L_{1,\mathrm{loc}}(J, E_0), C(J, E_0)\bigr) \tag{1.5.8}$$

and that

$$u(\cdot, 0, x, A, f) = R_A x + K_A f . \tag{1.5.9}$$

After these preparations we obtain the following characterization of pairs of maximal regularity.

1.5.2 Theorem *Let (E_0, E_1) be a densely injected Banach couple, let $(\mathbb{E}_0, \mathbb{E}_1)$ be a pair of Banach spaces satisfying* (1.5.1) *and* (1.5.2), *respectively, and let $\mathcal{A}$ be a nonempty subset of $\mathcal{H}(E_1, E_0)$. Then $(\mathbb{E}_0, \mathbb{E}_1)$ is a pair of maximal regularity for $\mathcal{A}$ iff*

$$\partial + A \in \mathcal{L}(\mathbb{E}_1, \mathbb{E}_0) , \quad R_A \in \mathcal{L}(\gamma\mathbb{E}_1, \mathbb{E}_1) , \quad K_A \in \mathcal{L}(\mathbb{E}_0, \mathbb{E}_1) ,$$

and

$$R_A x + K_A f$$

is a $W^1_{1,\mathrm{loc}}$-solution of $(1.5.5)_{(x,A,f)}$ for $(x, A, f) \in \gamma\mathbb{E}_1 \times \mathcal{A} \times \mathbb{E}_0$.

Proof This is an immediate consequence of Lemma 1.5.1, Proposition 1.3.1, and (1.5.9). ■

Let E and $\mathbb{E}$ be Banach spaces such that $\mathbb{E} \hookrightarrow L_{1,\mathrm{loc}}(J,E)$. Given $\alpha \in \mathbb{R}$, put

$$e^{\alpha}\mathbb{E} := \{ u \in L_{1,\mathrm{loc}}(J,E) \ ; \ e^{-\alpha t}u \in \mathbb{E} \} ,$$

equipped with the norm

$$u \mapsto \|u\|_{e^{\alpha}\mathbb{E}} := \|e^{-\alpha t}u\|_{\mathbb{E}}$$

where, as usual, $e^{-\alpha t}u := [t \mapsto e^{-\alpha t}u(t)]$.

The following observation is almost trivial, though useful in applications.

1.5.3 Proposition *Let $\mathcal{A}$ be a nonempty subset of $\mathcal{H}(E_1,E_0)$ and let $\alpha \in \mathbb{R}$. Suppose that $(\mathbb{E}_0,\mathbb{E}_1)$ is a pair of maximal regularity for $\alpha+\mathcal{A}$. Then $\gamma(e^{\alpha}\mathbb{E}_1) = \gamma\mathbb{E}_1$ and $(e^{\alpha}\mathbb{E}_0, e^{\alpha}\mathbb{E}_1)$ is a pair of maximal regularity for $\mathcal{A}$. Moreover,*

$$\|(\partial + A,\gamma)^{-1}\|_{\mathcal{L}(e^{\alpha}\mathbb{E}_0 \times \gamma\mathbb{E}_1, e^{\alpha}\mathbb{E}_1)} = \|(\partial + \alpha + A,\gamma)^{-1}\|_{\mathcal{L}(\mathbb{E}_0 \times \gamma\mathbb{E}_1, \mathbb{E}_1)}$$

for $A \in \mathcal{A}$.

Proof It is obvious that the map $(u \mapsto e^{\alpha t}u)$ is an isometric isomorphism from $\mathbb{E}_1$ onto $e^{\alpha}\mathbb{E}_1$. This readily implies $\gamma(e^{\alpha}\mathbb{E}_1) = \gamma\mathbb{E}_1$.

Let $(f,x) \in e^{\alpha}\mathbb{E}_0 \times \gamma\mathbb{E}_1$ be given. It follows from Remark II.2.1.2(e) and from Theorem 1.5.2 that

$$\begin{aligned}(\partial + A,\gamma)^{-1}(f,x) &= u(\cdot,x,A,f) = e^{\alpha t}u(\cdot,x,\alpha + A, e^{-\alpha t}f) \\ &= e^{\alpha t}(\partial + \alpha + A,\gamma)^{-1}(e^{-\alpha t}f,x) \ .\end{aligned}$$

Now the assertion is obvious. ■

1.6 Stability

For further considerations we take advantage of the following simple lemma that we give in slightly greater generality than actually needed.

1.6.1 Lemma *Let F_j be LCSs for $j = 0,1,2$ and suppose*

$$(B_1,B_2) \in \mathcal{L}\mathrm{is}(F_0, F_1 \times F_2) \ . \tag{1.6.1}$$

Then

$$F_0 = \ker(B_1) \oplus \ker(B_2) \tag{1.6.2}$$

and

$$B_j | \ker(B_k) \in \mathcal{L}\mathrm{is}(\ker(B_k), F_j) \ , \qquad j = 1,2 \ , \quad j \neq k \ . \tag{1.6.3}$$

Proof Clearly, (1.6.1) implies (1.6.3). Hence $P := \big[B_2|\ker(B_1)\big]^{-1}B_2 \in \mathcal{L}(F_0)$ and $B_2\big[B_2|\ker(B_1)\big]^{-1} = 1_{F_2}$. This implies that $P^2 = P$, that is, P is a continuous projection onto F_0. Consequently, $F_0 = \operatorname{im}(P) \oplus \ker(P)$ and, since $\operatorname{im}(P) = \ker(B_1)$ and $\ker(P) = \ker(B_2)$, the assertion follows. ■

1.6.2 Corollary *Suppose that $(\mathbb{E}_0, \mathbb{E}_1)$ is a pair of maximal regularity for $\mathcal{A} \subset \mathcal{H}(E_1, E_0)$. Put*

$$\mathbb{E}_{1,\gamma} := \ker(\gamma) \ , \quad \mathbb{E}_{1,A} := \ker(\partial + A)$$

and

$$(\partial + A)_\gamma := (\partial + A, \gamma)|\mathbb{E}_{1,\gamma} \ , \quad r_A := \gamma|\mathbb{E}_{1,A}$$

for $A \in \mathcal{A}$. Then

$$\mathbb{E}_1 = \mathbb{E}_{1,\gamma} \oplus \mathbb{E}_{1,A} \simeq \mathbb{E}_0 \times \gamma\mathbb{E}_1$$

and

$$\operatorname{diag}\big[(\partial + A)_\gamma, r_A\big] \in \mathcal{L}\mathrm{is}(\mathbb{E}_{1,\gamma} \times \mathbb{E}_{1,A}, \mathbb{E}_0 \times \gamma\mathbb{E}_1)$$

with

$$\big(\operatorname{diag}\big[(\partial + A)_\gamma, r_A\big]\big)^{-1} = \operatorname{diag}[K_A, R_A] \ .$$

Now it is easy to show that the property of being a pair of maximal regularity for $B \in \mathcal{H}(E_1, E_0)$ is stable under small time-dependent perturbations of B.

1.6.3 Proposition *Suppose that $(\mathbb{E}_0, \mathbb{E}_1)$ is a pair of maximal regularity for $B \in \mathcal{H}(E_1, E_0)$ and that $B \in \mathcal{L}(\mathbb{E}_1, \mathbb{E}_0)$. Then $(\mathbb{E}_1, \mathbb{E}_0)$ is a pair of maximal regularity for each $A \in \mathcal{L}(\mathbb{E}_1, \mathbb{E}_0)$ satisfying*

$$\|A - B\|_{\mathcal{L}(\mathbb{E}_1, \mathbb{E}_0)} \le \lambda \Big/ \|K_B\|_{\mathcal{L}(\mathbb{E}_0, \mathbb{E}_1)} \tag{1.6.4}$$

for some $\lambda \in (0, 1)$. If this is the case,

$$\|(\partial + A, \gamma)^{-1}\| \le c(1 - \lambda)^{-1} \, \|(\partial + B, \gamma)^{-1}\| \ ,$$

where c is independent of A, B, and λ.

Proof Note that $(\partial + A, \gamma)u = (f, x)$ for $u \in \mathbb{E}_1$ and $(f, x) \in \mathbb{E}_0 \times \gamma\mathbb{E}_1$ iff

$$(\partial + B)u + (A - B)u = f \ , \quad \gamma u = x \ . \tag{1.6.5}$$

Thanks to Corollary 1.6.2 we have the unique decomposition

$$u = v + w \in \mathbb{E}_{1,\gamma} \oplus \mathbb{E}_{1,B} \ .$$

Hence (1.6.5) is equivalent to

$$(\partial + B)_\gamma v + (A - B)v + (A - B)w = f \ , \quad r_B w = x \ ,$$

which, in turn, is equivalent to

$$\bigl[1 + (A - B)K_B\bigr](\partial + B)_\gamma v = f - (A - B)R_B x \ , \quad w = R_B x \ .$$

Note that (1.6.4) implies $\bigl[1 + (A - B)K_B\bigr] \in \mathcal{L}\mathrm{aut}(\mathbb{E}_0)$ and

$$\bigl\| \bigl[1 + (A - B)K_B\bigr]^{-1} \bigr\| \le (1 - \lambda)^{-1} \ .$$

Thus

$$v = K_B\bigl[1 + (A - B)K_B\bigr]^{-1}\bigl(f - (A - B)R_B x\bigr) \ , \quad w = R_B x$$

and

$$\|v\| \le (1 - \lambda)^{-1} \, \|K_B\| \, \bigl(\|f\| + \|A - B\| \, \|R_B\| \, \|x\|\bigr) \ , \quad \|w\| \le \|R_B\| \, \|x\| \ .$$

Hence, given $(f, x) \in \mathbb{E}_0 \times \gamma\mathbb{E}_1$, there exists a unique

$$u = v + w \in \mathbb{E}_{1,\gamma} \oplus \mathbb{E}_{1,B} = \mathbb{E}_1$$

satisfying $(\partial + A, \gamma)u = (f, x)$ and

$$\|v\| \le (1 - \lambda)^{-1}\bigl(\|K_B\| \vee \|R_B\|\bigr)\bigl(\|f\| + \|x\|\bigr) \ , \qquad \|w\| \le \|R_B\| \, \|x\| \ .$$

On the other hand, $B \in \mathcal{L}(\mathbb{E}_1, \mathbb{E}_0)$ and (1.5.4) imply $\partial \in \mathcal{L}(\mathbb{E}_1, \mathbb{E}_0)$. From this we deduce that $\partial + A \in \mathcal{L}(\mathbb{E}_1, \mathbb{E}_0)$ so that $(\partial + A, \gamma) \in \mathcal{L}(\mathbb{E}_1, \mathbb{E}_0 \times \gamma\mathbb{E}_1)$. This proves the assertion. ■

2 Maximal Hölder Regularity

In this section we show that maximal regularity occurs in Hölder spaces, provided we allow the Hölder seminorm to become singular at $t = 0$ at a controlled rate. We begin by introducing the needed function spaces.

2.1 Singular Hölder Spaces

Let J be a perfect subinterval of $\mathbb{R}^+$ containing 0 and let E be a Banach space. Given $\mu \in \mathbb{R}^+$, put

$$\|u\|_{C_\mu} := \|u\|_{C_\mu(J,E)} := \sup_{t \in \dot{J}} (1 \wedge t)^\mu \, \|u(t)\|$$

for $u \in E^J$. Then

$$BC_\mu(J,E) := \left(\left\{ u \in C(\dot{J},E) \ ; \ \|u\|_{C_\mu} < \infty \right\},\ \|\cdot\|_{C_\mu}\right)$$

is a Banach space. Note that

$$BC_0(J,E) = BC(\dot{J},E) \ .$$

If $\mu > 0$, put

$$\begin{aligned} &BUC_\mu(J,E) \\ &\quad := \left\{ u \in C(\dot{J},E) \ ; \ \left[t \mapsto (1 \wedge t)^\mu u(t)\right] \in BUC(\dot{J},E), \lim\nolimits_{t\to 0} t^\mu \left\|u(t)\right\| = 0 \right\} , \end{aligned}$$

whereas

$$BUC_0(J,E) := BUC(J,E) \ .$$

Observe that $BUC_\mu(J,E)$ is a closed linear subspace of $BC_\mu(J,E)$ for $\mu \in \mathbb{R}^+$. Thus it is a Banach space as well.

Given $\rho \in (0,1)$ and a nonempty subinterval I of $\mathbb{R}$, let

$$[u]_\rho^* := [u]_{\rho,I}^* := [u]_{C^\rho(I,E)}^* := \sup\left\{ \frac{\|u(s)-u(t)\|}{|s-t|^\rho} \ ; \ s,t \in I,\ 0 < t-s \le 1 \right\} .$$

For $\mu \in \mathbb{R}^+$ put

$$[\![u]\!]_{\rho,\mu} := [\![u]\!]_{\rho,\mu,J} := [\![u]\!]_{C^\rho_{\rho,\mu}(J,E)} := \sup_{2\varepsilon \in \dot{J}} (1 \wedge \varepsilon)^{\rho+\mu} [u]^*_{\rho,[\varepsilon,2\varepsilon]}$$

and

$$\|\cdot\|_{C^\rho_{\rho,\mu}} := \|\cdot\|_{C_\mu} + [\![\cdot]\!]_{\rho,\mu} \ .$$

Then

$$BC^\rho_{\rho,\mu}(J,E) := \left(\left\{ u \in C^\rho(\dot{J},E) \ ; \ \|\cdot\|_{C^\rho_{\rho,\mu}} < \infty \right\},\ \|\cdot\|_{C^\rho_{\rho,\mu}}\right)$$

is a Banach space. It is convenient to put

$$BC^\rho_\rho(J,E) := BC^\rho_{\rho,0}(J,E) \ , \quad [\![\cdot]\!]_\rho := [\![\cdot]\!]_{\rho,0} \ ,$$

and

$$BC^0_{0,\mu}(J,E) := BC_\mu(J,E) \ .$$

Lastly,

$$BUC^\rho_{\rho,\mu}(J,E) := \left\{ u \in BUC_\mu(J,E) \cap BC^\rho_{\rho,\mu}(J,E) \ ; \ \lim\nolimits_{\varepsilon\to 0} \varepsilon^{\rho+\mu} [u]_{\rho,[\varepsilon,2\varepsilon]} = 0 \right\}$$

and, of course,

$$BUC^\rho_\rho(J,E) := BUC^\rho_{\rho,0}(J,E) \ .$$

Clearly, $BUC^\rho_{\rho,\mu}(J,E)$ is a closed linear subspace of $BC^\rho_{\rho,\mu}(J,E)$, hence a Banach space.

If $\rho > 0$ and J is compact, the spaces $BC^{\rho}_{\rho,\mu}$ coincide essentially with the spaces $Z_{\mu,\rho}$ introduced in [AcT87]. These authors use the weights t^{μ} and $t^{\mu+\rho}$ instead of $(1 \wedge t)^{\mu}$ and $(1 \wedge t)^{\mu+\rho}$, respectively. Our choice is more convenient if $J = \mathbb{R}^+$, a situation that has not been studied in [AcT87]. Related spaces also occur in [Sob64]. The spaces BUC^{ρ}_{ρ} for $\rho > 0$ correspond to the spaces z^{ρ} used in [Lun87] (again for compact intervals). Also see [Lun95].

In the following proposition, in which we collect some embedding properties of the spaces introduced above, we let $BUC^{\rho} := BUC^{\rho}(J, E)$, etc.

2.1.1 Proposition *Suppose that*

$$0 \le \tau \le \sigma \le \rho < 1 \quad \textit{and} \quad 0 \le \mu \le \nu \le \lambda < \infty \ .$$

Then

$$BUC^{\rho} \hookrightarrow BUC^{\rho}_{\rho,\mu} \hookrightarrow BUC^{\sigma}_{\sigma,\nu} \hookrightarrow BC^{\sigma}_{\sigma,\nu} \hookrightarrow BC^{\tau}_{\tau,\lambda} \hookrightarrow L_{p,\mathrm{loc}}$$

for $0 < p < 1/\lambda$,

$$BUC^{\rho}_{\rho,\mu} \hookrightarrow C^{\rho}(\dot{J}, E)$$

and

$$BUC^{\rho}_{\rho} \hookrightarrow BUC(J, E) \cap C^{\rho}(\dot{J}, E) \ .$$

Moreover,

$$BUC \overset{d}{\hookrightarrow} BUC_{\mu}$$

In the Banach space cases, the norms of these injections are bounded by 1.

Proof Note that $(1 \wedge t)^{\nu} \le (1 \wedge t)^{\mu}$ for $t \in \mathbb{R}^+$. Hence

$$\|\cdot\|_{C_{\nu}} \le \|\cdot\|_{C_{\mu}} \ . \tag{2.1.1}$$

Moreover, if $\sigma > 0$,

$$\frac{\|u(s) - u(t)\|}{|s - t|^{\sigma}} \le |s - t|^{\rho - \sigma} \, [u]^*_{\rho,[\varepsilon,2\varepsilon]} \le (1 \wedge \varepsilon)^{\rho - \sigma} [u]^*_{\rho,[\varepsilon,2\varepsilon]}$$

for $0 < \varepsilon \le s < t \le 2\varepsilon$ and $t - s \le 1$, so that

$$(1 \wedge \varepsilon)^{\sigma} [\cdot]^*_{\sigma,[\varepsilon,2\varepsilon]} \le (1 \wedge \varepsilon)^{\rho} [\cdot]^*_{\rho,[\varepsilon,2\varepsilon]} \ , \qquad 2\varepsilon \in \dot{J} \ .$$

Consequently, $[\![\cdot]\!]_{\sigma,\nu} \le [\![\cdot]\!]_{\rho,\mu}$ for $0 < \sigma \le \rho < 1$ so that (2.1.1) implies

$$\|\cdot\|_{C^{\sigma}_{\sigma,\nu}} \le \|\cdot\|_{C^{\rho}_{\rho,\mu}} \ , \qquad 0 \le \sigma \le \rho < 1 \ , \quad 0 \le \mu \le \nu < \infty \ .$$

On the other hand, since $(1 \wedge \varepsilon)^{\rho+\mu} \le 1$ for $\varepsilon > 0$, it follows that $[\![\cdot]\!]_{\rho,\mu} \le [\cdot]_{\rho,J}$ for $0 < \rho < 1$. Since $\|\cdot\|_{C_\mu} \le \|\cdot\|_{C_0}$ by (2.1.1), we see that $\|\cdot\|_{C^\rho_{\rho,\mu}} \le \|\cdot\|_{C^\rho}$. Lastly,

$$\int_0^a \|u(t)\|^p \, dt \le \|u\|^p_{C_\lambda} \int_0^a t^{-p\lambda} \, dt \ , \qquad a \in \dot{J} \ , \quad a \le 1 \ ,$$

and the fact that u is bounded and continuous on $J \cap [a, \infty)$ imply $BC_\lambda \hookrightarrow L_{p,\mathrm{loc}}$ for $0 < p < 1/\lambda$. Now everything, except the density assertion, is obvious.

As for the density of BUC in BUC_μ, we can assume that $\mu > 0$. Let $u \in BUC_\mu$ and $\varepsilon > 0$ be given. Then there exists $\tau \in (0,1] \cap J$ such that $t^\mu \|u(t)\| \le \varepsilon/2$ for $0 < t \le \tau$. Define $v \in BUC$ by $v|[0,\tau] := u(\tau)$ and $v|[\tau,\infty) \cap J := u|[\tau,\infty) \cap J$, respectively. Then

$$t^\mu \|u(t) - v(t)\| \le t^\mu \|u(t)\| + t^\mu \|u(\tau)\| \le \varepsilon \ , \qquad 0 < t \le \tau \ ,$$

implies $\|u - v\|_{C_\mu} \le \varepsilon$. This proves the proposition. ■

Let S denote either one of the symbols BC or BUC. Then, given $\rho \in [0,1)$ and $\mu \in \mathbb{R}^+$,

$$S^{1+\rho}_{\rho,\mu}(J,E) := \left(\left\{ u \in S^\rho_{\rho,\mu}(J,E) \ ; \ \partial u \in S^\rho_{\rho,\mu}(J,E) \right\}, \ \|\cdot\|_{C^{1+\rho}_{\rho,\mu}} \right) \ ,$$

where

$$\|\cdot\|_{C^{1+\rho}_{\rho,\mu}} := \|\cdot\|_{C^\rho_{\rho,\mu}} + \|\partial\cdot\|_{C^\rho_{\rho,\mu}} \ .$$

Of course, $S^{1+\rho}_{\rho,\mu}(J,E)$ is a Banach space and

$$\partial \in \mathcal{L}\big(S^{1+\rho}_{\rho,\mu}(J,E), S^\rho_{\rho,\mu}\big) \ . \tag{2.1.2}$$

Note that ∂ is the classical derivative in this case.

Let J be bounded. If

$$u \in S^\rho_{\rho,\mu}(J,E) \cup BUC_\mu(J,E) \ , \qquad 0 < \rho < 1 \ , \quad \mu \ge 0 \ ,$$

then u is uniformly continuous near $\sup(J)$. Hence it has a unique continuous extension $\overline{u}$ over $\overline{J}$ and it is clear that $\overline{u}$ belongs to the set above with J replaced by $\overline{J}$ and has the same norm as u. Thus in the case of the spaces above we can assume without loss of generality that J is closed. The same is true for the spaces $S^{1+\rho}_{\rho,\mu}(J,E)$ and $BUC^1_\mu(J,E)$ with $0 < \rho < 1$ and $\mu \ge 0$.

2.1.2 Lemma *Let $T_0 > 0$ be fixed. Then, given $\rho \in [0,1)$ and $\mu \ge 0$, the space*

$$S^\rho_{\rho,\mu}(\mathbb{R}^+,E) \ [\textit{resp. } S^{1+\rho}_{\rho,\mu}(\mathbb{R}^+,E)]$$

is a regular extension of

$$S^\rho_{\rho,\mu}(J,E) \quad [\textit{resp. } S^{1+\rho}_{\rho,\mu}(J,E)] \ ,$$

uniformly with respect to any closed interval J containing $[0,T_0]$.

Proof Let $J := [0,T]$ for some $T \geq T_0$ and let $S \in \{S^{\rho}_{\rho,\mu}, S^{1+\rho}_{\rho,\mu}\}$ and $r := (u \mapsto u|J)$. Then it is clear that $r \in \mathcal{L}\big(S(\mathbb{R}^+,E), S(J,E)\big)$. Given $u \in S(J,E)$, put

$$\widetilde{u}(t) := \begin{cases} u(t) \ , & t \in J =: [0,T] \ , \\ 3u(2T-t) - 2u(3T-2t) \ , & t \in [T,(3/2)T] \ . \end{cases}$$

Let $\varphi \in C^{\infty}(\mathbb{R}^+)$ be equal to one on $[0, T+T_0/6]$ and vanish for $t \geq T + T_0/3$, and put $r^c u := \varphi\widetilde{u}$. It is easily verified that $r^c \in \mathcal{L}\big(S(J,E), S(\mathbb{R}^+,E)\big)$ and that we can choose φ such that r^c is bounded independently of J with $J \supset [0,T_0]$. Since $rr^c = \mathrm{id}$, the assertion follows. ■

2.1.3 Lemma *The spaces* $BC^j_{\mu}(\mathbb{R}^+,E)$ [*resp.* $BUC^j_{\mu}(\mathbb{R}^+,E)$] *are* [*resp. continuously*] *translation invariant for* $j = 0,1$ *and* $\mu \geq 0$.

Proof Note that $\{\, \lambda_s \ ; \ s \geq 0 \,\}$ is a contraction semigroup on $BC_{\mu}(\mathbb{R}^+,E)$ and on $BUC_{\mu}(\mathbb{R}^+,E)$. It is, trivially, strongly continuous on $BUC(\mathbb{R},E)$. Since, thanks to Proposition 2.1.1, $BUC(\mathbb{R},E)$ is dense in $BUC_{\mu}(\mathbb{R}^+,E)$, the semigroup of left translations is strongly continuous on $BUC_{\mu}(\mathbb{R}^+,E)$. Now the assertion for the spaces BC^1_{μ} and BUC^1_{μ} follows from the fact that translations and differentiation commmute. ■

Let (E_0,E_1) be a densely injected Banach couple and suppose that

$$\rho, \mu \in [0,1) \ . \tag{2.1.3}$$

Put

$$\mathbb{E}_0 := \mathbb{E}_0(J) := BUC^{\rho}_{\rho,\mu}(J,E_0) \tag{2.1.4}$$

and

$$\mathbb{E}_1 := \mathbb{E}_1(J) := BUC^{\rho}_{\rho,\mu}(J,E_1) \cap BUC^{1+\rho}_{\rho,\mu}(J,E_0) \ . \tag{2.1.5}$$

Note that (2.1.3) and Proposition 2.1.1 imply that

$$\mathbb{E}_0 \hookrightarrow L_{1,\mathrm{loc}}(J,E_0)$$

and

$$\mathbb{E}_1 \hookrightarrow L_{1,\mathrm{loc}}(J,E_1) \cap W^1_{1,\mathrm{loc}}(J,E_0) \ .$$

Thus we are in the situation of (1.5.1) and (1.5.2), respectively.

2.2 Semigroup Estimates

Now we prove some technical lemmas that will ultimately lead to maximal regularity results for the pair $(\mathbb{E}_0,\mathbb{E}_1)$, given suitable additional assumptions. For this we assume that

$$\left.\begin{aligned} &\mathcal{A} \subset \mathcal{H}(E_1, E_0) \text{ and there are constants } M > 0 \\ &\text{and } \vartheta \in (0, \pi/2) \text{ such that } \Sigma_\vartheta \subset \rho(-A) \text{ and} \\ &\|A\|_{\mathcal{L}(E_1,E_0)} + (1+|\lambda|)^{1-j}\, \|(\lambda+A)^{-1}\|_{\mathcal{L}(E_0,E_j)} \le M \\ &\text{for } (\lambda, A) \in \Sigma_\vartheta \times \mathcal{A} \text{ and } j = 0, 1. \end{aligned}\right\} \tag{2.2.1}$$

Note that (2.2.1) is a special case of (II.4.2.1). Thus (II.4.2.2) and Lemma II.4.2.1 are valid. However, we need the following improved estimates. As usual, $\|\cdot\|_j$ denotes the norm in E_j.

2.2.1 Lemma *Let assumption* (2.2.1) *be satisfied. Then*

$$M^{-1}\, \|x\|_1 \le \|Ax\|_0 \le M\, \|x\|_1 , \qquad (x, A) \in E_1 \times \mathcal{A} , \tag{2.2.2}$$

and there exists a constant $\omega > 0$ *such that*

$$\|(tA)^k e^{-tA}\|_{\mathcal{L}(E_0)} \le c(k) e^{-\omega t} , \qquad (t, A) \in \dot{\mathbb{R}}^+ \times \mathcal{A} , \quad k \in \mathbb{N} . \tag{2.2.3}$$

Proof The first estimate is a restatement of (II.4.2.2). Since (2.2.1) implies $\|A^{-1}\|_{\mathcal{L}(E_0,E_j)} \le M$ for $A \in \mathcal{A}$, it follows from

$$(\lambda + A)^{-1} = A^{-1}(1 + \lambda A^{-1})^{-1} \qquad \text{for} \quad |\lambda|\, \|A^{-1}\|_{\mathcal{L}(E_0)} < 1$$

that $[\,|z| < 1/M] \subset \rho(A)$ and

$$\|(\lambda + A)^{-1}\|_{\mathcal{L}(E_0,E_1)} \le 2M , \qquad \lambda \in [\,|z| \le r] , \quad A \in \mathcal{A} ,$$

where $r := 1/(2M)$. From this and from (2.2.1) we deduce, letting $\omega := r \sin\vartheta > 0$, that

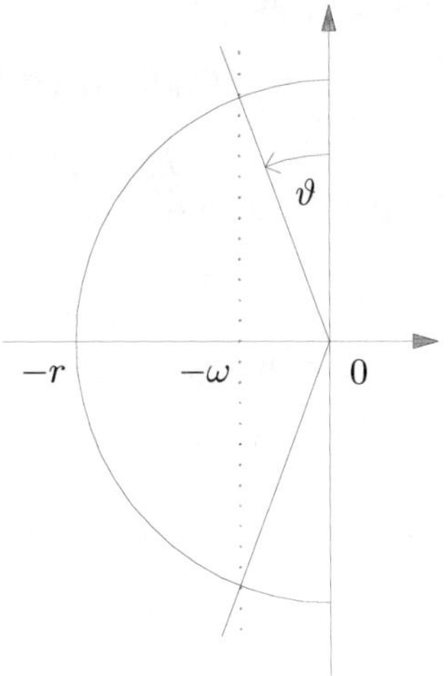

$$[\operatorname{Re} z \ge \omega] \subset \rho(2\omega - A)$$

and

$$\|(\lambda - 2\omega + A)^{-1}\|_{\mathcal{L}(E_0,E_1)} \le 2M$$

for $\operatorname{Re}\lambda \ge \omega$ and $A \in \mathcal{A}$. Thus we infer from Remark I.1.2.1(a) that there exists $\kappa \ge 1$ such that

$$-2\omega + \mathcal{A} \subset \mathcal{H}(E_1, E_0, \kappa, \omega) .$$

Consequently, Corollary I.1.4.3 implies the existence of constants $M' > 0$ and $\vartheta' \in (0, \pi/2)$ such that

$$\| -\omega + A\|_{\mathcal{L}(E_1,E_0)} + (1+|\lambda|)^{1-j}\, \|(\lambda - \omega + A)^{-1}\|_{\mathcal{L}(E_0,E_j)} \le M' \tag{2.2.4}$$

for $\lambda \in \Sigma_{\vartheta'}$ and $A \in \mathcal{A}$. Thus we deduce from Lemma II.4.2.1 that

$$\big\| \big[t(-\omega + A)\big]^k e^{-t(-\omega+A)} \big\|_{\mathcal{L}(E_0)} \leq c(k) \tag{2.2.5}$$

for $(t, A) \in \dot{\mathbb{R}}^+ \times \mathcal{A}$ and $k \in \mathbb{N}$. Now (2.2.3) follows (with any positive ω that is strictly smaller than the one of (2.2.5)). ■

In the following we put

$$E_\theta := (E_0, E_1)^0_{\theta,\infty} \ , \qquad 0 < \theta < 1 \ ,$$

and denote by $\|\cdot\|_\theta$ the norm in E_θ. Then we prove the following fundamental mapping result.

2.2.2 Lemma *Let assumption* (2.2.1) *be satisfied. Then*

$$(A \mapsto R_A) \in B\big(\mathcal{A}, \mathcal{L}\big(E_{1-\mu}, BUC^\rho_{\rho,\mu}(\mathbb{R}^+, E_1)\big)\big) \ .$$

Proof Choose $\omega > 0$ so that (2.2.5) is satisfied. Then

$$\|x\|_0 = \|(-\omega + A)^{-1}(-\omega + A)x\|_0 \leq c\,\|(-\omega + A)x\|_0$$

and (2.2.2) imply

$$\|x\|_1 \leq M\big(\|(-\omega + A)x\|_0 + \omega\,\|x\|_0\big) \leq c\,\|(-\omega + A)x\|_0$$

for $x \in E_1$ and $A \in \mathcal{A}$. Hence

$$t^\mu\,\|R_{-\omega+A}x(t)\|_1 \leq ct^{\mu-1}\,\|t(-\omega + A)e^{-t(-\omega+A)}x\|_0$$

for $t \in \dot{\mathbb{R}}^+$, $x \in E_0$, and $A \in \mathcal{A}$. Thus, if $\mu > 0$, we deduce from (2.2.5) and (I.2.10.6)–(I.2.10.10) that

$$\|e^{\omega\cdot} R_A x\|_{C_\mu(\mathbb{R}^+,E_1)} \leq c\,\|x\|_{1-\mu} \ , \qquad x \in E_{1-\mu} \ , \quad A \in \mathcal{A} \ ,$$

and that $t^\mu R_A x(t) \to 0$ in E_1 as $t \to 0$ for $x \in E_{1-\mu}$. If $\mu = 0$,

$$e^{\omega t}\,\|R_A x(t)\|_1 \leq M\,\|e^{-t(-\omega+A)}Ax\|_0 \leq c\,\|x\|_1 \ , \qquad x \in E_1 \ , \quad A \in \mathcal{A} \ ,$$

and $R_A x \in C(\mathbb{R}^+, E_1)$ for $x \in E_1$. This shows that

$$(A \mapsto R_A) \in B\big(\mathcal{A}, \mathcal{L}\big(E_{1-\mu}, BUC_\mu(\mathbb{R}^+, E_1)\big)\big) \ . \tag{2.2.6}$$

By interpolation we obtain from (2.2.2) and (2.2.3) that

$$\|(tA)^k e^{-tA}\|_{\mathcal{L}(E_{1-\mu},E_1)} \leq c(k,\mu)t^{-\mu}e^{-\omega t} \tag{2.2.7}$$

for $(t, A) \in \mathbb{R}^+ \times \mathcal{A}$ and $k \in \mathbb{N}$. Since

$$e^{-tA} - e^{-sA} = -\int_s^t Ae^{-\tau A}\, d\tau\ , \qquad 0 < s < t < \infty\ , \tag{2.2.8}$$

we deduce from (2.2.7) that

$$\|e^{-tA} - e^{-sA}\|_{\mathcal{L}(E_{1-\mu},E_1)} \le c\int_s^t \tau^{-\mu-1}e^{-\omega\tau}\, d\tau \le c(1\wedge\varepsilon)^{-\mu}\log(t/s)$$

for $0 < \varepsilon \le s < t < \infty$. Thanks to

$$\log(t/s) = \log\bigl(1 + (t-s)/s\bigr) \le (t-s)/s \le (t-s)^\rho(1\wedge\varepsilon)^{-\rho} \tag{2.2.9}$$

for $0 < \varepsilon \le s < t \le 2\varepsilon$, we see that

$$[\![R_A x]\!]_{C^\rho_{\rho,\mu}(\mathbb{R}^+,E_1)} \le c\,\|x\|_{1-\mu}\ , \qquad (x, A) \in E_{1-\mu} \times \mathcal{A}\ . \tag{2.2.10}$$

Given $y \in \mathrm{dom}(A^2)$, we infer from (2.2.2), (2.2.7), and (2.2.8) that

$$\|e^{-tA}y - e^{-sA}y\|_1 \le c\int_s^t \tau^{-\mu}e^{-\omega\tau}\, d\tau\, \|Ay\|_{1-\mu} \le c\varepsilon^{-\mu}(t-s)\,\|Ay\|_{1-\mu}$$

for $0 < \varepsilon \le s < t < \infty$. Thus

$$\varepsilon^{\rho+\mu}[R_A y]_{C^\rho([\varepsilon,2\varepsilon],E_1)} \le c\varepsilon\,\|Ay\|_{1-\mu}\ , \qquad \varepsilon > 0\ . \tag{2.2.11}$$

Given $x \in E_1 \doteq D(A)$ and $\eta > 0$, there exists $y \in E_1$ such that $\|y - Ax\|_0 < \eta/M$, thanks to the density of E_1 in E_0. Thus

$$z := A^{-1}y \in \mathrm{dom}(A^2)$$

and

$$\|z - x\|_1 \le M\,\|Az - Ax\|_0 = M\,\|y - Ax\|_0 < \eta\ .$$

This shows that $\mathrm{dom}(A^2)$ is dense in E_1. Since $E_1 \overset{d}{\hookrightarrow} E_{1-\mu}$, it follows that $\mathrm{dom}(A^2)$ is dense in $E_{1-\mu}$. Thus, given $x \in E_{1-\mu}$, we infer from (2.2.10) that there exists $y \in \mathrm{dom}(A^2)$ such that

$$\varepsilon^{\rho+\mu}[R_A y - R_A x]_{C^\rho([\varepsilon,2\varepsilon],E_1)} < \eta/2\ , \qquad 0 < \varepsilon \le 1\ . \tag{2.2.12}$$

On the other hand, (2.2.11) implies the existence of $\varepsilon_0 \in (0, 1]$ such that

$$\varepsilon^{\rho+\mu}[R_A y]_{C^\rho([\varepsilon,2\varepsilon],E_1)} < \eta/2\ , \qquad 0 < \varepsilon \le \varepsilon_0\ . \tag{2.2.13}$$

Hence we see from (2.2.12) and (2.2.13) that

$$\lim_{\varepsilon\to 0} \varepsilon^{\rho+\mu}[R_A x]_{C^\rho([\varepsilon,2\varepsilon],E_1)} = 0$$

for $x \in E_{1-\mu}$ and $A \in \mathcal{A}$. By combining this with (2.2.6) and (2.2.10), the assertion follows. ■

For the following corollary recall that $\mathbb{E}_1$ has been defined in (2.1.5).

2.2.3 Corollary *Let assumption* (2.2.1) *be satisfied. Then*

$$(A \mapsto R_A) \in B\big(\mathcal{A}, \mathcal{L}(E_{1-\mu}, \mathbb{E}_1)\big) ,$$

uniformly with respect to J, that is, the norm of this map can be bounded independently of J.

Proof Given $x \in E_{1-\mu}$, it follows that $\partial(R_A x) = -AR_A x$ in E_0. Hence we infer from (2.2.2) and Lemma 2.2.2 that

$$\|\partial(R_A x)\|_{C^\rho_{\rho,\mu}(J,E_0)} \le \|AR_A x\|_{C^\rho_{\rho,\mu}(\mathbb{R}^+,E_0)} \le M\,\|R_A x\|_{C^\rho_{\rho,\mu}(\mathbb{R}^+,E_1)} \le c\,\|x\|_{1-\mu}$$

for $(x, A) \in E_{1-\mu} \times \mathcal{A}$. Now the assertion is obvious. ■

2.2.4 Remarks **(a)** Put

$$\mathbb{F}_0 := BC^\rho_{\rho,\mu}(J, E_0)$$

and

$$\mathbb{F}_1 := BC^\rho_{\rho,\mu}(J, E_1) \cap BC^{1+\rho}_{\rho,\mu}(J, E_0)$$

if $0 < \mu < 1$, and let $\mathbb{F}_j := \mathbb{E}_j$ if $\mu = 0$. Then

$$\mathbb{E}_0 \hookrightarrow \mathbb{F}_0 \hookrightarrow L_{1,\mathrm{loc}}(J, E_0) \tag{2.2.14}$$

and

$$\mathbb{E}_1 \hookrightarrow \mathbb{F}_1 \hookrightarrow L_{1,\mathrm{loc}}(J, E_1) \cap W^1_{1,\mathrm{loc}}(J, E_0) , \tag{2.2.15}$$

respectively. Moreover, given assumption (2.2.1),

$$(A \mapsto R_A) \in B\big(\mathcal{A}, \mathcal{L}((E_0, E_1)_{1-\mu,\infty}, \mathbb{F}_1)\big)$$

uniformly with respect to J.

Proof This follows by replacing in the first part of the proof of Lemma 2.2.2 (up to (2.2.10)) the space $E_{1-\mu}$ by $(E_0, E_1)_{1-\mu,\infty}$. ■

(b) Since $\partial(R_A x) = -AR_A x$, it follows from (a) that $R_A x$ is a $W^1_{1,\mathrm{loc}}$-solution of $(1.5.5)_{(x,A,0)}$ for $x \in (E_0, E_1)_{1-\mu,\infty}$. ■

2.3 Trace Spaces

Thanks to (2.2.15) the trace spaces $\gamma\mathbb{E}_1$ and $\gamma\mathbb{F}_1$ are well-defined. The following theorem gives a characterization of these spaces. It is convenient to put

$$(E_0, E_1)^0_{1,\infty} := (E_0, E_1)_{1,\infty} := E_1 \tag{2.3.1}$$

to simplify the presentation.

2.3.1 Theorem *Let (E_0, E_1) be a densely injected Banach couple and suppose $\mathcal{H}(E_1, E_0) \neq \emptyset$. Then*

$$\gamma\mathbb{E}_1 \doteq (E_0, E_1)^0_{1-\mu,\infty} \quad \textit{and} \quad \gamma\mathbb{F}_1 \doteq (E_0, E_1)_{1-\mu,\infty}$$

for $\rho, \mu \in [0, 1)$.

Proof Let $\mathcal{A} \subset \mathcal{H}(E_1, E_0)$ satisfy assumption (2.2.1). Note that we can assume that $\mathcal{A} \neq \emptyset$. Indeed, $A \in \mathcal{H}(E_1, E_0)$ iff $\alpha + A \in \mathcal{H}(E_1, E_0)$ for $\alpha \in \mathbb{R}$. Thus there exists $A_0 \in \mathcal{H}(E_1, E_0)$ with $\text{type}(-A_0) < 0$. Now Corollary I.1.4.3 implies the existence of a neighborhood $\mathcal{A}$ of A in $\mathcal{L}(E_1, E_0)$ satisfying (2.2.1).

Let $x \in E_{1-\mu}$ (resp. $x \in (E_0, E_1)_{1-\mu,\infty}$) and $A \in \mathcal{A}$, and put $u := R_A x | J$. Then Corollary 2.2.3 (resp. Remark 2.2.4(a)) and $\gamma u = x$ imply

$$E_{1-\mu} \hookrightarrow \gamma\mathbb{E}_1 \quad (\text{resp. } (E_0, E_1)_{1-\mu,\infty} \hookrightarrow \gamma\mathbb{F}_1) \ .$$

Conversely, given $u \in \mathbb{F}_1$, put $x := u(0) \in E_0$. We deduce from (2.2.15) and Theorem 1.2.2 that

$$x = u(t) - \int_0^t \dot{u}(\tau)\, d\tau \ , \qquad t \in \dot{J} \ ,$$

in E_0. Thus, given $A \in \mathcal{A}$,

$$Ae^{-tA}x = Ae^{-tA}u(t) - Ae^{-tA} \int_0^t \dot{u}(\tau)\, d\tau \ , \qquad t \in \dot{J} \ .$$

From this we obtain by means of Lemma 2.2.1 that

$$\begin{aligned} t^{\mu-1} \|tAe^{-tA}x\|_0 &\leq c\Big(t^\mu \|u(t)\|_1 + t^{\mu-1} \int_0^t \|\dot{u}(\tau)\|_0\, d\tau\Big) e^{-\omega t} \\ &\leq c\Big\{(1 \wedge t)^\mu \|u(t)\|_1 + \sup_{0<\tau\leq t} (1 \wedge \tau)^\mu \|\dot{u}(\tau)\|_0\Big\} \\ &\leq c\, \|u\|_{\mathbb{F}_1} \end{aligned} \tag{2.3.2}$$

for $t \in \dot{J}$. Thus, if $\mu > 0$, we infer from (2.2.3) and (I.2.10.6)–(I.2.10.9) that

$$\|x\|_{(E_0,E_1)_{1-\mu,\infty}} \leq c\, \|u\|_{\mathbb{F}_1} \ , \qquad A \in \mathcal{A} \ .$$

Since this is true for every $u \in \mathbb{F}_1$ with $\gamma u = x$, it follows that

$$\gamma\mathbb{F}_1 \hookrightarrow (E_0, E_1)_{1-\mu,\infty} \ .$$

If, in addition, $u \in \mathbb{E}_1$, we obtain from (2.3.2) that

$$\lim_{t\to 0} t^{\mu-1} \|tAe^{-tA}x\|_0 = 0 \ .$$

Hence (I.2.10.10) implies $\gamma\mathbb{E}_1 \hookrightarrow E_{1-\mu}$.

Lastly, suppose that $\mu = 0$. Then we deduce from (2.2.2) and (2.3.2) that

$$\|x\|_1 \le M \, \|Ax\|_0 \le M \sup_{0<t\le 1} \|e^{-tA}Ax\|_0 \le c \, \|u\|_{\mathbb{E}_1} \ ,$$

thanks to the fact that $\mathbb{F}_1 = \mathbb{E}_1$ in this case. Thus $\gamma\mathbb{E}_1 \hookrightarrow E_1$ if $\mu = 0$. This proves the theorem. ■

2.3.2 Remark The above proof shows that, given $\rho, \mu \in [0,1)$, there exists a constant $\alpha > 0$ such that

$$\alpha^{-1} \, \|\cdot\|_{(E_0,E_1)_{1-\mu,\infty}} \le \|\cdot\|_{\gamma\mathbb{F}_1} \le \alpha \, \|\cdot\|_{(E_0,E_1)_{1-\mu,\infty}}$$

and

$$\alpha^{-1} \, \|\cdot\|_{E_{1-\mu}} \le \|\cdot\|_{\gamma\mathbb{E}_1} \le \alpha \, \|\cdot\|_{E_{1-\mu}} \ ,$$

respectively, uniformly with respect to J. ■

The fact, contained in Remark 2.2.4(a), that $R_A \in \mathcal{L}\big((E_0,E_1)_{1-\mu,\infty}, \mathbb{F}_1\big)$ can be found — even in greater generality — in [AcT87].

If $\rho = 0$, Theorem 2.3.1 is essentially due to Da Prato and Grisvard [DaPG79] (also cf. [Lun85]), if one takes into consideration that it has been shown by Grisvard in the appendix to [Lun85] that the spaces $C(\mu, E_1, E_0)$, used in [DaPG79], can be replaced by $\mathbb{E}_1\big([0,1]\big)$ (for $\rho = 0$). To be more precise, in [DaPG79] it is shown that $\gamma\mathbb{E}_1$ is an interpolation space between E_0 and E_1 of exponent $1-\mu$ if $0 < \mu < 1$ and $\rho = 0$, even if it is not assumed that $\mathcal{H}(E_1, E_0) \ne \emptyset$ and without the assumption that E_1 is dense in E_0. In [DorF87] it has then been shown that this 'continuous interpolation method' is equivalent to $(\cdot,\cdot)^0_{\mu,\infty}$, which explains the name 'continuous interpolation functor'.

It should be remarked that, given $A \in \mathcal{H}(E_1, E_0)$, the notations

$$D_A(\theta, \infty) := \big(E_0, D(A)\big)_{\theta,\infty} \doteq (E_0, E_1)_{\theta,\infty}$$

and

$$D_A(\theta) := \big(E_0, D(A)\big)^0_{\theta,\infty} \doteq (E_0, E_1)^0_{\theta,\infty}$$

are often used in the literature. Note that the second equivalences follow from Lemma I.1.1.2.

Now it is easy to prove an important embedding theorem for the spaces $\mathbb{E}_1$ and $\mathbb{F}_1$, respectively. Here we again use the convention (2.3.1).

2.3.3 Theorem *Let J be closed in $\mathbb{R}^+$, let (E_0, E_1) be a densely injected Banach couple with $\mathcal{H}(E_1, E_0) \ne \emptyset$, and suppose that $\rho, \mu \in [0,1)$. Then*

$$BC^{\rho}_{\rho,\mu}(J, E_1) \cap BC^{1+\rho}_{\rho,\mu}(J, E_0) \hookrightarrow B\big(J, (E_0, E_1)_{1-\mu,\infty}\big)$$

and

$$BUC^{\rho}_{\rho,\mu}(J,E_1)\cap BUC^{1+\rho}_{\rho,\mu}(J,E_0)\hookrightarrow BUC\big(J,(E_0,E_1)^0_{1-\mu,\infty}\big)\ .$$

Proof It follows from Lemmas 2.1.2 and 2.1.3 that $\mathbb{F}_1(\mathbb{R}^+)$ and $\mathbb{E}_1(\mathbb{R}^+)$ are translation invariant regular extensions of $\mathbb{F}_1(J)$ and $\mathbb{E}_1(J)$, respectively, and that $\mathbb{E}_1(\mathbb{R}^+)$ is continuously translation invariant, provided $\rho=0$. Hence the assertion follows from Proposition 1.4.2, Theorem 2.3.1, and the embedding results of Proposition 2.1.1. ■

The above theorem generalizes [Sim92, Satz 5.1], where — by a different method — the second part of the assertion has been proven for a compact interval and $\rho=0$. It should be noted that such a result does not imply the above embedding theorem if J is not bounded. For this it is important that the elements of $BUC_\mu(J,E)$ are uniformly continuous near $\sup(J)$. For this reason we have introduced the spaces $BUC_\mu(J,E)$ and do not work with the simpler spaces

$$\big\{\,u\in C(\dot J,E)\ ;\ \big[t\mapsto t^\mu u(t)\big]\in B(\dot J,E),\ \lim_{t\to 0}t^\mu u(t)=0\,\big\}\ ,$$

that are used in the literature most frequently (e.g., [ClH⁺87]).

2.4 Estimates for K_A

In the following lemma we study the behavior of the linear operator K_A in the spaces $S^\rho_{\rho,\mu}$ introduced above.

2.4.1 Lemma *Let assumption* (2.2.1) *be satisfied and suppose that* ρ *is positive. Then*

$$(A\mapsto K_A)\in B\big(\mathcal{A},\mathcal{L}\big(S^\rho_{\rho,\mu}(J,E_0),S^\rho_{\rho,\mu}(J,E_1)\big)\big)\ ,$$

uniformly with respect to J.

Proof Given $A\in\mathcal{A}$ and $f\in BC^\rho_{\rho,\mu}(J,E_0)$, define $v_j:=v_{A,j}$ by

$$\begin{aligned}K_Af(t)=&\int_0^{t/2}e^{-(t-\tau)A}f(\tau)\,d\tau+\int_{t/2}^t e^{-(t-\tau)A}\big[f(\tau)-f(t)\big]\,d\tau\\&+\int_{t/2}^t e^{-(t-\tau)A}f(t)\,d\tau=:v_1(t)+v_2(t)+v_3(t)\end{aligned}\tag{2.4.1}$$

for $t\in J$. Then Lemma 2.2.1 implies

$$\begin{aligned}(1\wedge t)^\mu\,\|v_1(t)\|_1&\le c(1\wedge t)^\mu\int_0^{t/2}(t-\tau)^{-1}(1\wedge\tau)^{-\mu}e^{-\omega(t-\tau)}\,d\tau\\&\quad\cdot\sup_{0<\tau\le t}(1\wedge\tau)^\mu\,\|f(\tau)\|_0\\&\le c\,\|f\|_{C_\mu([0,t],E_0)}\le c\,\|f\|_{C_\mu(J,E_0)}\end{aligned}\tag{2.4.2}$$

for $t \in \dot{J}$ and $A \in \mathcal{A}$, uniformly with respect to J.

Note that, given $0 < t/2 \leq \tau < t$ with $t \in J$,

$$\|f(\tau) - f(t)\|_0 \leq (t-\tau)^\rho [f]^*_{\rho,[t/2,t]} \leq c(t-\tau)^\rho (1 \wedge \tau)^{-\rho-\mu} [\![f]\!]_{C^\rho_{\rho,\mu}([0,t],E_0)}$$

if $t - \tau \leq 1$, and that

$$\|f(\tau) - f(t)\|_0 \leq \|f(\tau)\|_0 + \|f(t)\|_0 \leq c(1 \wedge \tau)^{-\mu} \|f\|_{C_\mu([0,t],E_0)}$$

otherwise. From this it easily follows that

$$\begin{aligned}(1 \wedge t)^\mu \|v_2(t)\|_1 &\leq c(1 \wedge t)^\mu \int_{t/2}^t (t-\tau)^{-1} e^{-\omega(t-\tau)} \|f(t) - f(\tau)\|_0 \, d\tau \\ &\leq c \|f\|_{C^\rho_{\rho,\mu}([0,t],E_0)} \leq c \|f\|_{C^\rho_{\rho,\mu}(J,E_0)}\end{aligned} \tag{2.4.3}$$

for $(t, A) \in \dot{J} \times \mathcal{A}$, uniformly with respect to J.

Lastly, the closedness of A in E_0 implies

$$Av_3(t) = \int_{t/2}^t Ae^{-(t-\tau)A} f(t) \, d\tau = (1 - e^{-(t/2)A}) f(t) \; . \tag{2.4.4}$$

Thus, by Lemma 2.2.1,

$$(1 \wedge t)^\mu \|v_3(t)\|_1 \leq c \|f\|_{C_\mu([0,t],E_0)} \leq c \|f\|_{C_\mu(J,E_0)} \tag{2.4.5}$$

for $(t, A) \in \dot{J} \times \mathcal{A}$, uniformly with respect to J. Since it is easily verified that $v_j \in C(\dot{J}, E_1)$ (or see the estimates below),

$$(A \mapsto K_A) \in B\big(\mathcal{A}, \mathcal{L}\big(BC^\rho_{\rho,\mu}(J, E_0), BC_\mu(J, E_1)\big)\big) \; , \tag{2.4.6}$$

uniformly with respect to J.

If $f \in BUC^\rho_{\rho,\mu}(J, E_0)$ and $\mu > 0$, we see from (2.4.2), (2.4.3), and (2.4.5) that

$$\lim_{t \to 0} t^\mu \|v_j(t)\|_1 = 0 \; , \qquad 1 \leq j \leq 3 \; . \tag{2.4.7}$$

If $f \in BUC^\rho_\rho(J, E_0)$, write

$$K_A f = K_A\big(f - f(0)\big) + K_A f(0) \; . \tag{2.4.8}$$

By replacing in the estimates above f by $f - f(0)$ (and μ by 0, of course), we see that

$$K_A\big(f - f(0)\big)(t) \to 0 \quad \text{in } E_1 \qquad \text{as} \quad t \to 0 \; .$$

Moreover,

$$AK_A\big(f(0)\big)(t) = (1 - e^{-tA}) f(0) \to 0 \quad \text{in } E_0 \qquad \text{as} \quad t \to 0 \; .$$

This shows that $K_A f(t) \to 0$ in E_1 as $t \to 0$, that is, that $K_A f$ is continuous at $t = 0$ in the topology of E_1.

In the following, it is always understood that $0 < \varepsilon \le s < t \le 2\varepsilon$ with $2\varepsilon \in \dot{J}$, and that $(f, A) \in BC^{\rho}_{\rho,\mu}(J, E_0) \times \mathcal{A}$, and it should be noted that the estimates below hold uniformly with respect to J. Then, using (2.2.8), the closedness of A, and the fact that the estimates below justify the following calculations:

$$Av_1(t) - Av_1(s) = \int_{s/2}^{t/2} Ae^{-(t-\tau)A} f(\tau)\, d\tau - \int_0^{s/2}\int_s^t A^2 e^{-(\sigma-\tau)A} f(\tau)\, d\sigma\, d\tau\ .$$

Hence, thanks to Lemma 2.2.1,

$$\begin{aligned}&\|v_1(t) - v_1(s)\|_1\\ &\qquad \le c\Big[(1\wedge\varepsilon)^{-\mu}\log(2-s/t) + \int_0^{s/2}\big[(s-\tau)^{-1} - (t-\tau)^{-1}\big](1\wedge\tau)^{-\mu}\, d\tau\Big]\\ &\qquad\quad \cdot \|f\|_{C_\mu([0,\varepsilon],E_0)}\ .\end{aligned}$$

Observe that

$$\log(2-s/t) \le 1 - s/t = (t-s)^{\rho}(t-s)^{1-\rho}/t \le (1\wedge\varepsilon)^{-\rho}(t-s)^{\rho}$$

and

$$\begin{aligned}\int_0^{s/2}\big[(s-\tau)^{-1} - (t-\tau)^{-1}\big](1\wedge\tau)^{-\mu}\, d\tau &\le c(t-s)s^{-1}(1\wedge s)^{-\mu}\\ &\le c(1\wedge\varepsilon)^{-\rho-\mu}(t-s)^{\rho}\ .\end{aligned}$$

Thus

$$(1\wedge\varepsilon)^{\rho+\mu}[v_1]^{*}_{C^{\rho}([\varepsilon,2\varepsilon],E_1)} \le c\,\|f\|_{C_\mu([0,2\varepsilon],E_0)} \le c\,\|f\|_{C_\mu(J,E_0)}\ . \tag{2.4.9}$$

Similarly,

$$\begin{aligned}Av_2(t) - Av_2(s) &= \Big(\int_{t/2}^{t} - \int_{s/2}^{s}\Big) Ae^{-(t-\tau)A}\big(f(\tau) - f(t)\big)\, d\tau\\ &\quad + \int_{s/2}^{s} A\big[e^{-(t-\tau)A} - e^{-(s-\tau)A}\big]\big(f(\tau) - f(s)\big)\, d\tau\\ &\quad + \int_{s/2}^{s} Ae^{-(t-\tau)A}\big(f(s) - f(t)\big)\, d\tau\\ &= \Big(\int_{s}^{t} - \int_{s/2}^{t/2}\Big) Ae^{-(t-\tau)A}\big(f(\tau) - f(t)\big)\, d\tau\\ &\quad - \int_{s/2}^{s}\int_s^t A^2 e^{-(\sigma-\tau)A}\big[f(\tau) - f(s)\big]\, d\sigma\, d\tau\\ &\quad + \big[e^{-(t-s)A} - e^{-(t-s/2)A}\big]\big(f(s) - f(t)\big)\ .\end{aligned}$$

Note that, given $s/2 \le \tau \le \xi \le t$ with $\xi \ge s$,

$$\|f(\xi) - f(\tau)\|_0 \le (\xi - \tau)^\rho [f]^*_{\rho,[\varepsilon/2,2\varepsilon]} \le c(\xi - \tau)^\rho (1 \wedge \varepsilon)^{-\rho-\mu} [\![f]\!]_{C^\rho_{\rho,\mu}([0,2\varepsilon],E_0)}$$

if $\xi - \tau \le 1$, whereas

$$\|f(\xi) - f(\tau)\|_0 \le \|f(\xi)\|_0 + \|f(\tau)\|_0 \le 2(1 \wedge \tau)^{-\mu} \|f\|_{C_\mu([0,2\varepsilon],E_0)}$$

otherwise. Hence we easily deduce from Lemma 2.2.1 that

$$\begin{aligned}
&\|v_2(t) - v_2(s)\|_1 \\
&\quad \le c(1 \wedge \varepsilon)^{-\rho-\mu} \Big\{ (t-s)^\rho \|f\|_{C^\rho_{\rho,\mu}([0,2\varepsilon],E_0)} \\
&\qquad\qquad + \int_{s/2}^s \big[(s-\tau)^{-1} - (t-\tau)^{-1}\big](s-\tau)^\rho \, d\tau \, [\![f]\!]_{C^\rho_{\rho,\mu}([0,2\varepsilon],E_0)} \Big\} \\
&\quad \le c(1 \wedge \varepsilon)^{-\rho-\mu} (t-s)^\rho \|f\|_{C^\rho_{\rho,\mu}([0,2\varepsilon],E_0)}
\end{aligned}$$

for $t - s \le 1$, thanks to

$$\begin{aligned}
\int_{s/2}^s \big[(s-\tau)^{-1} - (t-\tau)^{-1}\big](s-\tau)^\rho \, d\tau &= (t-s) \int_{s/2}^s (s-\tau)^{\rho-1}(t-\tau)^{-1} \, d\tau \\
&\le (t-s)^\rho \int_0^\infty \xi^{\rho-1}(1+\xi)^{-1} \, d\xi ,
\end{aligned}$$

where we used the substitution $\xi := (s-\tau)(t-s)^{-1}$. Consequently,

$$(1 \wedge \varepsilon)^{\rho+\mu} [v_2]^*_{C^\rho([\varepsilon,2\varepsilon],E_1)} \le c \|f\|_{C^\rho_{\rho,\mu}([0,2\varepsilon],E_0)} \le c \|f\|_{C^\rho_{\rho,\mu}(J,E_0)} . \tag{2.4.10}$$

Lastly, (2.4.4) implies

$$Av_3(t) - Av_3(s) = \big(1 - e^{-(t/2)A}\big)\big(f(t) - f(s)\big) + \int_{s/2}^{t/2} Ae^{-\tau A} f(s) \, d\tau .$$

Hence we obtain from (2.2.9) and Lemma 2.2.1 that

$$(1 \wedge \varepsilon)^{\rho+\mu} [v_3]^*_{C^\rho_{\rho,\mu}([\varepsilon,2\varepsilon],E_1)} \le c \|f\|_{C^\rho_{\rho,\mu}([0,2\varepsilon],E_0)} \le c \|f\|_{C^\rho_{\rho,\mu}(J,E_0)} . \tag{2.4.11}$$

Estimates (2.4.9)–(2.4.11) show that

$$[\![K_A f]\!]_{C^\rho_{\rho,\mu}(J,E_1)} \le c \|f\|_{C^\rho_{\rho,\mu}(J,E_0)}$$

for $(f,A) \in BC^\rho_{\rho,\mu}(J,E_0) \times \mathcal{A}$, uniformly with respect to J. This, together with (2.4.6), proves the assertion if $S = BC$.

Now let $f \in BUC^{\rho}_{\rho,\mu}(J, E_0)$. If $\mu > 0$, we infer from (2.4.9)–(2.4.11) that

$$\lim_{\varepsilon \to 0} \varepsilon^{\rho+\mu} [K_A f]_{C^{\rho}([\varepsilon,2\varepsilon],E_1)} = 0 \ .$$

The same is true if $\mu = 0$ and $f(0) = 0$, since $f \in BUC(J, E_0)$ in this case. Thus suppose that $f \in BUC^{\rho}_{\rho}(J, E_0)$ and put $x := f(0) \in E_0$. Then, thanks to (2.4.8) and the preceding observation, it suffices to show that

$$\lim_{\varepsilon \to 0} \varepsilon^{\rho} [K_A x]_{C^{\rho}([\varepsilon,2\varepsilon],E_1)} = 0 \ . \tag{2.4.12}$$

Given $y \in E_0$, thanks to (2.2.9) and Lemma 2.2.1,

$$\begin{aligned} \|K_A y(s) - K_A y(t)\|_1 &\leq c \Big\| \int_s^t A e^{-\tau A} y \, d\tau \Big\|_0 \\ &\leq c \log(t/s) \, \|y\|_0 \leq c\varepsilon^{-\rho}(t-s)^{\rho} \, \|y\|_0 \end{aligned} \tag{2.4.13}$$

for $0 < \varepsilon \leq s < t \leq 2\varepsilon$ and $\varepsilon \leq 1$. If $y \in E_1$, we obtain instead

$$\|K_A y(s) - K_A y(t)\|_1 \leq c(t-s) \, \|y\|_1 \leq c\varepsilon^{1-\rho}(t-s)^{\rho} \, \|y\|_1 \ . \tag{2.4.14}$$

Since E_1 is dense in E_0, we deduce from (2.4.13) that, given $\eta > 0$, there exists $y \in E_1$ such that

$$\varepsilon^{\rho} \big[K_A(x-y)\big]_{C^{\rho}([\varepsilon,2\varepsilon],E_1)} \leq \eta/2 \ , \qquad \varepsilon \leq 1 \ .$$

Then (2.4.14) implies the existence of $\varepsilon_0 \in (0,1)$ such that

$$\varepsilon^{\rho} [K_A y]_{C^{\rho}([\varepsilon,2\varepsilon],E_1)} \leq \eta/2 \ , \qquad 0 < \varepsilon \leq \varepsilon_0 \ .$$

This proves (2.4.12). Note that $K_A f \in BUC^{\rho}_{\rho,\mu}(J, E_1)$ implies

$$K_A f \in BUC\big(J \cap [s,\infty), E_1\big) \ , \qquad s \in \dot{J} \ .$$

Hence we deduce from (2.4.7) for $\mu \in (0,1)$ that $\big[t \mapsto t^{\mu} K_A f(t)\big] \in BUC(J, E_1)$ so that $K_A f \in BUC^{\rho}_{\rho,\mu}(J, E_1)$ if $f \in BUC^{\rho}_{\rho,\mu}(J, E_0)$ for $0 \leq \mu < 1$. ■

The above proof uses ideas from the proof of [AcT87, Proposition 2.1(v)]. The case $\mu = 0$ has also been considered in [Lun87]. In either reference E_1 is not supposed to be dense in E_0, but J is compact.

2.5 Maximal Regularity

Now we shall show that $(\mathbb{E}_0, \mathbb{E}_1)$ and $(\mathbb{F}_0, \mathbb{F}_1)$ are pairs of maximal regularity for $\mathcal{H}(E_1, E_0)$, provided $\rho > 0$. For this we prepare the following technical result.

2.5.1 Lemma *Let assumption* (2.2.1) *be satisfied and suppose that* $\rho > 0$. *Then* $K_A f \in W^1_{1,\mathrm{loc}}(J, E_0)$ *and*

$$\partial(K_A f) = f - AK_A f$$

for $f \in BC^\rho_{\rho,\mu}(J, E_0)$ *and* $A \in \mathcal{A}$.

Proof Let $(f, A) \in BC^\rho_{\rho,\mu}(J, E_0) \times \mathcal{A}$ be fixed and put $T := \sup(J)$. Note that

$$f, K_A f \in BC_\mu(J, E_0) \hookrightarrow L_{1,\mathrm{loc}}(J, E_0) \hookrightarrow \mathcal{D}'(J, E_0) . \tag{2.5.1}$$

Thus, given $\varphi \in \mathcal{D}(\mathring{J})$,

$$(K_A f)(\varphi) = \int_0^T \varphi(t) K_A f(t)\, dt = \int_0^T \int_\tau^T \varphi(t) e^{-(t-\tau)A} f(\tau)\, dt\, d\tau . \tag{2.5.2}$$

Note that

$$-\int_\tau^T \partial\varphi(t) e^{-(t-\tau)A} f(\tau)\, dt = \varphi(\tau) f(\tau) - \int_\tau^T \varphi(t) A e^{-(t-\tau)A} f(\tau)\, dt , \tag{2.5.3}$$

since the last integral exists in E_0, thanks to

$$\begin{aligned}\int_\tau^T \varphi(t) A e^{-(t-\tau)A} f(\tau)\, dt = &\int_\tau^T \big[\varphi(t) - \varphi(\tau)\big] A e^{-(t-\tau)A}\, dt\, f(\tau)\\ &+ \varphi(\tau)[1 - e^{-(T-\tau)A}] f(\tau) .\end{aligned}$$

Hence we obtain from (2.5.2) and (2.5.3) that

$$\begin{aligned}\partial(K_A f)(\varphi) = -K_A f(\partial\varphi) &= \int_0^T \varphi(t) \Big[f(t) - \int_0^t A e^{-(t-\tau)A} f(\tau)\, d\tau \Big]\, dt\\ &= (f - AK_A f)(\varphi) ,\end{aligned}$$

since the proof of Lemma 2.4.1 shows that the inner integral in the second to last term exists, that A can be interchanged with integration, and that

$$AK_A f \in BC_\mu(J, E_0) \hookrightarrow L_{1,\mathrm{loc}}(J, E_0) .$$

This proves the assertion. ■

2.5.2 Corollary *Let assumption* (2.2.1) *be satisfied and suppose that* $\rho > 0$. *Then*

$$(A \mapsto K_A) \in B\big(\mathcal{A}, \mathcal{L}(\mathbb{F}_0, \mathbb{F}_1) \cap \mathcal{L}(\mathbb{E}_0, \mathbb{E}_1)\big) ,$$

uniformly with respect to J.

Proof By Lemma 2.5.1,

$$\partial(K_A f) = f - AK_A f \ , \qquad f \in BC^{\rho}_{\rho,\mu}(J, E_0) = \mathbb{F}_0 \ .$$

Since $A \in \mathcal{L}(E_1, E_0)$, it follows from Lemma 2.4.1 that

$$\partial(K_A f) \in \mathbb{F}_0 \ [\text{resp. } \mathbb{E}_0] \ , \qquad f \in \mathbb{F}_0 \ [\text{resp. } \mathbb{E}_0] \ ,$$

and that the norms of these elements can be bounded independently of

$$(f, A) \in \mathbb{F}_0 \times \mathcal{A} \ [\text{resp. } \mathbb{E}_0 \times \mathcal{A}] \text{ and } J \ .$$

Now the assertion is a consequence of Lemma 2.4.1. ∎

After these preparations it is easy to prove the following maximal regularity result.

2.5.3 Theorem *Let assumption* (2.2.1) *be satisfied and let* $\rho \in (0,1)$ *and* $\mu \in [0,1)$. *Put*

$$(\mathbb{E}_0, \mathbb{E}_1) := \big(BUC^{\rho}_{\rho,\mu}(J, E_0), BUC^{\rho}_{\rho,\mu}(J, E_1) \cap BUC^{1+\rho}_{\rho,\mu}(J, E_0)\big) \ . \tag{2.5.4}$$

Then $(\mathbb{E}_0, \mathbb{E}_1)$ *is a pair of maximal regularity for* $\mathcal{A}$, *and*

$$\gamma\mathbb{E}_1 \doteq E_{1-\mu} := (E_0, E_1)^0_{1-\mu,\infty} \ ,$$

where $(E_0, E_1)^0_{1,\infty} := E_1$. *Moreover,*

$$\|(\partial + A, \gamma)^{-1}\|_{\mathcal{L}(\mathbb{E}_0 \times E_{1-\mu}, \mathbb{E}_1)} \leq c$$

for $A \in \mathcal{A}$, *uniformly with respect to* J.

Proof It follows from (2.1.2) that $\partial \in \mathcal{L}(\mathbb{E}_1, \mathbb{E}_0)$. Since $A \in \mathcal{L}(\mathbb{E}_1, \mathbb{E}_0)$, we see that $\partial + A \in \mathcal{L}(\mathbb{E}_1, \mathbb{E}_0)$. Theorem 2.3.1 guarantees that $\gamma\mathbb{E}_1 \doteq E_{1-\mu}$, and Corollaries 2.2.3 and 2.5.2 imply

$$\big[A \mapsto (R_A, K_A)\big] \in B\big(\mathcal{A}, \mathcal{L}(E_{1-\mu}, \mathbb{E}_1) \times \mathcal{L}(\mathbb{E}_0, \mathbb{E}_1)\big) \ ,$$

uniformly with respect to J. Note that $R_A x + K_A f \in \mathbb{E}_1$ and (2.2.15) together with Remark 2.2.4(b) and Lemma 2.5.1 imply that $R_A x + K_A f$ is a $W^1_{1,\mathrm{loc}}$-solution of the Cauchy problem $(1.5.5)_{(x,A,f)}$ for $(x, A, f) \in E_{1-\mu} \times \mathcal{A} \times \mathbb{E}_0$. Now the assertion follows from $(\partial + A, \gamma)^{-1} = R_A x + K_A f$ and Theorem 1.5.2. ∎

2.5.4 Remark Observe that, by replacing in the above proof $(\mathbb{E}_0, \mathbb{E}_1)$ by $(\mathbb{F}_0, \mathbb{F}_1)$ and $E_{1-\mu}$ by $(E_0, E_1)_{1-\mu,\infty}$, respectively, it follows that $(\mathbb{F}_0, \mathbb{F}_1)$ is a pair of maximal regularity for $\mathcal{A}$ if $\rho > 0$. ∎

The above maximal regularity theorem for the pair $(\mathbb{F}_0, \mathbb{F}_1)$ is (implicitly) contained in [AcT87] and, without proof, in [Sob64]. The maximal regularity for the pair

$$\left(BUC_\rho^\rho(J, E_0), BUC_\rho^\rho(J, E_1) \cap BUC_\rho^{1+\rho}(J, E_0)\right),$$

that is, for the pair $(\mathbb{E}_0, \mathbb{E}_1)$ with $\mu = 0$, has been proven in [Lun87] (cf. [Lun95]).

In the following, we are mainly interested in the pair $(\mathbb{E}_0, \mathbb{E}_1)$. Reasons being, for example, that $(\cdot,\cdot)^0_{1-\mu,\infty}$ is an admissible interpolation functor whereas $(\cdot,\cdot)_{1-\mu,\infty}$ is not and, most importantly, the embedding $\mathbb{E}_1 \hookrightarrow BUC(J, E_{1-\mu})$. Hence we restrict our considerations to this pair and leave it to the interested reader to extend the following results to the pair $(\mathbb{F}_0, \mathbb{F}_1)$ (if they are valid in that case).

Using results from Section I.1, it is not difficult to give sufficient conditions for assumption (2.2.1) to be satisfied. A particularly important case is contained in the following statement.

2.5.5 Theorem *Suppose that (E_0, E_1) is a densely injected Banach couple, let $\rho \in (0,1)$ and $\mu \in [0,1)$, and define $(\mathbb{E}_0, \mathbb{E}_1)$ by (2.5.4) with $J := \mathbb{R}^+$. Also suppose that $\mathcal{A}_0$ is a compact subset of $\mathcal{H}(E_1, E_0)$ and that there exists $\omega_0 > 0$ such that the spectral bounds satisfy*

$$s(-A) \le -\omega_0 , \qquad A \in \mathcal{A}_0 .$$

Then, given $\omega \in [0, \omega_0)$, there exists a neighborhood $\mathcal{A}$ of $\mathcal{A}_0$ in $\mathcal{L}(E_1, E_0)$ such that $(e^{-\omega}\mathbb{E}_0, e^{-\omega}\mathbb{E}_1)$ is a pair of maximal regularity for $\mathcal{A}$, and

$$\gamma(e^{-\omega}\mathbb{E}_1) \doteq E_{1-\mu} := (E_0, E_1)^0_{1-\mu,\infty} ,$$

where $(E_0, E)^0_{1,\infty} := E_1$. Moreover,

$$\|(\partial + A, \gamma)^{-1}\|_{\mathcal{L}(e^{-\omega}\mathbb{E}_0 \times E_{1-\mu}, e^{-\omega}\mathbb{E}_1)} \le c$$

for $A \in \mathcal{A}$.

Proof It is an immediate consequence of Corollary I.1.4.3 that there exists a neighborhood $\mathcal{A}$ of $\mathcal{A}_0$ in $\mathcal{L}(E_1, E_0)$ such that $-\omega + \mathcal{A}$ satisfies condition (2.2.1). Now the assertion follows from Theorem 2.5.3 and Proposition 1.5.3. ■

If we restrict our consideration to bounded intervals, the next theorem shows that the pair (2.5.4) is a pair of maximal regularity for all of $\mathcal{H}(E_1, E_0)$.

2.5.6 Theorem *Let J be bounded, (E_0, E_1) be a densely injected Banach couple, $\rho \in (0,1)$ and $\mu \in [0,1)$, and*

$$(\mathbb{E}_0, \mathbb{E}_1) := \left(BUC^\rho_{\rho,\mu}(J, E_0), BUC^\rho_{\rho,\mu}(J, E_1) \cap BUC^{1+\rho}_{\rho,\mu}(J, E_0)\right) . \qquad (2.5.5)$$

Then

$$\gamma\mathbb{E}_1 \doteq E_{1-\mu} := (E_0, E_1)^0_{1-\mu,\infty} ,$$

where $(E_0, E_1)^0_{1,\infty} := E_1$, *and* $(\mathbb{E}_0, \mathbb{E}_1)$ *is a pair of maximal regularity for all of* $\mathcal{H}(E_1, E_0)$. *Given* $\kappa \geq 1$ *and* $\omega > 0$,

$$\|(\partial + A, \gamma)^{-1}\|_{\mathcal{L}(\mathbb{E}_0 \times E_{1-\mu}, \mathbb{E}_1)} \leq c , \qquad A \in \mathcal{H}(E_1, E_0, \kappa, \omega) .$$

Proof Assumption (2.2.1) is satisfied for $\omega + \mathcal{H}(E_1, E_0, \kappa, \omega)$, thanks to Corollary I.1.4.3. Hence Theorem 2.5.3 and Proposition 1.5.3 imply that $(e^\omega\mathbb{E}_0, e^\omega\mathbb{E}_1)$ is a pair of maximal regularity for $\mathcal{H}(E_1, E_0, \kappa, \omega)$. Since J is bounded, it is easily verified that $e^\omega\mathbb{E}_j \doteq \mathbb{E}_j$ for $j = 0, 1$. Now the assertion is obvious. ■

2.6 Nonautonomous Problems

Let F and F_j be Banach spaces and let

$$F_1 \times F_2 \to F , \quad (x_1, x_2) \mapsto x_1 \bullet x_2 \tag{2.6.1}$$

be a multiplication. It is an easy consequence of (II.1.1.6) that point-wise multiplication induced by (2.6.1) is bilinear and continuous:

$$BUC^\rho_\rho(J, F_1) \times BUC^\rho_{\rho,\mu}(J, F_2) \to BUC^\rho_{\rho,\mu}(J, F) \tag{2.6.2}$$

for $\rho, \mu \in [0, 1)$. From this we infer that, given Banach spaces E_j,

$$BUC^\rho_\rho\big(J, \mathcal{L}(E_1, E_0)\big) \hookrightarrow \mathcal{L}\big(BUC^\rho_{\rho,\mu}(J, E_1), BUC^\rho_{\rho,\mu}(J, E_0)\big) \tag{2.6.3}$$

via the natural injection

$$A \mapsto (u \mapsto Au) .$$

After these preparations it is possible to extend Theorem 2.5.6 to the nonautonomous case.

2.6.1 Theorem *Let* (E_0, E_1) *be a densely injected Banach couple and, given* $\rho \in (0, 1)$ *and* $\mu \in [0, 1)$, *put*

$$(\mathbb{E}_0, \mathbb{E}_1) := \big(BUC^\rho_{\rho,\mu}(J, E_0), BUC^\rho_{\rho,\mu}(J, E_1) \cap BUC^{1+\rho}_{\rho,\mu}(J, E_0)\big) . \tag{2.6.4}$$

Suppose that $A \in BUC^\rho_\rho\big(J, \mathcal{H}(E_1, E_0)\big)$ *and that either* J *is bounded or there exists* $T \geq 0$ *such that* $s\big(-A(T)\big) < 0$ *and*

$$\|A(T + \cdot) - A(T)\|_{\mathcal{L}(\mathbb{E}_1(\mathbb{R}^+), \mathbb{E}_0(\mathbb{R}^+))} \leq \lambda \Big/ \|K_{A(T)}\|_{\mathcal{L}(\mathbb{E}_0(\mathbb{R}^+), \mathbb{E}_1(\mathbb{R}^+))} \tag{2.6.5}$$

for some $\lambda \in (0,1)$. *Then*

$$(\partial + A, \gamma) \in \mathcal{L}\mathrm{is}(\mathbb{E}_1, \mathbb{E}_0 \times E_{1-\mu}) \ ,$$

where $E_{1-\mu} := (E_0, E_1)^0_{1-\mu,\infty}$ *with* $(E_0, E_1)^0_{1,\infty} := E_1$.

Proof First assume that J is bounded. Then we can assume that it is compact. Hence $A(J)$ is compact in $\mathcal{H}(E_1, E_0)$. Thus it follows from Corollary I.1.3.2 that there exist $\kappa \geq 1$ and $\omega > 0$ such that

$$A(J) \subset \mathcal{H}(E_1, E_0, \kappa, \omega) \ . \tag{2.6.6}$$

Given $s \in J$, denote by B_s the constant map $t \mapsto A(s)$. Theorem 2.5.6 implies

$$(\partial + B_s, \gamma) \in \mathcal{L}\mathrm{is}(\mathbb{E}_1, \mathbb{E}_0 \times E_{1-\mu}) \tag{2.6.7}$$

and the existence of a constant α such that

$$\|(\partial + B_s, \gamma)^{-1}\|_{\mathcal{L}(\mathbb{E}_0 \times E_{1-\mu}, \mathbb{E}_1)} \leq \alpha \ , \qquad s \in J \ . \tag{2.6.8}$$

Let $A_s(t) := A(s+t)$ for $s \in J$ and $t \in J - s$. Observe that, given $s \in J =: [0,T]$, there exists $\delta(s) \in (0,1)$ such that

$$\begin{aligned} \|A_s - B_s\|_{C^\rho_\rho([0,\delta]\cap(J-s), \mathcal{L}(E_1,E_0))} = \max_{0 \leq t \leq \delta \wedge (T-s)} &\|A(s+t) - A(s)\|_{\mathcal{L}(E_1,E_0)} \\ + \sup_{0<2\varepsilon<\delta\wedge(T-s)} &\varepsilon^\rho \big[A(s+\cdot)\big]_{\rho,[\varepsilon,2\varepsilon]} \leq 1/(2\alpha) \end{aligned}$$

for $0 < \delta \leq \delta(s)$. Note that there exists $\delta_0 > 0$ such that $\delta(s) \geq \delta_0$ for $s \in J$. From this and the compactness of J we easily deduce the existence of a number $\delta > 0$ and of finitely many points $0 =: s_0 < s_1 \cdots < s_m < s_{m+1} := T$ such that $s_{j+2} - s_j \leq \delta$ for $j = 0, 1, \ldots, m-1$ and such that

$$\|A_{s_j} - B_{s_j}\|_{C^\rho_\rho([0,\delta]\cap(J-s_j), \mathcal{L}(E_1,E_0))} \leq 1/(2\alpha) \ , \qquad j = 0, 1, \ldots, m \ . \tag{2.6.9}$$

Put

$$A_j(t) := \begin{cases} A(s_j + t) \ , & 0 \leq t \leq \delta \wedge (T - s_j) \ , \\ A\big(s_j + \delta \wedge (T - s_j)\big) \ , & \delta \wedge (T - s_j) < t \leq T \ , \end{cases}$$

for $j = 0, 1, \ldots, m$. Then (2.6.9) and (2.6.3) imply

$$\|A_j - B_{s_j}\|_{\mathcal{L}(\mathbb{E}_1,\mathbb{E}_0)} \leq \|A_j - B_{s_j}\|_{C^\rho_\rho([0,\delta]\cap(J-s_j), \mathcal{L}(E_1,E_0))} \leq 1/(2\alpha) \ .$$

Hence we infer from (2.6.7), (2.6.8), and Proposition 1.6.3 that

$$(\partial + A_j, \gamma) \in \mathcal{L}\mathrm{is}(\mathbb{E}_1, \mathbb{E}_0 \times E_{1-\mu}) \tag{2.6.10}$$

and that

$$\|(\partial + A_j, \gamma)^{-1}\|_{\mathcal{L}(\mathbb{E}_0 \times E_{1-\mu}, \mathbb{E}_1)} \leq c \ , \qquad j = 0, 1, \ldots, m \ .$$

Let $(f, x) \in \mathbb{E}_0 \times E_{1-\mu}$ be given. It follows from (2.6.10) that the Cauchy problem

$$\dot{u} + A(t)u = f(t) \ , \quad 0 < t \leq s_2 \ , \qquad u(0) = x$$

has a unique solution

$$u_0 \in BUC^{\rho}_{\rho,\mu}\big([0, s_2], E_1\big) \cap BUC^{1+\rho}_{\rho,\mu}\big([0, s_2], E_0\big) \ .$$

Put

$$f_1(t) := \begin{cases} f(s_1 + t) \ , & 0 \leq t \leq T - s_1 \ , \\ f(T) \ , & T - s_1 \leq t \leq T \ , \end{cases}$$

and $x_1 := u_0(s_1)$. Then $(f_1, x_1) \in \mathbb{E}_0 \times E_1$, and (2.6.10) guarantees that the Cauchy problem

$$\dot{u} + A(t)u = f(t) \ , \quad s_1 < t \leq s_3 \ , \qquad u(s_1) = u_0(s_1)$$

has a unique solution

$$u_1 \in BUC^{\rho}_{\rho,\mu}\big([s_1, s_3], E_1\big) \cap BUC^{1+\rho}_{\rho,\mu}\big([s_1, s_3], E_0\big) \ .$$

Since $A \in C^{\rho}\big([t_0, T], \mathcal{H}(E_1, E_0)\big)$ for $0 < t_0 < T$, it follows from that part of Theorem II.1.2.1 which has already been proven in Section II.4.5 that the Cauchy problem

$$\dot{u} + A(t)u = f(t) \ , \quad t_0 < t \leq t_1 \ , \qquad u(t_0) = x \tag{2.6.11}$$

has for each $x \in E_1$ at most one solution on $[t_0, t_1]$, where $0 < t_0 < t_1 \leq T$. This easily implies that

$$u_0(t) = u_1(t) \ , \qquad s_1 \leq t \leq s_2 \ .$$

Thus, letting

$$u(t) := \begin{cases} u_0(t) \ , & 0 \leq t \leq s_1 \ , \\ u_1(t) \ , & s_1 \leq t \leq s_3 \ , \end{cases}$$

it follows that

$$u \in BUC^{\rho}_{\rho,\mu}\big([0, s_3], E_1\big) \cap BUC^{1+\rho}_{\rho,\mu}\big([0, s_3], E_0\big)$$

and that u is the unique solution of the Cauchy problem (2.6.11) on this interval. By repeating this argument a finite number of times we see that $(\partial + A, \gamma)$ is surjective. Since (2.6.3) implies $(\partial + A, \gamma) \in \mathcal{L}(\mathbb{E}_1, \mathbb{E}_0 \times E_{1-\mu})$, and since this map is injective by what we just have shown, the assertion follows from the open mapping theorem.

Now suppose that $J = \mathbb{R}^+$ and let $(f, x) \in \mathbb{E}_0(\mathbb{R}^+) \times E_{1-\mu}$ be given. Let

$$u_T \in BUC^{\rho}_{\rho,\mu}\big([0, T+1], E_1\big) \cap BUC^{1+\rho}_{\rho,\mu}\big([0, T+1], E_0\big)$$

be the unique solution of

$$\dot{u} + A(t)u = f(t) \ , \quad 0 < t \leq T+1 \ , \qquad u(0) = x \ ,$$

whose existence has been proven above. By Theorem 2.5.5

$$\big(\partial + A(T), \gamma\big) \in \mathcal{L}\mathrm{is}\big(\mathbb{E}_1(\mathbb{R}^+), \mathbb{E}_0(\mathbb{R}^+) \times E_{1-\mu}\big) \ .$$

Assumption (2.6.5) and Proposition 1.6.3 guarantee that

$$\big(\partial + A(T + \cdot), \gamma\big) \in \mathcal{L}\mathrm{is}\big(\mathbb{E}_1(\mathbb{R}^+), \mathbb{E}_0(\mathbb{R}^+) \times E_{1-\mu}\big) \ .$$

Thus the Cauchy problem

$$\dot{v} + A(T+t)v = f(T+t) \ , \quad t > 0 \ , \qquad v(0) = u_T(T)$$

has a unique solution $v \in \mathbb{E}_1(\mathbb{R}^+)$. Letting

$$u(t) := \begin{cases} u_T(t) & 0 \leq t \leq T+1 \ , \\ v(t-T) & T \leq t < \infty \ , \end{cases}$$

it follows from uniqueness that $u \in \mathbb{E}_1(\mathbb{R}^+)$ and that it is the unique solution in $\mathbb{E}_1(\mathbb{R}^+)$ of $\dot{u} + A(t)u = f(t)$ on $\dot{\mathbb{R}}^+$ satisfying $u(0) = x$. Now the assertion follows similarly as above. ■

Theorem 2.6.1 seems to be new in this generality. The assumption that $A \in BUC^{\rho}_{\rho}\big(J, \mathcal{H}(E_1, E_0)\big)$ is natural in this setting. In previous maximal regularity results in singular Hölder spaces, the condition $A \in BUC^{\rho}\big(J, \mathcal{H}(E_1, E_0)\big)$ has been imposed (e.g., [AcT87], [Lun87]).

Now it is easy to give the missing part of the proof of the main existence theorem of Subsection II.1.2.1.

Proof of Theorem II.1.2.1: Part 2 Thanks to part 1 of the proof of this theorem we know already that, given

$$(x, A, f) \in E_0 \times C^{\rho}\big(J, \mathcal{H}(E_1, E_0)\big) \times C^{\rho}(J, E) \ ,$$

the Cauchy problem $(\mathrm{II}.1.2.2)_{(s,x,A,f)}$ has a unique solution

$$u \in C\big(J \backslash \{s\}, E_1\big) \cap C^1\big(J \backslash \{s\}, E_0\big) \cap C(J, E_0) \ .$$

Let $s < s_0 < s_1$ with $s_1 \in J$ and put $J_1 := [0, s_1 - s_0]$. Define

$$(x_1, A_1, f_1) \in E_1 \times C^\rho\big(J_1, \mathcal{H}(E_1, E_0)\big) \times C^\rho(J_1, E_0) \tag{2.6.12}$$

by $x_1 := u(s_0)$ and

$$(A_1, f_1)(t) := (A, f)(s_0 + t) , \qquad t \in J_1 , \tag{2.6.13}$$

respectively. Then (2.6.11), Proposition 2.1.1, and Theorem 2.6.1 imply that the Cauchy problem

$$\dot v + A_1(t)v = f_1(t) , \quad 0 < t \le s_1 - s_0 , \qquad v(0) = x_1$$

has a unique solution

$$v \in BUC_\rho^\rho(J_1, E_1) \cap BUC_\rho^{1+\rho}(J_1, E_0) .$$

It is obvious, letting $w_0(t) := v(t - s_0)$ for $s_0 \le t \le s_1$, that

$$w_0 \in BUC_\rho^\rho\big([s_0, s_1], E_1\big) \cap BUC_\rho^{1+\rho}\big([s_0, s_1], E_0\big)$$

and that w_0 is a solution of the Cauchy problem

$$\dot w + A(t)w = f(t) , \quad s_0 < t \le s_1 , \qquad w(s_0) = u(s_0) .$$

Thus $w = u|[s_0, s_1]$ by uniqueness. Since this is true for every admissible choice of s_0 and s_1, it follows that

$$u \in C^\rho\big(J\backslash\{s\}, E_1\big) \cap C^{1+\rho}\big(J\backslash\{s\}, E_0\big)$$

which proves the theorem. ∎

3 Maximal Continuous Regularity

Let J be a perfect subinterval of $\mathbb{R}^+$ containing 0 and let (E_0, E_1) be a densely injected Banach couple. In the preceding section we have proven maximal regularity results for the spaces $BUC_{\rho,\mu}^\rho$ under the assumption that $\rho > 0$. In this section we investigate the case $\rho = 0$, that is, the spaces BUC_μ.

3.1 Necessary Conditions

First we consider the case $\mu = 0$, that is, we consider the pair

$$(\mathbb{E}_0, \mathbb{E}_1) := \big(BUC(J, E_0), BUC(J, E_1) \cap BUC^1(J, E_0)\big) . \tag{3.1.1}$$

We also assume that J is bounded, for simplicity.

The following proposition shows that maximal regularity can occur in this case in the 'parabolic case' only.

3.1.1 Proposition *Suppose* $A \in \mathcal{G}(E_0) \cap \mathcal{L}(E_1, E_0)$ *and* $(\mathbb{E}_0, \mathbb{E}_1)$ *is a pair of maximal regularity for* A. *Then* $A \in \mathcal{H}(E_1, E_0)$.

Proof Note that the proof of Theorem 1.5.2 implies $K_A \in \mathcal{L}(\mathbb{E}_0, \mathbb{E}_1)$. For this it suffices to observe that the latter proof did not use all the properties of a parabolic evolution operator but only those that are valid for

$$\{ e^{-(t-s)A} \; ; \; 0 \le s < t < \infty \} .$$

Fix $x_0 \in E_0$ and put $f(t) := e^{-tA}x$ for $t \in J$. Then

$$K_A f(t) = te^{-tA}x .$$

Thus

$$\operatorname{im}(e^{-tA}) \subset \operatorname{dom}(A) , \quad \|tAe^{-tA}\|_{\mathcal{L}(E_0)} \le c , \qquad 0 < t \le 1 , \tag{3.1.2}$$

where we used Lemma I.1.1.2 for the last estimate. Since it is well-known that (3.1.2) implies that $-A$ generates a strongly continuous analytic semigroup, the assertion follows. ■

The following theorem, due to Baillon, shows that the assumption that $(\mathbb{E}_0, \mathbb{E}_1)$ is a pair of maximal regularity implies serious restrictions on the Banach spaces E_0 and E_1.

3.1.2 Theorem *Suppose that* $(\mathbb{E}_0, \mathbb{E}_1)$ *is a pair of maximal regularity for some* $A \in \mathcal{H}(E_1, E_0)$. *Then either* $E_1 = E_0$ *or* E_0 *contains a closed subspace which is toplinearly isomorphic to the space* c_0 *of null sequences.*

Proof For this we refer to Baillon's note [Bai80] (also cf. [EbG91]). ■

3.1.3 Remarks **(a)** Clearly, $E_1 = E_0$ iff $A \in \mathcal{L}(E_0)$. Also observe that E_1 is toplinearly isomorphic to E_0 if $\mathcal{H}(E_1, E_0) \ne \emptyset$. Hence E_1 contains a copy of c_0 if E_0 does.

(b) Every closed linear subspace of a reflexive [resp. weakly sequentially complete] Banach space is reflexive [resp. weakly sequentially complete]. It is well-known that c_0 is neither reflexive nor weakly sequentially complete. Thus, if $E_1 \ne E_0$, the pair $(\mathbb{E}_0, \mathbb{E}_1)$ cannot be a pair of maximal regularity for any $A \in \mathcal{H}(E_1, E_0)$ if E_0 is reflexive or weakly sequentially complete (an L_1-space, for example). ■

3.2 Higher Order Interpolation Spaces

Suppose that

$$A \in \mathcal{H}(E_1, E_0) \quad \text{with} \quad s(-A) < 0 \ ,$$

where s denotes the spectral bound. Then, letting

$$\|\cdot\|_{E_2(A)} := \|A^2 \cdot\|_0 \ ,$$

it follows that

$$E_2(A) := \big(\mathrm{dom}(A^2);\ \|\cdot\|_{E_2(A)}\big)$$

is a Banach space such that

$$E_2(A) \overset{d}{\hookrightarrow} E_1 \overset{d}{\hookrightarrow} E_0 \ , \tag{3.2.1}$$

since the density of $\mathrm{dom}(A^2)$ in E_1 has already been shown in the proof of Lemma 2.2.2.

Throughout the remainder of this section we put

$$E_\theta := (E_0, E_1)^0_{\theta,\infty} \ , \quad E_{1+\theta}(A) := \big(E_1, E_2(A)\big)^0_{\theta,\infty} \ , \qquad 0 < \theta < 1 \ .$$

Since $(\cdot,\cdot)^0_{\theta,\infty}$ is an admissible interpolation functor, it is a consequence of (3.2.1) that $\big(E_\theta, E_{1+\theta}(A)\big)$ is a densely injected Banach couple for $0 \le \theta \le 1$, where we put $E_1(A) := E_1$. Let A_θ be the E_θ-realization of A for $0 \le \theta \le 1$. Since A commutes with its resolvent, it is obvious that $\rho(A_1) = \rho(A)$. Moreover, using on E_1 the norm $\|A\cdot\|$,

$$\frac{\|(\lambda + A_1)x\|_1}{|\lambda|\,\|x\|_1 + \|x\|_{E_2(A)}} = \frac{\|(\lambda + A)Ax\|_0}{|\lambda|\,\|Ax\|_0 + \|Ax\|_1} \ , \qquad x \in \dot{E}_2(A) \ ,$$

shows that

$$A \in \mathcal{H}(E_1, E_0, \kappa, \omega) \quad \Longrightarrow \quad A_1 \in \mathcal{H}\big(E_2(A), E_1, \kappa, \omega\big) \ .$$

Now again we suppose that assumption (2.2.1) is satisfied, that is,

$$\left.\begin{array}{l} \mathcal{A} \subset \mathcal{H}(E_1, E_0) \text{ and there are constants } M > 0 \\ \text{and } \vartheta \in (0, \pi/2) \text{ such that } \Sigma_\vartheta \subset \rho(-A) \text{ and} \\ \|A\|_{\mathcal{L}(E_1,E_0)} + (1 + |\lambda|)^{1-j}\, \|(\lambda + A)^{-1}\|_{\mathcal{L}(E_0,E_j)} \le M \\ \text{for } (\lambda, A) \in \Sigma_\vartheta \times \mathcal{A} \text{ and } j = 0, 1. \end{array}\right\} \tag{3.2.2}$$

Then, similarly as above, we see that

$$\|A_1\|_{\mathcal{L}(E_2(A),E_1)} + (1 + |\lambda|)^{1-j}\, \|(\lambda + A_1)^{-1}\|_{\mathcal{L}(E_1,E_{1+j}(A))} \le c$$

for $(\lambda, A) \in \Sigma_\vartheta \times \mathcal{A}$ and $j = 0, 1$. Thus, by interpolation,

$$\|A_\theta\|_{\mathcal{L}(E_{1+\theta}(A),E_\theta)} + (1+|\lambda|)^{1-j}\, \|(\lambda + A_\theta)^{-1}\|_{\mathcal{L}(E_\theta,E_{1+\theta}(A))} \le c(\theta) \tag{3.2.3}$$

for $0 \le \theta \le 1$, $(\lambda, A) \in \Sigma_\vartheta \times \mathcal{A}$, and $j = 0, 1$. Now Lemma 2.2.1 guarantees the existence of positive constants $M(\theta)$ and ω such that

$$M(\theta)^{-1}\, \|x\|_{E_{1+\theta}(A)} \le \|Ax\|_\theta \le M(\theta)\, \|x\|_{E_{1+\theta}(A)}\ , \qquad x \in E_{1+\theta}(A)\ , \tag{3.2.4}$$

and

$$\|(tA_\theta)^k e^{-tA_\theta}\|_{\mathcal{L}(E_\theta)} \le c(k,\theta)e^{-\omega t}\ , \qquad t > 0\ , \quad k \in \mathbb{N}\ , \tag{3.2.5}$$

for $\theta \in [0, 1]$ and $A \in \mathcal{A}$. Note that ω is independent of θ.

It should be remarked that the above construction of the higher order interpolation spaces $E_{1+\theta}(A)$, $0 < \theta < 1$, is a very particular case of the much more general situation discussed in detail in Chapter V. It should also be observed that, unlike the spaces E_θ, $0 \le \theta \le 1$, the higher order spaces $E_{1+\theta}(A)$, $0 < \theta \le 1$, depend on $A \in \mathcal{A}$, in general.

3.3 Estimates for K_A

Now we study the behavior of the map K_A in the densely injected Banach couples $\big(E_\theta, E_{1+\theta}(A)\big)$, $0 < \theta < 1$. For this we begin with the following simple technical

3.3.1 Lemma *Suppose that $0 \le \mu < 1$ and $0 < \theta < 1$. Then*

$$s^{1-\theta}(1 \wedge t)^\mu \int_0^t (s + t - \tau)^{\theta-2}(1 \wedge \tau)^{-\mu}\, d\tau \le c$$

for $s, t \in \dot{\mathbb{R}}^+$.

Proof If $t \le 2$,

$$\begin{aligned} I_1(s,t) := \int_0^{t/2} (s+t-\tau)^{\theta-2}(1\wedge\tau)^{-\mu}\, d\tau &= \int_0^{t/2} (s+t-\tau)^{\theta-2}\tau^{-\mu}\, d\tau \\ \le (s+t)^{\theta-1-\mu} \int_0^{1/2} (1-\sigma)^{\theta-2}\sigma^{-\mu}\, d\sigma &\le c(s+t)^{\theta-1-\mu}\ . \end{aligned}$$

If $t \ge 2$,

$$\begin{aligned} I_1(s,t) &= \int_0^1 (s+t-\tau)^{\theta-2}\tau^{-\mu}\, d\tau + \int_1^{t/2} (s+t-\tau)^{\theta-2}\, d\tau \\ &\le c\big\{(s+t)^{\theta-1-\mu} + (s+t/2)^{\theta-1}\big\} \le c(s+1)^{\theta-1}\ . \end{aligned}$$

Similarly, if $t \geq 2$,

$$I_2(s,t) := \int_{t/2}^{t} (s+t-\tau)^{\theta-2}(1 \wedge \tau)^{-\mu}\, d\tau = \int_{t/2}^{t} (s+t-\tau)^{\theta-2}\, d\tau \leq cs^{\theta-1} .$$

If $t \leq 1$,

$$I_2(s,t) = \int_{t/2}^{t} (s+t-\tau)^{\theta-2}\tau^{-\mu}\, d\tau \leq ct^{-\mu}s^{\theta-1} .$$

Lastly, if $1 < t < 2$,

$$I_2(s,t) = \int_{t/2}^{1} (s+t-\tau)^{\theta-2}\tau^{-\mu}\, d\tau + \int_{1}^{t} (s+t-\tau)^{\theta-2}\, d\tau \leq cs^{\theta-1} .$$

Now the assertion is obvious. ■

After these preparations we can derive the mapping properties of K_A in the interpolation spaces introduced above.

3.3.2 Proposition *Let assumption* (3.2.2) *be satisfied and suppose that* $0 \leq \mu < 1$ *and* $0 < \theta < 1$. *Then*

$$K_{A_\theta} \in \mathcal{L}\big(BUC_\mu(J,E_\theta), BUC_\mu(J,E_{1+\theta}(A)) \cap BUC_\mu^1(J,E_\theta)\big)$$

and

$$\|K_{A_\theta}\| \leq c\,, \qquad A \in \mathcal{A}\,,$$

uniformly with respect to J.

Proof If $0 \leq \alpha < \beta \leq 1$, it is easily verified that $A_\alpha \supset A_\beta$. Thus, without fearing confusion, we simply write A for A_α, $0 \leq \alpha \leq 1$. Note that

$$Ae^{-sA}K_A f(t) = \int_0^t Ae^{-(s+t-\tau)A} f(\tau)\, d\tau\,, \qquad t \in J\,, \quad s > 0\,, \tag{3.3.1}$$

for $f \in BUC_\mu(J,E_\theta)$. Indeed, it follows from (2.2.7) and the closedness of A that (3.3.1) is true in E_1 and that, thanks to Lemma 3.3.1,

$$\begin{aligned}\|Ae^{-sA}K_A f(t)\|_1 &\leq c\int_0^t (s+t-\tau)^{-2+\theta}(1\wedge\tau)^{-\mu}\, d\tau\, \|f\|_{C_\mu([0,t],E_\theta)} \\ &\leq cs^{\theta-1}(1\wedge t)^{-\mu}\, \|f\|_{C_\mu([0,t],E_\theta)} \\ &\leq cs^{\theta-1}(1\wedge t)^{-\mu}\, \|f\|_{C_\mu(J,E_\theta)}\,,\end{aligned}$$

uniformly with respect to J. Consequently,

$$\sup_{t>0}(1\wedge t)^\mu \sup_{s>0} s^{-\theta}\, \|sAe^{-sA}K_A f(t)\|_1 \leq c\, \|f\|_{C_\mu(J,E_\theta(A))}$$

for $f \in BUC_\mu(J, E_\theta)$, uniformly with respect to J. From this, from (3.2.5), and from (I.2.10.6)–(I.2.10.9) we deduce that

$$K_A \in \mathcal{L}\big(BUC_\mu(J, E_\theta), BC_\mu\big(J, (E_1, E_2(A))_{\theta,\infty}\big)\big) \tag{3.3.2}$$

and that

$$\|K_A\|_{\mathcal{L}(BUC_\mu(J,E_\theta), BC_\mu(J,(E_1,E_2(A))_{\theta,\infty}))} \le c\,, \qquad A \in \mathcal{A}\,, \tag{3.3.3}$$

uniformly with respect to J.

Now suppose that $f \in BUC^\rho(J, E_\theta)$ for some $\rho \in (0,1)$. Then it follows from Proposition 2.1.1, estimate (3.2.3), and Theorem 2.5.3 that

$$K_A f \in BUC^\rho_{\rho,\mu}\big(J, E_{1+\theta}(A)\big) \hookrightarrow BUC_\mu\big(J, E_{1+\theta}(A)\big)\ .$$

Since

$$BUC^\rho(J, E_\theta) \overset{d}{\hookrightarrow} BUC(J, E_\theta) \overset{d}{\hookrightarrow} BUC_\mu(J, E_\theta)\ , \tag{3.3.4}$$

where the density of the first inclusion follows by mollifying, for example, and where the density of the second injection is proven in Proposition 2.1.1, we deduce from (3.3.2) and (3.3.3) that

$$K_A \in \mathcal{L}\big(BUC_\mu(J, E_\theta), BUC_\mu(J, E_{1+\theta}(A))\big)$$

and that

$$\|K_A\|_{\mathcal{L}(BUC_\mu(J,E_\theta), BUC_\mu(J,E_{1+\theta}(A)))} \le c\,, \qquad A \in \mathcal{A}\,, \tag{3.3.5}$$

uniformly with respect to J.

If $f \in BUC^\rho(J, E_\theta)$ for some $\rho \in (0,1)$, we deduce from Theorem 2.5.3 that

$$K_A f \in BUC^{1+\rho}_{\rho,\mu}(J, E_\theta) \hookrightarrow BUC^1_\mu(J, E_\theta)$$

and that $\partial(K_A f) = f - AK_A f$. Hence, thanks to (3.2.3), (3.3.4), and (3.3.5),

$$\|\partial(K_A f)\|_{C_\mu(J,E_\theta)} \le \|f\|_{C_\mu(J,E_\theta)} + c\,\|K_A f\|_{C_\mu(J,E_{1+\theta}(A))} \le c\,\|f\|_{C_\mu(J,E_\theta)}\ ,$$

uniformly with respect to $(f, A) \in BUC_\mu(J, E_\theta) \times \mathcal{A}$ and J. Now the assertion follows from (3.3.4). ■

3.4 Maximal Regularity

After these preparations it is now easy to prove the main result of this section, namely the following maximal regularity theorem.

3.4.1 Theorem *Let condition* (3.2.2) *be satisfied and suppose that* $0 \le \mu < 1$ *and* $0 < \theta < 1$. *Then*

$$\begin{aligned}(\mathbb{E}_\theta, \mathbb{E}_{1+\theta}(A)) \\ := \big(BUC_\mu(J, E_\theta), BUC_\mu(J, E_{1+\theta}(A)) \cap BUC_\mu^1(J, E_\theta)\big)\end{aligned} \tag{3.4.1}$$

is a pair of maximal regularity for $\mathcal{A}_\theta := \{\, A_\theta \ ; \ A \in \mathcal{A} \,\}$, *and*

$$\gamma\mathbb{E}_{1+\theta}(A) \doteq \big(E_\theta, E_{1+\theta}(A)\big)^0_{1-\mu,\infty} , \tag{3.4.2}$$

where $\big(E_\theta, E_{1+\theta}(A)\big)^0_{1,\infty} := E_{1+\theta}(A)$. *Moreover,*

$$\|(\partial + A_\theta, \gamma)^{-1}\|_{\mathcal{L}(\mathbb{E}_\theta \times \gamma\mathbb{E}_{1+\theta}(A), \mathbb{E}_{1+\theta}(A))} \le c , \qquad A_\theta \in \mathcal{A}_\theta ,$$

uniformly with respect to J.

Proof Since $\big(E_\theta, E_{1+\theta}(A)\big)$ is a densely injected Banach couple and $\mathcal{A}_\theta$ satisfies (3.2.3), it follows from Corollary 2.2.3 that

$$(A_\theta \mapsto R_{A_\theta}) \in B\big(\mathcal{A}_\theta, \mathcal{L}\big((E_\theta, E_{1+\theta}(A))^0_{1-\mu,\infty}, \mathbb{E}_{1+\theta}(A)\big)\big) ,$$

uniformly with respect to J. Similarly, it follows from Theorem 2.3.1 that (3.4.2) is true. Now the assertion is a consequence of Proposition 3.3.2, Theorem 1.5.2, and the fact that $R_{A_\theta}x + K_{A_\theta}f$ is a $W^1_{1,\mathrm{loc}}(J, E_\theta)$-solution of $\dot u + A_\theta u = f$ with $u(0) = x$ whenever $(x, f) \in \gamma\mathbb{E}_{1+\theta}(A) \times \mathbb{E}_\theta$. ∎

3.4.2 Remarks **(a)** Theorem V.1.5.9 and Remark V.1.5.11 imply

$$\big(E_\theta, E_{1+\theta}(A)\big)^0_{1-\mu,\infty} \doteq \begin{cases} E_{\theta+1-\mu} , & \theta < \mu , \\ E_{\theta+1-\mu}(A) , & \theta > \mu . \end{cases}$$

Note however that, in general, $\big(E_\theta, E_{1+\theta}(A)\big)^0_{1-\theta,\infty} \ne E_1$.

(b) It is clear that one can deduce — by arguments similar to the ones used in Section 2 — from Theorem 3.4.1 results that correspond to Theorems 2.5.5 and 2.5.6, respectively.

(c) Suppose that J is bounded, $A\colon J \to \mathcal{H}(E_1, E_0)$ such that $A(J)$ is relatively compact in $\mathcal{H}(E_1, E_0)$ and, given $\theta \in (0,1)$,

$$E_{1+\theta}\big(A(t)\big) \doteq E_{1+\theta}\big(A(0)\big) ,$$

uniformly with respect to $t \in J$, and

$$A_\theta \in BUC\big(J, \mathcal{H}\big(E_{1+\theta}(A(0)), E_\theta\big)\big) ,$$

where $A_\theta(t) := \big[A(t)\big]_\theta$ for $t \in J$. Then the proof of Theorem 2.6.1 shows that

$$(\partial + A, \gamma) \in \mathcal{L}\mathrm{is}\Big(BUC_\mu\big(J, E_{1+\theta}(A(0))\big) \cap BUC^1_\mu(J, E_\theta),$$
$$BUC_\mu(J, E_\theta) \times \big(E_\theta, E_{1+\theta}(A(0))\big)^0_{1-\mu,\infty}\Big) .$$

Of course, the analogue to Theorem 2.6.1 is valid for $J = \mathbb{R}^+$ in the present case as well. ■

Theorem 3.4.1 is due to Da Prato and Grisvard [DaPG79, Théorème 3.1] for the case $\mu = 0$. That paper also contains the result of Remark 3.4.2(b) (in the case $\mu = 0$, of course, and for J bounded, cf. [DaPG79, Théorème 3.6]). The extension to the case $\mu > 0$ has been carried out in [Ang90]. Our proof of it, more precisely: our proof of Proposition 3.3.2, is closer to [Sim92, Theorem 5.4].

4 Maximal Sobolev Regularity

In this section we investigate the problem of maximal regularity in Sobolev spaces, that is, the maximal regularity of pairs of the form

$$\big(L_p(J, E_0), L_p(J, E_1) \cap W^1_p(J, E_0)\big)$$

for $1 < p < \infty$ and a densely injected Banach couple (E_0, E_1). We shall see that we need restrictions for the underlying Banach spaces E_j as well as for the class of admissible generators.

For the proof of maximal Sobolev regularity we need a considerable amount of preparatory material that is of interest for its own sake. Below we give almost self-contained treatments of the Banach-space-valued Hilbert transform and UMD spaces as well as of parts of the theory of fractional powers of linear operators of positive type.

4.1 Temperate Distributions

First we recall some simple facts about vector-valued distributions and Fourier transforms. The proofs are straightforward extensions of the corresponding ones for the scalar case. We assume that the reader has a working knowledge of scalar distributions and Fourier transforms and we refer for details to [Hör83, vol. 1], [Hor66], [Pet83], and [Schw66], for example. For later purposes we consider the general n-dimensional case though, in this section, only distributions on $\mathbb{R}$ are of interest.

Let $E := (E, |\cdot|)$ be a Banach space. If X is a nonempty open subset of $\mathbb{R}^n$, we denote by

$$C^k(X, E) \ , \qquad k \in \bar{\mathbb{N}} := \mathbb{N} \cup \{\infty\} \ ,$$

the Fréchet spaces of all E-valued functions on X whose derivatives of order $\leq k$ are continuous, equipped with the topology induced by the family of seminorms

$$u \mapsto p_{m,K}(u) := \max_{|\alpha| \leq m} \|\partial^\alpha u\|_{\infty,K} \ , \qquad m < k+1 \ , \quad K \subset\subset X \ . \tag{4.1.1}$$

and $C^k(X) := C^k(X, \mathbb{K})$. We also put $\mathcal{E}(X, E) := C^\infty(X, E)$ and $\mathcal{E}(X) := \mathcal{E}(X, \mathbb{K})$. Moreover,

$$\mathcal{E}'(X, E) := \mathcal{L}\big(\mathcal{E}(X), E\big)$$

(equipped, as always, with the bounded convergence topology) so that

$$\mathcal{E}'(X) = \mathcal{E}(X)' \ .$$

Standard truncation and mollification arguments show that

$$\mathcal{D}(X, E) \overset{d}{\hookrightarrow} \mathcal{E}(X, E) \ . \tag{4.1.2}$$

From this it follows easily that

$$\mathcal{E}'(X, E) \hookrightarrow \mathcal{D}'(X, E) \ , \tag{4.1.3}$$

by restriction, of course. Hence $\mathcal{E}'(X, E)$ is a space of distributions and the standard 'scalar proof' applies to show that $u \in \mathcal{E}'(X, E)$ iff u is a distribution with compact support, where the **support**, supp, of $u \in \mathcal{D}'(X, E)$ is given by

$$\begin{aligned} \operatorname{supp}(u) := X \setminus \big\{ x \in X \ ; \ & \text{there exists a neighborhood } U \text{ of } x \in X \\ & \text{such that } u(\varphi) = 0 \text{ for } \varphi \in \mathcal{D}(U) \big\} \ . \end{aligned}$$

Recall that $\mathcal{E}(X)$ and $\mathcal{E}'(X)$ are Montel spaces. Hence they are reflexive.

We denote by $\mathcal{S}(\mathbb{R}^n, E)$ the Schwartz space of smooth rapidly decreasing E-valued functions on $\mathbb{R}^n$. Thus $u \in \mathcal{S}(\mathbb{R}^n, E)$ iff $u \in \mathcal{E}(\mathbb{R}^n, E)$ and

$$q_{k,m}(u) := \sup_{\substack{x \in \mathbb{R}^n \\ |\alpha| \leq m}} (1 + |x|^2)^k \, |\partial^\alpha u(x)| < \infty \ , \qquad k, m \in \mathbb{N} \ . \tag{4.1.4}$$

Then $\mathcal{S}(\mathbb{R}^n, E)$ is a Fréchet space with the topology induced by the family of seminorms (4.1.4), and $\mathcal{S}(\mathbb{R}^n) := \mathcal{S}(\mathbb{R}^n, \mathbb{K})$. By standard arguments

$$\mathcal{D}(\mathbb{R}^n, E) \overset{d}{\hookrightarrow} \mathcal{S}(\mathbb{R}^n, E) \overset{d}{\hookrightarrow} \mathcal{E}(\mathbb{R}^n, E) \ . \tag{4.1.5}$$

We define the space $\mathcal{S}'(\mathbb{R}^n, E)$ of E-valued temperate distributions by

$$\mathcal{S}'(\mathbb{R}^n, E) := \mathcal{L}\big(\mathcal{S}(\mathbb{R}^n), E\big) \ ,$$

and $\mathcal{S}'(\mathbb{R}^n) := \mathcal{S}'(\mathbb{R}^n, \mathbb{K})$. Note that (4.1.5) implies

$$\mathcal{E}'(\mathbb{R}^n, E) \hookrightarrow \mathcal{S}'(\mathbb{R}^n, E) \hookrightarrow \mathcal{D}'(\mathbb{R}^n, E) . \tag{4.1.6}$$

Moreover, $\mathcal{S}(\mathbb{R}^n)$ and $\mathcal{S}'(\mathbb{R}^n)$ are Montel spaces, thus reflexive.

Lastly, we introduce the space $\mathcal{O}_M(\mathbb{R}^n, E)$ of E-valued slowly increasing smooth functions on $\mathbb{R}^n$. Namely, $u \in \mathcal{O}_M(\mathbb{R}^n, E)$ iff $u \in \mathcal{E}(\mathbb{R}^n, E)$ and, given $\alpha \in \mathbb{N}^n$, there exist $m_\alpha \in \mathbb{N}$ and $c_\alpha > 0$ such that

$$|\partial^\alpha u(x)| \le c_\alpha (1 + |x|^2)^{m_\alpha} , \qquad x \in \mathbb{R}^n .$$

The space $\mathcal{O}_M(\mathbb{R}^n, E)$ is a *LCS* with respect to the topology induced by the family of seminorms

$$u \mapsto \|\varphi \partial^\alpha u\|_\infty , \qquad \varphi \in \mathcal{S}(\mathbb{R}^n) , \quad \alpha \in \mathbb{N}^n .$$

It follows that

$$\mathcal{S}(\mathbb{R}^n, E) \hookrightarrow \mathcal{O}_M(\mathbb{R}^n, E) \hookrightarrow \mathcal{S}'(\mathbb{R}^n, E) . \tag{4.1.7}$$

Of course, $\mathcal{O}_M(\mathbb{R}^n) := \mathcal{O}_M(\mathbb{R}^n, \mathbb{K})$. It is easily verified that the map

$$\mathcal{O}_M(\mathbb{R}^n) \times \mathcal{S}(\mathbb{R}^n, E) \to \mathcal{S}(\mathbb{R}^n, E) , \quad (m, u) \mapsto mu \tag{4.1.8}$$

is well-defined and bilinear. Moreover, if $m \in \mathcal{E}(\mathbb{R}^n)$ then

$$m \in \mathcal{O}_M(\mathbb{R}^n) \quad \text{iff} \quad (\varphi \mapsto m\varphi) \in \mathcal{L}\big(\mathcal{S}(\mathbb{R}^n, E)\big) . \tag{4.1.9}$$

Clearly, the rôles of E and $\mathbb{K}$ can be interchanged on the left-hand side of (4.1.8). Similarly, $\mathcal{O}_M(\mathbb{R}^n)$ can be replaced by $\mathcal{O}_M(\mathbb{R}^n, E)$ if $\mathcal{S}(\mathbb{R}^n, E)$ is replaced by $\mathcal{S}(\mathbb{R}^n)$.

4.2 Fourier Transforms and Convolutions

Let $E := (E, |\cdot|)$ be a Banach space. Given $u \in L_1(\mathbb{R}^n, E)$,

$$\mathcal{F}u(\xi) := \widehat{u}(\xi) := \int_{\mathbb{R}^n} e^{-i\langle \xi, x\rangle} u(x)\, dx , \qquad \xi \in \mathbb{R}^n ,$$

is the **Fourier transform** of u, where

$$\langle \xi, x\rangle := \textstyle\sum_{j=1}^n \xi^j x^j .$$

The RIEMANN-LEBESGUES LEMMA asserts that

$$\mathcal{F} \in \mathcal{L}\big(L_1(\mathbb{R}^n, E), C_0(\mathbb{R}^n, E)\big) ,$$

where, in general, given any locally compact metric space M,

$$C_0(M, E) := \big(C_0(M, E), \|\cdot\|_\infty\big)$$

is the closed subspace of $BUC(M, E)$ consisting of the continuous functions **vanishing at infinity**. Recall that this means that, given any $\varepsilon > 0$, there exists a compact subset K of M such that $|u(x)| < \varepsilon$ for $x \in M \backslash K$. Thus $C_0(M, E)$ is a Banach space.

The FOURIER INVERSION THEOREM guarantees that

$$\mathcal{F} \in \mathcal{L}\mathrm{aut}\big(\mathcal{S}(\mathbb{R}^n, E)\big) \tag{4.2.1}$$

and

$$\mathcal{F}^{-1}u = (2\pi)^{-n}\check{\widehat{u}} = (2\pi)^{-n}\widehat{\check{u}}\ , \qquad u \in \mathcal{S}(\mathbb{R}^n, E)\ , \tag{4.2.2}$$

where

$$\check{u}(x) := u(-x)\ , \qquad x \in \mathbb{R}^n\ , \quad u \in E^{\mathbb{R}^n}\ ,$$

is the reflection of u.

The Fourier transform $\widehat{u} := \mathcal{F}u$ of the temperate distribution $u \in \mathcal{S}'(\mathbb{R}^n, E)$ is defined by

$$\widehat{u}(\varphi) := u(\widehat{\varphi})\ , \qquad \varphi \in \mathcal{S}(\mathbb{R}^n)\ .$$

Define the reflection of $u \in \mathcal{D}'(\mathbb{R}^n, E)$ by

$$\check{u}(\varphi) := u(\check{\varphi})\ , \qquad \varphi \in \mathcal{D}(\mathbb{R}^n)\ .$$

Then the FOURIER-SCHWARTZ THEOREM guarantees that

$$\mathcal{F} \in \mathcal{L}\mathrm{aut}\big(\mathcal{S}'(\mathbb{R}^n, E)\big) \tag{4.2.3}$$

and that (4.2.2) is true for $u \in \mathcal{S}'(\mathbb{R}^n, E)$. Moreover, if $u \in L_1(\mathbb{R}^n, E) \subset \mathcal{S}'(\mathbb{R}^n, E)$, the new definition of $\widehat{u}$ coincides with the original one.

It is a fundamental property of the Fourier transform that

$$(D^\alpha u)\widehat{\ } := \widehat{D^\alpha u} = \xi^\alpha \widehat{u}\ , \qquad u \in \mathcal{S}'(\mathbb{R}^n, E)\ , \quad \alpha \in \mathbb{N}^n\ , \tag{4.2.4}$$

where $D_j := -i\partial_j$ for $j = 1, \ldots, n$.

Given $a \in \mathbb{R}^n$, we denote by $\tau_a u$ the (right) **translation** of $u \in E^{\mathbb{R}^n}$ by a, that is,

$$\tau_a u(x) := u(x - a)\ , \qquad x \in \mathbb{R}^n\ , \quad u \in E^{\mathbb{R}^n}\ .$$

Moreover,

$$\tau_a u(\varphi) := u(\tau_{-a}\varphi)\ , \qquad u \in \mathcal{D}'(\mathbb{R}^n, E)\ , \quad \varphi \in \mathcal{D}(\mathbb{R}^n)\ .$$

It is not difficult to verify that $\{\, \tau_a\ ;\ a \in \mathbb{R}^n \,\}$, the **group of translations** on $\mathbb{R}^n$, is strongly continuous on each one of the spaces $\mathfrak{F}(\mathbb{R}^n, E)$ for

$$\mathfrak{F} \in \{\mathcal{D}, \mathcal{D}', \mathcal{S}, \mathcal{S}', \mathcal{E}, \mathcal{E}'\}\ . \tag{4.2.5}$$

It is easy to see that

$$\mathcal{F}\tau_a = e^{-i\langle a,\xi\rangle}\mathcal{F}\ , \qquad a \in \mathbb{R}^n\ , \tag{4.2.6}$$

where $e^{-i\langle a,\xi\rangle}$ stands for multiplication with the function $\xi \mapsto e^{-i\langle a,\xi\rangle}$.

Given $\alpha \in \dot{\mathbb{R}}^+$, we define the **dilation** $\sigma_\alpha u$ of $u \in E^{\mathbb{R}^n}$ by α by

$$\sigma_\alpha u(x) := u(\alpha x) , \qquad x \in \mathbb{R}^n . \tag{4.2.7}$$

Moreover,

$$\sigma_\alpha u(\varphi) := \alpha^{-n} u(\sigma_{\alpha^{-1}}\varphi) , \qquad u \in \mathcal{D}'(\mathbb{R}^n, E) , \quad \varphi \in \mathcal{D} . \tag{4.2.8}$$

Note that (4.2.8) is consistent with (4.2.7) if u is a regular distribution. It is easily verified that $\{ \sigma_\alpha \ ; \ \alpha \in \dot{\mathbb{R}}^+ \}$, the **group of dilations**, is strongly continuous on $\mathfrak{F}(\mathbb{R}^n, E)$, where $\mathfrak{F}$ satisfies (4.2.5). Moreover, it is not difficult to see that $\widehat{\sigma_\alpha u} = \alpha^{-n}\sigma_{\alpha^{-1}}\widehat{u}$ for $u \in \mathcal{S}(\mathbb{R}^n, E)$ and $\alpha > 0$, that is,

$$\mathcal{F}\sigma_\alpha = \alpha^{-n}\sigma_{\alpha^{-1}}\mathcal{F} , \qquad \alpha > 0 . \tag{4.2.9}$$

Given

$$\begin{array}{lccc} & u \in \mathcal{D}'(\mathbb{R}^n, E) & \text{and} & \varphi \in \mathcal{D}(\mathbb{R}^n) , \\ \text{or} & u \in \mathcal{E}'(\mathbb{R}^n, E) & \text{and} & \varphi \in \mathcal{E}(\mathbb{R}^n) , \\ \text{or} & u \in \mathcal{S}'(\mathbb{R}^n, E) & \text{and} & \varphi \in \mathcal{S}(\mathbb{R}^n) , \end{array}$$

respectively, the **convolution**, $u * \varphi$, of u and φ is defined by

$$u * \varphi(x) := u(\tau_x \check{\varphi}) , \qquad x \in \mathbb{R}^n .$$

It follows that

$$u * \varphi \in \mathcal{E}(\mathbb{R}^n, E)$$

and that

$$\partial^\alpha (u * \varphi) = (\partial^\alpha u) * \varphi = u * \partial^\alpha \varphi , \qquad \alpha \in \mathbb{N}^n . \tag{4.2.10}$$

In fact, convolution is a bilinear and separately continuous (indeed, hypocontinuous) map:

$$\mathcal{D}'(\mathbb{R}^n, E) \times \mathcal{D}(\mathbb{R}^n) \to \mathcal{E}(\mathbb{R}^n, E) , \tag{4.2.11}$$

$$\mathcal{E}'(\mathbb{R}^n, E) \times \mathcal{E}(\mathbb{R}^n) \to \mathcal{E}(\mathbb{R}^n, E) , \tag{4.2.12}$$

$$\mathcal{E}'(\mathbb{R}^n, E) \times \mathcal{D}(\mathbb{R}^n) \to \mathcal{D}(\mathbb{R}^n, E) , \tag{4.2.13}$$

$$\mathcal{S}'(\mathbb{R}^n, E) \times \mathcal{S}(\mathbb{R}^n) \to \mathcal{O}_M(\mathbb{R}^n, E) . \tag{4.2.14}$$

It is not difficult to verify that

$$\tau_a(u * \varphi) = (\tau_a u) * \varphi = u * \tau_a \varphi , \qquad a \in \mathbb{R}^n . \tag{4.2.15}$$

Note that

$$u(\tau_x \check{\varphi}) = u\big(\check{\varphi}(\cdot - x)\big) = u\big(\varphi(x - \cdot)\big) = \check{u}\big(\varphi(\cdot + x)\big) = \check{u}(\tau_{-x}\varphi) = \tau_x \check{u}(\varphi) .$$

Hence

$$\varphi * u(x) := \tau_x \check{u}(\varphi) = u * \varphi(x) , \qquad x \in \mathbb{R}^n . \tag{4.2.16}$$

Of course, if $u \in L_{1,\mathrm{loc}}(\mathbb{R}^n, E)$ is a regular distribution and $\varphi \in \mathcal{D}(\mathbb{R}^n)$,

$$u * \varphi(x) = \int_{\mathbb{R}^n} u(x-y)\varphi(y)\,dy = \int_{\mathbb{R}^n} u(y)\varphi(x-y)\,dy , \qquad x \in \mathbb{R}^n , \tag{4.2.17}$$

(where, as usual, $e\alpha := \alpha e$ for $\alpha \in \mathbb{K}$ and $e \in E$).

Literally as in the classical case (e.g., [Fol84, Section 8.2]), direct estimates of the integrals in (4.2.17) combined with density arguments show that convolution is bilinear and continuous:

$$BC(\mathbb{R}^n, E) \times L_1(\mathbb{R}^n) \to BC(\mathbb{R}^n, E) , \tag{4.2.18}$$
$$BUC(\mathbb{R}^n, E) \times L_1(\mathbb{R}^n) \to BUC(\mathbb{R}^n, E) , \tag{4.2.19}$$
$$L_p(\mathbb{R}^n, E) \times L_1(\mathbb{R}^n) \to L_p(\mathbb{R}^n, E) , \tag{4.2.20}$$
$$L_\infty(\mathbb{R}^n, E) \times L_1(\mathbb{R}^n) \to BUC(\mathbb{R}^n, E) , \tag{4.2.21}$$
$$L_q(\mathbb{R}^n, E) \times L_{q'}(\mathbb{R}^n) \to C_0(\mathbb{R}^n, E) \tag{4.2.22}$$

for $1 \le p \le \infty$ and $1 < q < \infty$. Moreover, the norms of these bilinear maps are bounded by 1. Observe that in situation (4.2.20), for example, this means that

$$\|u * \varphi\|_p \le \|u\|_p \, \|\varphi\|_1 , \qquad u \in L_p(\mathbb{R}^n, E) , \quad \varphi \in L_1(\mathbb{R}^n) ,$$

which is **Young's inequality** for convolutions. Note that (4.2.16) implies that E and $\mathbb{K}$ can be interchanged on the left-hand sides of (4.2.11)–(4.2.14) and (4.2.18)–(4.2.22).

Suppose that $\varphi \in \mathcal{L}_1(\mathbb{R}^n)$ and put

$$\varphi_\varepsilon(x) := \varepsilon^{-n}\varphi(x/\varepsilon) , \qquad x \in \mathbb{R}^n , \quad \varepsilon > 0 .$$

Also let $a := \int \varphi\,dx$. If $a = 1$ then $\{\, \varphi_\varepsilon \ ; \ \varepsilon > 0 \,\}$ is said to be an **approximate identity** and if, in addition, $\varphi \in \mathcal{D}(\mathbb{R}^n)$, $\varphi \ge 0$, and $\operatorname{supp}\varphi = \bar{\mathbb{B}}^n$, it is a **mollifier**. Classical arguments (e.g., [Fol84, Section 8.2]) show that

$$\varphi_\varepsilon * u \to au \qquad \text{as} \quad \varepsilon \to 0 \tag{4.2.23}$$

in

$$L_p(\mathbb{R}^n, E) \qquad \text{if} \quad u \in L_p(\mathbb{R}^n, E) , \qquad 1 \le p < \infty , \tag{4.2.24}$$

and in

$$BUC(\mathbb{R}^n, E) \qquad \text{if} \quad u \in BUC(\mathbb{R}^n, E) , \tag{4.2.25}$$

respectively. If there exist $c > 0$ and $\delta > 0$ such that

$$|\varphi(x)| \le c(1 + |x|)^{-n-\delta} , \qquad x \in \mathbb{R}^n , \tag{4.2.26}$$

then the proof of [Fol84, Theorem (8.15)] carries over to the E-valued situation to show that, given $u \in L_p(\mathbb{R}^n, E)$ for some $p \in [1, \infty]$,

$$\varphi_\varepsilon * u(x) \xrightarrow[\varepsilon \to 0]{} au(x) , \qquad \text{a.a. } x \in \mathbb{R}^n , \tag{4.2.27}$$

in fact: for every x in the Lebesgue set of u.

It follows from (4.1.7) and (4.2.14) that $u * \varphi \in \mathcal{S}'(\mathbb{R}^n, E)$ for $u \in \mathcal{S}'(\mathbb{R}^n, E)$ and $\varphi \in \mathcal{S}(\mathbb{R}^n)$. Hence the Fourier transform of $u * \varphi$ is well-defined and the CONVOLUTION THEOREM states that

$$(u * \varphi)\widehat{\ } = \widehat{u}\widehat{\varphi} .$$

Lastly, suppose that $E := \big(E, (\cdot|\cdot)\big)$ is a Hilbert space. Then $L_2(\mathbb{R}^n, E)$ is also a Hilbert space with respect to the inner product

$$(u|v)_2 := (u|v)_{L_2(\mathbb{R}^n, E)} := \int_{\mathbb{R}^n} (u|v)\, dx .$$

Given $u, v \in \mathcal{S}(\mathbb{R}^n, E)$, PLANCHEREL'S THEOREM guarantees that

$$(\widehat{u}|\widehat{v})_2 = (2\pi)^n (u|v)_2 . \tag{4.2.28}$$

Since $\mathcal{S}(\mathbb{R}^n, E)$ is dense in $L_2(\mathbb{R}^n, E)$, it follows from (4.2.28) and Fourier's inversion theorem that $(2\pi)^{-n/2}\mathcal{F}$ is a unitary operator on $L_2(\mathbb{R}^n, E)$.

4.2.1 Remark The formulas (4.2.11)–(4.2.14) and (4.2.17)–(4.2.22) exhibit a nonsymmetry due to the fact that we did not define multiplication and convolutions if both factors are vector-valued. For this suppose that E_j, $1 \le j \le 3$, are Banach spaces and

$$E_1 \times E_2 \to E_3 , \quad (e_1, e_2) \mapsto e_1 \bullet e_2$$

is a multiplication (cf. (II.1.1.4)). Then, given $u_j \in \mathcal{D}'(\mathbb{R}^n, E_j)$ for $j = 1, 2$, such that at least one has compact support, there exists a unique

$$u_1 *_\bullet u_2 \in \mathcal{D}'(\mathbb{R}^n, E_3) ,$$

the **convolution** of u_1 and u_2 (with respect to the multiplication $\bullet$), such that

$$(v_1 \otimes e_1) *_\bullet (v_2 \otimes e_2) = (v_1 * v_2) \otimes (e_1 \bullet e_2) , \qquad v_j \in \mathcal{D}'(\mathbb{R}^n) , \quad e_j \in E_j ,$$

provided v_1 or v_2 has compact support. Of course, $*$ is the 'scalar' convolution and $(v \otimes e_j)(\varphi) = \langle v, \varphi \rangle e_j$ for $(v, e_j) \in \mathcal{D}'(\mathbb{R}^n) \times E_j$ and $\varphi \in \mathcal{D}(\mathbb{R}^n)$. Moreover, letting

$$\mathfrak{F}_j \in \{ \mathcal{D}, \mathcal{D}', \mathcal{E}, \mathcal{E}', \mathcal{S}, \mathcal{S}', \mathcal{O}_M, BC, BUC, C_0, L_p \ ; \ 1 \le p < \infty \} ,$$

it follows that convolution with respect to $\bullet$ is a bilinear map

$$\mathfrak{F}_1(\mathbb{R}^n, E_1) \times \mathfrak{F}_2(\mathbb{R}^n, E_2) \to \mathfrak{F}_3(\mathbb{R}^n, E_3) ,$$

having the same continuity properties as the maps (4.2.11)–(4.2.14) and (4.2.18)–(4.2.22), respectively, provided the triple $(\mathfrak{F}_1, \mathfrak{F}_2, \mathfrak{F}_3)$ has the same meaning as in the corresponding formulas (thus $(\mathfrak{F}_1, \mathfrak{F}_2, \mathfrak{F}_3) = (\mathcal{D}', \mathcal{D}, \mathcal{E})$ in the analogue to (4.2.11) etc.). This follows from the general theory of vector-valued distributions for which we refer to [Schw57b] and [Schw57a]. Since we hardly have occasions to use this general setting we do not go into details. ■

4.3 The Hilbert Transform

Unless explicitly stated otherwise, we assume from now on that $n = 1$ and simply write $\mathfrak{F}$ for $\mathfrak{F}(\mathbb{R})$, where $\mathfrak{F}$ stands for any one of the spaces of distributions introduced so far. Thus $\mathcal{D} := \mathcal{D}(\mathbb{R})$, $\mathcal{S}' := \mathcal{S}'(\mathbb{R})$, etc. Of course, $E := (E, |\cdot|)$ is a Banach space.

Given $\varepsilon > 0$, define $(1/t)_\varepsilon \in L_{1,\text{loc}}$ by

$$(1/t)_\varepsilon(\tau) := \tau^{-1}\chi_{[\,|\tau| \geq \varepsilon]}(\tau) , \qquad \tau \in \mathbb{R} .$$

Then

$$\langle (1/t)_\varepsilon, \varphi \rangle = \int_{|\tau| \geq \varepsilon} \varphi(\tau)\, d\tau/\tau , \qquad \varphi \in \mathcal{S} .$$

It is well-known and easily seen that there exists a unique temperate distribution PV$(1/t)$, the **principal value** of $1/t$, such that

$$\langle \mathrm{PV}(1/t), \varphi \rangle = \lim_{\varepsilon \to 0} \langle (1/t)_\varepsilon, \varphi \rangle = \lim_{\varepsilon \to 0} \int_{|\tau| \geq \varepsilon} \varphi(\tau)\, d\tau/\tau , \qquad \varphi \in \mathcal{S} . \tag{4.3.1}$$

Given $u \in \mathcal{S}(\mathbb{R}, E)$, we define the **Hilbert transform** Hu of u by

$$Hu := H_E u := \pi^{-1}\,\mathrm{PV}(1/t) * u ,$$

and the **truncated Hilbert transform**, H_ε, $\varepsilon > 0$, by

$$H_\varepsilon u := H_{E,\varepsilon} u := \pi^{-1}(1/t)_\varepsilon * u , \qquad \varepsilon > 0 ,$$

so that

$$H_\varepsilon u(t) = \frac{1}{\pi} \int_{|\tau| \geq \varepsilon} u(t - \tau)\, d\tau/\tau = \frac{1}{\pi} \int_{|t-\tau| \geq \varepsilon} \frac{u(\tau)}{t - \tau}\, d\tau , \qquad t \in \mathbb{R} , \quad u \in \mathcal{S}(\mathbb{R}, E) .$$

The following proposition shows that H_ε is an approximation of H.

4.3.1 Proposition *If $u \in \mathcal{S}(\mathbb{R}, E)$ then $H_\varepsilon u \to Hu$ in $\mathcal{O}_M(\mathbb{R}, E)$ as $\varepsilon \to 0$.*

Proof By (4.3.1) we know that $(1/t)_\varepsilon \to \mathrm{PV}(1/t)$ in the w^*-topology of $\mathcal{S}'$. Since $\mathcal{S}$ is a Montel space, $(1/t)_\varepsilon \to \mathrm{PV}(1/t)$ in $\mathcal{S}'$ as $\varepsilon \to 0$, by the Banach-Steinhaus theorem. Now the assertion follows from (4.2.14). ■

Since $\mathcal{O}_M(\mathbb{R},E) \hookrightarrow \mathcal{E}(\mathbb{R},E)$, this proposition justifies the notation

$$Hu(t) := H_E u(t) = \frac{1}{\pi}\int_{-\infty}^{\infty} u(t-\tau)\,d\tau/\tau = \frac{1}{\pi}\int_{-\infty}^{\infty} \frac{u(\tau)}{t-\tau}\,d\tau\ , \qquad t \in \mathbb{R}\ ,$$

for the Hilbert transform of $u \in \mathcal{S}(\mathbb{R},E)$, where the integral is a principal value integral.

Now we show that H extends to a bounded linear operator on $L_2(\mathbb{R},E)$, provided E is a Hilbert space. For this we need the following Fourier transform:

4.3.2 Lemma $\big(\mathrm{PV}(1/t)\big)\hat{} = -i\pi\,\mathrm{sign}$.

Proof Given $0<\varepsilon<R<\infty$, let

$$(1/t)_\varepsilon^R(\tau) := \tau^{-1}\chi_{[\varepsilon\le|\tau|\le R]}(\tau)\ , \qquad \tau\in\mathbb{R}\ .$$

Then $(1/t)_\varepsilon^R \in L_1 \hookrightarrow \mathcal{S}'$, and

$$\big\langle (1/t)_\varepsilon^R, \varphi\big\rangle = \int_{\varepsilon\le|t|\le R} \varphi(\tau)\,d\tau/\tau \xrightarrow[R\to\infty]{} \big\langle (1/t)_\varepsilon,\varphi\big\rangle\ , \qquad \varphi\in\mathcal{S}\ .$$

Thus

$$(1/t)_\varepsilon^R \to (1/t)_\varepsilon \quad \text{in } \mathcal{S}'_{w^*} \qquad \text{as} \quad R\to\infty\ . \tag{4.3.2}$$

Note that

$$\begin{aligned}\big((1/t)_\varepsilon^R\big)\hat{}(s) &= \int_{\varepsilon\le|\tau|\le R} \tau^{-1}e^{-is\tau}\,d\tau = -2i\int_\varepsilon^R \frac{\sin(s\tau)}{\tau}\,d\tau\\ &= -2i\,\mathrm{sign}(s)\int_{\varepsilon|s|}^{R|s|}\frac{\sin\xi}{\xi}\,d\xi\end{aligned}$$

for $s\in\mathbb{R}$. Hence

$$\big((1/t)_\varepsilon^R\big)\hat{} \to -2i\,\mathrm{sign}(\cdot)\int_{\varepsilon|\cdot|}^{\infty}\frac{\sin\xi}{\xi}\,d\xi\ , \qquad R\to\infty\ ,$$

in $L_{1,\mathrm{loc}} \hookrightarrow \mathcal{S}' \hookrightarrow \mathcal{S}_{w^*}$. Since, trivially, $\mathcal{F}\in\mathcal{L}(\mathcal{S}'_{w^*})$, we deduce from (4.3.2) that

$$\widehat{(1/t)_\varepsilon} := \big((1/t)_\varepsilon\big)\hat{} = -2i\,\mathrm{sign}(\cdot)\int_{\varepsilon|\cdot|}^{\infty}\frac{\sin\xi}{\xi}\,d\xi$$

Hence, given $\varphi\in\mathcal{S}$,

$$\big\langle \widehat{(1/t)_\varepsilon},\varphi\big\rangle \to -i\pi\langle \mathrm{sign},\varphi\rangle\ , \qquad \varepsilon\to 0\ ,$$

that is, $\widehat{(1/t)_\varepsilon} \to -i\pi\,\mathrm{sign}$ in $\mathcal{S}'_{w^*}$ as $\varepsilon\to 0$. Since $(1/t)_\varepsilon \to \mathrm{PV}(1/t)$ in $\mathcal{S}'$ by the proof of Proposition 4.3.1 and (4.2.3), the assertion follows. ■

Suppose that, given $p \in [1, \infty)$,

$$\|Hu\|_{L_p(\mathbb{R},E)} \le c\,\|u\|_{L_p(\mathbb{R},E)} , \qquad u \in \mathcal{S}(\mathbb{R}, E) .$$

Then, since $\mathcal{S}(\mathbb{R}, E)$ is dense in $L_p(\mathbb{R}, E)$, there exists a unique $\overline{H} \in \mathcal{L}\big(L_p(\mathbb{R}, E)\big)$ extending H. Of course, $\overline{H}$ is again denoted by H and said to be the **Hilbert transform on** $L_p(\mathbb{R}, E)$.

4.3.3 Theorem *Let E be a Hilbert space. Then the Hilbert transform is a unitary skew-adjoint linear operator on $L_2(\mathbb{R}, E)$.*

Proof Given $u \in \mathcal{S}(\mathbb{R}, E)$, the convolution theorem and Lemma 4.3.2 imply

$$\widehat{Hu} = -i\,\mathrm{sign}(\cdot)\widehat{u} . \tag{4.3.3}$$

Thus, by Plancherel's theorem,

$$(Hu|Hv)_2 = (2\pi)^{-1}(\widehat{Hu}|\widehat{Hv})_2 = (2\pi)^{-1}(\widehat{u}|\widehat{v})_2 = (u|v)_2 , \qquad u, v \in \mathcal{S}(\mathbb{R}, E) .$$

From this and the density of $\mathcal{S}(\mathbb{R}, E)$ in $L_2(\mathbb{R}, E)$ we see that $H \in \mathcal{L}\mathrm{aut}\big(L_2(\mathbb{R}, E)\big)$ and $H^{-1}v = -Hv$ for $v \in L_2(\mathbb{R}, E)$. Now the assertion is obvious. ■

Now we show that, given any Banach space E, the Hilbert transform is bounded on $L_p(\mathbb{R}, E)$ for $1 < p < \infty$, provided this is true for at least one $p \in (1, \infty)$. For this we need some preparations.

4.3.4 Lemma *Let E be a Banach space. Then $Hw \in L_1\big([\,|t| > 2T], E\big)$ and*

$$\int_{|t|>2T} |Hw|\,dt \le (2/\pi)\,\|w\|_1$$

for $T > 0$ and $w \in L_1(\mathbb{R}, E)$ satisfying $\int w\,dt = 0$ and $\mathrm{supp}(w) \subset [-T, T]$.

Proof Let $T > 0$ and suppose that $v \in \mathcal{D}(\mathbb{R}, E)$ with $\int v\,dt = 0$ and

$$\mathrm{supp}(v) \subset [-T - \delta, T + \delta]$$

for some $\delta \in (0, T/2)$. Given $\varepsilon \in (0, T/2)$ it follows that

$$\begin{aligned}
\int_{|t|>2T} |H_\varepsilon v|\,dt &= \frac{1}{\pi}\int_{|t|>2T}\Big|\int_{|t-\tau|\ge\varepsilon} \frac{v(\tau)}{t-\tau}\,d\tau\Big|\,dt \\
&= \frac{1}{\pi}\int_{|t|>2T}\Big|\int_{-T-\delta}^{T+\delta} v(\tau)\big[(t-\tau)^{-1} - t^{-1}\big]\,d\tau\Big|\,dt \\
&\le \frac{1}{\pi}\int_{|t|\ge 2T} \frac{T+\delta}{(|t|-T-\delta)\,|t|}\,dt\,\|v\|_1 \le \frac{2}{\pi}\frac{T+\delta}{T-\delta}\int_{|t|\ge 2T}\frac{T\,dt}{t^2}\,\|v\|_1 \\
&\le \frac{2}{\pi}\frac{T+\delta}{T-\delta}\,\|v\|_1 .
\end{aligned}$$

It follows from Proposition 4.3.1 and (4.1.7) that

$$H_\varepsilon v(\varphi) \to Hv(\varphi) \qquad \text{as} \quad \varepsilon \to 0 \;, \qquad \varphi \in \mathcal{S} \;.$$

If $\operatorname{supp}(\varphi) \subset [\,|t| \geq 2T]$, similarly as above,

$$H_\varepsilon v(\varphi) = \frac{1}{\pi} \int_{|t|>2T} \varphi(t) \int_{-\infty}^{\infty} v(\tau)\big[(t-\tau)^{-1} - t^{-1}\big]\, d\tau\, dt \tag{4.3.4}$$

for $0 < \varepsilon < T/2$, that is, $H_\varepsilon v(\varphi)$ is independent of $\varepsilon \in (0, T/2)$ for $\varphi \in \mathcal{D}$ with support in the set $[\,|t| \geq 2T]$. Hence

$$H_\varepsilon v|[\,|t| > 2T] = Hv|[\,|t| > 2T] \;. \tag{4.3.5}$$

This shows that $Hv \in L_1\big([\,|t| > 2T], E\big)$ and that

$$\int_{|t|>2T} |Hv|\, dt \leq \frac{2}{\pi} \frac{T+\delta}{T-\delta} \|v\|_1 \;. \tag{4.3.6}$$

Let $\{\, \varphi_\varepsilon \;;\; \varepsilon > 0 \,\}$ be a mollifier. Then $\varphi_\varepsilon * w \in \mathcal{D}(\mathbb{R}, E)$ with support in the interval $[-T-\varepsilon, T+\varepsilon]$, and, by Fubini's theorem,

$$\int \varphi_\varepsilon * w\, dt = \iint \varphi_\varepsilon(t-\tau) w(\tau)\, d\tau\, dt = \int w\, d\tau\, \|\varphi_\varepsilon\|_1 = 0 \;.$$

Since $\varphi_\varepsilon * w \to w$ in $L_1(\mathbb{R}, E)$, it follows from (4.3.4) and (4.3.5) that (4.3.6) is true with v replaced by w. Since $\delta \in (0, T/2)$ was arbitrary, the assertion follows. ■

In the next lemma we recall the CALDERÓN-ZYGMUND DECOMPOSITION THEOREM, that we state in greater generality than presently needed.

4.3.5 Lemma *Let E be a Banach space and let $u \in L_1(\mathbb{R}^n, E)$ and $\alpha > 0$ be given. Then there exist $v, w_k \in L_1(\mathbb{R}^n, E)$ for $k \in \mathbb{N}$ such that*

$$u = v + \sum w_k \tag{4.3.7}$$

and such that

$$\|v\|_\infty \leq 2^n \alpha \;, \qquad \int w_k\, dx = 0 \tag{4.3.8}$$

and

$$\|v\|_1 + \sum \|w_k\|_1 \leq 3\, \|u\|_1 \;. \tag{4.3.9}$$

Furthermore, there exists a sequence of pair-wise disjoint cubes Q_k, with sides parallel to the coordinate-hyperplanes, such that

$$\operatorname{supp}(w_k) \subset \overline{Q}_k \tag{4.3.10}$$

and

$$\alpha \sum \lambda_n(Q_k) \leq \|u\|_1 \;, \tag{4.3.11}$$

where $\lambda_n(\cdot)$ denotes the n-dimensional Lebesgue measure.

Proof The 'scalar proof' given in [Hör83, Lemma 4.5.5], for example, applies literally to the E-valued case (also see [Tri78, Lemma 2.2.2]). ■

After these preparations we can prove the following important theorem concerning the boundedness of the Hilbert transform.

4.3.6 Theorem *Let E be a Banach space and suppose that the Hilbert transform is bounded on $L_p(\mathbb{R}, E)$ for some $p \in (1, \infty)$. Then it is bounded on $L_q(\mathbb{R}, E)$ and on $L_q(\mathbb{R}, E')$ for each $q \in (1, \infty)$.*

Proof (i) First we show that H is of 'weak type' $(1, \infty)$ in the sense that

$$\sigma\lambda_1\bigl([\,|Hu| > \sigma]\bigr) \leq c\,\|u\|_1 \ , \qquad u \in L_1(\mathbb{R}, E) \ , \quad \sigma > 0 \ . \tag{4.3.12}$$

By assumption there exist $p \in (1, \infty)$ and $\mu > 0$ such that

$$\|Hu\|_p \leq \mu\,\|u\|_p \ , \qquad u \in L_p(\mathbb{R}, E) \ . \tag{4.3.13}$$

Fix $u \in \mathcal{S}(\mathbb{R}, E)$. Then, given $\sigma > 0$, we have the decomposition of Lemma 4.3.5 with $\alpha := \sigma/\mu$. Hence, thanks to (4.3.8),

$$\sigma^p\lambda_1\bigl([\,|Hv| > \sigma]\bigr) \leq \|Hv\|_p^p \leq \mu^p\,\|v\|_p^p \leq \mu^p 2^{p-1}(\sigma/\mu)^{p-1}\,\|v\|_1$$

so that

$$\sigma\lambda_1\bigl([\,|Hv| > \sigma]\bigr) \leq 2^{p-1}\mu\,\|v\|_1 \ . \tag{4.3.14}$$

Let Q_k^* be the 'double cube' (that is, interval) with the same center as Q_k and twice its length, and let $V := \bigcup Q_k^*$. Then, thanks to (4.3.11)

$$\sigma\lambda_1(V) \leq \sigma\textstyle\sum\lambda_1(Q_k^*) \leq 2\sigma\sum\lambda_1(Q_k) \leq 2\mu\,\|u\|_1 \ . \tag{4.3.15}$$

Note that H is translation invariant. Thus

$$\begin{aligned}\sigma\lambda_1\bigl(V^c \cap \bigl[\textstyle\sum|Hw_k| > \sigma\bigr]\bigr) &\leq \int_{V^c} \textstyle\sum|Hw_k|\,dt \\ &= \sum \int_{\bigcap (Q_k^*)^c} |Hw_k|\,dt \leq \sum \int_{|t|\geq 2T_k} |Hw_k^*|\,dt\end{aligned}$$

for suitable $T_k > 0$, where $w_k^* \in L_1(\mathbb{R}, E)$, has its support in $[-T_k, T_k]$, and satisfies $\int w_k^*\,dt = 0$. Now we deduce from Lemma 4.3.4 that

$$\sigma\lambda_1\bigl(V^c \cap \bigl[\textstyle\sum|Hw_k| > \sigma\bigr]\bigr) \leq (2/\pi)\textstyle\sum\|w_k\|_1 \ . \tag{4.3.16}$$

Lastly, observe that by (4.3.7)

$$[\,|Hu| > \sigma] \subset [\,|Hv| > \sigma/2] \cup \bigl(V^c \cap \bigl[\textstyle\sum|Hw_k| > \sigma/2\bigr]\bigr) \cup V \ .$$

Thus we deduce from (4.3.14)–(4.3.16) that

$$\begin{aligned}
&\sigma\lambda_1\big([\,|Hu|>\sigma]\big)\\
&\quad\le 2\Big\{(\sigma/2)\lambda_1\big([\,|Hv|>\sigma/2]\big)+(\sigma/2)\lambda_1\big(V^c\cap\big[\textstyle\sum|Hw_k|>\sigma/2\big]\big)\Big\}+\sigma\lambda_1(V)\\
&\quad\le \mu 2^p\,\|v\|_1+(4/\pi)\textstyle\sum\|w_k\|_1+2\mu\,\|u\|_1\ .
\end{aligned}$$

Now (4.3.9) and the density of $\mathcal{S}(\mathbb{R},E)$ in $L_1(\mathbb{R},E)$ imply the validity of (4.3.12).

(ii) From (4.3.12), (4.3.13), and the vector-valued version of the interpolation theorem of Marcinkiewicz ([Tri78, Theorem 1.18.3]) we see that $H_E\in\mathcal{L}\big(L_q(\mathbb{R},E)\big)$ for $1<q<p$.

(iii) Now let $E:=\mathbb{K}$. Then Theorem 4.3.3 implies $H_{\mathbb{K}}\in\mathcal{L}(L_2)$. Consequently, we deduce from (ii) that $H_{\mathbb{K}}\in\mathcal{L}(L_q)$, $\;1<q\le 2$. Theorem 4.3.3 also implies that $(H_{\mathbb{K}})'=-H_{\mathbb{K}}\in\mathcal{L}(L_2)$. Hence we deduce from (ii) that $(H_{\mathbb{K}})'\in\mathcal{L}(L_{q'})$ for $1<q'\le 2$, since it is trivially verified that the dual $H_{\mathbb{K}}'$ of $H_{\mathbb{K}}$ on $L_{q'}$ is again given by $-H_{\mathbb{K}}$. Thus, by the reflexity of L_q for $1<q<\infty$, $\;H_{\mathbb{K}}=(H_{\mathbb{K}}')'\in\mathcal{L}(L_q)$ for $2\le q<\infty$. That is, $H_{\mathbb{K}}\in\mathcal{L}(L_q)$ for $1<q<\infty$.

(iv) Suppose that $1<r<\infty$. Then we claim that

$$H_E\in\mathcal{L}\big(L_r(\mathbb{R},E)\big)\quad\Longleftrightarrow\quad H_{E'}\in\mathcal{L}\big(L_{r'}(\mathbb{R},E')\big)\ . \tag{4.3.17}$$

For this we recall first that

$$\left.\begin{aligned}
&L_{r'}(\mathbb{R},E')\text{ is (identified with) a closed linear subspace}\\
&\text{of } L_r(\mathbb{R},E)'\text{ such that}\\
&\langle v',u\rangle_{L_r(\mathbb{R},E)}=\int_{\mathbb{R}}\big\langle v'(t),u(t)\big\rangle\,dt\\
&\text{for } v'\in L_{r'}(\mathbb{R},E')\text{ and } u\in L_r(\mathbb{R},E).
\end{aligned}\right\} \tag{4.3.18}$$

Next, given $\varphi\in L_r$ and $e\in E$, put $\varphi\otimes e:=\varphi e$. Moreover,

$$L_r\otimes E:=\big\{\textstyle\sum_{j=0}^m\varphi_j\otimes e_j\ ;\ \varphi_j\in L_r,\ e_j\in E,\ m\in\mathbb{N}\big\}\ .$$

Then

$$L_r\otimes E\overset{d}{\subset}L_r(\mathbb{R},E)\ , \tag{4.3.19}$$

thanks to the fact that the simple functions are dense in $L_r(\mathbb{R},E)$ and contained in $L_r\otimes E$. Observe that (iii) and the obvious fact that $H_E(\varphi\otimes e)=(H_{\mathbb{K}}\varphi)\otimes e$ imply

$$H_E(L_r\otimes E)\subset L_r\otimes E\ . \tag{4.3.20}$$

Since

$$\big\langle\varphi'\otimes e',H_E(\varphi\otimes e)\big\rangle_{L_r(\mathbb{R},E)}=\langle\varphi',H_{\mathbb{K}}\varphi\rangle_{L_r}\langle e',e\rangle_E$$

for $(\varphi',\varphi)\in L_{r'}\times L_r$ and $(e',e)\in E'\times E$, it follows from (iii) that

$$\langle v',H_Eu\rangle_{L_r(\mathbb{R},E)}=-\langle H_{E'}v',u\rangle_{L_r(\mathbb{R},E)} \tag{4.3.21}$$

for $u\in L_r\otimes E$ and $v'\in L_{r'}\otimes E'$.

Suppose that $H_E \in \mathcal{L}\big(L_r(\mathbb{R}, E)\big)$. Then (4.3.19) and (4.3.21) imply

$$(H_E)' \supset -H_{E'} | (L_{r'} \otimes E') .$$

Thus, using (4.3.18)–(4.3.20) and the fact that $(H_E)' \in \mathcal{L}\big(L_r(\mathbb{R}, E)'\big)$, we see that

$$H_{E'} \in \mathcal{L}\big(L_{r'}(\mathbb{R}, E')\big) . \tag{4.3.22}$$

Conversely, let (4.3.22) be satisfied. Then an analogous argument, combined with the fact that E is a closed linear subspace of E'', proves that $H_E \in \mathcal{L}\big(L_r(\mathbb{R}, E)\big)$.

(v) From (ii) and (iv) it follows that $H_{E'}$ is bounded on $L_{q'}(\mathbb{R}, E')$ for $1 < q' \leq p'$. Hence (iv) implies $H_E \in \mathcal{L}\big(L_q(\mathbb{R}, E)\big)$ for $p \leq q < \infty$. This proves the theorem. ■

The proofs of this section are adaptions of well-known arguments from harmonic analysis used in the study of the scalar Hilbert transform (e.g., [Ste70a]).

4.4 UMD Spaces and Fourier Multipliers

Let $E := (E, |\cdot|)$ be a Banach space. Then, in general, the Hilbert transform is not bounded on $L_p(\mathbb{R}, E)$ for any $p \in (1, \infty)$. This justifies the following definition: a Banach space E is a **UMD space** if the Hilbert transform is bounded on $L_p(\mathbb{R}, E)$ for some $p \in (1, \infty)$.

4.4.1 Theorem *Let E be a UMD space. Then the Hilbert transform is bounded on $L_p(\mathbb{R}, E)$ for $1 < p < \infty$.*

Proof This is a consequence of Theorem 4.3.6. ■

4.4.2 Remarks **(a)** Let $(\Omega, \mathcal{A}, P)$ be a probability space and let $(\mathcal{A}_k)$ be an increasing sequence of sub-σ-fields of $\mathcal{A}$. A sequence (u_k) of E-valued P-integrable random variables on $(\Omega, \mathcal{A}, P)$ is said to be a **martingale** on E if

$$\int_A u_k \, dP = \int_A u_j \, dP , \qquad A \in \mathcal{A}_k , \quad 0 \leq k < j < \infty ,$$

that is, if

$$u_k = \mathbb{E}(u_j | \mathcal{A}_k) , \qquad 0 \leq k < j < \infty ,$$

where $\mathbb{E}(\cdot\, | \mathcal{A}_k)$ is the conditional expectation.

The Banach space E is said to possess the property of **unconditionality of martingale differences** (UMD) if for some $p \in (1, \infty)$

$$\Big\| \sum_{k=0}^{n} \varepsilon_k (u_k - u_{k-1}) \Big\|_{L_p(\Omega, \mathcal{A}, P; E)} \leq c \Big\| \sum_{k=0}^{n} (u_k - u_{k-1}) \Big\|_{L_p(\Omega, \mathcal{A}, P; E)}$$

for all $n \in \mathbb{N}$, $\varepsilon_k \in \{\pm 1\}$, and all E-valued martingales (u_k), where $u_{-1} := 0$. It has been shown by Bourgain [Bour83] that E possesses the property UMD if E is a UMD space. Conversely, Burkholder [Bur83] proved that the validity of property UMD implies that E is a UMD space. This explains the name.

(b) A Banach space E is **ζ-convex** if there exists a symmetric, biconvex function $\zeta : E \times E \to \mathbb{R}$ satisfying $\zeta(0,0) > 0$ and $\zeta(x,y) \leq \|x+y\|$ for $x, y \in E$ with $\|x\| = \|y\| = 1$. Burkholder [Bur81] proved that a Banach space is ζ-convex iff it is a UMD space.

(c) UMD spaces are reflexive. The converse is not true (e.g., [Rub86]). ■

In order to establish some important properties of UMD spaces we need a few basic facts about Fourier multipliers.

Let $m \in \mathcal{O}_M(\mathbb{R}^n)$ be given. It follows from (4.1.8), the convolution theorem, (4.2.1), and (4.2.3) that, given a Banach space E,

$$\big[u \mapsto \mathcal{F}^{-1} m \mathcal{F} u := \mathcal{F}^{-1}(m\widehat{u}) = \mathcal{F}^{-1} m * u\big] \in \mathcal{L}\big(\mathcal{S}(\mathbb{R}^n, E)\big) . \tag{4.4.1}$$

We also put

$$m(D)u := \mathcal{F}^{-1} m \mathcal{F} u , \qquad u \in \mathcal{S}(\mathbb{R}^n, E) ,$$

and say that $m(D)$ is a translation invariant **pseudodifferential operator** with **symbol** m. This is justified by (4.2.4) and (4.2.15).

Now one can ask what conditions have to be imposed on m so that (4.4.1) is true if $\mathcal{S}(\mathbb{R}^n, E)$ is replaced by $L_p(\mathbb{R}^n, E)$ or some other Banach space, $\mathfrak{F}(\mathbb{R}^n, E)$, satisfying

$$\mathcal{S}(\mathbb{R}^n, E) \hookrightarrow \mathfrak{F}(\mathbb{R}^n, E) \hookrightarrow \mathcal{S}'(\mathbb{R}^n, E) . \tag{4.4.2}$$

If $E = \mathbb{K}$, the famous Mikhlin multiplier theorem gives a satisfactory answer for the L_p-case. It is interesting and of great importance that this theorem essentially carries over to the E-valued case, provided E is a UMD space.

To be more precise, let $\mathfrak{F}(\mathbb{R}^n, E)$ be a Banach space satisfying (4.4.2). Then an element $m \in L_\infty(\mathbb{R}^n)$ is said to be a **Fourier multiplier** for $\mathfrak{F}(\mathbb{R}^n, E)$ (an $\mathfrak{F}(\mathbb{R}^n, E)$-multiplier), if

$$m(D) = \mathcal{F}^{-1} m \mathcal{F} = \mathcal{F}^{-1} m \, * \in \mathcal{L}\big(\mathfrak{F}(\mathbb{R}^n, E)\big) . \tag{4.4.3}$$

This means that $m\widehat{u} \in \mathcal{S}'(\mathbb{R}^n, E)$ for $u \in \mathfrak{F}(\mathbb{R}^n, E)$ so that $\mathcal{F}^{-1}(m\widehat{u})$ is well-defined, $\mathcal{F}^{-1}(m\widehat{u}) \in \mathfrak{F}(\mathbb{R}^n, E)$, and there exists $c > 0$ such that

$$\|m(D)u\|_{\mathfrak{F}(\mathbb{R}^n,E)} = \|\mathcal{F}^{-1}(m\widehat{u})\|_{\mathfrak{F}(\mathbb{R}^n,E)} \leq c\, \|u\|_{\mathfrak{F}(\mathbb{R}^n,E)} \tag{4.4.4}$$

for $u \in \mathfrak{F}(\mathbb{R}^n, E)$. If $\mathcal{S}(\mathbb{R}^n, E)$ is dense in $\mathfrak{F}(\mathbb{R}^n, E)$, it suffices, of course, to verify (4.4.4) for all $u \in \mathcal{S}(\mathbb{R}^n, E)$. (Note that $\mathcal{F}^{-1} m * u$ is well-defined for $m \in L_\infty(\mathbb{R}^n)$ and $u \in \mathcal{S}(\mathbb{R}, E)$. In the general case the middle term in (4.4.3) can be taken to define $\mathcal{F}^{-1} m \, *$.)

The space of all $L_p(\mathbb{R}^n, E)$-multipliers is denoted by

$$M_p(E) := M_p(\mathbb{R}^n, E) , \qquad 1 \leq p \leq \infty .$$

It is a Banach algebra with respect to the norm

$$\|m\|_{M_p(E)} := \|m(D)\|_{\mathcal{L}(L_p(\mathbb{R}^n,E))} .$$

We write $\mathcal{M}_M := \mathcal{M}_M(\mathbb{R}^n)$ for the vector space of all $m \in L_\infty(\mathbb{R}^n)$ such that the distributional derivatives of order α, where $\alpha \leq (1,\dots,1) \in \mathbb{N}^n$, are represented on $\dot{\mathbb{R}}^n$ by functions and such that

$$\|m\|_{\mathcal{M}_M} := \max_{\alpha \leq (1,\dots,1)} \sup_{\xi \in \dot{\mathbb{R}}^n} |\xi|^{|\alpha|} \, |\partial^\alpha m(\xi)| < \infty . \tag{4.4.5}$$

Then

$$\mathcal{M}_M := \mathcal{M}_M(\mathbb{R}^n) := (\mathcal{M}_M, \|\cdot\|_{\mathcal{M}_M}) \tag{4.4.6}$$

is a Banach space and a continuous multiplication algebra, that is, point-wise multiplication

$$\mathcal{M}_M \times \mathcal{M}_M \to \mathcal{M}_M , \quad (m_1, m_2) \mapsto m_1 m_2$$

is well-defined and continuous.

After these preparations we can formulate the following vector-valued extension of the MIKHLIN MULTIPLIER THEOREM.

4.4.3 Theorem *Let E be a UMD space. Then*

$$\mathcal{M}_M \hookrightarrow M_p(E) , \qquad 1 < p < \infty ,$$

and this injection is an algebra homomorphism.

Proof For the first assertion we refer to [Zim89], the second one is obvious. ■

It should be remarked that an earlier, somewhat weaker result has been given in [McC84].

4.4.4 Remark If $E = \mathbb{C}$, Hörmander's extension of Mikhlin's theorem guarantees that $\mathcal{M}_M \hookrightarrow M_p := M_p(\mathbb{C})$ for $1 < p < \infty$, provided the norm (4.4.5) is replaced by

$$m \mapsto \max_{|\alpha| \leq [n/2]+1} \sup_{\xi \in \dot{\mathbb{R}}^n} |\xi|^{|\alpha|} \, |\partial^\alpha m(\xi)| ,$$

where $[s]$ denotes the largest integer not exceeding $s \in \mathbb{R}$. ■

4.5 Properties of UMD Spaces

As a first application of the generalized Mikhlin multiplier theorem we show that, given $u \in L_p(\mathbb{R}^n, E)$, the truncated Hilbert transform $H_\varepsilon u$ converges in $L_p(\mathbb{R}^n, E)$, and almost everywhere, towards Hu if E is a UMD space.

4.5.1 Theorem *Suppose that E is a UMD space, $1 < p < \infty$, and $u \in L_p(\mathbb{R}, E)$. Then $H_\varepsilon u \to Hu$ in $L_p(\mathbb{R}, E)$ and almost everywhere as $\varepsilon \to 0$.*

Proof Let $\psi(\xi) := e^{-|\xi|}$ for $\xi \in \mathbb{R}$. Then

$$\begin{aligned}\varphi(x) := \mathcal{F}^{-1}\psi(x) &= \frac{1}{2\pi}\int_0^\infty e^{(ix-1)\xi}\,d\xi + \frac{1}{2\pi}\int_\infty^0 e^{(ix+1)\xi}\,d\xi \\ &= \frac{1}{2\pi}\Big[\frac{1}{1-ix} + \frac{1}{1+ix}\Big] = \frac{1}{\pi}\frac{1}{1+x^2}\end{aligned}$$

for $x \in \mathbb{R}$. Hence $\varphi \in \mathcal{L}_1(\mathbb{R})$ and $\int \varphi\,dx = \widehat{\varphi}(0) = \psi(0) = 1$. Thus $\{\,\varphi_\varepsilon\ ;\ \varepsilon > 0\,\}$ is an approximate identity and it follows from (4.2.23), (4.2.24), (4.2.26), and (4.2.27) and the fact that $Hu \in L_p(\mathbb{R}, E)$ that

$$\varphi_\varepsilon * Hu \to Hu \qquad \text{in } L_p(\mathbb{R}, E) \text{ and a.e.} \tag{4.5.1}$$

as $\varepsilon \to 0$.

Let

$$m_\varepsilon := -\frac{2i\,\mathrm{sign}(\cdot)}{\pi}\int_{\varepsilon|\cdot|}^\infty \frac{\sin\xi}{\xi}\,d\xi\ , \qquad \varepsilon > 0\ .$$

It is easily verified that $m_\varepsilon \in \mathcal{M}_M$ and that $\|m_\varepsilon\|_{\mathcal{M}_M} \le c$ for $\varepsilon > 0$. Hence, since by the proof of Lemma 4.3.2 and the convolution theorem, $H_\varepsilon u = m_\varepsilon(D)u$ for $\varepsilon > 0$, Theorem 4.4.3 implies

$$H_\varepsilon u \in L_p(\mathbb{R}, E) \quad \text{and} \quad \|H_\varepsilon\|_{\mathcal{L}(L_p(\mathbb{R},E))} \le c\ , \qquad \varepsilon > 0\ . \tag{4.5.2}$$

Since $\varphi_\varepsilon = \varepsilon^{-1}\sigma_{\varepsilon^{-1}}\varphi$, we infer from (4.2.9) that $\widehat{\varphi_\varepsilon} = \sigma_\varepsilon\widehat{\varphi} = \sigma_\varepsilon\psi$. Note that ψ and $m := -i\,\mathrm{sign}(\cdot)$ belong to $\mathcal{M}_M$. Hence Theorem 4.4.3, the convolution theorem, and Lemma 4.3.2 imply

$$\varphi_\varepsilon * Hu = \mathcal{F}^{-1}\big((\sigma_\varepsilon\psi)m\big)\mathcal{F}u = \mathcal{F}^{-1}\big((\sigma_\varepsilon\psi)m\big) * u\ , \qquad \varepsilon > 0\ .$$

Observe that $(\sigma_\varepsilon\psi)m = \sigma_\varepsilon(\psi m)$ so that (4.2.9) gives

$$\mathcal{F}^{-1}\big((\sigma_\varepsilon\psi)m\big) = \varepsilon^{-1}\sigma_{\varepsilon^{-1}}\mathcal{F}^{-1}(\psi m) = \chi_\varepsilon\ ,$$

where

$$\begin{aligned}\chi(x) := \mathcal{F}^{-1}(\psi m)(x) &= \frac{-i}{2\pi}\int_0^\infty e^{-\xi(1-ix)}\,d\xi + \frac{i}{2\pi}\int_{-\infty}^0 e^{\xi(1+ix)}\,d\xi \\ &= \frac{i}{2\pi}\Big[\frac{-1}{1-ix} + \frac{1}{1+ix}\Big] = \frac{1}{\pi}\frac{x}{1+x^2}\end{aligned}$$

for $x \in \mathbb{R}$. Consequently,

$$\varphi_\varepsilon * Hu = \chi_\varepsilon * u \ , \qquad \varepsilon > 0 \ .$$

Lastly, let

$$k(x) := \begin{cases} \dfrac{1}{\pi x(1+x^2)} \ , & |x| > 1 \ , \\ \dfrac{-x}{\pi(1+x^2)} \ , & |x| < 1 \ . \end{cases}$$

Then $k \in \mathcal{L}_1(\mathbb{R})$ and $\int k\,dx = 0$ since k is odd. Moreover, $|k(x)| \le c(1+|x|)^2$ for $x \in \mathbb{R}$. Hence $k_\varepsilon * u \to 0$ in $L_p(\mathbb{R}, E)$ and a.e. as $\varepsilon \to 0$, thanks to (4.2.23), (4.2.24), (4.2.26), and (4.2.27). Thus

$$H_\varepsilon u - \varphi_\varepsilon * Hu = k_\varepsilon * u \to 0$$

in $L_p(\mathbb{R}, E)$ and a.e. as $\varepsilon \to 0$ which, together with (4.5.1) and (4.5.2), implies the assertion. ■

Theorem 4.5.1 was first proven by Burkholder [Bur83]. Our proof has been motivated by the 'classical' proof of M. Riesz [Rie27] based on conjugate functions (also cf. [Tit48] and [Prü88]).

In the next theorem we collect the most important properties of UMD spaces. It follows from this result that most of the reflexive Banach spaces of distributions commonly used in the theory of partial differential equations are UMD spaces.

4.5.2 Theorem

(i) *Every Banach space isomorphic to a UMD space is a UMD space.*

(ii) *Every Hilbert space is a UMD space.*

(iii) *Every finite-dimensional Banach space is a UMD space.*

(iv) *Finite products of UMD spaces are UMD spaces.*

(v) *If E is a UMD space, E' is one as well.*

(vi) *If E is a UMD space and (X, μ) is a σ-finite measure space, the Lebesgue space $L_p(X, \mu; E)$ is a UMD space for $1 < p < \infty$.*

(vii) *If (E_0, E_1) is an interpolation couple of UMD spaces, the interpolation spaces $[E_0, E_1]_\theta$ and $(E_0, E_1)_{\theta,p}$ are UMD spaces for $0 < \theta < 1$ and $1 < p < \infty$.*

(viii) *Closed linear subspaces of UMD spaces are UMD spaces.*

(ix) *Quotients of UMD spaces modulo closed linear subspaces are UMD spaces.*

Proof (i) Let E be a UMD space and let F be a Banach space such that there exists $T \in \mathcal{L}\mathrm{is}(E, F)$. Then, by 'point-wise multiplication',

$$T \in \mathcal{L}\mathrm{is}\big(L_p(\mathbb{R}, E), L_p(\mathbb{R}, F)\big) \ , \qquad 1 < p < \infty \ ,$$

and $H_F = T H_E T^{-1}$. This proves the assertion.

(ii) follows from Theorem 4.3.3.

(iii) is a consequence of (i) and (ii).

(iv) Let $E := \prod E_j$. Then, using obvious notation, $H_E = \prod H_{E_j}$. Now the assertion is evident.

(v) is implied by Theorem 4.3.6.

(vi) Given $\varepsilon > 0$ and $v \in L_p\big(\mathbb{R}, L_p(X,\mu;E)\big)$, Fubini's theorem implies

$$\begin{aligned}
\|H_{L_p(X,\mu;E),\varepsilon} v\|_{L_p(\mathbb{R},L_p(X,\mu;E))} &= \frac{1}{\pi}\Big\{\int_{\mathbb{R}}\int_X \Big|\int_{|\tau|\geq\varepsilon} v(x,t-\tau)\,d\tau/\tau\Big|^p\,d\mu\,dt\Big\}^{1/p} \\
&= \frac{1}{\pi}\Big\{\int_X\int_{\mathbb{R}} \Big|\int_{|\tau|\geq\varepsilon} v(x,t-\tau)\,d\tau/\tau\Big|^p\,dt\,d\mu\Big\}^{1/p} \\
&\leq \|H_{E,\varepsilon}\|_{\mathcal{L}(L_p(\mathbb{R},E))}\Big\{\int_X\int_{\mathbb{R}} |v(x,t)|^p\,dt\,d\mu\Big\}^{1/p} \\
&= \|H_{E,\varepsilon}\|_{\mathcal{L}(L_p(\mathbb{R},E))}\,\|v\|_{L_p(\mathbb{R},L_p(X,\mu;E))}\ .
\end{aligned}$$

Let (ε_j) be a positive null sequence. Since, thanks to (4.5.2),

$$\|H_{E,\varepsilon_j}\|_{\mathcal{L}(L_p(\mathbb{R},E))} \leq c\ , \qquad j \in \mathbb{N}\ ,$$

it follows that, letting $F := L_p(X,\mu;E)$,

$$\|H_{F,\varepsilon_j} v\|_{L_p(\mathbb{R},F)} \leq c\,\|v\|_{L_p(\mathbb{R},F)}\ , \qquad j \in \mathbb{N}\ . \tag{4.5.3}$$

Thus, given $\varphi \in \mathcal{S}(\mathbb{R},F)$, the sequence $(H_{F,\varepsilon_j}\varphi)$ is bounded in $L_p(\mathbb{R},F)$. Note that $1 < p < \infty$ and Remark 4.4.2(c) imply that F is reflexive. Hence $L_p(\mathbb{R},F)$ is also reflexive. Consequently, $(H_{F,\varepsilon_j}\varphi)$ is weakly relatively compact. Thus there exists $w \in L_p(\mathbb{R},F)$ and a subsequence converging weakly towards w. By the continuous injection of $L_p(\mathbb{R},F)$ in $\mathcal{S}'(\mathbb{R},F)$ it also converges weakly in $\mathcal{S}'(\mathbb{R},F)$ towards w. On the other hand, we infer from Proposition 4.3.1 and (4.1.7) that the sequence $(H_{F,\varepsilon_j}\varphi)$ converges strongly, hence weakly, in $\mathcal{S}'(\mathbb{R},F)$ towards $H_F\varphi$. Thus $H_F\varphi = w$ and (4.5.3) implies

$$\|H_F\varphi\|_{L_p(\mathbb{R},F)} \leq c\,\|\varphi\|_{L_p(\mathbb{R},F)}\ , \qquad \varphi \in \mathcal{S}(\mathbb{R},E)\ .$$

Now we infer from the density of $\mathcal{S}(\mathbb{R},E)$ in $L_p(\mathbb{R},E)$ that

$$H_{L_p(X,\mu;E)} \in \mathcal{L}\big(L_p(\mathbb{R},L_p(X,\mu;E))\big)\ .$$

(vii) Recall that

$$\big[L_p(\mathbb{R};E_0),L_p(\mathbb{R};E_1)\big]_\theta = L_p\big(\mathbb{R};[E_0,E_1]_\theta\big) \tag{4.5.4}$$

and

$$\big(L_p(\mathbb{R};E_0),L_p(\mathbb{R};E_1)\big)_{\theta,p} = L_p\big(\mathbb{R};(E_0,E_1)_{\theta,p}\big) \tag{4.5.5}$$

for $0 < \theta < 1$ and $1 < p < \infty$ (e.g., [Tri78, Theorem 1.18.4]). Thus, if

$$H \in \mathcal{L}\big(L_p(\mathbb{R}, E_j)\big) , \qquad 1 < p < \infty , \quad j = 0, 1 ,$$

it follows that

$$H \in \mathcal{L}\big(L_p(\mathbb{R}, E_\theta)\big) , \qquad 1 < p < \infty ,$$

where E_θ equals either $[E_0, E_1]_\theta$ or $(E_0, E_1)_{\theta,p}$.

(viii) Let F be a closed linear subspace of a UMD space E. Given

$$u = \sum_{j=0}^{m} \varphi_j \otimes e_j \in L_p \otimes F ,$$

it follows from (iii) and (vi) that

$$H_E u = \sum_{j=0}^{m} (H_{\mathbb{K}} \varphi_j) \otimes e_j \in L_p \otimes F ,$$

that is,

$$H_E(L_p \otimes F) \subset L_p \otimes F \subset L_p(\mathbb{R}, F) .$$

Now the assertion follows from the density of $L_p \otimes F$ in $L_p(\mathbb{R}, F)$, the closedness of $L_p(\mathbb{R}, F)$ in $L_p(\mathbb{R}, E)$, and the continuity of H_E on $L_p(\mathbb{R}, E)$.

(ix) Let F be a closed linear subspace of a UMD space. Then $(E/F)'$ is toplinearly isomorphic to $F^\perp := \big\{\, x' \in E' \,;\, \langle x', x\rangle = 0,\ x \in F \,\big\}$. Since $F^\perp$ is a closed linear subspace of E', it follows from (viii) that $(E/F)'$ is a UMD space. Now the assertion is a consequence of (v) and Remark 4.4.2(c). ■

It should be remarked that the proof of the preceding Theorem 4.5.2 follows essentially [Rub86].

4.6 Fractional Powers

Let E be a Banach space. A linear operator A in E is said to be of **positive type** K if it is closed, densely defined, $\mathbb{R}^+ \subset \rho(-A)$, and

$$(1 + s)\, \|(s + A)^{-1}\| \le K , \qquad s \in \mathbb{R}^+ , \tag{4.6.1}$$

where $K \ge 1$. We denote by

$$\mathcal{P}_K := \mathcal{P}_K(E)$$

the set of all operators of positive type K, and A is of **positive type** if it belongs to

$$\mathcal{P} := \mathcal{P}(E) := \bigcup_{K \ge 1} \mathcal{P}_K(E) .$$

Throughout the remainder of this section we assume that $A \in \mathcal{P}_K$ for some $K \geq 1$.

Given $s \in \mathbb{R}^+$ and $\lambda \in \mathbb{C}$ satisfying

$$|\lambda - s| \leq (1+s)/(2K) , \tag{4.6.2}$$

it follows from $\lambda + A = (s+A)\big(1 + (\lambda - s)(s+A)^{-1}\big)$ that $\lambda \in \rho(-A)$ and

$$\begin{aligned}\|(\lambda + A)^{-1}\| &\leq \big\|\big[1 + (\lambda - s)(s+A)^{-1}\big]^{-1}\big\| \, \|(s+A)^{-1}\| \leq 2K(1+s)^{-1} \\ &\leq \frac{2K}{1+|\lambda|}\frac{1+s+|\lambda - s|}{1+s} \leq \frac{2K}{1+|\lambda|}\Big(1 + \frac{1}{2K}\Big) - \frac{2K+1}{1+|\lambda|}\end{aligned} \tag{4.6.3}$$

From this we deduce that

$$S(K) := \big[|\arg z| \leq \arcsin 1/(2K)\big] \cup \big[|z| \leq 1/(2K)\big] \subset \rho(-A) \tag{4.6.4}$$

and that

$$(1 + |\lambda|) \, \|(\lambda + A)^{-1}\| \leq 2K + 1 , \qquad \lambda \in S(K) . \tag{4.6.5}$$

We denote by Φ the set of holomorphic functions $\varphi \colon \mathbb{C}\backslash(-\mathbb{R}^+) \to \mathbb{C}$ such that there exists $\delta > 0$ (depending on φ) with

$$|\lambda|^\delta \, \varphi(\lambda) \to 0 \qquad \text{as} \quad |\lambda| \to \infty , \tag{4.6.6}$$

uniformly in $\big[|\arg \lambda| \leq \pi - \varepsilon\big]$ for each $\varepsilon \in (0, \pi)$. It is obvious that Φ is a commutative algebra without unit with respect to point-wise multiplication. Given $\varphi \in \Phi$, we put

$$\varphi(A) := \frac{1}{2\pi i}\int_\Gamma \varphi(-\lambda)(\lambda + A)^{-1} \, d\lambda = \frac{1}{2\pi i}\int_{-\Gamma} \varphi(\lambda)(\lambda - A)^{-1} \, d\lambda , \tag{4.6.7}$$

where Γ is any piece-wise smooth simple curve in $S(K)\backslash\mathbb{R}^+$ running from $\infty e^{-i\vartheta}$ to $\infty e^{i\vartheta}$ for some $\vartheta \in \big(0, \arcsin 1/(2K)\big]$. Of course, $-\Gamma := \big[-\lambda \in \Gamma\big]$. It follows from (4.6.4)–(4.6.6) and Cauchy's theorem that $\varphi(A)$ is well-defined in $\mathcal{L}(E)$ and independent of the particular choice of Γ. In fact, more is true.

4.6.1 Lemma *The map $\Phi \to \mathcal{L}(E)$, $\varphi \mapsto \varphi(A)$ is an algebra homomorphism.*

Proof The linearity is obvious. Given $\varphi_1, \varphi_2 \in \Phi$, choose admissible contours Γ_1

and Γ_2 such that Γ_1 lies to the left of Γ_2. Then

$$\begin{aligned}
&\varphi_1(A)\varphi_2(A)\\
&\quad= \frac{1}{(2\pi i)^2}\int_{\Gamma_1}\int_{\Gamma_2}\varphi_1(-\lambda)\varphi_2(-\mu)(\lambda+A)^{-1}(\mu+A)^{-1}\,d\mu\,d\lambda\\
&\quad= \frac{1}{(2\pi i)^2}\int_{\Gamma_1}\int_{\Gamma_2}\varphi_1(-\lambda)\varphi_2(-\mu)(\lambda-\mu)^{-1}\big[(\mu+A)^{-1}-(\lambda+A)^{-1}\big]\,d\mu\,d\lambda\\
&\quad= \frac{1}{2\pi i}\int_{\Gamma_2}\varphi_2(-\mu)(\mu+A)^{-1}\Big(\frac{1}{2\pi i}\int_{\Gamma_1}\frac{\varphi_1(-\lambda)}{\lambda-\mu}\,d\lambda\Big)\,d\mu\\
&\qquad+\frac{1}{2\pi i}\int_{\Gamma_1}\varphi_1(-\lambda)(\lambda+A)^{-1}\Big(\frac{1}{2\pi i}\int_{\Gamma_2}\frac{\varphi_2(-\mu)}{\mu-\lambda}\,d\mu\Big)\,d\lambda\;.
\end{aligned}$$

The map $\lambda \mapsto (\lambda-\mu)^{-1}\varphi_1(-\lambda)$ is for each $\mu \in \Gamma_2$ holomorphic on Γ_1 and to the left of it. Hence it follows from (4.6.6) and Cauchy's theorem that the integral in the first pair of parentheses equals zero. Similarly, Cauchy's integral formula implies that the integral in the second pair of parentheses equals $\varphi_2(-\lambda)$. Consequently,

$$\varphi_1(A)\varphi_2(A) = \frac{1}{2\pi i}\int_{\Gamma_1}\varphi_1(-\lambda)\varphi_2(-\lambda)(\lambda+A)^{-1}\,d\lambda = \varphi_1\varphi_2(A)\;,$$

which proves the assertion. ■

Given $z \in \mathbb{C}$, put

$$\varphi_z(\lambda) := \lambda^z = e^{z\log\lambda}\;,\qquad \lambda \in \mathbb{C}\backslash(-\mathbb{R}^+)\;,$$

where the principal branch of the logarithm is used. Note that $\varphi_z \in \Phi$ for $\operatorname{Re} z < 0$. If $k \in \dot{\mathbb{N}}$ then φ_{-k} is holomorphic in $\dot{\mathbb{C}}$ and we can deform Γ to the positively oriented circle centered at 0 of radius r, where $0 < r < 1/\|A^{-1}\|$. Since

$$(\lambda - A)^{-1} = -A^{-1}(1-\lambda A^{-1})^{-1} = -\sum_{j=0}^{\infty} A^{-j-1}\lambda^j\;,\qquad |\lambda| = r\;,$$

it follows that

$$\varphi_{-k}(A) = \frac{-1}{2\pi i}\int_{|\lambda|=r}\lambda^{-k}(\lambda-A)^{-1}\,d\lambda = \sum_{j=0}^{\infty}A^{-j-1}\operatorname{Res}(\lambda^{j-k},0) = A^{-k}\;. \tag{4.6.8}$$

This justifies the following definition of the **fractional powers** of A:

$$A^z := \varphi_z(A)\;,\qquad \operatorname{Re} z < 0\;,$$

(cf. (I.2.9.4)).

Suppose that $0 < \operatorname{Re} z < 1$. Then we can contract Γ to $\mathbb{R}^+$. Hence

$$\begin{aligned} A^{-z} &= \frac{1}{2\pi i}\int_{\infty - i0}^{0-i0} (-\lambda)^{-z}(\lambda + A)^{-1}\, d\lambda + \frac{1}{2\pi i}\int_{0+i0}^{\infty + i0} (-\lambda)^{-z}(\lambda + A)^{-1}\, d\lambda \\ &= -\frac{e^{-i\pi z}}{2\pi i}\int_0^\infty s^{-z}(s+A)^{-1}\, ds + \frac{e^{i\pi z}}{2\pi i}\int_0^\infty s^{-z}(s+A)^{-1}\, ds\ , \end{aligned}$$

that is,

$$A^{-z} = \frac{\sin \pi z}{\pi}\int_0^\infty s^{-z}(s+A)^{-1}\, ds\ , \qquad 0 < \operatorname{Re} z < 1\ . \tag{4.6.9}$$

Applying formula (4.6.9) to the case $E := \mathbb{C}$ and $A := 1$, in particular, it follows that

$$\int_0^\infty s^{-z}(1+s)^{-1}\, ds = \pi / \sin \pi z\ , \qquad 0 < \operatorname{Re} z < 1\ . \tag{4.6.10}$$

Hence we deduce from (4.6.9) and from $A \in \mathcal{P}_K$ that

$$\|A^{-z}\| \le K\frac{|\sin \pi z|}{\pi}\int_0^\infty s^{-\operatorname{Re} z}(1+s)^{-1}\, ds = K\frac{|\sin \pi z|}{\sin(\pi \operatorname{Re} z)} \tag{4.6.11}$$

for $0 < \operatorname{Re} z < 1$. Now it is not difficult to prove the following continuity result:

4.6.2 Theorem $\{\, A^z\ ;\ \operatorname{Re} z < 0\,\} \cup \{A^0 = 1_E\}$ *is a strongly continuous holomorphic semigroup on* E.

Proof It is an easy consequence of the theorem on the differentiation of parameter integrals that the map $z \mapsto A^z$ is holomorphic on $[\operatorname{Re} z < 0]$. Thus, thanks to Lemma 4.6.1, it remains to show that it is strongly continuous at $z = 0$.

Note that

$$(s+A)^{-1} - (1+s)^{-1} \supset (s+A)^{-1}\bigl(1 - (s+A)(1+s)^{-1}\bigr) = (1+s)^{-1}(s+A)^{-1}(1-A)$$

for $s > 0$. Thus, given $x \in \operatorname{dom}(A)$ and z with $0 < \operatorname{Re} z < 1$, it follows from (4.6.9) and (4.6.10) that

$$\begin{aligned} A^{-z}x - x &= \frac{\sin \pi z}{\pi}\int_0^\infty s^{-z}(s+A)^{-1}x\, ds - \frac{\sin \pi z}{\pi}\int_0^\infty s^{-z}(1+s)^{-1}x\, ds \\ &= \frac{\sin \pi z}{\pi}\int_0^\infty \frac{s^{-z}}{(1+s)}(s+A)^{-1}(1-A)x\, ds\ . \end{aligned}$$

Consequently,

$$\|A^{-z}x - x\| \le K\frac{|\sin \pi z|}{\pi}\int_0^\infty \frac{s^{-\operatorname{Re} z}}{(1+s)^2}\, ds\, \|(1-A)x\|\ , \qquad 0 < \operatorname{Re} z < 1\ .$$

Since the integral converges towards 1 as $\operatorname{Re} z \to 0$, we see that $A^{-z}x \to x$ as $z \to 0$ in $[\,|\arg z| \le \alpha]$ for each $\alpha \in (0, \pi/2)$. Since A^{-z} is uniformly bounded for z

in $[\,|\arg z| \le \alpha] \cap [0 < \operatorname{Re} z < 1]$ for each $\alpha \in (0, \pi/2)$, thanks to (4.6.11), it follows that $A^z \to 1$ in $\mathcal{L}_s(E)$ as $z \to 0$ in $[\,|\arg z| \ge \pi/2 + \varepsilon]$ for each $\varepsilon \in (0, \pi/2)$. This proves the theorem. ■

Suppose that $A^z x = 0$ for some $x \in E$ and $z \in \mathbb{C}$ with $\operatorname{Re} z < 0$. Then it follows that $A^{z+w}x = A^w A^z x = 0$ for $\operatorname{Re} w < 0$. Thus $A^w x = 0$ for $\operatorname{Re} w < \operatorname{Re} z$. In particular, $A^{-k}x = 0$ for $k \in \mathbb{N}$ with $k > -\operatorname{Re} z$. Consequently, $x = 0$. This shows that A^z is injective for $\operatorname{Re} z < 0$. Hence we can define the fractional powers for $\operatorname{Re} z > 0$ by

$$A^z := (A^{-z})^{-1} , \qquad \operatorname{Re} z > 0 . \tag{4.6.12}$$

Clearly, $A^z \in \mathcal{C}(E)$. Given $z, w \in \mathbb{C}$ with $0 < \operatorname{Re} z < \operatorname{Re} w$ and $x \in \operatorname{dom}(A^w)$, it follows from

$$x = A^{-w}A^w x = A^{-z-(w-z)}A^w x = A^{-z}A^{-(w-z)}A^w x$$

that $x \in \operatorname{dom}(A^z)$, that is,

$$\operatorname{dom}(A^w) \subset \operatorname{dom}(A^z) , \qquad 0 < \operatorname{Re} z < \operatorname{Re} w . \tag{4.6.13}$$

Given $x \in \operatorname{dom}(A)$, put $y := Ax$. Since $\operatorname{dom}(A)$ is dense in E, we can find for each $\varepsilon > 0$ an element $u \in \operatorname{dom}(A)$ such that $\|u - y\| \le \varepsilon / \|A^{-1}\|$. Thus, letting $v := Au$,

$$\|A^{-2}v - x\| = \|A^{-1}u - A^{-1}y\| \le \|A^{-1}\|\,\|u - y\| \le \varepsilon .$$

This shows that $\overline{\operatorname{dom}(A^2)} \supset \operatorname{dom}(A)$. Hence $\overline{\operatorname{dom}(A^2)} \supset \overline{\operatorname{dom}(A)} = E$ which guarantees that $\operatorname{dom}(A^2)$ is dense in E. By induction we see that $\operatorname{dom}(A^k)$ is dense in E for each $k \in \mathbb{N}$. Hence we infer from (4.6.13) that

$$\overline{\operatorname{dom}(A^z)} = E , \qquad \operatorname{Re} z > 0 . \tag{4.6.14}$$

Now suppose that $\operatorname{Re} z > 0$ and $\operatorname{Re} w > 0$. Given

$$x \in \operatorname{dom}(A^{z+w}) \subset \operatorname{dom}(A^w) \cap \operatorname{dom}(A^z) ,$$

put $y := A^{z+w}x$. Then $x = A^{-(z+w)}y = A^{-w}A^{-z}y$ implies $A^w x = A^{-z}y$ which, in turn, shows that $y = A^z A^w x$, that is,

$$A^{z+w}x = A^z A^w x = A^w A^z x , \qquad x \in \operatorname{dom}(A^{z+w}) . \tag{4.6.15}$$

If $\operatorname{Re} z > \operatorname{Re} w$ and $x \in \operatorname{dom}(A^w)$ then

$$A^{-z}A^w x = A^{-(z-w)}A^{-w}A^w x = A^{-(z-w)}x = A^{w-z}x .$$

Moreover, if $x \in \operatorname{dom}(A^z)$ then, thanks to (4.6.15),

$$A^{-w}A^z x = A^{-w}A^w A^{z-w}x = A^{z-w}x .$$

This proves that, given $z, w \in \mathbb{C}$ with $\operatorname{Re} z, \operatorname{Re} w, \operatorname{Re}(z+w) \ne 0$,

$$A^z A^w x = A^{z+w}x , \qquad x \in \operatorname{dom}(A^u) , \tag{4.6.16}$$

where $u \in \{z, w, z+w\}$ with $\operatorname{Re} u = \max\bigl\{\operatorname{Re} z, \operatorname{Re} w, \operatorname{Re}(z+w)\bigr\}$.

In order to be able to define A^z for $\operatorname{Re} z = 0$ we need the following extension of (4.6.9). As usual, the empty product is given the value 1.

4.6.3 Proposition *Suppose that* $m \in \mathbb{N}$. *Then*

$$A^{-z} = \frac{\sin \pi z}{\pi} \frac{m!}{(1-z)(2-z)\cdots(m-z)} \int_0^\infty s^{m-z}(s+A)^{-m-1}\, ds \qquad (4.6.17)$$

for $0 < \operatorname{Re} z < m+1$.

Proof Suppose z satisfies $0 < \operatorname{Re} z < 1$. Then we infer from (4.6.9) by integration by parts that

$$\begin{aligned} A^{-z} &= \frac{\sin \pi z}{\pi(1-z)} \Big[\big(s^{1-z}(s+A)^{-1}\big)\Big|_0^\infty + \int_0^\infty s^{1-z}(s+A)^{-2}\, ds \Big] \\ &= \frac{\sin \pi z}{\pi(1-z)} \int_0^\infty s^{1-z}(s+A)^{-2}\, ds \ . \end{aligned}$$

Now (4.6.17) follows by induction for $0 < \operatorname{Re} z < 1$. Thanks to (4.6.1) it is easily verified that the integral in (4.6.17) converges absolutely for $0 < \operatorname{Re} z < m+1$ and that the right-hand side of (4.6.17) is a holomorphic map from $[0 < \operatorname{Re} z < m+1]$ into $\mathcal{L}(E)$. Now the assertion follows from Theorem 4.6.2 and the identity theorem for holomorphic functions. ■

It is a consequence of Theorem 4.6.2 that $\{\, A^{-t} \ ; \ t \geq 0 \,\}$ is a strongly continuous semigroup on E. We denote its infinitesimal generator by

$$-\log A$$

which defines the logarithm of $A \in \mathcal{P}(E)$. Then the intuitive formula

$$A^{-t} = e^{-t \log A} \ , \qquad t \geq 0 \ ,$$

is valid. This also shows that the definition is consistent with the one for $A \in \mathcal{L}(E)$ obtained by the Dunford calculus.

Recall that an analytic semigroup $\{\, e^{-tB} \ ; \ t \geq 0 \,\}$ is said **to be of angle** α, where $0 < \alpha \leq \pi$, if there exists a holomorphic function

$$T \colon [\, |\arg z| < \alpha] \to \mathcal{L}(E)$$

extending $\{\, e^{-tB} \ ; \ t \geq 0 \,\}$, such that T is strongly continuous on the closed sector $[\, |\arg z| \leq \alpha - \varepsilon] \cup \{0\}$ for each $\varepsilon \in (0, \alpha)$. Then T is a semigroup on E and we put $e^{-zB} := T(z)$ for $z \in [\, |\arg z| < \alpha] \cup \{0\}$. This notation is justified since $\partial T(z) = -BT(z)$ for $|\arg z| < \alpha$.

4.6.4 Theorem *Suppose $A \in \mathcal{P}_K(E)$. Then $\{ A^{-t} \; ; \; t \geq 0 \}$ is an analytic semigroup of angle $\pi/2$. Moreover,*

$$\|A^{-t}\| \leq K^m , \qquad 0 \leq t \leq m , \quad m \in \mathbb{N} ,$$

and $A^{-z} = e^{-z \log A}$ for $\operatorname{Re} z < 0$.

Proof Thanks to Theorem 4.6.2 it remains to prove the asserted bound. By applying Proposition 4.6.3 to $E := \mathbb{C}$ and $A := 1$ we see that

$$\frac{\pi(1-t)(2-t)\cdots(m-t)}{m! \, \sin \pi t} = \int_0^\infty s^{m-t}(1+s)^{-m-1} \, ds > 0$$

for $0 < s < m+1$ and $m \in \mathbb{N}$. Now Proposition 4.6.3 and (4.6.1) imply

$$\|A^{-t}\| \leq K^{m+1} \frac{\sin \pi t}{\pi} \frac{m!}{(1-t)(2-t)\cdots(m-t)} \int_0^\infty s^{m-t}(1+s)^{-m-1} \, dt = K^{m+1}$$

for $0 < t < m+1$ and $m \in \mathbb{N}$. ■

Now suppose that $-1 < \operatorname{Re} z < 1$. Then we put

$$A_z x := \frac{\sin \pi z}{\pi z} \int_0^\infty s^z (s+A)^{-2} A x \, ds , \qquad x \in \operatorname{dom}(A) .$$

Observe that

$$A_0 x = \int_0^\infty (s+A)^{-2} \, ds \, Ax = -(s+A)^{-1} Ax \Big|_0^\infty = x , \qquad x \in \operatorname{dom}(A) . \quad (4.6.18)$$

Moreover, if $\operatorname{Re} z \neq 0$, it follows from (4.6.9) and (4.6.17) that

$$A^z x = A^{z-1} A x = \frac{\sin \pi(1-z)}{\pi z} \int_0^\infty s^z (s+A)^{-2} A x \, ds = A_z x \qquad (4.6.19)$$

for $x \in \operatorname{dom}(A)$. Note that

$$A^{-1} A_z \subset B_z := \frac{\sin \pi z}{\pi z} \int_0^\infty s^z (s+A)^{-2} \, ds \in \mathcal{L}(E) . \qquad (4.6.20)$$

Let (x_j) be a sequence in $\operatorname{dom}(A)$ such that $x_j \to 0$ and $A_z x_j \to y$ in E. Then, thanks to (4.6.20), $B_z x_j \to 0$ and $B_z x_j \to A^{-1} y$, which implies $y = 0$. Hence A_z is closable. Motivated by (4.6.18) and (4.6.19) we put

$$A^z := \text{closure of } A_z , \qquad \operatorname{Re} z = 0 . \qquad (4.6.21)$$

By these considerations we have already proven most of the following statement.

4.6.5 Theorem *Suppose that $A \in \mathcal{P}(E)$. Then the* **fractional power** A^z *is for each $z \in \mathbb{C}$ a densely defined closed linear operator in E. If* $\operatorname{Re} z < 0$ *then $A^z \in \mathcal{L}(E)$ and is given by the Dunford integral*

$$A^z = \frac{1}{2\pi i} \int_\Gamma (-\lambda)^z (\lambda + A)^{-1}\, d\lambda \ , \tag{4.6.22}$$

where Γ is any piece-wise smooth simple curve running in $\mathbb{C}\backslash\mathbb{R}^+$ from $\infty e^{-i\varphi}$ to $\infty e^{i\psi}$ for some $\varphi, \psi \in (0,\pi)$ such that $\sigma(-A)$ lies strictly to the left of Γ. Moreover,

(i) *A^z is the 'usual power' of A if $z \in \mathbb{Z}$.*

(ii) $A^z x = \frac{\sin \pi z}{\pi z} \int_0^\infty s^z (s+A)^{-2} Ax\, ds, \quad x \in \operatorname{dom}(A), \quad -1 < \operatorname{Re} z < 1.$

(iii) *Suppose that either $m \in \dot{\mathbb{N}}$, $x \in \operatorname{dom}(A^{2m})$, and* $\operatorname{Re} z \vee \operatorname{Re} w < m$, *or* $\operatorname{Re} z$, $\operatorname{Re} w$, *and* $\operatorname{Re}(z+w)$ *are distinct from zero and $x \in \operatorname{dom}(A^u)$, where u belongs to $\{z, w, z+w\}$ and satisfies* $\operatorname{Re} u = \max\{\operatorname{Re} z, \operatorname{Re} w, \operatorname{Re}(z+w)\}$. *Then*

$$A^z A^w x = A^{z+w} x \ .$$

(iv) $A^z A^w = A^{z+w}, \quad \operatorname{Re} z, \operatorname{Re} w > 0.$

(v) $D(A^w) \overset{d}{\hookrightarrow} D(A^z) \overset{d}{\hookrightarrow} E, \quad 0 < \operatorname{Re} z < \operatorname{Re} w.$

(vi) $A^z \in \mathcal{L}\mathrm{is}\big(D(A^{z+w}), D(A^w)\big) \cap \mathcal{L}\mathrm{is}\big(D(A^z), E\big), \quad \operatorname{Re} z, \operatorname{Re} w > 0.$

(vii) *Given $m \in \mathbb{N}$, the map*

$$[\operatorname{Re} z < m] \to \mathcal{L}\big(D(A^m), E\big) \ , \quad z \mapsto A^z$$

is holomorphic.

Proof The first part of the assertion follows from the investigations preceding this theorem.

(i) follows from (4.6.8) and (4.6.12).

(ii) If $\operatorname{Re} z \neq 0$, this has been shown in (4.6.19), and it follows from the definition of A^z if $\operatorname{Re} z = 0$.

(iii) If $\operatorname{Re} z$, $\operatorname{Re} w$, and $\operatorname{Re}(z+w)$ are all distinct from zero, this is a consequence of (4.6.16) and (4.6.13). From (ii) and (4.6.1) we infer that

$$(z \mapsto A^z) \in C^1\Big([-1 < \operatorname{Re} z < 1], \mathcal{L}(D(A), E) \cap \mathcal{L}(D(A^2), D(A))\Big) \ . \tag{4.6.23}$$

Thus suppose that $z, w \in [-1 < \operatorname{Re} z < 1]$. Choose sequences (z_j), (w_j) in

$$[-1 < \operatorname{Re} z < 1] \backslash [\operatorname{Re} z = 0] =: Z$$

such that $z_j + w_j \in Z$ and such that $z_j \to z$ and $w_j \to w$. Then, by what we already know,

$$A^{z_j} A^{w_j} x = A^{z_j + w_j} x \ , \qquad x \in \operatorname{dom}(A^2) \ .$$

Hence, letting $j \to \infty$, we infer from (4.6.23) that (iii) is true if $-1 < \operatorname{Re} z, \operatorname{Re} w < 1$.

Suppose that $\operatorname{Re} z = 0$ and $|\operatorname{Re} w| \geq 1$. Fix $\alpha \in \mathbb{R}$ with $0 < \alpha - \operatorname{Re} w < 1$. Then

$$A^z A^w x = A^z A^{w-\alpha} A^\alpha x = A^{z+(w-\alpha)} A^\alpha x = A^{(z+w-\alpha)+\alpha} x = A^{z+w} x$$

for $x \in D(A^{2m})$ with $m \in \dot{\mathbb{N}}$ and $\operatorname{Re} w < m$, since $-1 < \operatorname{Re}(w-\alpha) < 0$ and $\alpha \neq 0$.

Lastly, let $\operatorname{Re} z \leq -1 < 1 \leq \operatorname{Re} w$ and $\operatorname{Re}(z+w) = 0$. Write $z = r + s$ with $-1 < \operatorname{Re} r < 0$. Since the real parts of r, w, and $r+w$ are distinct from zero and z, r, and s have negative real parts, it follows that $A^z = A^r A^s$ and $A^s A^w x = A^{s+w} x$ for $x \in \operatorname{dom}(A^w)$. Thus

$$A^z A^w x = A^r A^{s+w} x \ , \qquad x \in \operatorname{dom}(A^w) \subset \operatorname{dom}(A^{2m}) \ .$$

So we can assume that $-1 < \operatorname{Re} z < 0$. Then $\operatorname{Re}(z+w) = 0$ implies $0 < \operatorname{Re} w < 1$, so that we are back to a situation already considered. Consequently, (iii) has been completely proven.

(iv) By Theorem 4.6.2 and (iii) it suffices to prove that $x \in \operatorname{dom}(A^w)$ and $A^w x \in \operatorname{dom}(A^z)$ imply $x \in \operatorname{dom}(A^{w+z})$ if $\operatorname{Re} z > 0$ and $\operatorname{Re} w > 0$. Let $y := A^z(A^w x)$. Then it follows from (iii) that $x = A^{-w}(A^{-z} y) = A^{-(w+z)} y \in \operatorname{dom}(A^{w+z})$.

(v) From (4.6.13) and (iii) we deduce that

$$\|A^z x\| = \|A^{z-w} A^w x\| \leq \|A^{z-w}\| \, \|A^w x\| \ , \qquad x \in D(A^w) \ .$$

Since $x \mapsto \|A^u x\|$ is an equivalent norm on $D(A^u)$ for $\operatorname{Re} u > 0$, thanks to the boundedness of A^{-u}, it follows that $D(A^w) \hookrightarrow D(A^z) \hookrightarrow E$.

Given $x \in D(A^z)$, put $y := A^z x \in E$. Since $D(A^{w-z})$ is dense in E by (4.6.14), given $\varepsilon > 0$, we can find $u \in D(A^{w-z})$ such that $\|u - y\| < \varepsilon$. Hence

$$v := A^{-z} u \in D(A^w) \quad \text{and} \quad \|A^z(v - x)\| = \|u - y\| < \varepsilon \ .$$

This shows that $D(A^w)$ is dense in $D(A^z)$ which, together with (4.6.14), implies the assertion.

(vi) The first assertion follows from (iv), the second one is trivial.

(vii) Thanks to Theorem 4.6.2 and (4.6.23), we can assume that $m \geq 2$. Since (v) implies

$$\mathcal{L}\big(D(A), E\big) \hookrightarrow \mathcal{L}\big(D(A^m), E\big) \ ,$$

we infer that

$$(z \mapsto A^z) \in C^1\big([\operatorname{Re} z < 1], \mathcal{L}(D(A^m), E)\big) \ . \tag{4.6.24}$$

If $0 < \operatorname{Re} z < m$ then (iii) implies $A^z x = A^{z-m} A^m x$ for $x \in D(A^m)$. Hence Theorem 4.6.2 guarantees that

$$(z \mapsto A^z) \in C^1\big([0 < \operatorname{Re} z < m], \mathcal{L}(D(A^m), E)\big) \ .$$

This, together with (4.6.24), proves the theorem. ■

Observe that $\mathcal{G}(E, M, -\sigma) \subset \mathcal{P}(E)$ for $\sigma > 0$, that is, if $-A$ is the infinitesimal generator of a strongly continuous exponentially decaying semigroup on E then A is of positive type. In this case we can obtain another useful representation formula for A^z with $\operatorname{Re} z > 0$.

4.6.6 Theorem *Suppose that $A \in \mathcal{G}(E, M, -\sigma)$ for some $M \geq 1$ and $\sigma > 0$. Then*

$$A^{-z} = \frac{1}{\Gamma(z)} \int_0^\infty t^{z-1} e^{-tA}\, dt\ , \qquad \operatorname{Re} z > 0\ .$$

Proof It is an easy consequence of

$$\Big\| \int_0^\infty t^{z-1} e^{-tA}\, dt \Big\|_{\mathcal{L}(E)} \leq M \int_0^\infty t^{\operatorname{Re} z - 1} e^{-\sigma t}\, dt$$

and the known properties of the Euler Γ-function that the map

$$[\operatorname{Re} z > 0] \to \mathcal{L}(E)\ , \quad z \mapsto \frac{1}{\Gamma(z)} \int_0^\infty t^{z-1} e^{-tA}\, dt$$

is holomorphic. Thus, thanks to Theorem 4.6.2 and the identity theorem for holomorphic functions, it suffices to prove the asserted equality for $0 < z < 1$.

Given $z \in (0,1)$,

$$A^{-z} = \frac{\sin \pi z}{\pi} \int_0^\infty s^{-z}(s+A)^{-1}\, ds$$

by Proposition 4.6.3. On the other hand, we know from semigroup theory that

$$(s+A)^{-1} = \int_0^\infty e^{-st} e^{-tA}\, dt\ , \qquad s > 0\ .$$

Thus, by Fubini's theorem,

$$\begin{aligned} A^{-z} &= \frac{\sin \pi z}{\pi} \int_0^\infty s^{-z} \int_0^\infty e^{-st} e^{-tA}\, dt\, ds = \frac{\sin \pi z}{\pi} \int_0^\infty e^{-tA} \int_0^\infty s^{-z} e^{-ts}\, ds\, dt \\ &= \frac{\sin \pi z}{\pi} \Gamma(1-z) \int_0^\infty t^{z-1} e^{-tA}\, dt\ . \end{aligned}$$

Hence the assertion follows from the well-known 'complementing' formula

$$\Gamma(z)\Gamma(1-z) = \pi / \sin \pi z \tag{4.6.25}$$

(e.g., [Schw65, VIII,1;20]). ■

Now we assume that H is a Hilbert space and A is a positive definite self-adjoint linear operator in H, that is, $A = A^* \geq \alpha > 0$ for some $\alpha > 0$. Let

$\{E_\lambda \,;\, \lambda \in \mathbb{R}\}$ be the spectral resolution of A. Then, given $z \in \mathbb{C}$, we can define A^z by

$$A^z := \int_0^\infty \lambda^z \, dE_\lambda \,, \qquad z \in \mathbb{C} \,. \tag{4.6.26}$$

The following theorem shows that this definition coincides with the former one.

4.6.7 Theorem *Let H be a Hilbert space and let A be a positive definite self-adjoint linear operator in H. Then $A \in \mathcal{P}(H)$ and the fractional powers defined in* (4.6.26) *by means of the spectral resolution coincide with the fractional powers of Theorem* 4.6.5.

Proof First note that $\sigma(-A) \subset (-\infty, -\alpha]$ if $A = A^* \geq \alpha > 0$. Moreover,

$$(s+\alpha)\,\|x\|^2 \leq \big((s+A)x|x\big) \leq \|(s+A)x\|\,\|x\| \,, \qquad x \in \mathrm{dom}(A) \,,$$

implies

$$\|(s+A)^{-1}\| \leq (s+\alpha)^{-1} \leq K(1+s)^{-1} \,, \qquad s \geq 0 \,.$$

Hence $A \in \mathcal{P}(H)$.

Let Γ be the contour consisting of the two rays $-\beta + \mathbb{R}^+ e^{\pm i\varphi}$ for some $\beta \in (0,\alpha)$ and $\varphi \in (0,\pi)$, and oriented so that the imaginary parts increase along Γ. Then, given $z \in \mathbb{C}$ with $\mathrm{Re}\, z < 0$ and $\mu \geq \alpha$, Cauchy's integral formula implies

$$\frac{1}{2\pi i} \int_\Gamma \frac{(-\lambda)^z}{\lambda + \mu}\, d\lambda = \mu^z \,.$$

Hence, by Fubini's theorem and the spectral calculus for A,

$$\begin{aligned}\frac{1}{2\pi i} \int_\Gamma (-\lambda)^z (\lambda + A)^{-1}\, d\lambda &= \frac{1}{2\pi i} \int_\Gamma (-\lambda)^z \int_0^\infty (\lambda+\mu)^{-1}\, dE_\mu \, d\lambda \\ &= \int_0^\infty \Big[\frac{1}{2\pi i} \int_\Gamma \frac{(-\lambda)^z}{z+\mu}\, d\lambda\Big]\, dE_\mu = \int_0^\infty \mu^z \, dE_\mu\end{aligned}$$

in $\mathcal{L}(H)$, thanks to the fact that the support of the spectral resolution is contained in $[\alpha,\infty)$. This proves the assertion for $\mathrm{Re}\, z < 0$. Now the theorem follows from the spectral calculus for self-adjoint linear operators and the definition of the fractional powers for $A \in \mathcal{P}(H)$ given above. ∎

4.6.8 Remarks **(a)** Observe that $s(s+A)^{-1} \supset 1 - (s+A)^{-1}A$, and thus,

$$s^{-1}(s+A)^{-1} = s^{-1}A^{-1} - (s+A)^{-1}A^{-1} \,.$$

Thus, by splitting the integral in (4.6.9) at $s = 1$ and using these formulas, we see that

$$\begin{aligned}A^z x = \frac{\sin \pi z}{\pi} \Big\{ \frac{x}{z} - \frac{1}{1+z} A^{-1}x &+ \int_0^1 s^{z+1}(s+A)^{-1}A^{-1}x\, ds \\ &+ \int_1^\infty s^{z-1}(s+A)^{-1}Ax\, ds \Big\}\end{aligned}$$

for $x \in \operatorname{dom}(A)$ and $-1 < \operatorname{Re} z < 0$. By analytic continuation this representation is also valid in $[-1 < \operatorname{Re} z < 1]$.

(b) Let $0 < \alpha < 1$. Given $A \in \mathcal{P}(E)$, it is not difficult to see that $\mathbb{R}^+ \subset \rho(-A^\alpha)$ and

$$(\lambda + A^\alpha)^{-1} = \frac{1}{2\pi i} \int_\Gamma \frac{1}{\lambda + (-\mu)^\alpha} (\mu + A)^{-1}\, d\mu\ , \qquad \lambda \geq 0\ ,$$

where Γ is a contour as in (4.6.22). By contracting Γ to $\mathbb{R}^+$ it follows that

$$(\lambda + A^\alpha)^{-1} = \frac{\sin \pi\alpha}{\pi} \int_0^\infty \frac{s^\alpha (s + A)^{-1}}{s^{2\alpha} + 2\lambda s^\alpha \cos \pi\alpha + \lambda^2}\, ds \tag{4.6.27}$$

for $\lambda \geq 0$. Assume that

$$[\,|\arg z| < \vartheta] \cup \{0\} \subset \rho(-A)$$

for some $\vartheta \in (0, \pi)$ and that, given $\varepsilon \in (0, \vartheta)$, there exists K_ε such that

$$(1 + |\lambda|)\, \|(\lambda + A)^{-1}\| \leq K_\varepsilon\ , \qquad |\arg \lambda| \leq \vartheta - \varepsilon\ .$$

Then by means of the above formulas it can be shown that

$$\big[|\arg z| < (1 - \alpha)\pi + \alpha\vartheta\big] \cup \{0\} \subset \rho(-A^\alpha)$$

and that, given $\varepsilon \in \big(0, (1 - \alpha)\pi + \alpha\vartheta\big)$, there exists M_ε such that

$$(1 + |\lambda|)\, \|(\lambda + A^\alpha)^{-1}\| \leq M_\varepsilon\ , \qquad |\arg \lambda| < (1 - \alpha)\pi + \alpha\vartheta - \varepsilon\ .$$

This implies, in particular, that A^α is the negative generator of a strongly continuous analytic semigroup on E whenever $A \in \mathcal{G}(E, M, -\sigma)$ for some $\sigma > 0$. Details can be found in [Kre72] and [Tan79].

(c) Let $A \in \mathcal{C}(E)$ be densely defined and assume that $\dot{\mathbb{R}}^+ \subset \rho(-A)$ and

$$\|(s + A)^{-1}\| \leq K/s\ , \qquad s > 0\ , \tag{4.6.28}$$

for some $K \geq 1$. Note that we no longer require that A has a bounded inverse. Then, given $\alpha \in (0, 1)$, we put

$$A^\alpha := \big[I_\alpha(\lambda)\big]^{-1} - \lambda\ , \qquad \lambda > 0\ , \tag{4.6.29}$$

where $I_\alpha(\lambda)$ denotes the right-hand side of (4.6.27). It can be shown that this definition is meaningful and that A^α is a closed and densely defined linear operator. Put $A_\varepsilon := \varepsilon + A$ for $\varepsilon \in (0, 1)$ and note that (4.6.28) implies $\mathbb{R}^+ \subset \rho(-A_\varepsilon)$ and

$$\|(s + A_\varepsilon)^{-1}\| \leq \frac{K}{s + \varepsilon} < \frac{K}{\varepsilon} \frac{1}{1 + s}\ , \qquad s \geq 0\ .$$

Hence A_ε is of positive type K/ε and, consequently, the fractional power A_ε^α is well-defined. It can be shown that $\operatorname{dom}(A_\varepsilon^\alpha) = \operatorname{dom}(A^\alpha)$ and that there exists a constant c depending on K and α only such that

$$\|A_\varepsilon^\alpha x - A^\alpha x\| \le c\varepsilon^\alpha \|x\| \ , \qquad x \in \mathrm{dom}(A^\alpha) \ . \tag{4.6.30}$$

Using these facts it is not difficult to prove that the **fractional powers** A^α enjoy the following properties in this case in which 0 may belong to $\sigma(A)$:

(i) if $0 < \alpha < \beta \le 1$ then $D(A^\beta) \hookrightarrow D(A^\alpha)$;

(ii) if $\alpha, \beta > 0$ and $\alpha + \beta \le 1$ then $x \in D(A^{\alpha+\beta})$ implies $x \in D(A^\alpha A^\beta)$ and the relation $A^\alpha A^\beta x = A^{\alpha+\beta} x$.

For details we again refer to [Kre72] and [Tan79]. ■

Let H be a Hilbert space and let A be a positive semidefinite self-adjoint linear operator in H, that is, $A = A^* \ge 0$. Then definition (4.6.26) is meaningful. On the other hand, A obviously satisfies (4.6.28) with $K = 1$. Hence A^α can also be defined by (4.6.29). The following theorem shows that the two definitions coincide.

4.6.9 Theorem *Let H be a Hilbert space and let A be a positive semidefinite self-adjoint linear operator in H. Then the two definitions* (4.6.26) *and* (4.6.29) *coincide for $z = \alpha \in (0,1)$.*

Proof Given $\varepsilon > 0$, the operator $A_\varepsilon := \varepsilon + A$ is positive definite and self-adjoint. Hence, denoting by $\{\, E_\lambda(A_\varepsilon) \ ; \ \lambda \in \mathbb{R} \,\}$ the spectral resolution of A_ε, we infer from Theorem 4.6.7 that

$$(A_\varepsilon)^\alpha = \int_0^\infty \lambda^\alpha \, dE_\lambda(A_\varepsilon) = \int_0^\infty (\mu + \varepsilon)^\alpha \, dE_\mu \ , \qquad 0 < \alpha < 1 \ ,$$

where the last equality follows from the spectral calculus for self-adjoint operators. Given $x \in \mathrm{dom}(A^\alpha)$, it follows from (4.6.30) that $(A_\varepsilon)^\alpha x \to A^\alpha x$ in H as $\varepsilon \to 0$. On the other hand,

$$\int_0^\infty (\mu + \varepsilon)^\alpha \, dE_\mu x \to \int_0^\infty \mu^\alpha \, dE_\mu x$$

in H as $\varepsilon \to 0$ by Lebesgue's theorem. This implies the assertion. ■

We close this subsection by showing that we can 'raise powers to powers', a result that is needed in Chapter V. For this we prove first that $A \in \mathcal{P}(E)$ entails $A^\alpha \in \mathcal{P}(E)$ for $0 < \alpha < 1$. Indeed, we derive the following more precise statement:

4.6.10 Proposition *Suppose that $A \in \mathcal{P}(E)$ and $\alpha \in (0,1)$. Then $A^\alpha \in \mathcal{P}(E)$. In fact, if there exists $\vartheta \in (0,\pi)$ such that* (4.6.1) *is satisfied for $s \in \mathbb{C}$ with $|\arg s| \le \vartheta$ then $\big[|\arg z| < \pi - (\pi - \vartheta)\alpha\big] \cup \{0\} \subset \rho(-A^\alpha)$ and, given $\vartheta' \in (0,\vartheta)$,*

$$(1 + |s|) \, \|(s + A^\alpha)^{-1}\| \le c \ , \qquad |\arg s| \le \pi - (\pi - \vartheta')\alpha \ .$$

Proof Fix $s \in \mathbb{R}^+$ and define $\varphi := \varphi_s \in \Phi$ by $\varphi(\lambda) := (s + \lambda^\alpha)^{-1}$. Then

$$\psi(\lambda) := \lambda^{\alpha-1}\varphi(\lambda) = \lambda^{-1} - s\lambda^{-1}\varphi(\lambda)$$

defines $\psi \in \Phi$. Since $(\lambda \mapsto \lambda^{-\beta}) \in \Phi$ for $\beta > 0$, it follows from Lemma 4.6.1 that

$$\psi(A) = \varphi(A)A^{\alpha-1} = \bigl(1 - s\varphi(A)\bigr)A^{-1} \ . \tag{4.6.31}$$

Consequently, $\varphi(A)(s + A^\alpha)A^{-1} = A^{-1}$, that is,

$$\varphi(A)(s + A^\alpha)x = x \ , \qquad x \in \mathrm{dom}(A) \ . \tag{4.6.32}$$

Given $x \in \mathrm{dom}(A^\alpha)$, by Theorem 4.6.5(v) and (vi) we find $x_n \in \mathrm{dom}(A)$ such that $x_n \to x$ and $A^\alpha x_n \to A^\alpha x$ in E as $n \to \infty$. This shows that (4.6.32) is valid for $x \in \mathrm{dom}(A^\alpha)$. On the other hand, by multiplying (4.6.31) by $A^{-\alpha}$ we obtain

$$\varphi(A)A^{-1} = A^{-\alpha}\bigl(1 - s\varphi(A)\bigr)A^{-1} \ ,$$

hence $\varphi(A) = A^{-\alpha}\bigl(1 - s\varphi(A)\bigr)$. This gives $(s + A^\alpha)\varphi(A) = 1$, which proves that $\mathbb{R}^+ \subset \rho(-A^\alpha)$ and $(s + A^\alpha)^{-1} = \varphi_s(A)$ for $s \in \mathbb{R}^+$. By contracting the contour Γ in (4.6.7) to $\mathbb{R}^+$ it follows that

$$\begin{aligned}(s + A^\alpha)^{-1} &= \frac{1}{2\pi i}\int_0^\infty \bigl[(s + r^\alpha e^{-i\pi\alpha})^{-1} - (s + r^\alpha e^{i\pi\alpha})^{-1}\bigr](r + A)^{-1}\,dr \\ &= \frac{\sin\pi\alpha}{\pi}\int_0^\infty r^\alpha(s + r^\alpha e^{-i\pi\alpha})^{-1}(s + r^\alpha e^{i\pi\alpha})^{-1}(r + A)^{-1}\,dr \\ &= \frac{\sin\pi\alpha}{\pi}\int_0^\infty r^\alpha(r^\alpha + se^{i\pi\alpha})^{-1}(r^\alpha + se^{-i\pi\alpha})^{-1}(r + A)^{-1}\,dr\end{aligned} \tag{4.6.33}$$

for $s \in \mathbb{R}^+$. Observe that the last integral is an analytic function of s in the sector $\bigl[|\arg z| < \pi(1-\alpha)\bigr]$. Hence this sector belongs to $\rho(-A^\alpha)$ and $(s + A^\alpha)^{-1}$ is given by the integral above. Now suppose that $s = |s|\,e^{i\psi}$ with $|\psi| < \pi - (\pi - \vartheta')\alpha$ and $0 < \vartheta' < \vartheta$. Then we can again deform the path of integration in the last integral to the ray $\Gamma_\pm := \{re^{\pm i\vartheta} \ ; \ r \geq 0\}$, where we choose the positive [resp. negative] sign if $\psi \geq 0$ [resp. $\psi < 0$], thanks to the estimate $\|(\lambda + A)^{-1}\| \leq K(1 + |\lambda|)^{-1}$ for $|\arg\lambda| \leq \vartheta$. Letting $\beta := \psi + (\pi \mp \vartheta)\alpha$ and $\gamma := \psi - (\pi \pm \vartheta)\alpha$, where we choose the upper [resp. lower] sign if we integrate over Γ_+ [resp. Γ_-], we obtain the estimate

$$\begin{aligned}\|(s + A^\alpha)^{-1}\| &\leq K\frac{\sin\pi\alpha}{\pi}\int_0^\infty \frac{r^\alpha\,dr}{(1 + r)\,\bigl|r^\alpha + |s|\,e^{i\beta}\bigr|\,\bigl|r^\alpha + |s|\,e^{i\gamma}\bigr|} \\ &\leq \frac{K}{|s|}\frac{\sin\pi\alpha}{\pi\alpha}\int_0^\infty \frac{d\tau}{|\tau + e^{i\beta}|\,|\tau + e^{i\gamma}|} \ .\end{aligned}$$

This proves that $(1 + |s|)\,\|(s + A^\alpha)^{-1}\| \leq c(\alpha, \vartheta')$ for $0 < \vartheta' < \vartheta$ and s in the sector $\bigl[|\arg z| \leq \pi - (\pi - \vartheta')\alpha\bigr] \cup \{0\}$, hence the assertion. ■

4.6.11 Corollary *Suppose that $A \in \mathcal{P}(E)$ and there exists $\vartheta \in (0,\pi)$ such that* (4.6.1) *is satisfied for $s \in \mathbb{C}$ with $|\arg s| \le \vartheta$. If $\alpha \in (0,1)$ satisfies $\alpha < \pi/\big[2(\pi-\vartheta)\big]$ then $-A^\alpha$ generates a strongly continuous analytic semigroup on E.*

4.6.12 Remark Corollary 4.6.11 and (4.6.5) imply that $-A^\alpha$ generates a strongly continuous analytic semigroup on E whenever $A \in \mathcal{P}(E)$ and $0 < \alpha \le 1/2$. Moreover, if $-A$ generates a strongly continuous semigroup on E, this is also true for $-A^\alpha$ for each $\alpha \in (0,1)$. ∎

It follows from Proposition 4.6.10 and Theorem 4.6.5 that the fractional powers $(A^\alpha)^z$ are well-defined for $z \in \mathbb{C}$ and $\alpha \in (0,1)$. For simplicity, we restrict ourselves in the following theorem to the case $z \in \mathbb{R}$.

4.6.13 Theorem *Suppose that $A \in \mathcal{P}(E)$ and $0 < \alpha < 1$. Then $(A^\alpha)^\beta = A^{\alpha\beta}$ for $\beta \in \mathbb{R}$.*

Proof Thanks to Proposition 4.6.10 we can find $K \ge 1$ such that A and A^α belong to $\mathcal{P}_K(E)$. Then it follows from the proof of that proposition that

$$(\mu + A^\alpha)^{-1} = \frac{1}{2\pi i}\int_\Gamma \frac{(\lambda + A)^{-1}}{\mu + (-\lambda)^\alpha}\,d\lambda\,, \qquad \mu \in S(K)\,,$$

where Γ is a piece-wise smooth curve running in $S(K)\backslash\mathbb{R}^+$ from $\infty e^{-i\vartheta}$ to $\infty e^{i\vartheta}$ for a sufficiently small positive ϑ. Thus, by (4.6.22) and Cauchy's integral formula,

$$\begin{aligned}(A^\alpha)^{-\beta} &= \frac{1}{(2\pi i)^2}\int_{\Gamma'}\int_\Gamma \frac{(-\mu)^{-\beta}}{\mu + (-\lambda)^\alpha}(\lambda + A)^{-1}\,d\lambda\,d\mu \\ &= \frac{1}{(2\pi i)^2}\int_\Gamma (\lambda + A)^{-1}\int_{\Gamma'} \frac{(-\mu)^{-\beta}}{\mu + (-\lambda)^\alpha}\,d\mu\,d\lambda \\ &= \frac{1}{2\pi i}\int_\Gamma (-\lambda)^{-\alpha\beta}(\lambda + A)^{-1}\,d\lambda = A^{-\alpha\beta}\end{aligned}$$

for $\beta > 0$, where Γ' is a contour with the same properties as Γ and lying to the right of Γ. Moreover, $(A^\alpha)^\beta = \big[(A^\alpha)^{-\beta}\big]^{-1} = [A^{-\alpha\beta}]^{-1} = A^{\alpha\beta}$ for $\beta > 0$. This proves the theorem. ∎

The proofs of this subsection follow essentially [Kre72] and [KraZ76]. By means of Proposition 4.6.3 it is not difficult to derive a representation formula similar to the one of Theorem 4.6.5(ii) for arbitrary $z \in \mathbb{C}$. This is done in [Tri78], where such a formula is used to define A^z for $A \in \mathcal{P}(E)$ without reference to the Dunford integral representation (4.6.22). For a detailed study of fractional powers of operators that are not necessarily invertible, we refer to [Kom66]. The representation formula of Theorem 4.6.6 is often used to define the fractional powers of generators of strongly continuous semigroups (for real exponents) (e.g., [Gols85], [Hen81], [Paz83]).

Lemma 4.6.1 shows that one can define a good functional calculus for a class of holomorphic functions that is larger than the class of powers. By mimicking the definition of fractional powers for $\operatorname{Re} z \geq 0$ one can also establish a reasonable functional calculus for suitable unbounded holomorphic functions. This has been elaborated by McIntosh [McI86] (also cf. [deL87], [deL94] for holomorphic functional calculi for unbounded operators).

4.7 Bounded Imaginary Powers

Let E be a Banach space. A linear operator A in E is said to have **bounded imaginary powers**, in symbols,

$$A \in \mathcal{BIP} := \mathcal{BIP}(E) ,$$

provided $A \in \mathcal{P}(E)$ and there exist $\varepsilon > 0$ and $M \geq 1$ such that

$$A^{it} \in \mathcal{L}(E) \quad \text{and} \quad \|A^{it}\| \leq M , \qquad -\varepsilon \leq t \leq \varepsilon . \tag{4.7.1}$$

The following theorem shows that this assumption has far-reaching consequences.

4.7.1 Theorem *Suppose that $A \in \mathcal{BIP}$. Then $\{ A^z \ ; \ \operatorname{Re} z \leq 0 \}$ is a strongly continuous semigroup on $\mathcal{L}(E)$, that is, the map*

$$[\operatorname{Re} z \leq 0] \to \mathcal{L}_s(E) , \quad z \mapsto A^z$$

is a continuous representation of the additive semigroup $[\operatorname{Re} z \leq 0]$. Moreover, $\{ A^{it} \ ; \ t \in \mathbb{R} \}$ is a strongly continuous group on E whose infinitesimal generator is $i \log A$.

Proof Suppose that $A \in \mathcal{P}_K$. If $|t| \in \big[n\varepsilon, (n+1)\varepsilon\big)$ for some $n \in \mathbb{N}$, it follows from Theorem 4.6.4, Theorem 4.6.5(iii), and (4.7.1) that

$$\begin{aligned} \|A^{-s+it}x\| &= \|A^{-s}(A^{i\,\operatorname{sign}(t)\varepsilon})^n A^{i\,\operatorname{sign}(t)(|t|-n\varepsilon)}x\| \\ &\leq K^m M^{n+1}\,\|x\| = K^m M e^{n \log M}\,\|x\| \leq K^m M e^{\theta |t|}\,\|x\| \end{aligned} \tag{4.7.2}$$

for $0 \leq s \leq m$ and $x \in \operatorname{dom}(A^2)$, where $\theta := \varepsilon^{-1} \log M \geq 0$. Thus, by the density of $\operatorname{dom}(A^2)$ in E,

$$\|A^z\| \leq K^{1-\operatorname{Re} z} M e^{\theta |\operatorname{Im} z|} , \qquad \operatorname{Re} z \leq 0 . \tag{4.7.3}$$

From this and Theorem 4.6.5(v) and (vii) it follows that the map $z \mapsto A^z$ is strongly continuous on $[\operatorname{Re} z \leq 0]$. Now, using Theorem 4.6.5(iii) and the density of $\operatorname{dom}(A^2)$ in E, we see that $\{ A^z \ ; \ \operatorname{Re} z \leq 0 \}$ is a strongly continuous semigroup on E. Consequently, $\{ A^{it} \ ; \ t \in \mathbb{R} \}$ is a strongly continuous group on E. Let B denote the infinitesimal generator of this group and recall that

$$Bx = \lim_{t \to 0} t^{-1}[A^{it}x - x]$$

iff $x \in \mathrm{dom}(B)$. Since

$$\tau^{-1}[A^{-s+i(t+\tau)}x - A^{-s+it}x] = \tau^{-1}(A^{i\tau} - 1)A^{-s+it}x = A^{-s+it}\tau^{-1}[A^{i\tau} - 1]x$$

for $x \in E$, $s \geq 0$, and $t, \tau \in \mathbb{R}$ with $\tau \neq 0$, we see that

$$BA^{-s+it}x = A^{-s+it}Bx = \partial_t A^{-s+it}x \tag{4.7.4}$$

for $x \in \mathrm{dom}(B)$, $s \geq 0$, and $t \in \mathbb{R}$. On the other hand, the analyticity of A^{-z} for $\mathrm{Re}\, z > 0$ implies

$$\partial_s A^{-s+it}x = i\partial_t A^{-s+it}x \ , \qquad x \in E \ , \quad s > 0 \ , \quad t \in \mathbb{R} \ . \tag{4.7.5}$$

Since

$$\partial_s A^{-s+it}x = \partial_s A^{-s}A^{it}x = (-\log A)A^{-s}A^{it}x = (-\log A)A^{-s+it}x$$

for $x \in E$, $s > 0$, and $t \in \mathbb{R}$, thanks to $\mathrm{im}(A^{-s}) \subset \mathrm{dom}(\log A)$ by Theorem 4.6.2, we deduce from (4.7.4) and (4.7.5) that

$$(i \log A)A^{-s+it}x = BA^{-s+it}x = A^{-s+it}Bx \ , \qquad s > 0 \ , \quad t \in \mathbb{R} \ , \tag{4.7.6}$$

for $x \in \mathrm{dom}(B)$. Thus

$$(i \log A)A^{-s}x = BA^{-s}x = A^{-s}Bx \to Bx \ , \qquad x \in \mathrm{dom}(B) \ ,$$

as $s \to 0+$. Since $i \log A$ is closed and $A^{-s}x \to x$ for $s \to 0+$ we find $i \log A \supset B$. On the other hand, since the argument leading to (4.7.4) implies

$$BA^{-s+it}x = \partial_t A^{-s+it}x \ , \qquad x \in E \ , \quad s > 0 \ , \quad t \in \mathbb{R} \ ,$$

it follows from (4.7.5) that

$$i\,BA^{-s}x = A^{-s}(-\log A)x \to (-\log A)x \ , \qquad s \to 0+ \ , \quad x \in \mathrm{dom}(\log A) \ .$$

Since B is closed and $A^{-s}x \to x$ we see that $i\,Bx = (-\log A)x$ for $x \in \mathrm{dom}(\log A)$, that is, $B \supset i \log A$. This proves the theorem. ■

4.7.2 Corollary *Suppose that $A \in \mathcal{BIP}$. Then there exist constants $M \geq 1$ and $\theta \geq 0$ such that*

$$\|A^{it}\| \leq Me^{\theta|t|} \ , \qquad t \in \mathbb{R} \ . \tag{4.7.7}$$

Proof This follows from the proof of (4.7.3) by letting $s = 0$ in (4.7.2). ■

The last part of the proof of Theorem 4.7.1 follows [HiP57, Theorem 17.9.2], where 'boundary values' of analytic semigroups of angle $\pi/2$ have been studied in detail.

Suppose that $M \geq 1$ and $\theta \geq 0$. Then we write

$$A \in \mathcal{BIP}(M,\theta) := \mathcal{BIP}(E;M,\theta)$$

iff $A \in \mathcal{BIP}$ and estimate (4.7.7) is valid. Moreover,

$$\mathcal{BIP}(\theta) := \mathcal{BIP}(E;\theta) := \bigcup_{M\geq 1} \mathcal{BIP}(M,\theta)$$

for $\theta \geq 0$, so that

$$\mathcal{BIP} = \bigcup_{\theta\geq 0} \mathcal{BIP}(\theta) .$$

In general, $\mathcal{BIP}$ is a proper subset of $\mathcal{P}$, that is, there exist operators of positive type which do not possess bounded imaginary powers. In fact, Venni [Ven93] has shown that in every Banach space E with basis, thus in every separable Hilbert space, there exists $A \in \mathcal{P}(E)$ such that $-A$ generates a strongly continuous analytic semigroup, $A^{it} = 1$ for $t \in 2\pi\mathbb{Z}$, whereas A^{it} is unbounded for $t \in \pi(2\mathbb{Z}+1)$. Earlier examples showing that $\mathcal{BIP}(E) \neq \mathcal{P}(E)$, in general, are contained in [Kom66, part I, section 14] and [BaiC91].

Although a general characterization of the class $\mathcal{BIP}$ is not available at present, it is known that certain families of operators belong to $\mathcal{BIP}$. Next we collect some known cases:

4.7.3 Examples **(a)** Let E be a Hilbert space and suppose that $A = A^* \geq \alpha > 0$. Then $A \in \mathcal{BIP}(1,0)$.

Proof This is an immediate consequence of Theorem 4.6.7. ■

(b) Let E be a Hilbert space and let A be maximal accretive with $0 \in \rho(A)$. Then $A \in \mathcal{BIP}(1,\pi/2)$.

This is proven in [Kat62].

(c) Let (Ω,μ) be a σ-finite positive measure space and let $1 < p < \infty$. Suppose that A is the negative infinitesimal generator of a strongly continuous contraction semigroup of negative type on $L_p(\Omega,\mu)$. Also suppose that A is resolvent positive with respect to the natural order of L_p induced by the positive cone $L_p^+ := L_p(\Omega,\mu;\mathbb{R}^+)$. Then there exists a constant $M \geq 1$, depending on p only, but not on A, such that

$$\|A^{it}\| \leq M(1+t^2)e^{\pi|t|/2} , \qquad t \in \mathbb{R} .$$

Thus, given any $\varepsilon > 0$, there exists $M \geq 1$ such that

$$A \in \mathcal{BIP}\big(L_p(\Omega,\mu); M, (\pi+\varepsilon)/2\big)$$

for each A satisfying the above conditions.

This is due to Coifman and Weiss [CoiW77] (also cf. [Ste70b]).

(d) Let the hypotheses of (c) be true for every $p \in (1,\infty)$ and let A be self-adjoint in $L_2(\Omega,\mu)$. Then, given $q \in (1,\infty)$ and $\theta > \pi\,|q^{-1} - 2^{-1}|$, there is $K \geq 1$, independently of A, such that $A \in \mathcal{BIP}\big(L_q(\Omega,\mu); K, \theta\big)$.

Proof Thanks to (a) we can assume that $q \neq 2$. Fix any $p \in (1,\infty)$ with

$$\vartheta := (q^{-1} - 2^{-1})/(p^{-1} - 2^{-1}) \in (0,1)\ .$$

Then fix $\varepsilon \in \big(0, (\pi/\vartheta) - \pi\big)$. Since $L_q \doteq [L_2, L_p]_\vartheta$ (e.g., [Tri78, 1.18.6/2]), it follows from (a) and (c) that $A \in \mathcal{BIP}\big(L_q(\Omega,\mu); K, \theta\big)$, where $K := M^\vartheta$ and $2\theta := \vartheta(\pi+\varepsilon)$. Since we can choose p such that $|p^{-1} - 2^{-1}|$ is arbitrarily close to $1/2$ and ε arbitrarily close to 0, the assertion follows. ■

We also refer to [ClP90, Theorem 5.8] for bounded imaginary powers of operators on $L_p(\Omega,\mu;E)$ that are 'tensor product' extensions of operators on $L_p(\Omega,\mu)$ for some $p \in (1,\infty)$.

The following lemma is almost trivial but is useful in proving certain deeper results about bounded imaginary powers.

4.7.4 Lemma *Suppose that $K, M \geq 1$ and $\theta \geq 0$, and let $A \in \mathcal{P}$.*

(i) *If there exists $\varepsilon > 0$ such that*

$$\|A^z\| \leq M e^{\theta|\operatorname{Im} z|}\ , \qquad -\varepsilon < \operatorname{Re} z < 0\ ,$$

then $A \in \mathcal{BIP}(M,\theta)$.

(ii) *If $A \in \mathcal{BIP}(M,\theta) \cap \mathcal{P}_K$ then*

$$\|A^z\| \leq KM e^{\theta|\operatorname{Im} z|}\ , \qquad -1 < \operatorname{Re} z < 0\ .$$

Proof (i) Given $x \in D(A)$, it follows from Theorem 4.6.5(vii) that $A^z x \to A^{it}x$ in E as $z \to it$ and $\operatorname{Re} z < 0$. Since $D(A)$ is dense in E, by Theorem 4.6.5(v), the assertion is now an obvious consequence of the assumption.

(ii) is obtained from $A^{-s+it} = A^{-s}A^{it}$ for $s, t \in \mathbb{R}$, and from Theorem 4.6.4. ■

For $K \geq 1$ and $\vartheta \in [0, \pi)$, a linear operator A in E is said **to be of type** (K, ϑ), in symbols:

$$A \in \mathcal{P}(K, \vartheta) := \mathcal{P}(E; K, \vartheta) ,$$

if it is densely defined, if

$$S_\vartheta := [\,|\arg z| \leq \vartheta] \cup \{0\} \subset \rho(-A) ,$$

and if

$$(1 + |\lambda|)\, \|(\lambda + A)^{-1}\| \leq K , \qquad \lambda \in S_\vartheta . \tag{4.7.8}$$

Put

$$\mathcal{P}(\vartheta) := \mathcal{P}(E; \vartheta) := \bigcup_{K \geq 1} \mathcal{P}(K, \vartheta)$$

and note that, trivially

$$\mathcal{P}(K, \vartheta) \subset \mathcal{P}(L, \theta) , \qquad 1 \leq K \leq L , \quad 0 \leq \theta \leq \vartheta < \pi . \tag{4.7.9}$$

Observe that $\mathcal{P} = \mathcal{P}(0)$, that is, every operator of type (K, ϑ) is of positive type.

Let

$$\vartheta_K := \vartheta + \frac{\pi - \vartheta}{2} \wedge \arcsin \frac{1}{2K} . \tag{4.7.10}$$

An obvious modification of (4.6.3) shows that, given $A \in \mathcal{P}(K, \vartheta)$,

$$\begin{aligned} &\|(\lambda + A)^{-1}\| \leq 2K(1 + |\lambda_0|)^{-1} \leq (2K + 1)/(1 + |\lambda|) \\ &\text{for } |\lambda - \lambda_0| \leq (1 + |\lambda_0|)/(2K) \text{ and } \lambda_0 \in S_\vartheta. \end{aligned} \tag{4.7.11}$$

Hence

$$S_{\vartheta_K} \cup \big[|z| \leq 1/(2K)\big] \subset \rho(-A) \tag{4.7.12}$$

and

$$\mathcal{P}(K, \vartheta) \subset \mathcal{P}(2K + 1, \vartheta_K) . \tag{4.7.13}$$

In the following, we denote by

$$\Gamma := \Gamma(K, \vartheta)$$

the negatively oriented boundary of $S_{\vartheta_K} \cup \big[|z| \leq 1/(2K)\big] \subset \rho(-A)$. Then

$$A^z = \frac{1}{2\pi i} \int_\Gamma (-\lambda)^z (\lambda + A)^{-1}\, d\lambda , \qquad \operatorname{Re} z < 0 , \quad A \in \mathcal{P}(K, \vartheta) , \tag{4.7.14}$$

by Theorem 4.6.5.

After these preparations we can prove the following surprising theorem due to Dore [Dor93a], which shows that, given any $A \in \mathcal{P}(E)$, its $\big(E, D(A)\big)_\alpha$-realization (that we again denote by A) belongs to $\mathcal{BIP}\big((E, D(A))_\alpha\big)$, where $(\cdot, \cdot)_\alpha$ is any admissible real interpolation functor of exponent $\alpha \in (0, 1)$.

4.7.5 Theorem *Suppose that* $0 < \alpha < 1$ *and*

$$(\cdot,\cdot)_\alpha \in \{\, (\cdot,\cdot)_{\alpha,p}, (\cdot,\cdot)^0_{\alpha,\infty} \ ;\ 1 \le p < \infty \,\} ,$$

and that $A \in \mathcal{P}(E;\vartheta)$ *for some* $\vartheta \in (0,\pi)$. *Then*

$$A \in \mathcal{BIP}\big((E, D(A))_\alpha; \pi - \vartheta\big) .$$

Proof Put $E_\alpha := \big(E, D(A)\big)_\alpha$ and let A_α be the E_α-realization of A. Then $A_\alpha \in \mathcal{C}(E_\alpha)$. Since $D(A^2) \subset \mathrm{dom}(A_\alpha)$ and since $D(A^2) \overset{d}{\hookrightarrow} D(A) \overset{d}{\hookrightarrow} E_\alpha$ by Theorem 4.6.5(v) and the admissibility of $(\cdot,\cdot)_\alpha$, it follows that A_α is densely defined. These assertions are also true for A_1, the $D(A)$-realization of A. Since A commutes with its resolvent, it follows that $A_1 \in \mathcal{P}\big(D(A), \vartheta\big)$. Hence, by interpolation, $A_\alpha \in \mathcal{P}(E_\alpha, \vartheta)$.

It follows from Theorem 4.6.5 that

$$A^z x = A^{z-1} A x = \frac{1}{2\pi i} \int_\Gamma (-\lambda)^{z-1} A(\lambda + A)^{-1} x \, d\lambda , \qquad \mathrm{Re}\, z = 0 , \quad x \in D(A) .$$

Note that

$$\begin{aligned} A(s+A)^{-1}(\lambda + A)^{-1} &= (s-\lambda)^{-1}\big[A(\lambda+A)^{-1} - A(s+A)^{-1}\big] \\ &= (s-\lambda)^{-1}\big(1 - \lambda(\lambda+A)^{-1} - \big[1 - s(s+A)^{-1}\big]\big) \\ &= (s-\lambda)^{-1}\big[s(s+A)^{-1} - \lambda(\lambda+A)^{-1}\big] \end{aligned}$$

for $s > 0$ and $\lambda \in \rho(-A)$. Hence, by Cauchy's theorem,

$$\begin{aligned} A(s+A)^{-1}A^z x &= \frac{1}{2\pi i}\Big[\int_\Gamma \frac{(-\lambda)^{z-1}}{s-\lambda}\, d\lambda\, sA(s+A)^{-1}x + \int_\Gamma \frac{(-\lambda)^z}{s-\lambda} A(\lambda+A)^{-1} x \, d\lambda\Big] \\ &= \frac{1}{2\pi i} \int_\Gamma \frac{(-\lambda)^z}{s-\lambda} A(\lambda + A)^{-1} x \, d\lambda \end{aligned}$$

for $s > 0$. In the last integral we can contract Γ to the negatively oriented boundary of S_ϑ. Then

$$\begin{aligned} \|A(s+A)^{-1}A^z x\| \le e^{(\pi-\vartheta)|\mathrm{Im}\, z|} \int_0^\infty \Big\{ &\frac{\|A(re^{i\vartheta} + A)^{-1}x\|}{|s - re^{i\vartheta}|} \\ &+ \frac{\|A(re^{-i\vartheta} + A)^{-1}x\|}{|s - re^{-i\vartheta}|} \Big\}\, dr \end{aligned} \tag{4.7.15}$$

for $\mathrm{Re}\, z = 0$ and $x \in D(A)$.

Note that $\mathrm{dom}(A) = \mathrm{dom}(e^{\pm i\vartheta}A)$ and $\|Ax\| = \|e^{\pm i\vartheta}Ax\|$ for $x \in \mathrm{dom}(A)$. Thus $D(A) \doteq D(e^{\pm i\vartheta}A)$ and, consequently, $\big(E, D(A)\big)_{\alpha,\infty} \doteq \big(E, D(e^{\pm i\vartheta}A)\big)_{\alpha,\infty}$

for $0<\alpha<1$. Now it follows from (I.2.9.12) that

$$x \mapsto \sup_{s>0} s^\alpha \left\|A(s+e^{\pm i\vartheta}A)^{-1}x\right\| , \quad x \mapsto \sup_{s>0} s^\alpha \left\|A(s+A)^{-1}x\right\|$$

are equivalent norms on $\big(E,D(A)\big)_{\alpha,\infty}$. Hence we infer from (4.7.15) that

$$\left\|A(s+A)^{-1}A^z x\right\| \le s^{-\alpha} M e^{(\pi-\vartheta)|\operatorname{Im} z|} \, \|x\|_{(E,D(A))_{\alpha,\infty}} \tag{4.7.16}$$

for $\operatorname{Re} z = 0$ and $x \in D(A)$, where

$$M := c(\alpha) \int_0^\infty t^{-\alpha} \big\{ |1-te^{i\vartheta}|^{-1} + |1-te^{-i\vartheta}|^{-1} \big\} \, dt \, .$$

By invoking (I.2.9.12) once more, we see that

$$\|A^z x\|_{(E,D(A))_{\alpha,\infty}} \le c e^{(\pi-\vartheta)|\operatorname{Im} z|} \, \|x\|_{(E,D(A))_{\alpha,\infty}}$$

for $\operatorname{Re} z = 0$ and $x \in D(A)$. Now the assertion follows from the density of $D(A)$ in $\big(E,D(A)\big)^0_{\alpha,\infty}$, provided $(\cdot,\cdot)_\alpha = (\cdot,\cdot)^0_{\alpha,\infty}$. If we have $(\cdot,\cdot)_\alpha = (\cdot,\cdot)_{\alpha,p}$ for some $p\in[1,\infty)$, we choose $0<\alpha_0<\alpha<\alpha_1<1$ and apply (I.2.5.2) and the reiteration theorem (I.2.8.1), (I.2.8.2) to deduce the assertion. ∎

4.8 Perturbation Theorems

Now we prove some perturbation theorems for the classes $\mathcal{P}$ and $\mathcal{BIP}$. They are useful for proving that L_p-realizations of rather general elliptic systems possess bounded imaginary powers.

In the following,

$$s(\vartheta) := 1 + \Theta(\vartheta-\pi/2)\Big[\frac{1}{\sin\vartheta} - 1\Big] , \qquad 0<\vartheta<\pi ,$$

where $\Theta(t) := 1$ for $t \ge 0$ and $\Theta(t) := 0$ for $t<0$ is the Heaviside function.

4.8.1 Lemma *Suppose that $A \in \mathcal{P}(K,\vartheta)$.*

(i) *If B is a linear operator in E satisfying* $\operatorname{dom}(B) \supset \operatorname{dom}(A)$ *and*

$$\|B(\lambda+A)^{-1}\| \le \beta < 1 , \qquad \lambda \in S_\vartheta , \tag{4.8.1}$$

then $A+B \in \mathcal{P}\big((1-\beta)^{-1}K,\vartheta\big)$.

(ii) *$\mu + A \in \mathcal{P}\big(Ks(\vartheta),\vartheta\big)$ for $\mu \ge 0$.*

Proof (i) It follows from (4.8.1) that $1+B(\lambda+A)^{-1} \in \mathcal{L}\mathrm{aut}(E)$ and

$$\left\|\big[1+B(\lambda+A)^{-1}\big]^{-1}\right\| \le (1-\beta)^{-1} \, .$$

Thus we deduce from

$$\lambda + A + B = \left[1 + B(\lambda + A)^{-1}\right](\lambda + A) \tag{4.8.2}$$

that $S_\vartheta \subset \rho\big(-(A+B)\big)$ and

$$\|(\lambda + A + B)^{-1}\| \le (1-\beta)^{-1} \|(\lambda + A)^{-1}\| \ , \qquad \lambda \in S_\vartheta \ .$$

Now the assertion is obvious.

(ii) follows from the fact that $\lambda \in S_\vartheta$ implies $|\lambda + \mu| \ge |\lambda| \sin(\pi - \vartheta)$ if ϑ lies in $(\pi/2, \pi)$, and $|\lambda + \mu| \ge |\lambda|$ if $0 < \vartheta \le \pi/2$. ∎

In the following, we put

$$E_0 := E \ , \quad E_1 := \big(\mathrm{dom}(A), \|A\cdot\|\big)$$

for $A \in \mathcal{P}(E)$, if no confusion seems likely. We also assume that $K \ge 1$, $\vartheta \in (0, \pi)$, and $\theta \ge 0$.

Now we can prove an important perturbation theorem for the class $\mathcal{P}$ concerning suitable 'subordinate' perturbations.

4.8.2 Theorem *Suppose that $A \in \mathcal{P}(K, \vartheta)$ and that $0 \le \alpha < 1$. If $\alpha > 0$, put $E_\alpha := \mathfrak{F}_\alpha(E_0, E_1)$, where $\mathfrak{F}$ is an exact interpolation functor of exponent α. Then, given $B \in \mathcal{L}(E_\alpha, E_0)$ and $\beta \in (0,1)$,*

$$\mu + A + B \in \mathcal{P}\big((1-\beta)^{-1} K s(\vartheta), \vartheta\big) \ ,$$

provided

$$\mu \ge \mu_0 := \Big[\big((1+K)\,\|B\|\,\beta^{-1}\big)^{1/(1-\alpha)} - 1\Big]^+ s(\vartheta) \ .$$

Proof Since $A(\lambda + A)^{-1} = 1 - \lambda(\lambda + A)^{-1}$ we see that

$$\|(\lambda + A)^{-1}\|_{\mathcal{L}(E_0, E_1)} \le 1 + K \ , \qquad \lambda \in S_\vartheta \ .$$

This and (4.7.8) imply

$$\|(\lambda + A)^{-1}\|_{\mathcal{L}(E_0, E_\alpha)} \le (1+K)(1+|\lambda|)^{\alpha-1} \ , \qquad \lambda \in S_\vartheta \ . \tag{4.8.3}$$

Hence

$$\|B(\lambda + \mu + A)^{-1}\| \le (1+K)\,\|B\|\,(1 + |\lambda + \mu|)^{\alpha - 1} \ , \qquad \lambda \in S_\vartheta \ , \quad \mu \ge 0 \ .$$

Since $|\lambda + \mu| \ge \mu$ if $0 < \vartheta \le \pi/2$, and $|\lambda + \mu| \ge \mu \sin(\pi - \vartheta)$ if $\pi/2 < \vartheta < \pi$, we see that

$$\|B(\lambda + \mu + A)^{-1}\| \le \beta < 1 \ , \qquad \lambda \in S_\vartheta \ , \quad \mu \ge \mu_0 \ .$$

Now the assertion follows from Lemma 4.8.1. ∎

Our next almost obvious lemma implies, in particular, that the classes $\mathcal{P}(\vartheta)$ and $\mathcal{BIP}(\vartheta)$ are invariant under similarity transformations.

4.8.3 Lemma *Let F be a Banach space, let A and B be densely defined linear operators in E and in F, respectively, and let $C \in \mathcal{L}(E,F)$ and $D \in \mathcal{L}(F,E)$. Suppose that $S_\vartheta \subset \rho(-A) \cap \rho(-B)$ and*

$$(\lambda + B)^{-1} = C(\lambda + A)^{-1} D\ , \qquad \lambda \in S_\vartheta\ . \tag{4.8.4}$$

Also suppose that $A \in \mathcal{P}(E; K, \vartheta)$. Then

(i) *$B \in \mathcal{P}(F; K_1, \vartheta)$ with $K_1 := \|C\|\,\|D\|\,K$.*

(ii) *If $A \in \mathcal{BIP}(E; M, \theta)$ then $B \in \mathcal{BIP}(F; M_1, \theta)$ with $M_1 := \|C\|\,\|D\|\,KM$.*

Proof (i) is obvious.

(ii) Thanks to (4.7.14),

$$B^z = \frac{1}{2\pi i} \int_\Gamma (-\lambda)^z C(\lambda + A)^{-1} D\, d\lambda = CA^z D\ , \qquad \operatorname{Re} z < 0\ .$$

Hence, by Lemma 4.7.4(ii),

$$\|B^z\|_{\mathcal{L}(F)} \le \|C\|\,\|D\|\,KMe^{\theta|\operatorname{Im} z|}\ , \qquad \operatorname{Re} z < 0\ .$$

Now the assertion follows from Lemma 4.7.4(i). ■

Next we prove a simple 'splitting lemma' that will greatly simplify our proofs that a given operator of positive type has bounded imaginary powers.

4.8.4 Lemma *Suppose that $A \in \mathcal{P}(K, \vartheta)$ and*

$$(\lambda + A)^{-1} = R(\lambda) + S(\lambda)\ , \qquad \lambda \in \Gamma := \Gamma(K, \vartheta)\ , \tag{4.8.5}$$

and put

$$R_z(\lambda) := (-\lambda)^z R(\lambda)\ , \qquad \lambda \in \Gamma\ , \quad \operatorname{Re} z < 0\ .$$

Also suppose that

$$R_z, S \in L_1\big(\Gamma, ds, \mathcal{L}(E)\big)$$

and

$$\Big\| \int_\Gamma R_z(\lambda)\, d\lambda \Big\|_{\mathcal{L}(E)} \le 2\pi M e^{\theta|\operatorname{Im} z|}\ , \qquad -\varepsilon < \operatorname{Re} z < 0\ , \tag{4.8.6}$$

for some $\varepsilon > 0$, ds denoting the 'arc-length measure'. Then

$$A \in \mathcal{BIP}(M + \|S\|_{L_1}, (\pi - \vartheta) \vee \theta)\ .$$

Proof It follows from (4.7.14) and (4.8.5) that

$$A^z = \frac{1}{2\pi i}\int_\Gamma R_z(\lambda)\,d\lambda + \frac{1}{2\pi i}\int_\Gamma (-\lambda)^z S(\lambda)\,d\lambda\ , \qquad \operatorname{Re} z < 0\ .$$

Thus, thanks to (4.8.6),

$$\|A^z\| \le \bigl(M + (2\pi)^{-1}(2K)^{|\operatorname{Re} z|}\,\|S\|_{L_1}\bigr)e^{((\pi-\vartheta)\vee\theta)|\operatorname{Im} z|}\ , \qquad -\varepsilon < \operatorname{Re} z < 0\ ,$$

and Lemma 4.7.4(i) implies the assertion. ∎

As a first application of this splitting lemma we show that $\mathcal{BIP}(\theta)$ is invariant under suitable 'lower order perturbations'.

4.8.5 Theorem *Suppose that $A \in \mathcal{P}(K,\vartheta) \cap \mathcal{BIP}(M,\theta)$ and $0 \le \beta < 1$. Fix $K_1 \ge K$ and put*

$$R(\lambda) := (\lambda + A)^{-1}\ , \qquad \lambda \in \Gamma := \Gamma\bigl((1-\beta)^{-1}K_1, \vartheta\bigr)\ .$$

Let B be a linear operator in E satisfying

(i) $\operatorname{dom}(B) \supset \operatorname{dom}(A)$;

(ii) $\|BR(\lambda)\| \le \beta < 1$ *for* $\lambda \in \Gamma \cup S_\vartheta$;

(iii) $\|RBR\|_{L_1(\Gamma, ds, \mathcal{L}(E))} \le \sigma < \infty$.

Then $A + B \in \mathcal{BIP}\bigl(KM + (1-\beta)^{-1}\sigma, (\pi-\vartheta)\vee\theta\bigr)$.

Proof Lemma 4.8.1 implies $A + B \in \mathcal{P}\bigl((1-\beta)^{-1}K, \vartheta\bigr)$. From (4.8.2) we deduce that

$$(\lambda + A + B)^{-1} = R(\lambda) + S(\lambda)\ , \qquad \lambda \in \Gamma\ ,$$

where

$$S := -RBR[1 + BR]^{-1}\ .$$

Hence (ii) and (iii) imply $S \in L_1\bigl(\Gamma, ds, \mathcal{L}(E)\bigr)$ and $\|S\|_{L_1} \le (1-\beta)^{-1}\sigma$. Now the assertion follows from Lemmas 4.7.4(ii) and 4.8.4 and the fact that $\Gamma(K,\vartheta)$ can be replaced by $\Gamma\bigl((1-\beta)^{-1}K_1, \vartheta\bigr)$. ∎

4.8.6 Corollary *Suppose that $A \in \mathcal{P}(K,\vartheta) \cap \mathcal{BIP}(M,\theta)$. Then, given $\nu > 0$, there exists N such that*

$$\mu + A \in \mathcal{BIP}\bigl(N, (\pi-\vartheta)\vee\theta\bigr)\ , \qquad 0 \le \mu \le \nu\ .$$

Proof Let $\mu_1 \ge 0$ be fixed and put $A_1 := \mu_1 + A$. Then Lemma 4.8.1 implies $A_1 \in \mathcal{P}(K_1, \vartheta)$ with $K_1 := Ks(\vartheta)$. Suppose that

$$A_1 \in \mathcal{BIP}(M_1, \theta)$$

for some $M_1 \ge 1$. Note that this is true if $\mu_1 = 0$.

Let $0 < \mu \le 1/(6K_1) =: \nu_1$ and put $B := \mu 1_E$ and

$$R_1(\lambda) := (\lambda + A_1)^{-1} , \qquad \lambda \in \Gamma_1 := \Gamma(2K_1, \vartheta) .$$

It follows from (4.7.11) that

$$\|R_1(\lambda)\| \le 3K_1(1 + |\lambda|)^{-1} , \qquad \lambda \in \Gamma_1 .$$

Hence B satisfies (i)–(iii) of Theorem 4.8.5 with $\beta := 1/2$ and $\sigma := 2K_1\rho$, where ρ is the $L_1(\Gamma_1, ds)$-norm of $(1 + |\cdot|)^{-2}$. Thus, thanks to the latter theorem,

$$\mu + \mu_1 + A = \mu + A_1 \in \mathcal{BIP}\big(M_2, (\pi - \vartheta) \vee \theta\big) ,$$

where $M_2 := K_1 M_1 + 2\sigma$. Now the assertion follows by induction starting with $\mu_1 := 0$, since ν can be reached in finitely many steps of length at most ν_1. ■

The following perturbation theorem is particularly important in applications since it allows to discard 'lower order terms'.

4.8.7 Theorem *Suppose $A \in \mathcal{P}(K, \vartheta) \cap \mathcal{BIP}(M, \theta)$. Also suppose $\alpha \in [0, 1)$ and $\mathfrak{F}_\alpha$ is an exact interpolation functor of exponent α, if $\alpha > 0$. Lastly, assume that $B \in \mathcal{L}(E_\alpha, E_0)$, where $E_\alpha := \mathfrak{F}_\alpha(E_0, E_1)$, and put*

$$\mu_B := \Big[\big(4K\,\|B\|\big)^{1/(1-\alpha)} - 1\Big]^+ .$$

Then $\mu + A + B \in \mathcal{BIP}\big((\pi - \vartheta) \vee \theta\big)$ for $\mu \ge \mu_B$.

Proof Put $\mu := \mu_B$ and let $A_0 := \mu + A$ and $R_0(\lambda) := (\lambda + A_0)^{-1}$. Since

$$\lambda \in M(K, \vartheta) := \Big[|z| \le \frac{\sin \vartheta_K}{2K}\Big] \cup S_{\vartheta_K}$$

implies $\lambda + \mu \in \Gamma(K, \vartheta) \cup S_{\vartheta_K}$, we deduce from (4.7.8) and (4.7.11) that

$$\|R_0(\lambda)\| \le \frac{3K}{1 + |\lambda + \mu|} , \qquad \lambda \in M(K, \vartheta) .$$

From this and from $AR_0(\lambda) = 1 - (\lambda + \mu)R_0(\lambda)$ it follows that

$$\|AR_0(\lambda)\| \le 4K , \qquad \lambda \in M(K, \vartheta) . \tag{4.8.7}$$

Thus, by interpolation,

$$\|R_0(\lambda)\|_{\mathcal{L}(E_0, E_\alpha)} \le \frac{3^{1-\alpha} 4^\alpha K}{(1 + |\lambda + \mu|)^{1-\alpha}} , \qquad \lambda \in M(K, \vartheta) ,$$

so that

$$\|BR_0(\lambda)\| \le \frac{3^{1-\alpha} 4^\alpha K\,\|B\|}{(1 + |\lambda + \mu|)^{1-\alpha}} , \qquad \lambda \in M(K, \vartheta) . \tag{4.8.8}$$

Of course, we can assume that $B \neq 0$. Hence

$$3^{1-\alpha}4^{\alpha}K\,\|B\| < 4K\,\|B\| \leq (1+\mu)^{1-\alpha}\ . \tag{4.8.9}$$

Note that $|\lambda+\mu| \geq \mu \wedge |\mu - r|$ for $r \in \bigl(0, \sin(\vartheta_K)/(2K)\bigr)$ and $\lambda \in [\,|z| = r] \cup S_\vartheta$. Thus it follows from (4.8.8) and (4.8.9) that we can find $K_1 > K$ and $\beta \in (0,1)$ such that

$$\|BR_0(\lambda)\| \leq \frac{3^{1-\alpha}4^{\alpha}K}{\bigl(1+\bigl|\mu - \frac{1-\beta}{2K_1}\bigr|\bigr)^{1-\alpha}} \leq \beta\ , \qquad \lambda \in \Sigma := \Gamma\bigl((1-\beta)^{-1}K_1, \vartheta\bigr) \cup S_\vartheta\ .$$

Lemma 4.8.1(ii) implies $A_0 \in \mathcal{P}\bigl(Ks(\vartheta), \vartheta\bigr)$. Hence we deduce from (4.7.8) and (4.7.11) that

$$(1+|\lambda|)\,\|R_0(\lambda)\| \leq 3Ks(\vartheta)\ , \qquad \lambda \in \Sigma\ .$$

Thus we infer from (4.8.8) that

$$\|R_0(\lambda)BR_0(\lambda)\| \leq c(1+|\lambda|)^{\alpha-2}\ , \qquad \lambda \in \Sigma\ .$$

Since we know from Corollary 4.8.6 that $A_0 \in \mathcal{BIP}\bigl((\pi-\vartheta) \vee \theta\bigr)$, the assertion is now a consequence of Theorem 4.8.5 and Corollary 4.8.6. ■

The results of this subsection are specializations of the corresponding assertions in [AmHS94] where the slightly more general situation of the bounded H_∞-calculus has been considered. The fact that we have given quantitative estimates in some cases will be useful in Chapter VII.

A closely related result to Corollary 4.8.6 has also been proven in [PrüS90, Theorem 3]. In that paper a functional calculus for operators of positive type possessing bounded imaginary powers has been developed. However, it is not assumed that $0 \in \rho(A)$, but only that A satisfies the assumptions of Remark 4.6.8(c) and that $\operatorname{im}(A)$ is dense in E. This generality requires rather more complicated proofs.

We also refer to [DorV90, Theorem 2.2 and Remark 2.3] and to [PrüS93, Proposition 3.1] for results related to Theorems 4.8.5 and 4.8.7.

4.9 Sums of Closed Operators

Let E be a Banach space. Given linear operators $A, B : E \to E$, their **commutator** is defined by

$$[A,B] := AB - BA\ .$$

Thus A and B commute iff $[A,B] = 0$. If $A : \operatorname{dom}(A) \subset E \to E$ and $B : E \to E$ are linear then A and B **commute** if $AB \supset BA$. Note that this means that $B\bigl(\operatorname{dom}(A)\bigr) \subset \operatorname{dom}(A)$ and $ABx = BAx$ for $x \in \operatorname{dom}(A)$.

For convenience, we collect some well-known facts about commuting linear operators. Observe that the first part of the following lemma has already been used in several of the preceding proofs.

4.9.1 Lemma *Suppose that $A \in \mathcal{C}(E)$. Then*

(i) $A(\lambda+A)^{-1} \supset (\lambda+A)^{-1}A, \quad \lambda \in \rho(-A)$.

(ii) *If $B \in \mathcal{L}(E)$ then*

$$\begin{aligned} AB \supset BA \iff & \big[(\lambda+A)^{-1}, B\big] = 0 \quad \textit{for some} \quad \lambda \in \rho(-A) \\ \iff & \big[(\lambda+A)^{-1}, B\big] = 0 \quad \textit{for all} \quad \lambda \in \rho(-A). \end{aligned}$$

(iii) *If $A, B \in \mathcal{G}(E)$ then*

$$\begin{aligned} & Ae^{-tB} \supset e^{-tB}A, \quad t \geq 0, \\ & \iff A(\mu+B)^{-1} \supset (\mu+B)^{-1}A, \quad \mu > \operatorname{type}(-B), \\ & \iff \big[(\lambda+A)^{-1}, (\mu+B)^{-1}\big] = 0, \quad \lambda > \operatorname{type}(-A), \quad \mu > \operatorname{type}(-B), \\ & \iff \big[(\lambda+A)^{-1}, e^{-tB}\big] = 0, \quad \lambda > \operatorname{type}(-A), \quad t \geq 0, \\ & \iff \big[e^{-sA}, e^{-tB}\big] = 0, \quad s, t \geq 0. \end{aligned}$$

Moreover, the assertions involving $\lambda > \operatorname{type}(-A)$ or $\mu > \operatorname{type}(-B)$ are already true if they hold for some $\lambda > \operatorname{type}(-A)$ or at least one $\mu > \operatorname{type}(-B)$, respectively.

Proof (i) follows from the obvious relation

$$A(\lambda+A)^{-1} = 1 - \lambda(\lambda+A)^{-1} \supset (\lambda+A)^{-1}A$$

for $\lambda \in \rho(-A)$.

(ii) If $AB \supset BA$ then $(\lambda+A)B \supset B(\lambda+A)$. Thus, if $\lambda \in \rho(-A)$, it follows that $B \supset (\lambda+A)^{-1}B(\lambda+A)$; hence $B(\lambda+A)^{-1} \supset (\lambda+A)^{-1}B$. Now the assertion is obvious.

(iii) is an easy consequence of (ii) and the representation formulas (II.6.3.3) and (II.6.3.4). ■

Two closed linear operators A, B in E are said to be **resolvent commuting** if there exist $\lambda \in \rho(-A)$, $\mu \in \rho(-B)$ such that

$$\big[(\lambda+A)^{-1}, (\mu+B)^{-1}\big] = 0\ . \tag{4.9.1}$$

It follows from Lemma 4.9.1(ii) that (4.9.1) then holds for all $\lambda \in \rho(-A)$ and all $\mu \in \rho(-B)$.

4.9.2 Lemma (i) *Suppose that $A \in \mathcal{P}(E)$ and $B \in \mathcal{L}(E)$. If $AB \supset BA$ then $A^zB \supset BA^z$ for $z \in \mathbb{C}$.*

(ii) *Suppose that* $A, B \in \mathcal{P}(E)$ *and are resolvent commuting. Then*

$$A^z B^w \supset B^w A^z \qquad \text{for} \quad z \in \mathbb{C} \text{ and } \operatorname{Re} w < 0 .$$

If $B \in \mathcal{BIP}(E)$ *then* $A^z B^w \supset B^w A^z$ *for* $z \in \mathbb{C}$ *and* $\operatorname{Re} w \leq 0$.

Proof (i) If $\operatorname{Re} z < 0$, the assertion follows from Lemma 4.9.1(ii) and from (4.6.22). Then $A^z = (A^{-z})^{-1}$ for $\operatorname{Re} z > 0$ implies the assertion for $\operatorname{Re} z > 0$ by again invoking Lemma 4.9.1(ii). Suppose that $\operatorname{Re} z = 0$. If $x \in \operatorname{dom}(A)$, the very same lemma and the representation formula of Theorem 4.6.5(ii) guarantee that $Bx \in \operatorname{dom}(A^z)$ and $BA^z x = A^z Bx$. By the definition of A^z as the closure of its restriction to $D(A)$, given $x \in \operatorname{dom}(A^z)$, there exists a sequence (x_j) in $D(A)$ such that $x_j \to x$ and $A^z x_j \to A^z x$ in E. Thus $A^z B x_j = B A^z x_j \to B A^z x$ by the boundedness of B. Since $Bx_j \to Bx$, the closedness of A^z shows that $Bx \in D(A^z)$ and $A^z Bx = BA^z x$. Hence $A^z B \supset BA^z$ in this case as well.

(ii) From (i) we infer $A^z(\lambda + B)^{-1} \supset (\lambda + B)^{-1} A^z$ for $z \in \mathbb{C}$ and $\lambda \in \rho(-B)$. Now the closedness of A^z and (4.6.22) imply $A^z B^w \supset B^w A^z$ for $z \in \mathbb{C}$ and $\operatorname{Re} w < 0$. Suppose that $B \in \mathcal{BIP}(E)$ and $\operatorname{Re} w = 0$. Then, given $x \in D(A^z)$, we see that

$$A^z B^{w-\varepsilon} x = B^{w-\varepsilon} A^z x , \qquad \varepsilon > 0 .$$

Thus, letting $\varepsilon \to 0$, we deduce from Theorem 4.7.1 and the closedness of A^z that $A^z B^w x = B^w A^z x$, which proves the lemma. ∎

Throughout the remainder of this subsection we presuppose that

$$\begin{aligned} &A, B \in \mathcal{P}_K(E) \text{ and are resolvent commuting and} \\ &A \in \mathcal{BIP}(E; M, \theta_A) , \quad B \in \mathcal{BIP}(E; M, \theta_B) \\ &\text{with } \theta_A + \theta_B < \pi. \end{aligned} \tag{4.9.2}$$

It follows from Theorem 4.6.2 that the map

$$[0 < \operatorname{Re} z < 1] \to \mathcal{L}(E) , \quad z \mapsto \frac{A^{-z} B^{z-1}}{\sin \pi z}$$

is holomorphic. Note that, thanks to (4.9.2) and Lemma 4.7.4(ii),

$$\frac{\|A^{-z} B^{z-1}\|}{|\sin \pi z|} \leq 4K^2 M^2 e^{-\pi |\operatorname{Im} z|} e^{(\theta_A + \theta_B)|\operatorname{Im} z|} \tag{4.9.3}$$

for $0 < \operatorname{Re} z < 1$ and $|\operatorname{Im} z| \geq 1$. Thus, given any $c \in (0,1)$,

$$S := \frac{1}{2i} \int_{c-i\infty}^{c+i\infty} \frac{A^{-z} B^{z-1}}{\sin \pi z} \, dz \in \mathcal{L}(E)$$

is well-defined and, thanks to Cauchy's theorem, independent of $c \in (0,1)$. Below we show, following [DorV87], that $S = (A+B)^{-1}$ provided E is a UMD space.

4.9.3 Lemma *If $x \in \operatorname{dom}(A+B)$ then $S(A+B)x = x$.*

Proof It follows from Lemma 4.9.2(ii) that

$$S(A+B)x = \frac{1}{2i}\int_{c-i\infty}^{c+i\infty} \frac{A^{-z}B^{z}x}{\sin \pi z}\,dz - \frac{1}{2i}\int_{c-1-i\infty}^{c-1+i\infty} \frac{B^{z}A^{-z}x}{\sin \pi z}\,dz\ . \tag{4.9.4}$$

Define $g\colon [-1 < \operatorname{Re} z < 1] \to E$ by

$$g(z) := \begin{cases} A^{-z}B^{z}x = A^{-z}B^{z-1}Bx\ , & 0 \le \operatorname{Re} z < 1\ , \\ B^{z}A^{-z}x = B^{z}A^{-z-1}Ax\ , & -1 < \operatorname{Re} z < 0\ . \end{cases}$$

Note that, thanks to Lemma 4.9.2(ii), g is well-defined. By Theorem 4.6.5(vii) the function g is holomorphic in $[-1 < \operatorname{Re} z < 1]\backslash i\mathbb{R}$ and, by Theorem 4.7.1, it is continuous on $[-1 < \operatorname{Re} z < 1]$. From this and from Cauchy's theorem we easily deduce that $\int_R g(z)\,dz = 0$ for every closed rectangle in $[-1 < \operatorname{Re} z < 1]$ whose sides are parallel to the coordinate axes. Hence g is holomorphic in $[-1 < \operatorname{Re} z < 1]$ by Morera's theorem (e.g., [Nar85, Theorem I.2.5]), which is obviously also valid in the E-valued case. Put $f(z) := g(z)/\sin \pi z$ for $-1 < \operatorname{Re} z < 1$. Then f is holomorphic in $[-1 < \operatorname{Re} z < 1]\backslash\{0\}$ and has a pole at $z = 0$ with residuum $\pi^{-1}x$. Lastly we obtain, similarly as in (4.9.3), that

$$\|f(z)\| \le c e^{-(\pi-\theta_A-\theta_B)|\operatorname{Im} z|}$$

for $-1 < \operatorname{Re} z < 1$ and $|\operatorname{Im} z| \ge 1$. Now (4.9.4) and the residue theorem imply the assertion. ■

4.9.4 Lemma *Suppose that $0 < \varepsilon < 1$ and $x \in \operatorname{dom}(A^{1-\varepsilon})$. Then $Sx \in \operatorname{dom}(A+B)$ and $(A+B)Sx = x$.*

Proof Since $A^{1-\varepsilon-it}B^{\varepsilon-1+it}x = B^{\varepsilon-1}B^{it}A^{-it}A^{1-\varepsilon}x$ by Theorem 4.7.1 and by Lemma 4.9.2(ii), the integral

$$\int_{\varepsilon-i\infty}^{\varepsilon+i\infty} \frac{A^{1-z}B^{z-1}x}{\sin \pi z}\,dz$$

is absolutely convergent. Hence $Sx \in \operatorname{dom}(A)$ and

$$ASx = \frac{1}{2i}\int_{\varepsilon-i\infty}^{\varepsilon+i\infty} \frac{A^{1-z}B^{z-1}x}{\sin \pi z}\,dz\ . \tag{4.9.5}$$

On the other hand, the map

$$z \mapsto B^{z-1}A^{-z}x = B^{z-1}A^{\varepsilon-1-z}A^{1-\varepsilon}x$$

is holomorphic in $[\varepsilon - 1 < \operatorname{Re} z < 1]$ by Theorem 4.6.2. Moreover, it follows from Lemma 4.7.4(ii) that

$$\|B^{z-1}A^{-z}x\| \le \|B^{z-1}\|\,\|A^{\varepsilon-1-z}\|\,\|A^{1-\varepsilon}x\| \le K^2M^2\,\|A^{1-\varepsilon}x\|\,e^{(\theta_A+\theta_B)|\operatorname{Im} z|}$$

for $\varepsilon - 1 \le \operatorname{Re} z \le 1$. Thus, by the residue theorem,

$$Sx = \frac{1}{2i}\int_{\varepsilon-1-i\infty}^{\varepsilon-1+i\infty} \frac{A^{-z}B^{z-1}x}{\sin \pi z}\,dz + \pi \operatorname{Res}(h,0)\ , \tag{4.9.6}$$

where $h(z) := A^{-z}B^{z-1}x/\sin \pi z$. Since $\operatorname{Res}(h,0) = \pi^{-1}B^{-1}x$ and since the integral in (4.9.6) is abolutely convergent, it follows that $Sx \in \operatorname{dom}(B)$ and

$$BSx = x + \frac{1}{2i}\int_{\varepsilon-1-i\infty}^{\varepsilon-1+i\infty} \frac{A^{-z}B^{z}x}{\sin \pi z}\,dz = x - ASx\ ,$$

thanks to (4.9.5). ■

4.9.5 Lemma *The operator $A+B$ is closable, its closure $\overline{A+B}$ is injective, and $S = \left(\overline{A+B}\right)^{-1}$.*

Proof Let (x_j) be a null sequence in $\operatorname{dom}(A+B)$ such that $(A+B)x_j \to y$. Then $x_j = S(A+B)x_j \to Sy$ by Lemma 4.9.3, so that $Sy = 0$. From Lemma 4.9.4 we know that $SA^{-1}y \in \operatorname{dom}(A+B)$ and $(A+B)SA^{-1}y = A^{-1}y$. Moreover, the definition of S and Lemma 4.9.2 imply $SA^{-1}y = A^{-1}Sy = 0$. Hence $A^{-1}y = 0$. Consequently, $y = 0$ and $A+B$ is closable.

Let $x \in \operatorname{dom}\left(\overline{A+B}\right)$. There exists a sequence (x_j) in $\operatorname{dom}(A+B)$ such that $x_j \to x$ and $(A+B)x_j \to \left(\overline{A+B}\right)x$. Since $x_j = S(A+B)x_j$ by Lemma 4.9.3, it follows that $x = S(\overline{A+B})x$. Conversely, given $x \in E$, choose a sequence (x_j) in $\operatorname{dom}(A)$ such that $x_j \to x$. Then $Sx_j \in \operatorname{dom}(A+B)$ by Lemma 4.9.4, and $Sx_j \to Sx$ and $(A+B)Sx_j = x_j \to x$. Hence $Sx \in \operatorname{dom}\left(\overline{A+B}\right)$ and $\left(\overline{A+B}\right)Sx = x$, which proves the assertion. ■

As a first application of these considerations we prove the following theorem concerning the spectrum of operators in $\mathcal{BIP}(E)$.

4.9.6 Theorem *If $A \in \mathcal{BIP}(E;\theta)$ then $\sigma(A) \subset \left[\,|\arg \lambda| \le \theta\right]$.*

Proof We can, of course, assume that $\theta < \pi$. Fix $\lambda \in \dot{\mathbb{C}}$ with $|\arg \lambda| < \pi - \theta$ and put $B := \lambda 1_E$. Then $B \in \mathcal{BIP}(E, |\arg \lambda|)$ and condition (4.9.2) is satisfied. Since $A + B = \lambda + A$ is closed, we infer from Lemma 4.9.5 that $(\lambda + A)^{-1} \in \mathcal{L}(E)$, that is, $\lambda \in \rho(-A)$. Hence $\left[\,|\arg z| > \theta\right] \subset \rho(A)$, which proves the theorem. ■

The preceding theorem can be improved to show that $(1 + |\lambda|)\ \|(\lambda + A)^{-1}\|$ is bounded in $\left[\,|\arg \lambda| \le \pi - \theta - \varepsilon\right]$ for each $\varepsilon \in (0, \pi - \theta)$ if $\theta < \pi$ (cf. [DorV90, Theorem 2.1]). This shows, in particular, that $-A$ generates a strongly continuous analytic semigroup on E if $A \in \mathcal{BIP}(E;\theta)$ for some $\theta < \pi/2$.

Now we can prove the main result of this subsection, namely the following DORE-VENNI THEOREM.

4.9.7 Theorem *Let E be a UMD space, let $A \in \mathcal{BIP}(E;\theta_A)$ and $B \in \mathcal{BIP}(E;\theta_B)$ with $\theta_A + \theta_B < \pi$, and let A and B be resolvent commuting. Then $A + B \in \mathcal{C}(E)$ and $0 \in \rho(A+B)$.*

Proof We shall prove that $Sx \in \mathrm{dom}(A+B)$ for $x \in E$. Then Lemma 4.9.5 implies $\mathrm{dom}\big(\overline{A+B}\big) \subset \mathrm{dom}(A+B)$. Hence $\overline{A+B} = A+B$ and $S = (A+B)^{-1}$.

By Theorems 4.6.2 and 4.7.1 the map $z \mapsto A^{-z}B^{z-1}/\sin\pi z$ is holomorphic on $[0 < \mathrm{Re}\, z < 1]$, continuous on $[0 \le \mathrm{Re}\, z \le 1]\backslash\{0,1\}$, and has poles at $z=0$ and $z=1$. Thus it follows from the validity of estimate (4.9.3) for the strip $[0 < \mathrm{Re}\, z < 1]$ that, given $\varepsilon > 0$,

$$Sx = \frac{1}{2i}\int_{\Gamma_{0,\varepsilon}} \frac{A^{-z}B^{z-1}x}{\sin\pi z}\,dz = \frac{1}{2i}\int_{\Gamma_{1,\varepsilon}} \frac{A^{-z}B^{z-1}x}{\sin\pi z}\,dz\ ,$$

where $\Gamma_{j,\varepsilon}$ are for $j=0,1$ the curves running from $j - i\infty$ to $j + i\infty$ and consisting of the half-lines $\{\, j + it\ ;\ |t| \ge \varepsilon \,\}$, and the circular arcs

$$\{\, j + \varepsilon e^{i(\varphi + j\pi)}\ ;\ -\pi/2 \le \varphi \le \pi/2 \,\}\ .$$

Thus

$$Sx = R_{0,\varepsilon}x + T_{0,\varepsilon}x = R_{1,\varepsilon}x + T_{1,\varepsilon}x\ ,$$

where

$$R_{j,\varepsilon}x = \frac{1}{2}\int_{|t|\ge\varepsilon} \frac{A^{-j-it}B^{j-1+it}x}{\sin\pi(j+it)}\,dt$$

and

$$T_{j,\varepsilon}x = \frac{(-1)^j}{2}\int_{j\pi-\pi/2}^{j\pi+\pi/2} \frac{\varepsilon e^{i\varphi}}{\sin\pi(j+\varepsilon e^{i\varphi})} A^{-j-\varepsilon e^{i\varphi}} B^{j-1+\varepsilon e^{i\varphi}} x\,d\varphi\ .$$

From Theorem 4.7.1 we easily deduce that $T_{0,\varepsilon}x \to 2^{-1}B^{-1}x$ and $T_{1,\varepsilon}x \to 2^{-1}A^{-1}x$ as $\varepsilon \to 0$. Hence

$$R_{0,\varepsilon}x \to Sx - 2^{-1}B^{-1}x\ ,\quad R_{1,\varepsilon}x \to Sx - 2^{-1}A^{-1}x$$

as $\varepsilon \to 0$. Since $A, B \in \mathcal{C}(E)$, it follows from Theorem 4.7.1 and Lemma 4.9.2(ii) that $R_{0,\varepsilon}x \in \mathrm{dom}(B)$ and $R_{1,\varepsilon}x \in \mathrm{dom}(A)$ and that

$$BR_{0,\varepsilon}x = \frac{1}{2}\int_{|t|\ge\varepsilon} \frac{A^{-it}B^{it}x}{\sin\pi it}\,dt = -AR_{1,\varepsilon}x\ .$$

Consequently, it remains to prove that $BR_{0,\varepsilon}x$ converges as $\varepsilon \to 0$. Then

$$Sx - 2^{-1}B^{-1}x \in \mathrm{dom}(B) \quad\text{and}\quad Sx - 2^{-1}A^{-1}x \in \mathrm{dom}(A)$$

by the closedness of B and A, respectively. Hence $Sx \in \mathrm{dom}(A) \cap \mathrm{dom}(B)$.

Observe that

$$\begin{aligned} BR_{0,\varepsilon} &= \frac{1}{2}\int_{|t|>1} \frac{A^{-it}B^{it}}{\sin \pi i t}\,dt + \frac{1}{2}\int_{\varepsilon\le|t|\le 1} \frac{A^{-it}B^{it}}{\pi i t}\,dt \\ &\quad + \frac{1}{2}\int_{\varepsilon\le|t|\le 1} A^{-it}B^{it}\Big(\frac{1}{\sin \pi i t} - \frac{1}{\pi i t}\Big)\,dt\ . \end{aligned} \tag{4.9.7}$$

The first integral is independent of $\varepsilon > 0$ and the last one is easily seen to converge in $\mathcal{L}_s(E)$ as $\varepsilon \to 0$. Given any $s \in (0,1)$, the middle integral can be written as

$$\begin{aligned} \int_{\varepsilon\le|t|\le 1} \frac{A^{-it}B^{it}}{\pi i t}\,dt &= A^{-is}B^{is}\int_{\varepsilon\le|t|\le 1} \frac{A^{i(s-t)}B^{-i(s-t)}}{\pi i t}\,dt \\ &= A^{-is}B^{is}\Big\{\int_{|t|\ge\varepsilon} \frac{A^{i(s-t)}B^{-i(s-t)}\chi_{[-1,1]}(s-t)}{\pi i t}\,dt \\ &\qquad - \Big(\int_1^{s+1} - \int_{-1}^{s-1}\Big)\frac{A^{i(s-t)}B^{-i(s-t)}}{\pi i t}\,dt\Big\}\ , \end{aligned} \tag{4.9.8}$$

provided $0 < \varepsilon < 1 - s$. It is obvious that

$$u := \big[t \mapsto A^{it}B^{-it}\chi_{[-1,1]}(t)\big]x \in L_2(\mathbb{R}, E)$$

for each $x \in E$ and that the first integral in the curly brackets equals $-iH_\varepsilon u(s)$, where H_ε is the truncated Hilbert transform. Since $H_\varepsilon u(s)$ converges in E for a.e. $s \in \mathbb{R}$ by Theorem 4.5.1, it follows from (4.9.8), by choosing $s \in (0,1)$ appropriately, that the middle integral in (4.9.7) also converges in $\mathcal{L}_s(E)$ as $\varepsilon \to 0$. ■

4.9.8 Corollary *Let the hypotheses of Theorem* 4.9.7 *be satisfied and suppose that*

$$A \in \mathcal{P}_K(E) \cap \mathcal{BIP}(E; M, \theta_A) \quad \textit{and} \quad B \in \mathcal{P}_K(E) \cap \mathcal{BIP}(E; M, \theta_B)\ .$$

Then

$$\|(A+B)^{-1}\| + \|A(A+B)^{-1}\| + \|B(A+B)^{-1}\| \le c\ ,$$

where c depends upon K, M, θ_A, and θ_B, but not upon the individual operators A and B.

Proof Since $(A+B)^{-1} = S$, the estimate for $(A+B)^{-1}$ follows easily from (4.9.3). The above proof shows that

$$-ASx + 2^{-1}x = BSx - 2^{-1}x = \frac{1}{2}\int_{-\infty}^{\infty} \frac{A^{-it}B^{it}x}{\sin \pi i t}\,dt$$

for $x \in E$, where the integral is a principal value integral. Now the assertion is a consequence of (4.9.7), (4.9.8), and Theorem 4.5.1. ■

The Dore-Venni theorem has been extended by Prüss and Sohr [PrüS90] to the case $0 \notin \rho(A) \cap \rho(B)$. In this case $A + B$ is not invertible, in general. In [PrüS90] it is also shown that $A + B \in \mathcal{BIP}(E, \theta_A \vee \theta_B)$ if the hypotheses of the Dore-Venni theorem are satisfied and $\theta_A \neq \theta_B$, which extends a slightly weaker result of [DorV90]

The sum of two resolvent commuting operators of positive type has been studied earlier by Sobolevskii [Sob75] and Da Prato and Grisvard [DaPG75]. In the latter paper it is shown that, given an arbitrary Banach space E and $A \in \mathcal{P}(E; \vartheta_A)$, $B \in \mathcal{P}(E; \vartheta_B)$ with $\vartheta_A + \vartheta_B > \pi$, it follows that the $\big(E, D(A)\big)_{\theta,q}$-realization of $A + B$ is closed and has a bounded inverse for any $\theta \in (0,1)$ and $1 \leq q \leq \infty$. In general, $A + B$ is not closed, even if E is a Hilbert space (cf. [BaiC91]).

4.10 Maximal Regularity

Let (E_0, E_1) be a densely injected Banach couple. We denote by J a perfect subinterval of $\mathbb{R}^+$ containing 0 and fix $p \in (1, \infty)$. Then we put

$$\mathbb{E}_0 := \mathbb{E}_0(J) := L_p(J, E_0)$$

and

$$\mathbb{E}_1 := \mathbb{E}_1(J) := L_p(J, E_1) \cap W_p^1(J, E_0)\ .$$

It is clear that $\mathbb{E}_0$ and $\mathbb{E}_1$ satisfy (1.5.1) and (1.5.2), respectively.

4.10.1 Lemma (i) *$\mathbb{E}_j(\mathbb{R}^+)$, $j = 0, 1$, are continuously translation invariant.*

(ii) *Let $T_0 > 0$ be fixed. Then $\mathbb{E}_j(J)$ is a regular extension of $\mathbb{E}_j\big([0, T_0)\big)$, $j = 0, 1$, uniformly with respect to any subinterval J of $\mathbb{R}^+$ containing $[0, T_0)$.*

Proof (i) Observe that $\{\, \lambda_s\ ;\ s \geq 0 \,\}$ is a contraction semigroup on $\mathbb{E}_j(\mathbb{R}^+)$ whose restriction to the subspace consisting of the smooth E_j-valued functions with compact supports in $\mathbb{R}^+$ is strongly continuous. Since the latter subspace is dense in $\mathbb{E}_j(\mathbb{R}^+)$, the assertion follows.

(ii) It is well-known and not difficult to see that the smooth functions with compact supports in $\overline{J}$ are dense in $\mathbb{E}_j(\mathbb{R}^+)$ (cf. Chapter VI). Using this fact it is easily seen that the coretraction r^c constructed in the proof of Lemma 2.1.2 extends to coretractions from $\mathbb{E}_j(J)$ onto $\mathbb{E}_j(\mathbb{R}^+)$ whose norms are bounded independently of $J \supset [0, T_0)$. ∎

Now we can characterize the trace space of $\mathbb{E}_1$ and prove an important embedding theorem.

4.10.2 Theorem $\gamma\mathbb{E}_1 \doteq (E_0, E_1)_{1-1/p,p}$ *and* $\mathbb{E}_1 \hookrightarrow BUC\big(J, (E_0, E_1)_{1-1/p,p}\big)$.

Proof The first assertion follows from the characterization of the real interpolation functors by means of the Lions-Peetre trace method (e.g., [BerL76, Corollary 3.12.3] or [Tri78, Theorem 1.8.2]). Now the second one is a consequence of Proposition 1.4.2 and Lemma 4.10.1. ■

As in Section 2 we assume that

$$\left.\begin{aligned}&\mathcal{A}\subset\mathcal{H}(E_1,E_0) \text{ and there are constants } M>0\\ &\text{and } \vartheta\in(0,\pi/2) \text{ such that } \Sigma_\vartheta\subset\rho(-A) \text{ and}\\ &\|A\|_{\mathcal{L}(E_1,E_0)}+(1+|\lambda|)^{1-j}\,\|(\lambda+A)^{-1}\|_{\mathcal{L}(E_0,E_j)}\leq M\\ &\text{for } (\lambda,A)\in\Sigma_\vartheta\times\mathcal{A} \text{ and } j=0,1.\end{aligned}\right\}\tag{4.10.1}$$

Also recall that R_A has been defined in (1.5.6).

4.10.3 Proposition *Let assumption* (4.10.1) *be satisfied. Then*

$$(A\mapsto R_A)\in B\big(\mathcal{A},\mathcal{L}\big((E_0,E_1)_{1-1/p,p},\mathbb{E}_1(J)\big)\big)\ ,$$

uniformly with respect to J.

Proof Lemma 2.2.1 and Remark II.5.1.2 imply the existence of $\omega>0$ such that $\mathrm{type}(-A)\leq-\omega$ for $A\in\mathcal{A}$. Now the assertion follows from (2.2.2) and (I.2.10.6)–(I.2.10.9). ■

4.10.4 Remark Given $T>0$, there exists a constant $\alpha\geq 1$ such that

$$\alpha^{-1}\,\|\cdot\|_{(E_0,E_1)_{1-1/p,p}}\leq\|\cdot\|_{\gamma\mathbb{E}_1(J)}\leq\alpha\,\|\cdot\|_{(E_0,E_1)_{1-1/p,p}}\ ,$$

uniformly with respect to $J\supset[0,T]$.

Proof Letting $J_0:=[0,T]$, we infer from Theorem 4.10.2 that

$$\|\cdot\|_{(E_0,E_1)_{1-1/p,p}}\leq c\,\|\cdot\|_{\gamma\mathbb{E}_1(J_0)}\ .$$

Next, Lemma 4.10.1(ii) and Proposition 1.4.1 imply that we can replace J_0 in the last inequality by any $J\supset J_0$. This gives the first of the asserted inequalities.

Note that $\gamma R_A=1$. Thus, thanks to the definition of the norm in $\gamma\mathbb{E}_1(J)$,

$$\|x\|_{\gamma\mathbb{E}_1(J)}\leq\|R_Ax\|_{\mathbb{E}(J)}\ ,\qquad x\in(E_0,E_1)_{1-1/p,p}\ .$$

Now we obtain the second asserted inequality from Proposition 4.10.3. ■

Given $A\in\mathcal{A}$, it is obvious that

$$\partial+A\in\mathcal{L}(\mathbb{E}_1,\mathbb{E}_0)\ .\tag{4.10.2}$$

Thus, letting $\mathbb{E}_{1,A} := \ker(\partial + A)$, it follows from Theorem 4.10.2, Proposition 4.10.3, and Remark 4.10.4 that, given $T > 0$,

$$r_A := \gamma | \mathbb{E}_{1,A} \in \mathcal{L}\mathrm{is}\big(\mathbb{E}_{1,A}, (E_0, E_1)_{1-1/p,p}\big) , \tag{4.10.3}$$

uniformly with respect to $J \supset [0,T]$. Hence Lemma 1.5.1 implies that $(\mathbb{E}_0, \mathbb{E}_1)$ is a pair of maximal regularity for $\mathcal{A}$ if we can show that

$$\partial + A \in \mathcal{L}\mathrm{is}(\mathbb{E}_{1,\gamma}, \mathbb{E}_0) , \qquad A \in \mathcal{A} , \tag{4.10.4}$$

where $\mathbb{E}_{1,\gamma} := \ker \gamma$. For this we put

$$W^1_{p,\gamma} := \big\{\, u \in W^1_p(J, E_0) \ ; \ \gamma u = 0 \,\big\}$$

and define a linear operator B in $\mathbb{E}_0$ by

$$B : W^1_{p,\gamma} \subset \mathbb{E}_0 \to \mathbb{E}_0 , \quad u \mapsto \partial u .$$

The following lemma shows that, given $\varepsilon \in (0, \pi)$, there exists $M \geq 1$, independently of J, such that

$$B \in \mathcal{P}_{1+T}(\mathbb{E}_0) \cap \mathcal{BIP}\big(\mathbb{E}_0; M, (\pi + \varepsilon)/2\big) , \tag{4.10.5}$$

provided J is bounded, $T := \sup J$, and E_0 is a UMD space.

4.10.5 Lemma *Suppose that J is bounded with $T := \sup J$ and E_0 is a UMD space. Then $B \in \mathcal{P}_{1+T}(\mathbb{E}_0)$ and there exists a constant M, independently of J, such that*

$$\|B^{it}\|_{\mathcal{L}(\mathbb{E}_0)} \leq M(1 + |t|)e^{\pi|t|/2} , \qquad t \in \mathbb{R} .$$

Proof Given $\lambda \in \mathbb{C}$, it is easily verified that $(\lambda + B)^{-1}$ exists and is given by

$$(\lambda + B)^{-1} v(t) = \int_0^t e^{-\lambda(t-\tau)} v(\tau)\, d\tau , \qquad t \in J , \quad v \in \mathbb{E}_0 .$$

Thus we deduce from Young's inequality for convolutions that

$$\|(\lambda + B)^{-1}\| \leq \|e^{-\lambda\cdot}\|_{L_1(J)} = \lambda^{-1}(1 - e^{-\lambda T}) \leq T , \qquad \lambda > 0 .$$

Hence $(1 + \lambda)\, \|(\lambda + B)^{-1}\| \leq 1 + T$ for $\lambda \geq 0$, that is, $B \in \mathcal{P}_{1+T}(\mathbb{E}_0)$.

From (4.6.9) we see that

$$\begin{aligned}(\varepsilon + B)^{-z} v(t) &= \frac{\sin \pi z}{\pi} \int_0^\infty s^{-z} \int_0^t e^{-(s+\varepsilon)(t-\tau)} v(\tau)\, d\tau\, ds \\ &= \frac{\sin \pi z}{\pi} \int_0^t v(\tau) \int_0^\infty s^{-z} e^{-(s+\varepsilon)(t-\tau)}\, ds\, d\tau \\ &= \frac{1}{\Gamma(z)} \int_0^t (t - \tau)^{z-1} e^{-\varepsilon(t-\tau)} v(\tau)\, d\tau\end{aligned}$$

for $v \in \mathbb{E}_0$, $t \in J$, $\varepsilon \geq 0$, and $0 < \operatorname{Re} z < 1$, thanks to the complementing formula (4.6.25). Denote by $r^c \in \mathcal{L}\big(\mathbb{E}_0, L_p(\mathbb{R}, E_0)\big)$ the trivial extension operator, that is, $r^c u$ is the extension of u over $\mathbb{R}$ by zero, and by $r \in \mathcal{L}\big(\mathcal{D}'(\mathbb{R}, E_0), \mathcal{D}'(J, E_0)\big)$ the restriction operator to J. Then

$$(\varepsilon + B)^{-z} v = r(g_{z,\varepsilon} * r^c v) , \qquad v \in \mathbb{E}_0 ,$$

in $\mathcal{D}'(J, E_0)$, where

$$g_{z,\varepsilon}(t) := \begin{cases} t^{z-1} e^{-\varepsilon t} / \Gamma(z) , & t > 0 , \\ 0 , & t \leq 0 , \end{cases}$$

for $0 < \operatorname{Re} z < 1$ and $\varepsilon \geq 0$. By the convolution theorem

$$g_{z,\varepsilon} * r^c v = \mathcal{F}^{-1} m_{z,\varepsilon} \mathcal{F} r^c v , \qquad v \in \mathbb{E}_0 ,$$

where $m_{z,\varepsilon} := \widehat{g}_{z,\varepsilon}$. Thus

$$m_{z,\varepsilon}(\tau) := \mathcal{F}(g_{z,\varepsilon})(\tau) = \frac{1}{\Gamma(z)} \int_0^\infty t^{z-1} e^{-t(\varepsilon + i\tau)} \, dt = \frac{(\varepsilon + i\tau)^{-z}}{\Gamma(z)} \int_S \zeta^{z-1} e^{-\zeta} \, d\zeta ,$$

where $S := \big\{ t(\varepsilon + i\tau) \; ; \; t \geq 0 \big\}$. Since $0 < \operatorname{Re} z < 1$, we can deform S to $\mathbb{R}^+$, thanks to Cauchy's theorem. Hence

$$m_{z,\varepsilon}(\tau) = (\varepsilon + i\tau)^{-z} , \qquad \varepsilon > 0 , \quad \tau \in \mathbb{R} .$$

Note that $g_{z,\varepsilon} \rightharpoonup g_{z,0}$ in $\mathcal{S}'$, as $\varepsilon \to 0$. Thus, by the continuity, hence the weak continuity, of the Fourier transform in $\mathcal{S}'$, and since

$$\big[\tau \mapsto (\varepsilon + i\tau)^{-z}\big] \to \big[\tau \mapsto (i\tau)^{z}\big] \quad \text{in } L_{1,\mathrm{loc}} \qquad \text{as} \quad \varepsilon \to 0 ,$$

it follows that $m_{z,0}(\tau) = (i\tau)^{-z}$ for $\tau \in \mathbb{R}$ and $0 < \operatorname{Re} z < 1$. This shows that

$$(\varepsilon + B)^{-z} v = r \mathcal{F}^{-1} m_{z,\varepsilon} \mathcal{F} r^c v , \qquad \varepsilon \geq 0 , \quad 0 < \operatorname{Re} z < 1 , \tag{4.10.6}$$

in $\mathcal{D}'(J, E_0)$ for $v \in \mathbb{E}_0$. It is not difficult to see that

$$|m_{z,\varepsilon}(\tau)| \leq \varepsilon^{-\operatorname{Re} z} e^{\pi |\operatorname{Im} z|/2} , \qquad \tau \in \mathbb{R} , \quad \varepsilon > 0 , \quad 0 < \operatorname{Re} z < 1 .$$

Since $\dot{m}_{z,\varepsilon}(\tau) = -iz(\varepsilon + i\tau)^{-1} m_{z,\varepsilon}(\tau)$, it follows that $m_{z,\varepsilon} \in \mathcal{M}_M$ and

$$\|m_{z,\varepsilon}\|_{\mathcal{M}_M} \leq (1 + |z|) \varepsilon^{-\operatorname{Re} z} e^{\pi |\operatorname{Im} z|/2} , \qquad \varepsilon > 0 , \quad 0 < \operatorname{Re} z < 1 . \tag{4.10.7}$$

Now we infer from (4.10.6), (4.10.7), Theorem 4.4.3, and Lemma 4.7.4 that

$$\|(\varepsilon + B)^{it}\| \leq c(1 + |t|) e^{\pi |t|/2} , \qquad 0 < \varepsilon < 1 , \quad t \in \mathbb{R} . \tag{4.10.8}$$

It is easily verified that $m_{z,\varepsilon} \to m_{it,\varepsilon}$ in $\mathcal{M}_M$ as $z \to it$ in $[0 < \operatorname{Re} z < 1]$. Hence we deduce from (4.10.6) and Theorem 4.4.3 that

$$(\varepsilon + B)^{-it} v = r\mathcal{F}^{-1} m_{it,\varepsilon} \mathcal{F} r^c v , \qquad 0 < \varepsilon < 1 , \quad t \in \mathbb{R} , \tag{4.10.9}$$

in $L_p(J, E_0)$ for $v \in \mathbb{E}_0$. It is not hard to see that $m_{it,\varepsilon} \rightharpoonup m_{it,0}$ in $\mathcal{S}'$ as $\varepsilon \to 0$. Thus, given $v \in \mathcal{D}(J, E_0)$, the continuity properties of the Fourier transform and (4.10.6) and (4.10.9) imply

$$(\varepsilon + B)^{-it} v \rightharpoonup r\mathcal{F}^{-1} m_{it,0} \mathcal{F} r^c v \quad \text{in } \mathcal{D}'(J, E_0) \qquad \text{as} \quad \varepsilon \to 0 \tag{4.10.10}$$

for $t \in \mathbb{R}$. On the other hand, it follows from (4.10.8) and the reflexivity of $\mathbb{E}_0$ that there exists a null sequence (ε_j) such that $\big((\varepsilon_j + B)^{-it} v\big)$ converges weakly in $\mathbb{E}_0$ to some $w \in \mathbb{E}_0$. Since $\mathbb{E}_0 \hookrightarrow \mathcal{D}'(J, E_0)$, we see that $(\varepsilon_j + B)^{it} v \rightharpoonup w$ in $\mathcal{D}'(J, E_0)$. Hence (4.10.10) implies $r\mathcal{F}^{-1} m_{it,0} \mathcal{F} r^c v = w \in \mathbb{E}_0$. Finally, it is an easy consequence of Theorem 4.6.5(ii) that $(\varepsilon + B)^{it} v \to B^{it} v$ in $\mathbb{E}_0$, which shows that $w = B^{it} v$. Now we infer from (4.10.8) that

$$\|B^{it} v\|_{\mathbb{E}_0} \le c(1 + |t|) e^{\pi |t|/2} \, \|v\|_{\mathbb{E}_0} , \qquad t \in \mathbb{R} , \quad v \in \mathcal{D}(J, E_0) ,$$

and the assertion follows from the density of $\mathcal{D}(J, E_0)$ in $\mathbb{E}_0$. ■

4.10.6 Remark The above theorem shows that the symbol of the pseudo-differential operator ∂^{is} equals $(i\tau)^{is}$ for $s \in \mathbb{R}$, where τ is the Fourier variable corresponding to t. ■

After these preparations we can prove the main result of this section.

4.10.7 Theorem *Suppose that E_0 is a UMD space. Let assumption* (4.10.1) *be satisfied and suppose that there exist constants $N \ge 1$ and $\theta \in [0, \pi/2)$ such that $\mathcal{A} \subset \mathcal{BIP}(E_0; N, \theta)$. Then*

$$(\mathbb{E}_0, \mathbb{E}_1) := \big(L_p(J, E_0), L_p(J, E_1) \cap W_p^1(J, E_0)\big)$$

is a pair of maximal regularity for $\mathcal{A}$, and

$$\gamma \mathbb{E}_1 \doteq E_{1-1/p,p} := (E_0, E_1)_{1-1/p,p} \ .$$

Moreover,

$$\|(\partial + A, \gamma)^{-1}\|_{\mathcal{L}(\mathbb{E}_0 \times E_{1-1/p,p}, \mathbb{E}_1)} \le c , \qquad A \in \mathcal{A} ,$$

uniformly with respect to J.

Proof (i) First suppose that J is bounded. We denote the 'multiplication operator', induced in $L_p(J, E_1)$ by A via $Au(t) := A\big(u(t)\big)$, again by A. Then

$$\mathcal{A} \subset \mathcal{L}\big(L_p(J, E_1), \mathbb{E}_0\big) , \tag{4.10.11}$$

and it follows from (2.2.2) that the graph norm of A, considered as a linear operator in $\mathbb{E}_0$, is equivalent to the $L_p(J, E_1)$-norm. Hence $A \in \mathcal{C}(\mathbb{E}_0)$ by Lemma I.1.1.2. It is an obvious consequence of (4.10.1) and the density of $L_p(J, E_1)$ in $L_p(J, E_0)$ that

$$\mathcal{A} \subset \mathcal{P}_M(\mathbb{E}_0) . \tag{4.10.12}$$

From Theorem 4.6.5(ii) we easily infer that A^{it} on $\mathbb{E}_0$ is given by the corresponding multiplication operator. Hence

$$\mathcal{A} \subset \mathcal{BIP}(\mathbb{E}_0; N, \theta) . \tag{4.10.13}$$

Fix $\theta_B \in (\pi/2, \pi - \theta)$. Then Lemma 4.10.5 guarantees that $B \in \mathcal{BIP}(\mathbb{E}_0; \theta_B)$. Of course, A and B are resolvent commuting for $A \in \mathcal{A}$. Since $\mathbb{E}_0$ is a UMD space by Theorem 4.5.2(vi), the Dore-Venni Theorem 4.9.7 guarantees that $A + B \in \mathcal{C}(\mathbb{E}_0)$ and $0 \in \rho(A + B)$. Note that

$$\mathrm{dom}(A + B) = \mathrm{dom}(A) \cap \mathrm{dom}(B) = W^1_{p,\gamma} \cap L_p(J, E_1) = \mathbb{E}_{1,\gamma} .$$

Hence $A + B \in \mathcal{L}\mathrm{is}(\mathbb{E}_{1,\gamma}, \mathbb{E}_0)$ and, thanks to (2.2.2),

$$\|u\|_{\mathbb{E}_1} = \|u\|_{L_p(J,E_1)} \vee \|\dot{u}\|_{L_p(J,E_0)} \le M \, \|Au\|_{L_p(J,E_0)} \vee \|Bu\|_{L_p(J,E_0)}$$

for $u \in \mathbb{E}_{1,\gamma}$. Now Corollary 4.9.8, (4.10.12), (4.10.13), Lemma 4.10.5, and (4.10.5) imply

$$\|(A + B)^{-1}\|_{\mathcal{L}(\mathbb{E}_0, \mathbb{E}_{1,\gamma})} \le c(J) , \qquad A \in \mathcal{A} . \tag{4.10.14}$$

Lastly, thanks to (4.10.14) and Proposition 4.10.3,

$$R_A x + (A + B)^{-1} f \in \mathbb{E}_1$$

and

$$\|R_A x + (A + B)^{-1} f\|_{\mathbb{E}_1} \le c(J) \, \|(f, x)\|_{\mathbb{E}_0 \times E_{1-1/p,p}} \tag{4.10.15}$$

for $(f, x) \in \mathbb{E}_0 \times E_{1-1/p,p}$ and $A \in \mathcal{A}$. Theorem 4.10.2 and (4.10.2) imply

$$(\partial + A, \gamma) \in \mathcal{L}(\mathbb{E}_1, \mathbb{E}_0 \times E_{1-1/p,p})$$

and it is clear that $(\partial + A, \gamma)\big(R_A x + (A + B)^{-1} f\big) = (f, x)$. Hence

$$(\partial + A, \gamma)^{-1}(f, x) = R_A x + (A + B)^{-1} f .$$

Now the assertion follows from (4.10.13) and Lemma 1.5.1.

(ii) Suppose that $J = \mathbb{R}^+$ and recall from (1.5.7) the definition of K_A. From (i) it follows, of course, that $R_A x + K_A f$ is a $W^1_{1,\mathrm{loc}}$-solution of $(1.5.5)_{(x,A,f)}$ for each $f \in \mathbb{E}_0$. Hence, thanks to Theorem 1.5.2, Proposition 4.10.3, and (4.10.2), it suffices to prove that

$$(A \mapsto K_A) \in B\big(\mathcal{A}, \mathcal{L}(\mathbb{E}_0, \mathbb{E}_1)\big) . \tag{4.10.16}$$

Given $v \in \mathbb{E}_0$, put $v_j := v\chi_{[j,j+1)}$ for $j \in \mathbb{N}$. Then

$$\begin{aligned}
\|K_A v\|_{L_p(\mathbb{R}^+,E_1)}^p &= \sum_{j=0}^{\infty} \int_j^{j+1} \|K_A v(t)\|_1^p \, dt \\
&= \int_0^2 \|K_A v(t)\|_1^p \, dt \\
&\quad + \sum_{j=2}^{\infty} \int_j^{j+1} \Big\| \int_0^{j-1} e^{-(t-\tau)A} v(\tau)\, d\tau + \int_{j-1}^t e^{-(t-\tau)A} v(\tau)\, d\tau \Big\|_1^p \, dt \\
&\le \int_0^2 \|K_A(v_0+v_1)(t)\|_1^p \, dt \\
&\quad + 2^{p-1} \sum_{j=2}^{\infty} \int_j^{j+1} \Big\{ \Big\| \int_0^{j-1} e^{-(t-\tau)A} v(\tau)\, d\tau \Big\|_1^p \\
&\qquad\qquad + \Big\| \int_{j-1}^t e^{-(t-\tau)A} \big[v_{j-1}(\tau) + v_j(\tau)\big]\, d\tau \Big\|_1^p \Big\} \, dt .
\end{aligned} \tag{4.10.17}$$

Part (i) of this proof guarantees that

$$\int_0^2 \|K_A(v_0+v_1)\|_1^p \, dt \le c\, \|v_0+v_1\|_{L_p((0,2),E_0)}^p . \tag{4.10.18}$$

From (4.10.1) and Lemma 2.2.1 we see that

$$\begin{aligned}
&\sum_{j=2}^{\infty} \int_j^{j+1} \Big\| \int_0^{j-1} e^{-(t-\tau)A} v(\tau)\, d\tau \Big\|_1^p \, dt \\
&\qquad \le c \sum_{j=2}^{\infty} \int_j^{j+1} \Big(\int_0^{j-1} (t-\tau)^{-1} e^{-\omega(t-\tau)} \, \|v(\tau)\|_0 \, d\tau \Big)^p \, dt \\
&\qquad \le c \int_2^{\infty} \Big(\int_0^{t-1} e^{-\omega(t-\tau)} \, \|v(\tau)\|_0 \, d\tau \Big)^p \, dt \\
&\qquad \le \omega^{-p} c \int_0^{\infty} \|v(t)\|_0^p \, dt \le c\, \|v\|_{L_p(\mathbb{R}^+,E_0)}^p ,
\end{aligned} \tag{4.10.19}$$

where for the second to the last estimate we used Young's inequality for convolutions. Lastly, using again (i) for the interval $[0,2]$, we can estimate the integrals

$$\int_j^{j+1} \Big\| \int_{j-1}^t e^{-(t-\tau)A} \big[v_{j-1}(\tau) + v_j(\tau)\big]\, d\tau \Big\|_1^p \, dt$$

from above by

$$\int_0^2 \Big\| \int_0^t e^{-(t-\tau)A} \big[v_{j-1}(\tau+j-1) + v_j(\tau+j-1) \big] \, d\tau \Big\|_1^p \, dt \qquad (4.10.20)$$
$$\leq c \, \| v_{j-1} + v_j \|_{L_p(\mathbb{R}^+, E_0)}^p \; .$$

Since the functions v_j have disjoint supports it follows that

$$\| v_0 + v_1 \|_{L_p((0,2),E_0)}^p + \sum_{j=2}^{\infty} \| v_{j-1} + v_j \|_{L_p(\mathbb{R}^+,E_0)}^p \leq 2 \sum_{j=0}^{\infty} \| v_j \|_{L_p(\mathbb{R}^+,E_0)}^p$$
$$= 2 \, \| v \|_{L_p(\mathbb{R}^+,E_0)}^p \; .$$

Thus (4.10.17)–(4.10.20) imply

$$(A \mapsto K_A) \in B\big(\mathcal{A}, \mathcal{L}(\mathbb{E}_0, L_p(\mathbb{R}^+, E_1))\big) \; . \qquad (4.10.21)$$

Now suppose that $v \in BUC^{\rho}(\mathbb{R}^+, E_0) \cap \mathbb{E}_0$ for some $\rho \in (0,1)$. Then we know from Proposition 2.1.1 and Theorem 2.5.3 that $w := K_A v \in C^1(\dot{\mathbb{R}}^+, E_0)$ and that $\dot{w}(t) = v(t) - A K_A v(t)$ for $t > 0$. Thus we infer from (2.2.2) and (4.10.21) that

$$\| (K_A v)^{\cdot} \|_{L_p(\mathbb{R}^+,E_0)} \leq c \, \| v \|_{\mathbb{E}_0} \qquad (4.10.22)$$

for $A \in \mathcal{A}$ and $v \in BUC^{\rho}(\mathbb{R}^+, E_0) \cap \mathbb{E}_0$. Since the latter set is dense in $\mathbb{E}_0$, estimate (4.10.22) is true for all $v \in \mathbb{E}_0$ and $A \in \mathcal{A}$, and (4.10.16) follows from this and from (4.10.21).

(iii) Lastly, suppose again that J is bounded. Then, by an obvious extension and restriction argument, we obtain from Lemma 4.10.2(ii) and step (ii) that $(\partial + A, \gamma)^{-1}$ can be estimated uniformly with respect to J. ■

The next theorem shows that we can weaken assumption (4.10.1) if we restrict our considerations to bounded intervals.

4.10.8 Theorem *Suppose that J is bounded and E_0 is a UMD space. Also suppose that $\kappa, N \geq 1$, $\omega > 0$, and $\theta \in [0, \pi/2)$, and that*

$$\mathcal{A} \subset \mathcal{H}(E_1, E_0, \kappa, \omega) \text{ satisfies } \omega + \mathcal{A} \subset \mathcal{BIP}(E_0; N, \theta) \; . \qquad (4.10.23)$$

Then

$$(\mathbb{E}_0, \mathbb{E}_1) := \big(L_p(J, E_0), L_p(J, E_1) \cap W_p^1(J, E_0) \big)$$

is a pair of maximal regularity for $\mathcal{A}$, and

$$\gamma \mathbb{E}_1 \doteq E_{1-1/p,p} := (E_0, E_1)_{1-1/p,p} \; . \qquad (4.10.24)$$

Moreover,

$$\| (\partial + A, \gamma)^{-1} \|_{\mathcal{L}(\mathbb{E}_0 \times E_{1-1/p}, \mathbb{E}_1)} \leq c \; , \qquad A \in \mathcal{A} \; . \qquad (4.10.25)$$

Proof Thanks to Corollary I.1.4.3 the set $\omega + \mathcal{A}$ satisfies assumption (4.10.1). Hence Theorem 4.10.7 guarantees that $(\mathbb{E}_0, \mathbb{E}_1)$ is a pair of maximal regularity for $\omega + \mathcal{A}$ and that (4.10.24) and (4.10.25) are satisfied (with $\mathcal{A}$ replaced by $\omega + \mathcal{A}$ in (4.10.25)). Now the assertion follows from Proposition 1.5.3 and the obvious fact that $e^{\omega}\mathbb{E}_j \doteq \mathbb{E}_j$ for $j = 0, 1$, thanks to the boundedness of J. ■

4.10.9 Remarks **(a)** It is clear that a result which is analogous to Theorem 2.5.5 holds in this case as well.

(b) By a result of Dore [Dor93b, Theorems 2.1 and 2.2] and Lemma I.1.1.2 the assumption:

$$\left.\begin{array}{l}\text{there exists } p \in (1,\infty) \text{ such that}\\ \big(L_p(J,E_0), L_p(J,E_1) \cap W_p^1(J,E_0)\big) \text{ is a pair of}\\ \text{maximal regularity for } A \in \mathcal{L}(E_1,E_0) \cap \mathcal{C}(E_0),\end{array}\right\} \tag{4.10.26}$$

implies that $A \in \mathcal{H}(E_1, E_0)$. Moreover, if $J = \mathbb{R}^+$ then $\operatorname{type}(-A) < 0$.

(c) It may be worthwhile to reformulate the assertion of Theorem 4.10.7 as:

Given any $(x, A, f) \in E_{1-1/p} \times \mathcal{A} \times L_p(J, E_0)$, the Cauchy problem

$$\dot{u} + Au = f(t)\ , \quad t \in \dot{J}\ , \qquad u(0) = x \tag{4.10.27}$$

has a unique solution $u \in L_p(J, E_1) \cap W_p^1(J, E_0)$, and

$$\begin{aligned}\int_J \|u(t)\|_0^p\,dt + \int_J \|\dot{u}(t)\|_0^p\,dt + \int_J \|Au(t)\|_0^p\,dt\\ \leq c\Big(\int_J \|f(t)\|_0^p\,dt + \|x\|_{1-1/p}^p\Big)\ ,\end{aligned} \tag{4.10.28}$$

where c is independent of $(x, A, f) \in E_{1-1/p} \times \mathcal{A} \times L_p(J, E_0)$. Moreover, Theorem 4.10.2 implies that the left-hand side of (4.10.28) can be estimated below by

$$c_0 \sup_{t \in J} \|u(t)\|_{E_{1-1/p}}^p$$

for a suitable $c_0 > 0$. ■

Now we extend the above maximal regularity results to the nonautonomous case. The proof of the following theorem is a modification of the one of Theorem 2.6.1.

4.10.10 Theorem *Let E_0 be a UMD space and let there exist $\kappa \geq 1$ and $\omega > 0$ such that*

$$A \in BUC\big(J, \mathcal{H}(E_1, E_0, \kappa, \omega)\big)\ . \tag{4.10.29}$$

Suppose that there are $N \geq 1$ and $\theta \in [0, \pi/2)$ such that

$$\omega + A(t) \in \mathcal{BIP}(E_0; N, \theta)\ , \qquad t \in J\ .$$

If $J = \mathbb{R}^+$, *also assume that there exists* $T \geq 0$ *such that* $s(-A(T)) < 0$ *and*

$$\sup_{t \geq T} \|A(t) - A(T)\|_{\mathcal{L}(E_1,E_0)} \leq \lambda \Big/ \|K_{A(T)}\|_{\mathcal{L}(\mathbb{E}_0,\mathbb{E}_1)} \tag{4.10.30}$$

for some $\lambda \in (0,1)$. *Then*

$$(\partial + A, \gamma) \in \mathcal{L}\mathrm{is}(\mathbb{E}_1, \mathbb{E}_0 \times E_{1-1/p}) ,$$

where

$$(\mathbb{E}_0, \mathbb{E}_1) := \big(L_p(J, E_0), L_p(J, E_1) \cap W_p^1(J, E_0)\big)$$

and $E_{1-1/p} := (E_0, E_1)_{1-1/p,p}$.

Proof It follows from (4.10.29) and Theorem 4.10.2 that

$$(\partial + A, \gamma) \in \mathcal{L}(\mathbb{E}_1, \mathbb{E}_0 \times E_{1-1/p}) .$$

Thus, by Banach's homomorphism theorem, it remains to show that $(\partial + A, \gamma)$ is bijective.

(i) First suppose that J is bounded. Then, thanks to (4.10.29), we can assume that $J = [0, T]$ for some $T > 0$. Theorem 4.10.8 implies

$$\big(\partial + A(s), \gamma\big) \in \mathcal{L}\mathrm{is}(\mathbb{E}_1, \mathbb{E}_0 \times E_{1-1/p}) , \qquad s \in J , \tag{4.10.31}$$

and the existence of a constant α such that

$$\big\|\big(\partial + A(s), \gamma\big)^{-1}\big\|_{\mathcal{L}(\mathbb{E}_0 \times E_{1-1/p}, \mathbb{E}_1)} \leq \alpha , \qquad s \in J .$$

By the compactness of J and by (4.10.29) we can find points

$$0 =: s_0 < s_1 \cdots < s_{m+2} := T$$

such that

$$\max_{s_j \leq t \leq s_{j+2}} \|A(t) - A(s_j)\|_{\mathcal{L}(E_1,E_0)} \leq 1/(2\alpha) , \qquad j = 0, \ldots, m .$$

Put

$$A_j(t) := A(s_j + t) , \qquad t \in J_j := [0, s_{j+2} - s_j] , \quad j = 0, \ldots, m .$$

It is an easy consequence of (4.10.29), Lemma 4.10.1, and Proposition 1.6.3 that

$$(\partial + A_j, \gamma) \in \mathcal{L}\mathrm{is}\big(\mathbb{E}_1(J_j), \mathbb{E}_0(J_j) \times E_{1-1/p}\big) , \qquad j = 0, \ldots, m . \tag{4.10.32}$$

Let $(f, x) \in \mathbb{E}_0 \times E_{1-1/p}$ be given and put

$$f_j(t) := f(s_j + t) , \qquad t \in J_j , \quad j = 0, \ldots, m .$$

Then (4.10.32) guarantees that the Cauchy problem

$$\dot{u} + A_0(t)u = f_0(t) \ , \quad t \in \dot{J}_0 \ , \qquad u(0) = x$$

has a unique solution $v_0 \in \mathbb{E}_1(J_0)$. From Lemma 4.10.1 and Theorem 4.10.2 we infer that $v_0 \in BUC(J_0, E_{1-1/p})$, so $v_0(s_1) \in E_{1-1/p}$. Hence, by again invoking (4.10.32), we see that the Cauchy problem

$$\dot{u} + A_1(t)u = f_1(t) \ , \quad t \in \dot{J}_1 \ , \qquad u(0) = v_0(s_1)$$

has a unique solution $v_1 \in \mathbb{E}_1(J_1)$. It is clear that

$$u_1(t) := \begin{cases} v_0(t) \ , & 0 \le t \le s_1 \ , \\ v_1(t - s_1) \ , & s_1 \le t \le s_3 \ , \end{cases}$$

is the unique solution of the Cauchy problem

$$\dot{u} + A(t)u = f(t) \ , \quad 0 \le t \le s_3 \ , \qquad u(0) = x \ .$$

Now the assertion follows by an obvious finite induction argument.

(ii) Suppose that $J = \mathbb{R}^+$ and fix $(f, x) \in \mathbb{E}_0 \times E_{1-1/p}$. Put $J_T := [0, T+1]$ and let u_T be the solution of the Cauchy problem

$$\dot{u} + A(t)u = f(t) \ , \qquad t \in \dot{J}_T \ , \qquad u(0) = x \ ,$$

whose existence and uniqueness are guaranteed by the first part of this proof. Since $s\big(-A(T)\big) < 0$ we infer from Corollary I.1.4.3 that $A(T)$ satisfies condition (4.10.1). Hence Theorem 4.10.7 shows that

$$\big(\partial + A(T), \gamma\big) \in \mathcal{L}\mathrm{is}(\mathbb{E}_1, \mathbb{E}_0 \times E_{1-1/p}) \ .$$

Assumption (4.10.30) and Proposition 1.6.3 imply

$$\big(\partial + A(T + \cdot), \gamma\big) \in \mathcal{L}\mathrm{is}(\mathbb{E}_1, \mathbb{E}_0 \times E_{1-1/p}) \ .$$

Thus the Cauchy problem

$$\dot{u} + A(T + t)u = f(T + t) \ , \quad t > 0 \ , \qquad u(0) = u_T(T)$$

has a unique solution $v \in \mathbb{E}_1$. Then

$$u(t) := \begin{cases} u_T(t) \ , & 0 \le t \le T \ , \\ v(t - T) \ , & t > T \ , \end{cases}$$

is the unique solution in $\mathbb{E}_1$ of $\dot{u} + A(t)u = f$ satisfying $u(0) = x$. Hence the assertion follows. ■

4.10.11 Remarks **(a)** The above proof shows that, in the case where $J = \mathbb{R}^+$, assumption (4.10.29) can be replaced by

$$A \in BC\big(\mathbb{R}^+, \mathcal{H}(E_1, E_0, \kappa, \omega)\big) .$$

The uniform continuity in (4.10.29) is only needed if J is bounded and not closed, in order to extend A over $\overline{J}$.

(b) Of course, Theorem 4.10.10 implies an estimate of the form (4.10.28). ■

Theorem 4.10.7 is due to Dore and Venni [DorV87, Theorem 3.2] in the case of a bounded interval (and for vanishing initial values). The extension to $\mathbb{R}^+$ given in part (ii) of the above proof follows [Dor93b, Theorem 2.4], where it is attributed to T. Kato. Other extensions of [DorV87, Theorem 3.2] are due to Prüss and Sohr [PrüS90] and Giga and Sohr [GiS91]. These authors admit operators A that are not required to have a bounded inverse. Thus the semigroup $\{\, e^{-tA} \ ;\ t \geq 0 \,\}$ is only bounded but not exponentially decaying. Hence, thanks to Remark 4.10.9(b), the pair (4.10.26) cannot be a pair of maximal regularity for A in this case. In fact, in [PrüS90] and [GiS91] it is only shown that, given $f \in L_p(\mathbb{R}^+, E_0)$ for some $p \in (1, \infty)$, the Cauchy problem

$$\dot{u} + Au = f(t) \ , \quad t > 0 \ , \qquad u(0) = 0$$

has a unique solution $u \in W^1_{p,\mathrm{loc}}(\mathbb{R}^+, E_0)$ such that $\dot{u}, Au \in L_p(\mathbb{R}^+, E_0)$ and

$$\|\dot{u}\|_{L_p(\mathbb{R}^+,E_0)} + \|Au\|_{L_p(\mathbb{R}^+,E_0)} \leq c\, \|f\|_{L_p(\mathbb{R}^+,E_0)} \ . \tag{4.10.33}$$

(In [GiS91] also nonzero initial values are admitted.) Note that the left-hand side of (4.10.33) is not an equivalent norm for $L_p(\mathbb{R}^+, E_1) \cap W^1_p(\mathbb{R}^+, E_0)$, since the map $u \mapsto \|Au\|$ is not an equivalent norm on E_1 if $0 \in \sigma(A)$.

Of course, the assumptions that E_0 is a UMD space and $A \in \mathcal{BIP}$ are sufficient conditions only for (4.10.26) to be a pair of maximal regularity for A. In fact, suppose that

$$A \in \mathcal{H}(E_1, E_0) \qquad \text{with} \quad \mathrm{type}(-A) < 0 \ . \tag{4.10.34}$$

Then a result of Benedek, Calderón, and Panzone [BCP62] implies that (4.10.26) is a pair of maximal regularity for A and any $p \in (1, \infty)$, provided this is true for some $p \in (1, \infty)$. In the case that (E_0, E_1) is a densely injected Hilbert couple, by means of Fourier transforms and Plancherel's theorem it is not difficult to see that (4.10.26) is a pair of maximal regularity for every A satisfying (4.10.34), provided $p = 2$. By combining these results one obtains a maximal Sobolev regularity result in the $L_p(J, E_0)$-setting if E_0 is a Hilbert space, as has first been proven in [DeSi64]. In a more concrete situation these observations have been applied in [vW82], for example, to derive 'L_p-L_q-estimates' for certain classes of parabolic differential equations. For further 'maximal Sobolev regularity' results we refer to [Sob64], [DaPG75], [CoL87], [Lam87], [Mie87], and [GiGS91].

Chapter IV

Variable Domains

So far we have studied the case of 'constant domains', that is, we have always assumed that $A(t) \in \mathcal{L}(E_1, E_0)$ for $t \in J$, where (E_0, E_1) is a fixed densely injected Banach couple. In this case the general theory is relatively easy and flexible. In addition, only mild continuity assumptions for the function $t \mapsto A(t)$ are needed.

We shall see in later chapters that a great many concrete situations lead to constant domain problems, either directly or via extrapolation techniques studied in detail in Chapter V. However, in order to develop a powerful theory of parabolic evolution equations we cannot restrict ourselves to the constant domain case for the following two main reasons. First, problems involving 'variable domains' occur naturally in many concrete applications. Second, if the extrapolation technique can be applied to transform a variable domain problem into a constant domain case, we arrive at a 'weak' formulation of the original setting. In order to prove that the solution of the weak constant domain problem is in fact a 'classical' solution of the original variable domain problem, we shall rely on bootstrapping arguments based on solvability results for a variable domain situation.

In Section 1 we prove regularity results for constant domain problems by lifting them into suitable interpolation spaces between E_0 and E_1. The lifted problem is then a variable domain problem, in general. In Section 2 we consider situations where the domain may be variable but a suitable interpolation space between E_0 and $D\big(A(t)\big)$ is independent of $t \in J$. The results of this section apply in particular to the constant domain case to give new proofs for some of the results of Chapter II. In Section 3 we briefly consider cases where no constancy assumption whatsoever is made. Here we rely on maximal regularity results and study abstract versions of initial-boundary value problems. The results of this last section provide a basis for more detailed studies in concrete situations that are carried out in Chapter VIII.

1 Higher Regularity

It is the main purpose of this section to prove Theorem II.1.2.2, that will be achieved in Subsection 1.5. For this we need some preparation. In Subsection 1.1 we collect a few simple, but useful, results about differentiable functions of one real variable. Then, in Theorem 1.2.1, we give a general sufficient condition for a mild solution to be, in fact, a solution. In addition, we prove quantitative estimates for the restriction of a parabolic evolution operator to suitable intermediate spaces.

1.1 Properties of Differentiable Functions

Let E be a Banach space. Given $-\infty < a < b < \infty$, we denote the right derivative of a function $u\colon [a,b) \to E$ by $\partial_+ u$. If $E = \mathbb{R}$,

$$d_+ f(x) := \liminf_{h\to 0+} \frac{f(x+h)-f(x)}{h}$$

is said to be the **lower right Dini derivative** of f at x. Observe that $d_+ f(x)$ exists in $\bar{\mathbb{R}}$.

1.1.1 Lemma *Suppose that $f \in C\big([a,b),\mathbb{R}\big)$ and $d_+ f(x) \le 0$ for $x \in [a,b)$. Then f is decreasing.*

Proof Given $\varepsilon > 0$, put $f_\varepsilon(x) := f(x) - \varepsilon x$ for $x \in [a,b)$. Then $d_+ f_\varepsilon \le -\varepsilon$. Suppose that f_ε is not decreasing. Then there exist two points $a \le x_1 < x_2 < b$ so that $f_\varepsilon(x_2) > f_\varepsilon(x_1)$. Let $x_0 := \max\{\, x \in [x_1,x_2] \ ;\ f_\varepsilon(x) = f_\varepsilon(x_1)\,\}$. The continuity of f implies that x_0 is well-defined, $x_0 < x_2$, and $f_\varepsilon(x) > f_\varepsilon(x_0)$ for $x_0 < x < x_2$. Consequently, $d_+ f_\varepsilon(x_0) \ge 0$, which is impossible. This shows that

$$f(x) - \varepsilon x \le f(y) - \varepsilon y \ , \qquad a \le y \le x < b \ .$$

Now the assertion follows by letting $\varepsilon \to 0$. ■

Suppose that $J \subset \mathbb{R}$ is a perfect interval and that $f \in E^J$ is right differentiable in $J\backslash\{\sup J\}$ (where $J\backslash\{\infty\} := J$). If J is bounded above and closed on the right, $\partial_+ f \in C(J,E)$ means that there exists a unique $g \in C(J,E)$ such that $g \supset \partial_+ f$.

1.1.2 Proposition *Suppose that $f \in C(J,E)$ is differentiable from the right and $\partial_+ f \in C(J,E)$. Then $f \in C^1(J,E)$ and $\partial f = \partial_+ f$.*

Proof Let $x_0 \in J$ be fixed and put

$$g(x) := f(x_0) + \int_{x_0}^x \partial_+ f(t)\,dt \ , \qquad x \in J \ .$$

Then $h := f - g \in C(J,E)$ and $h(x_0) = 0$. Moreover, h is differentiable from the right and $\partial_+ h = 0$. Let $e' \in E'$ be arbitrary and put $\varphi(x) := \mathrm{Re}\big\langle e', h(x)\big\rangle$, $x \in J$. Then $\varphi \in C(J,\mathbb{R})$ and φ is differentiable from the right with $\partial_+\varphi = \mathrm{Re}\langle e', \partial_+ h\rangle = 0$. Consequently, $\varphi(x) = \varphi(x_0)$ for $x \in J$, thanks to Lemma 1.1.1. This shows that $\mathrm{Re}\big\langle e', h(x)\big\rangle = 0$ for $x \in J$ and $e' \in E'$. If $\mathbb{K} = \mathbb{C}$, replace e' by ie' to find that $\mathrm{Im}\big\langle e', h(x)\big\rangle = 0$ for $x \in J$ and $e' \in E'$. Thus $\big\langle e', h(x)\big\rangle = 0$ for $x \in J$ and $e' \in E'$, which gives $h = 0$. Hence $f = g$ and, since $g \in C^1(J,E)$, the assertion follows. ■

Next we prove a simple continuity result for differentiable functions in situations where two topologies are involved. It is an elementary regularity theorem.

1.1.3 Proposition *Let $f : J \to E$ be differentiable, let F be a Banach space with $F \hookrightarrow E$, and suppose that $\partial f \in C(J,F)$ and $f(s) \in F$ for some $s \in J$. Then $f \in C^1(J,F)$ and the derivatives in E and in F coincide.*

Proof It follows from $F \hookrightarrow E$ that $\partial f \in C(J,E)$. Hence $f \in C^1(J,E)$. Thus

$$f(t) = f(s) + \int_s^t \partial f(\tau)\, d\tau \ , \qquad t \in J \ , \tag{1.1.1}$$

in E. Since $\partial f \in C(J,F)$, the integral in (1.1.1) converges in F. Now the assertion follows by differentiating the integral with respect to its upper limit. ■

1.2 General Solvability Results for Cauchy Problems

Let E be a Banach space, let J be a perfect subinterval of $\mathbb{R}$, and suppose that

$$(A, f) : J \to \mathcal{C}(E) \times E \ .$$

Given $s \in J$ and $x \in E$, consider the linear Cauchy problem

$$\dot{u} + A(t)u = f(t) \ , \quad t \in J \cap (s,\infty) \ , \qquad u(s) = x \ . \qquad (1.2.1)_{(s,x,A,f)}$$

Below we give sufficient conditions for this Cauchy problem to possess a solution. Observe that we do not require that $\mathrm{dom}\big(A(t)\big)$ be independent of $t \in J$.

1.2.1 Theorem *Let U be a parabolic evolution operator for A with regularity subspace F. Suppose that $F \hookrightarrow F_0 \hookrightarrow E$,*

$$f \in C(J, F_0) \ , \tag{1.2.2}$$

and there are $\alpha \in [0,1)$ and $c > 0$ such that

$$\|U(t,s)\|_{\mathcal{L}(F_0,F)} + \|AU(t,s)\|_{\mathcal{L}(F_0,E)} \le c(t-s)^{-\alpha} \ , \qquad (t,s) \in J_\Delta^* \ . \tag{1.2.3}$$

Then $(1.2.1)_{(s,x,A,f)}$ *has for each* $(s,x) \in J \times E$ *a unique solution* $u(\cdot,s,x)$ *and*

$$u(\cdot,\cdot,x) \in C(J_\Delta^*, F) \ .$$

If $U \in C\big(J_\Delta, \mathcal{L}_s(F)\big)$ *and* $x \in F$ *then*

$$u(\cdot,\cdot,x) \in C(J_\Delta, F) \ .$$

Proof It follows from (II.2.1.2), (II.2.1.6), and (II.2.1.7) that $U(\cdot,s)x$ is a solution of $(1.2.1)_{(s,x,A,0)}$ such that

$$U(\cdot,\cdot)x \in C(J_\Delta^*, F) \ .$$

It is an easy consequence of (II.2.1.2), (1.2.2), and (1.2.3) that

$$v := U \star f \in C(J_\Delta, F) \ . \tag{1.2.4}$$

Hence we infer from Remark II.2.1.2(b) that $u := U(\cdot,s)x + v(\cdot,s)$ has the desired properties, provided we show that $v(\cdot,s)$ is a solution of $(1.2.1)_{(s,0,A,f)}$.

Let $(t,s) \in J_\Delta^*$ and $h > 0$ with $t+h \in J$. Then

$$\begin{aligned} h^{-1}&\big[v(t+h,s) - v(t,s)\big] \\ &= h^{-1}\int_s^t \big[U(t+h,\tau) - U(t,\tau)\big] f(\tau)\, d\tau + h^{-1}\int_t^{t+h} U(t+h,\tau) f(\tau)\, d\tau \\ &=: I_h + II_h \ . \end{aligned}$$

Since (1.2.2) and (II.2.1.2) imply $Uf \in C(J_\Delta, E)$, it follows that

$$II_h \to f(t) \qquad \text{as} \quad h \to 0 \ . \tag{1.2.5}$$

Observe that, thanks to (II.2.1.4)–(II.2.1.7),

$$U(t+h,\tau) - U(t,\tau) = -\int_t^{t+h} AU(\sigma,\tau)\, d\sigma \ , \qquad (t,\tau) \in J_\Delta^* \ ,$$

in $\mathcal{L}(E)$. Hence (1.2.3) implies

$$\|U(t+h,\tau) - U(t,\tau)\|_{\mathcal{L}(F_0,E)} \le c(t-\tau)^{-\alpha} h \ .$$

Now we deduce from Lebesgue's theorem, (1.2.3), and (1.2.4) that

$$I_h \to -\int_s^t AU(t,\tau) f(\tau)\, d\tau = -A(t)v(t,s) \ , \qquad h \to 0+ \ , \tag{1.2.6}$$

where the last equality follows from the closedness of A. Thus (1.2.5) and (1.2.6) show that the right derivative $\partial_{1,+}v(t,s)$ exists on $J \cap (s,\infty)$ and satisfies

$$\partial_{1,+}v(t,s) = -A(t)v(t,s) + f(t) , \qquad t \in J \cap (s,\infty) . \tag{1.2.7}$$

Moreover, (1.2.2)–(1.2.4) and (1.2.7) imply that

$$\partial_{1,+}v(\cdot,s) \in C\big(J \cap [s,\infty), E\big) .$$

Now the first assertion follows from Proposition 1.1.2.

If $U \in C\big(J_\Delta, \mathcal{L}_s(F)\big)$ and $x \in F$ then $U(\cdot,\cdot)x \in C(J_\Delta, F)$. Hence (1.2.4) implies the second assertion. ■

The next theorem gives a sufficient condition for a solution of $(1.2.1)_{(s,x,A,f)}$ to be strict.

1.2.2 Theorem *Let the hypotheses of Theorem* 1.2.1 *be satisfied. Also assume that* $0 \in \rho\big(A(t)\big)$ *for* $t \in J$ *and*

$$AUA^{-1} \in C\big(J_\Delta, \mathcal{L}_s(E)\big) . \tag{1.2.8}$$

If $s \in J$ *and* $x \in \mathrm{dom}\big(A(s)\big)$ *then* $u(\cdot,s,x)$ *is a strict solution of the Cauchy problem* $(1.2.1)_{(s,x,A,f)}$.

Proof Observe that

$$Au(\cdot,s,x) = AU(\cdot,s)x + (AU \star f)(\cdot,s) \tag{1.2.9}$$

by (1.2.3) and the closedness of $A(t)$. Estimate (1.2.3) easily implies

$$AU \star f \in C(J_\Delta, E) . \tag{1.2.10}$$

Since $AU(t,s)x = AUA^{-1}(t,s)y$, where we have set $y := A(s)x \in E$, it follows from (1.2.8)–(1.2.10) that the assertion is true. ■

The above results are due to the author and were first published in [Ama88c].

1.3 Estimates for Evolution Operators

Let (E_0, E_1) be a densely injected Banach couple, J a perfect subinterval of $\mathbb{R}^+$ containing 0, and $\rho \in (0,1)$. Suppose that

$$\left.\begin{aligned}
&\mathcal{A} \subset C^\rho\big(J, \mathcal{H}(E_1, E_0)\big) \text{ and there are constants } M, \eta \in \mathbb{R}^+ \text{ and } \\
&\vartheta \in (0, \pi/2) \text{ such that} \\
&\qquad [A]_{\rho,J} \le \eta\ , \qquad A \in \mathcal{A}\ , \\
&\text{that } \Sigma_\vartheta \subset \rho\big(-A(s)\big), \text{ and that} \\
&\qquad \|A(s)\|_{\mathcal{L}(E_1,E_0)} + (1+|\lambda|)^{1-j}\, \big\|\big(\lambda + A(s)\big)^{-1}\big\|_{\mathcal{L}(E_0,E_j)} \le M \\
&\text{for } (s,\lambda,A) \in J \times \Sigma_\vartheta \times \mathcal{A} \text{ and } j = 0,1.
\end{aligned}\right\} \tag{1.3.1}$$

Note that this assumption coincides with (II.4.2.1). Hence Theorem II.4.4.1 guarantees the existence of a unique parabolic evolution operator U_A for $A \in \mathcal{A}$ possessing E_1 as regularity subspace. In the following, we fix μ as in (II.4.4.2) and suppose that

$$0 < \gamma < \rho \tag{1.3.2}$$

and that $(\cdot,\cdot)_\gamma$ is an admissible interpolation functor. Then we put $E_\gamma := (E_0, E_1)_\gamma$ and derive estimates for the restriction of A to E_γ.

1.3.1 Lemma *Given $A \in \mathcal{A}$,*

$$AU_A \in \mathfrak{K}(E_\gamma, 1) \cap \mathfrak{K}(E_\gamma, E_0, 1-\gamma)\ .$$

More precisely,

$$(t-s)\,\|AU_A(t,s)\|_{\mathcal{L}(E_\gamma)} + (t-s)^{1-\gamma}\,\|AU_A(t,s)\|_{\mathcal{L}(E_\gamma,E_0)} \le c(\varepsilon) e^{(\mu+\varepsilon)(t-s)}$$

for $\varepsilon > 0$, $(t,s) \in J_\Delta^*$, and $A \in \mathcal{A}$.

Proof Throughout this proof we use the notations of Section II.4 and omit the index A.

First we deduce from (II.4.2.3) by interpolation that

$$\big[(t,s) \mapsto [A(\tau)]^j e^{-(t-s)A(\tau)}\big] \in \mathfrak{K}_\infty(E_0, E_\gamma, \gamma + j) \tag{1.3.3}$$

and

$$\partial_1 a \in \mathfrak{K}_\infty(E_\gamma, 1)\ , \tag{1.3.4}$$

and that

$$(t-s)^{\gamma+j}\,\big\|\big[A(\tau)\big]^j e^{-(t-s)A(\tau)}\big\|_{\mathcal{L}(E_0,E_\gamma)} + (t-s)\,\|\partial_1 a(t,s)\|_{\mathcal{L}(E_\gamma)} \le c \tag{1.3.5}$$

for $(t,s) \in J_\Delta^*$, $\tau \in J$, and $A \in \mathcal{A}$. From (II.4.2.2) and (II.4.2.3) it also follows that

$$\|A(s)e^{-tA(s)}x\|_0 = \|e^{-tA(s)}A(s)x\|_0 \le cM\,\|x\|_1 \ , \qquad x \in E_1 \ , \quad A \in \mathcal{A} \ .$$

Thus, by using (II.4.2.2) again,

$$\|\partial_1 a(t,s)\|_{\mathcal{L}(E_1,E_0)} + (t-s)\,\|\partial_1 a(t,s)\|_{\mathcal{L}(E_0)} \le c$$

for $(t,s) \in J_\Delta^*$ and $A \in \mathcal{A}$. Hence, by interpolation,

$$\partial_1 a \in \mathfrak{K}_\infty(E_\gamma, E_0, 1-\gamma)$$

and

$$(t-s)^{1-\gamma}\,\|\partial_1 a(t,s)\|_{\mathcal{L}(E_\gamma,E_0)} \le c \ , \qquad A \in \mathcal{A} \ , \quad (t,s) \in J_\Delta^* \ . \tag{1.3.6}$$

Also by interpolating we obtain from Lemma II.4.3.1 that

$$w \in \mathfrak{K}(E_\gamma, E_0, 1-\gamma-\rho) \tag{1.3.7}$$

and that

$$(t-s)^{1-\gamma-\rho}\,\|w(t,s)\|_{\mathcal{L}(E_\gamma,E_0)} \le c(\varepsilon)e^{(\mu+\varepsilon)(t-s)} \tag{1.3.8}$$

for $\varepsilon > 0$, $(t,s) \in J_\Delta^*$, and $A \in \mathcal{A}$.

Put

$$I_1(t,s) := e^{-(t-s)A(t)}w(t,s) \ .$$

Then we deduce from (II.4.2.2), (II.4.2.3), (1.3.3), (1.3.5), and (1.3.7), (1.3.8) that

$$I_1 \in \mathfrak{K}(E_\gamma, 1-\rho) \cap \mathfrak{K}(E_\gamma, E_0, 1-\gamma-\rho) \tag{1.3.9}$$

and

$$(t-s)^{1-\rho}\,\|I_1(t,s)\|_{\mathcal{L}(E_\gamma)} + (t-s)^{1-\gamma-\rho}\,\|I_1(t,s)\|_{\mathcal{L}(E_\gamma,E_0)} \le c(\varepsilon)e^{(\mu+\varepsilon)(t-s)} \tag{1.3.10}$$

for $\varepsilon > 0$, $(t,s) \in J_\Delta^*$, and $A \in \mathcal{A}$.

By interpolation we obtain from (1.3.1) that

$$\big\|\big(\lambda + A(s)\big)^{-1}\big\|_{\mathcal{L}(E_0,E_\gamma)} \le M\,|\lambda|^{\gamma-1} \ , \qquad (s,\lambda,A) \in J \times \Sigma_\vartheta \times \mathcal{A} \ . \tag{1.3.11}$$

Using this estimate, an obvious modification of the proof of Lemma II.4.3.2 gives

$$e \in \mathfrak{K}(E_0, E_\gamma, 1+\gamma-\rho) \tag{1.3.12}$$

and

$$\|e(t,s)\|_{\mathcal{L}(E_0,E_\gamma)} \le c(t-s)^{\rho-\gamma-1} \tag{1.3.13}$$

for $(t,s) \in J_\Delta^*$ and $A \in \mathcal{A}$. Now we obtain from (1.3.7), (1.3.8), (1.3.12), (1.3.13), and (II.3.1.8), (II.3.1.9) on the one hand, and from (1.3.7), (1.3.8), Lemma II.4.3.2,

(II.3.1.8), (II.3.1.9) on the other hand, together with (II.3.1.1), (II.3.1.2), that

$$I_2 := e \star w \in \mathfrak{K}(E_\gamma, 1-\rho) \cap \mathfrak{K}(E_\gamma, E_0, 1-\gamma) \tag{1.3.14}$$

and

$$(t-s)^{1-\rho} \|I_2(t,s)\|_{\mathcal{L}(E_\gamma)} + (t-s)^{1-\gamma} \|I_2(t,s)\|_{\mathcal{L}(E_\gamma,E_0)} \leq c(\varepsilon) e^{(\mu+\varepsilon)(t-s)} \tag{1.3.15}$$

for $\varepsilon > 0$, $(t,s) \in J_\Delta^*$, and $A \in \mathcal{A}$.

By interpolation we infer from (II.4.3.19) that

$$(\tau - s)^{1-\gamma} \|k(t,s) - k(\tau,s)\|_{\mathcal{L}(E_\gamma,E_0)} \leq c(t-\tau)^\rho$$

for $0 \leq s < \tau < t$. Thus it is a consequence of (II.4.3.22), (II.4.3.21), (1.3.8), and (II.4.3.4) that

$$\begin{aligned} &\|w(t,s) - w(\tau,s)\|_{\mathcal{L}(E_\gamma,E_0)} \\ &\quad \leq c(\varepsilon,\beta)\Big[(t-\tau)^\rho(\tau-s)^{\gamma-1} + \int_s^\tau (t-\tau)^\beta(\tau-\sigma)^{\rho-\beta-1}(\sigma-s)^{\gamma+\rho-1}\,d\sigma \\ &\qquad + \int_\tau^t (t-\sigma)^{\rho-1}(\sigma-s)^{\gamma+\rho-1}\,d\sigma\Big] e^{(\mu+\varepsilon)(t-s)} \end{aligned}$$

for $\varepsilon > 0$, $0 < \beta < \rho$, and $0 \leq s < \tau < t$. Hence

$$\begin{aligned} &\|w(t,s) - w(\tau,s)\|_{\mathcal{L}(E_\gamma,E_0)} \\ &\quad \leq c(\varepsilon,\beta)\,\big[(t-\tau)^\beta(\tau-s)^{\gamma-1} + I(t,\tau,s)\big] e^{(\mu+\varepsilon)(t-s)} \end{aligned} \tag{1.3.16}$$

for $\varepsilon > 0$, $0 < \beta < \rho$, and $0 \leq s < \tau < t$, where

$$\begin{aligned} I(t,\tau,s) &:= \int_\tau^t (t-\sigma)^{\rho-1}(\sigma-s)^{\gamma+\rho-1}\,d\sigma \\ &\leq \begin{cases} c(t-\tau)^\rho(t-s)^{\gamma+\rho-1} & \text{if } \gamma+\rho-1 \geq 0\,, \\ c(t-\tau)^{2\rho+\gamma-1} & \text{if } \gamma+\rho-1 < 0\,. \end{cases} \end{aligned}$$

Fix $\beta \in (\gamma,\rho)$ and put

$$I_3(t,s) := \int_s^t A(t)e^{-(t-\tau)A(t)}\big[w(t,s) - w(\tau,s)\big]\,d\tau\,.$$

Then we infer from (1.3.3) and (1.3.16) that

$$\begin{aligned} \|I_3(t,s)\|_{\mathcal{L}(E_\gamma)} &\leq c(\varepsilon)\int_s^t (t-\tau)^{-1-\gamma}\big[(t-\tau)^\beta(\tau-s)^{\gamma-1} + I(t,\tau,s)\big]\,d\tau\, e^{(\mu+\varepsilon)(t-s)} \\ &\leq c(\varepsilon)\big[(t-s)^{\beta-1} + (t-s)^{2\rho-1}\big] e^{(\mu+\varepsilon)(t-s)} \end{aligned}$$

for $\varepsilon > 0$ and $(t,s) \in J_\Delta^*$. Similarly, (II.4.2.3) and (1.3.16) imply

$$\|I_3(t,s)\|_{\mathcal{L}(E_\gamma,E_0)} \leq c(\varepsilon)\big[(t-s)^{\gamma+\beta-1} + (t-s)^{\gamma+2\rho-1}\big]e^{(\mu+\varepsilon)(t-s)}$$

for $\varepsilon > 0$ and $(t,s) \in J_\Delta^*$. Hence

$$I_3 \in \mathfrak{K}(E_\gamma, 1-\gamma) \cap \mathfrak{K}(E_\gamma, E_0, 1-2\gamma) \tag{1.3.17}$$

and

$$(t-s)^{1-\gamma}\,\|I_3(t,s)\|_{\mathcal{L}(E_\gamma)} + (t-s)^{1-2\gamma}\,\|I_3(t,s)\|_{\mathcal{L}(E_\gamma,E_0)} \leq c(\varepsilon)e^{(\mu+\varepsilon)(t-s)} \tag{1.3.18}$$

for $\varepsilon > 0$ and $(t,s) \in J_\Delta^*$.

Note that Lemma II.4.3.4 and (II.4.3.26) and (II.4.3.30) imply

$$AU = -\partial_1 U = -\partial_1 a - I_1 - I_2 - I_3\ . \tag{1.3.19}$$

Hence the assertion follows from (1.3.4)–(1.3.6), (1.3.9), (1.3.10), (1.3.14), (1.3.15), and (1.3.17), (1.3.18). ■

1.3.2 Lemma *Given $A \in \mathcal{A}$,*

$$AU_A A^{-1} \in \mathfrak{K}(E_\gamma, 0) \cap C\big(J_\Delta, \mathcal{L}_s(E_\gamma)\big)\ .$$

Moreover,

$$\|AU_A A^{-1}(t,s)\|_{\mathcal{L}(E_\gamma)} \leq c(\varepsilon)e^{(\mu^{\rho/(\rho-\gamma)}+\varepsilon)(t-s)}$$

for $\varepsilon > 0$, $(t,s) \in J_\Delta^*$, and $A \in \mathcal{A}$.

Proof Put

$$b_A(t,s) := e^{-(t-s)A(t)}\ , \quad h_A(t,s) := b_A(t,s)\,\big[A(t) - A(s)\big]$$

for $(t,s) \in J_\Delta$ and $A \in \mathcal{A}$. In the following we again omit the index A. Then (II.4.2.3) and (1.3.3) imply

$$h \in \mathfrak{K}(E_1, 1-\rho) \tag{1.3.20}$$

and

$$Ah \in \mathfrak{K}(E_1, E_\gamma, 1+\gamma-\rho)\ , \tag{1.3.21}$$

respectively, together with the estimate

$$(t-s)^{1-\rho}\,\|h(t,s)\|_{\mathcal{L}(E_1)} + (t-s)^{1+\gamma-\rho}\,\|Ah(t,s)\|_{\mathcal{L}(E_1,E_\gamma)} \leq c\eta \tag{1.3.22}$$

for $(t,s) \in J_\Delta^*$ and $A \in \mathcal{A}$.

Note that, thanks to (II.2.2.3),

$$U = b + h \star U \; . \tag{1.3.23}$$

From (1.3.1) and (II.4.2.3) it follows that

$$AbA^{-1} \in \mathfrak{K}(E_0, 0) \tag{1.3.24}$$

and

$$\|AbA^{-1}(t,s)\|_{\mathcal{L}(E_0)} \le c \; , \qquad (t,s) \in J_\Delta^* \; , \quad A \in \mathcal{A}. \tag{1.3.25}$$

Using (1.3.21), (1.3.23), and (1.3.24) we find

$$AUA^{-1} = AbA^{-1} + A(h \star U)A^{-1} = AbA^{-1} + (AhA^{-1}) \star (AUA^{-1}) \; , \tag{1.3.26}$$

where the last 'convolution' is meaningful since $AhA^{-1} \in \mathfrak{K}(E_0, 1-\rho)$ by (1.3.1) and (1.3.20).

Observe that

$$AbA^{-1} = a + (Ab - aA)A^{-1} \tag{1.3.27}$$

and (cf. the proof of Lemma II.4.3.2)

$$\begin{aligned}(Ab - aA)A^{-1} \\ = \frac{1}{2\pi i} \int_\Gamma \lambda e^{\lambda(t-s)} \big(\lambda + A(t)\big)^{-1} \big[A(t) - A(s)\big] \big(\lambda + A(s)\big)^{-1} A^{-1}(s)\, d\lambda \; .\end{aligned} \tag{1.3.28}$$

From (1.3.1) and (II.4.2.2) we deduce that

$$\begin{aligned}\big\|\big(\lambda + A(s)\big)^{-1} x\big\|_1 &\le M \big\|A(s)\big(\lambda + A(s)\big)^{-1} x\big\|_0 = M \big\|\big(\lambda + A(s)\big)^{-1} A(s) x\big\|_0 \\ &\le (1 + |\lambda|)^{-1} M^2 \|x\|_1\end{aligned}$$

for $x \in E_1$ and $\lambda \in \Sigma_\vartheta$. Thus

$$\big\|\big(\lambda + A(s)\big)^{-1}\big\|_{\mathcal{L}(E_1)} \le (1+|\lambda|)^{-1} M^2 \; , \qquad s \in J \; , \quad \lambda \in \Sigma_\vartheta \; , \quad A \in \mathcal{A} \; . \tag{1.3.29}$$

By interpolation it follows from (1.3.1) and (1.3.29) that

$$(1 + |\lambda|) \big\|\big(\lambda + A(s)\big)^{-1}\big\|_{\mathcal{L}(E_\gamma)} + (1 + |\lambda|)^\gamma \big\|\big(\lambda + A(s)\big)^{-1}\big\|_{\mathcal{L}(E_\gamma, E_1)} \le c \tag{1.3.30}$$

for $s \in J$, $\lambda \in \Sigma_\vartheta$, and $A \in \mathcal{A}$. By means of (1.3.11) and (1.3.30) we can estimate the integrand in (1.3.28) in the norm of $\mathcal{L}(E_\gamma)$ by

$$c(t-s)^\rho \, |\lambda|^{\gamma - 1} \, e^{\operatorname{Re} \lambda (t-s)} \; .$$

Hence Lemma II.4.1.1, (II.3.1.1), and (II.3.1.5) imply

$$(Ab - aA)A^{-1} \in \mathfrak{K}(E_\gamma, \gamma - \rho) \hookrightarrow C\big(J_\Delta, \mathcal{L}(E_\gamma)\big) \cap \mathfrak{K}(E_\gamma, 0) \tag{1.3.31}$$

and

$$(t-s)^{\gamma-\rho}\,\|(Ab-aA)A^{-1}(t,s)\|_{\mathcal{L}(E_\gamma)} \le c \tag{1.3.32}$$

for $(t,s) \in J_\Delta^*$ and $A \in \mathcal{A}$. Lemma II.4.2.1 implies, by interpolation, that

$$a \in C\big(J_\Delta, \mathcal{L}_s(E_\gamma)\big) \cap \mathfrak{K}(E_\gamma, 0) \tag{1.3.33}$$

and

$$\|a(t,s)\|_{\mathcal{L}(E_\gamma)} \le c\,, \qquad (t,s) \in J_\Delta\,, \quad A \in \mathcal{A}\,. \tag{1.3.34}$$

Thus we obtain from (1.3.27) and (1.3.32), (1.3.34) that

$$AbA^{-1} \in C\big(J_\Delta, \mathcal{L}_s(E_\gamma)\big) \cap \mathfrak{K}(E_\gamma, 0) \tag{1.3.35}$$

and

$$\|AbA^{-1}(t,s)\|_{\mathcal{L}(E_\gamma)} \le c\big(1 \vee (t-s)^{\rho-\gamma}\big) \tag{1.3.36}$$

for $(t,s) \in J_\Delta$ and $A \in \mathcal{A}$. Thanks to (1.3.21), (1.3.22), and (1.3.30)

$$AhA^{-1} \in \mathfrak{K}(E_\gamma, 1+\gamma-\rho)\;. \tag{1.3.37}$$

and

$$(t-s)^{1+\gamma-\rho}\,\big\|AhA^{-1}(t,s)\big\|_{\mathcal{L}(E_\gamma)} \le c\eta\,, \qquad (t,s) \in J_\Delta^*\,, \quad A \in \mathcal{A}\,. \tag{1.3.38}$$

Using (1.3.26), (1.3.35)–(1.3.38), and Theorem II.3.2.2 we find

$$AUA^{-1} \in \mathfrak{K}(E_\gamma, 0) \tag{1.3.39}$$

and

$$AUA^{-1} = AbA^{-1} + r \star (AbA^{-1})\,, \tag{1.3.40}$$

where $r \in \mathfrak{K}(E_\gamma, 1+\gamma-\rho)$. Note that (1.3.38) and Lemma II.3.2.1 imply

$$(t-s)^{1+\gamma-\rho}\,\|r(t,s)\|_{\mathcal{L}(E_\gamma)} \le ce^{\mu^{\rho/(\rho-\gamma)}(t-s)}$$

for $(t,s) \in J_\Delta^*$ and $A \in \mathcal{A}$, provided we increase the constant $c(\rho)$ in (II.4.4.2) appropriately. Now it follows from (II.3.1.9) and (1.3.35), (1.3.36) that

$$r \star (AbA^{-1}) \in \mathfrak{K}(E_\gamma, \gamma-\rho) \hookrightarrow C(J_\Delta, E_\gamma) \tag{1.3.41}$$

and that

$$(t-s)^{\gamma-\rho}\,\|r \star (AbA^{-1})(t,s)\|_{\mathcal{L}(E_\gamma)} \le ce^{\mu^{\rho/(\rho-\gamma)}(t-s)} \tag{1.3.42}$$

for $(t,s) \in J_\Delta^*$ and $A \in \mathcal{A}$. Hence the assertion follows from (1.3.40), (1.3.35), (1.3.36), and (1.3.41), (1.3.42). ■

1.4 Evolution Operators on Interpolation Spaces

Let (E_0, E_1) be a densely injected Banach couple, J a perfect subinterval of $\mathbb{R}^+$ containing 0, and $\rho \in (0,1)$. Suppose that

$$0 < \gamma < \rho \tag{1.4.1}$$

and $(\cdot,\cdot)_\gamma$ is an admissible interpolation functor, and let $E_\gamma := (E_0, E_1)_\gamma$. Given a linear operator B in E_0, denote by B_γ its E_γ-realization.

Now it is easy to prove the following basic theorem which is the main result of this section.

1.4.1 Theorem *Let assumptions* (1.3.1) *and* (1.4.1) *be satisfied. Then, given $A \in \mathcal{A}$, the E_γ-realization, $U_{A,\gamma}$, of the parabolic evolution operator U_A for A is the parabolic evolution operator for the E_γ-realization A_γ of A. It possesses E_1 as regularity subspace and satisfies*

$$A_\gamma U_{A,\gamma} A_\gamma^{-1} \in C\big(J_\Delta, \mathcal{L}_s(E_\gamma)\big) . \tag{1.4.2}$$

Furthermore,

$$\begin{aligned}(t-s)^j \, \|A_\gamma^j U_{A,\gamma}(t,s)\|_{\mathcal{L}(E_\gamma)} + (t-s)^{1-\gamma} \, \|U_{A,\gamma}(t,s)\|_{\mathcal{L}(E_\gamma,E_1)} \\ \leq c(\varepsilon) e^{(\mu+\varepsilon)(t-s)}\end{aligned} \tag{1.4.3}$$

and

$$\|A_\gamma U_{A,\gamma} A_\gamma^{-1}(t,s)\|_{\mathcal{L}(E_\gamma)} \leq c(\varepsilon) e^{(\mu^{\rho/(\rho-\gamma)}+\varepsilon)(t-s)} \tag{1.4.4}$$

for $\varepsilon > 0$, $(t,s) \in J_\Delta^*$, $A \in \mathcal{A}$, and $j = 0, 1$, where we have put $\mu := c_0 \eta^{1/\rho}$ for a suitable positive constant c_0.

Proof Note that the E_γ-realization of U_A is simply the restriction of U_A to E_γ, considered as linear operator in E_γ. Moreover, $x \in \operatorname{dom}(A_\gamma)$ iff $x \in E_1$ and $Ax \in E_\gamma$. Using these facts, (1.4.2)–(1.4.4) with $j = 1$ follow immediately from Lemmas 1.3.1 and 1.3.2, from (II.4.2.2), and from $\operatorname{im}\big(U_A(t,s)\big) \subset E_1$ for $(t,s) \in J_\Delta^*$. By interpolation we deduce from (II.4.4.3) that (1.4.3) is true for $j = 0$ as well. From (II.4.4.1) and (1.4.3) we easily obtain — thanks to the density of E_1 in E_γ — that

$$U_{A,\gamma} \in C\big(J_\Delta, \mathcal{L}_s(E_\gamma)\big) \cap C\big(J_\Delta^*, \mathcal{L}_s(E_\gamma, E_1)\big) , \tag{1.4.5}$$

which is condition (II.2.1.2) with $E := E_\gamma$ and $F := E_1$.

By restricting $\partial_1 U_A(t,s) = -AU_A(t,s)$ for $(t,s) \in J_\Delta^*$ to E_γ we see that

$$\partial_1 U_A(t,s) | E_\gamma = -A_\gamma U_{A,\gamma}(t,s) , \qquad (t,s) \in J_\Delta^* . \tag{1.4.6}$$

Note that

$$U_A(t,s) = U_A(\tau,s) + \int_\tau^t \partial_1 U_A(\sigma,s)\, d\sigma , \qquad 0 \leq s < \tau < t , \quad t \in J , \tag{1.4.7}$$

in $\mathcal{L}(E)$. Thus, by restricting (1.4.7) to E_γ and using (1.4.6), we infer

$$U_{A,\gamma}(t,s) = U_{A,\gamma}(\tau,s) - \int_\tau^t A_\gamma U_{A,\gamma}(\sigma,s)\,d\sigma\ , \qquad 0 \le s < \tau < t\ , \quad t \in J\ ,$$

where, thanks to $A_\gamma U_{A,\gamma} \in \mathfrak{K}(E_\gamma, 1)$, the integral converges in $\mathcal{L}(E_\gamma)$. This implies

$$U_{A,\gamma}(\cdot,s) \in C^1\big(J \cap (s,\infty), \mathcal{L}(E_\gamma)\big) \quad \text{and} \quad \partial_1 U_{A,\gamma}(\cdot,s) = -A_\gamma U_{A,\gamma}(\cdot,s)$$

for $s \in J$, that is, conditions (II.2.1.6) and (II.2.1.7), respectively.

Recall that $U_A(t,\cdot) \in C^1\big([0,t), \mathcal{L}_s(E_1,E_0)\big)$ and that

$$\partial_2 U_A(t,\cdot)|E_1 = U_A A(t,\cdot)\ , \qquad t \in \dot{J}\ . \tag{1.4.8}$$

This shows that

$$\partial_2 U_A(t,\cdot)|E_1 \in C\big([0,t), \mathcal{L}_s(E_1,E_\gamma)\big)\ , \qquad t \in J\ ,$$

thanks to the continuity properties of A and U_A. Since, given $x \in E_1$,

$$U_A(t,\tau)x = U_A(t,\sigma)x + \int_\sigma^\tau \partial_2 U_A(t,r)x\,dr\ , \qquad 0 \le \sigma < \tau < t\ , \quad t \in \dot{J}\ , \tag{1.4.9}$$

we infer — by restricting (1.4.9) to E_γ — from (1.4.5), $E_1 \hookrightarrow E_\gamma$, and (1.4.8) that

$$U_{A,\gamma}(t,\cdot) \in C^1\big([0,t), \mathcal{L}_s(E_1,E_\gamma)\big) \quad \text{and} \quad \partial_2 U_{A,\gamma}(t,\cdot) \supset U_{A,\gamma} A_\gamma(t,\cdot)$$

for $t \in \dot{J}$, which are conditions (II.2.1.8) and (II.2.1.9), respectively. Conditions (II.2.1.3) and (II.2.1.4) are obvious, and (II.2.1.5) is a consequence of Lemma 1.3.1 as has been observed in the beginning of this proof. Hence $U_{A,\gamma}$ is a parabolic evolution operator for A_γ with regularity subspace E_1, and the theorem has been proven. ■

1.4.2 Corollary *Suppose that* $0 < \gamma < \rho < 1$ *and*

$$A \in C^\rho\big(J, \mathcal{H}(E_1,E_0)\big)\ .$$

Then the E_γ-realization of the parabolic evolution operator for A is the parabolic evolution operator for the E_γ-realization of A. It has E_1 as regularity subspace.

Proof This follows — similarly as in the proof of Corollary II.4.4.2 — by applying Theorem 1.4.1 to an arbitrary compact subinterval I of J containing 0 and to $\sigma + A$ for a suitable $\sigma \in \mathbb{R}$ (depending on I and A, of course). ■

1.4.3 Remark Given the hypotheses of Theorem 1.4.1, the E_γ-realization $A_\gamma(t)$ of $A(t)$ is densely defined for $t \in J$.

Indeed, let $E_2(t) := \big(\mathrm{dom}(A^2(t)), \|A^2(t)\cdot\|\big)$. Then $E_2(t)$ is a Banach space such that $E_2(t) \overset{d}{\hookrightarrow} E_1 \overset{d}{\hookrightarrow} E_0$. Let $A_1(t)$ be the E_1-realization of $A(t)$. It follows that $A_1(t) \in \mathcal{L}\mathrm{is}\big(E_2(t), E_1\big)$ and

$$A_1(t) \subset A(t) \in \mathcal{L}\mathrm{is}(E_1, E_0) .$$

From this we deduce by interpolation that $A_\gamma(t) \in \mathcal{L}\mathrm{is}\big(E_{1+\gamma}(t), E_\gamma\big)$, where

$$E_{1+\gamma}(t) := \big(E_1, E_2(t)\big)_\gamma \overset{d}{\hookrightarrow} E_1 \overset{d}{\hookrightarrow} E_\gamma .$$

Hence $\mathrm{dom}\big(A_\gamma(t)\big) = E_{1+\gamma}(t)$ is dense in E_γ (cf. Section V.1.5 for more details). ■

Now we give a variant of Theorem 1.4.1 in which the interpolation space E_γ is replaced by an intermediate space X between E_0 and E_1 that is not supposed to be an interpolation space. Of course, X has to satisfy some additional assumptions.

1.4.4 Theorem *Let assumption* (1.3.1) *be satisfied. Suppose that*

$$0 < \gamma_- \le \gamma_+ < \rho , \tag{1.4.10}$$

that $(\cdot,\cdot)_{\gamma_-}$ *and* $(\cdot,\cdot)_{\gamma_+}$ *are admissible interpolation functors, and that* X *is a Banach space such that*

$$E_{\gamma_+} \hookrightarrow X \hookrightarrow E_{\gamma_-} , \tag{1.4.11}$$

where $E_{\gamma_\pm} := (E_0, E_1)_{\gamma_\pm}$. *Let* $A_X(t)$ *be the* X*-realization of* $A(t)$ *for* $A \in \mathcal{A}$ *and* $t \in J$ *and suppose that* $A_X(t)$ *is densely defined and*

$$(1 + |\lambda|)\, \big\|\big(\lambda + A_X(t)\big)^{-1}\big\|_{\mathcal{L}(X)} \le M \tag{1.4.12}$$

for $A \in \mathcal{A}$, $t \in J$ *and* $\lambda \in \Sigma_\vartheta$. *Then* $U_{A,X}$, *the* X*-realization of* U_A, *is a parabolic evolution operator for* A_X. *It possesses* E_1 *as regularity subspace and satisfies*

$$A_X U_{A,X} A_X^{-1} \in C\big(J_\Delta, \mathcal{L}_s(X)\big) \tag{1.4.13}$$

and

$$\|A_X U_{A,X} A_X^{-1}(t,s)\|_{\mathcal{L}(X)} \le c(\varepsilon) e^{(\mu^{\rho/(\rho-\gamma_+)}+\varepsilon)(t-s)} \tag{1.4.14}$$

for $\varepsilon > 0$, $(t,s) \in J_\Delta^*$, *and* $A \in \mathcal{A}$. *Moreover,*

$$\begin{aligned}(t-s)^j\, \|A_X^j U_{A,X}(t,s)\|_{\mathcal{L}(X)} + (t-s)^{1-\gamma_-}\, \|U_{A,X}(t,s)\| \\ \le c(\varepsilon) e^{(\mu+\varepsilon)(t-s)}\end{aligned} \tag{1.4.15}$$

for $\varepsilon > 0$, $(t,s) \in J_\Delta^*$, $A \in \mathcal{A}$, *and* $j = 0, 1$.

Proof This follows by obvious modifications of the proof of Theorem 1.4.1 which we leave to the reader (cf. [Ama88c, Proof of Theorem 2.2] and [Ama90a, Appendix A]). ■

Theorem 1.4.1 — and, of course, Theorem 1.4.4 as well — guarantees the existence of a parabolic evolution operator for A_γ on E_γ, which is unique by Remark II.2.1.2(c). It is important to note that $\operatorname{dom}(A_\gamma(t))$ is, in general, not independent of t. Thus the situation is different from the one considered in Chapter II.

The above theorems show that the regularizing properties of analytic semigroups carry over, to some extent, to parabolic evolution operators. The amount of regularization is restricted by the smoothness with respect to t as follows from assumption (1.4.1).

The proof of Theorem 1.4.1 follows the proof of [Ama88c, Theorem 2.2] (also cf. [Ama90a, Appendix A]).

1.5 The Cauchy Problem

Let (E_0, E_1) be a densely injected Banach couple, let J be a perfect subinterval of $\mathbb{R}$ containing its left endpoint $s := \min J$, and suppose that

$$A \in C^\rho\big(J, \mathcal{H}(E_1, E_0)\big) \tag{1.5.1}$$

for some $\rho \in (0,1)$. Also suppose that $(\cdot,\cdot)_\theta$ is for each $\theta \in (0,1)$ an admissible interpolation functor, and put $E_\theta := (E_0, E_1)_\theta$. Lastly, suppose that $0 < \gamma < \rho$ and $0 < \varepsilon < 1 - \gamma$ and

$$f \in C^\varepsilon(J, E_\gamma) + C(J, E_{\gamma+\varepsilon}) \ . \tag{1.5.2}$$

Consider the Cauchy problem

$$\dot{u} + A(t)u = f(t) \ , \quad t \in J\backslash\{s\} \ , \qquad u(s) = x \ , \tag{1.5.3}$$

where $x \in E_0$.

Now we are ready to prove Theorem II.1.2.2 which, for the reader's convenience, we recall here:

1.5.1 Theorem *If f satisfies* (1.5.2), *the Cauchy problem* (1.5.3) *has a unique solution*

$$u \in C(J, E_0) \cap C\big(J\backslash\{s\}, E_1\big) \cap C^1\big(J\backslash\{s\}, E_\gamma\big) \ .$$

It is also a solution of the parabolic equation

$$\dot{v} + A_\gamma(t)v = f(t) \ , \qquad t \in J\backslash\{s\} \ , \tag{1.5.4}$$

in E_γ, where A_γ is the E_γ-realization of A. If $x \in E_\gamma$ then $u \in C(J, E_\gamma)$, and if $x \in \operatorname{dom}\big(A_\gamma(s)\big)$, it is a strict solution of (1.5.4).

Proof By an obvious translation argument we can assume that $s = 0$. Moreover, it suffices to consider the case that J is compact.

Let U be the parabolic evolution operator for A, whose existence is guaranteed, thanks to (1.5.1), by Corollary II.4.4.2. Since J is compact, Proposition I.1.4.2 guarantees the existence of constants $\omega > 0$, $\kappa \geq 1$, and $\sigma \in \mathbb{R}$ such that

$$\sigma + A \in C^\rho\big(J, \mathcal{H}(E_1, E_0, \kappa, \omega)\big) . \tag{1.5.5}$$

Since J is compact, A is uniformly ρ-Hölder continuous. Hence condition (II.5.0.1) is satisfied. Thus we deduce from Lemma II.5.1.3 that

$$(t-s)^{1-\gamma} \|U(t,s)\|_{\mathcal{L}(E_\gamma, E_1)} \leq c , \qquad (t,s) \in J_\Delta^* .$$

This implies, thanks to (1.5.1), that

$$\|U(t,s)\|_{\mathcal{L}(E_\gamma, E_1)} + \|AU(t,s)\|_{\mathcal{L}(E_\gamma, E_0)} \leq c(t-s)^{\gamma-1} , \qquad (t,s) \in J_\Delta^* .$$

Since $f \in C(J, E_\gamma)$ by (1.5.2), Theorem 1.2.1 guarantees that (1.5.3) has a unique solution

$$u \in C(J, E_0) \cap C^1(\dot{J}, E_0) \cap C(\dot{J}, E_\gamma) .$$

As $u \in C(\dot{J}, E_1)$ by (II.1.2.3), it remains to show that $u \in C^1(\dot{J}, E_\gamma)$ and that u is a solution of (1.5.4) which satisfies $u \in C(J, E_\gamma)$ if $x \in E_\gamma$ and which is strict if $x \in \text{dom}\big(A_\gamma(0)\big)$.

Thanks to (1.5.5), Corollary I.1.4.3, and the compactness of J there is a constant ω_0 such that $\mathcal{A} := \{\omega_0 + A\}$ satisfies condition (1.3.1). By Remark II.2.1.2(e) we can replace A by $\omega_0 + A$ in the remainder of this proof, that is, we can assume that $\mathcal{A} := \{A\}$ satisfies condition (1.3.1).

By (1.5.2) there exists a decomposition $f = f_1 + f_2$ with $f_1 \in C^\varepsilon(J, E_\gamma)$ and $f_2 \in C(J, E_{\gamma+\varepsilon})$. It follows from Remark II.2.1.2(b) that

$$u = v_0 + v_1 + v_2 , \tag{1.5.6}$$

where

$$v_0 := U(\cdot, 0)x \quad \text{and} \quad v_j := U \star f_j(\cdot, 0) , \qquad j = 1, 2 .$$

Note that $v_0(\tau) \in E_1 \hookrightarrow E_\gamma$ and that

$$v_0(t) = U_\gamma(t, \tau) v_0(\tau) , \qquad 0 < \tau < t , \quad t \in J ,$$

where U_γ denotes the E_γ-realization of U. From this and from Theorem 1.4.1 it follows that v_0 is a solution of the parabolic equation

$$\dot{v} + A_\gamma(t) v = 0 , \qquad t \in \dot{J} ,$$

which belongs to $C(J, E_\gamma)$ if $x \in E_\gamma$. From Theorems 1.2.2 and 1.4.1 we also deduce that v_0 is a strict solution of the Cauchy problem

$$\dot{v} + A_\gamma(t) v = 0 , \qquad t \in \dot{J} , \qquad v(0) = x ,$$

provided $x \in \text{dom}\big(A_\gamma(0)\big)$.

Observe that v_j is the solution of the Cauchy problem

$$\dot{v} + A(t)v = f_j(t) \ , \quad t \in \dot{J} \ , \qquad v(0) = 0 \tag{1.5.7$_j$}$$

for $j = 1, 2$. Since $f_1 \in C^\varepsilon(J, E_\gamma) \hookrightarrow C^\varepsilon(J, E_0)$, it follows from Theorem II.1.2.1 that v_1 is a strict solution of $(1.5.7)_1$. Hence

$$\dot{v}_1(t) = f_1(t) - A(t)v_1(t) \ , \qquad t \in J \ . \tag{1.5.8}$$

From Lemma 1.3.1 and $f_1 \in C^\varepsilon(J, E_\gamma)$ we deduce that

$$\big\| A(t)U(t,\tau)\big[f_1(t) - f_1(\tau)\big] \big\|_{E_\gamma} \le c(t-\tau)^{\varepsilon - 1} \ , \qquad (t,\tau) \in J_\Delta^* \ ,$$

and

$$\big((t,\tau) \mapsto A(t)U(t,\tau)\big[f_1(t) - f_1(\tau)\big]\big) \in C(J_\Delta^*, E_\gamma) \ .$$

This implies that the first term on the right-hand side of

$$A(t)v_1(t) = \int_0^t A(t)U(t,\tau)\big[f_1(\tau) - f_1(t)\big]\, d\tau + A(t)\int_0^t U(t,\tau)f_1(t)\, d\tau \tag{1.5.9}$$

is a continuous function of $t \in J$ with values in E_γ. Let $t \in \dot{J}$ be fixed. Since $U \in \mathfrak{K}(E_\gamma, E_1, 1-\gamma)$ by Lemma 1.3.1 and (1.3.1), we see that

$$g_\xi(t) := \int_0^{t-\xi} U(t,\tau)f_1(t)\, d\tau \quad \longrightarrow \quad g(t) := \int_0^t U(t,\tau)f_1(t)\, d\tau$$

in E_1 as $\xi \to 0$ in $(0,t)$. Hence $A(t) \in \mathcal{L}(E_1, E_0)$ implies $A(t)g_\xi(t) \to A(t)g(t)$ in E_0. Note that (1.3.9), (1.3.12), (1.3.14), (1.3.17), and (1.3.19) imply

$$AU(t,\tau) = A(t)e^{-(t-\tau)A(t)} + B(t,\tau) = \partial_\tau e^{-(t-\tau)A(t)} + B(t,\tau) \ , \qquad (t,\tau) \in J_\Delta^* \ ,$$

where

$$B \in \mathfrak{K}\big(E_\gamma, 1 - \gamma \wedge (\rho - \gamma)\big) \ . \tag{1.5.10}$$

Hence

$$\begin{aligned} A(t)g_\xi(t) &= \int_0^{t-\xi} AU(t,\tau)f_1(t)\, d\tau \\ &= [e^{-\xi A(t)} - e^{-tA(t)}]f_1(t) + \int_0^{t-\xi} B(t,\tau)f_1(t)\, d\tau \ , \end{aligned}$$

which shows that

$$A(t)g_\xi(t) \to [1 - e^{-tA(t)}]f_1(t) + \int_0^t B(t,\tau)f_1(t)\, d\tau$$

in E_0. Thus

$$A(t)\int_0^t U(t,\tau)f_1(t)\,d\tau = [1-e^{-tA(t)}]f_1(t) + \int_0^t B(t,\tau)f_1(t)\,d\tau \tag{1.5.11}$$

for $t \in \dot{J}$. From Lemma II.4.2.1 and the density of E_1 in E_γ we easily deduce that $[t \mapsto e^{-tA(t)}] \in C\big(J, \mathcal{L}_s(E_\gamma)\big)$. Now we infer from (1.5.10) and (1.5.11) that

$$\Big(t \mapsto A(t)\int_0^t U(t,\tau)f_1(t)\,d\tau\Big) \in C(J,E_\gamma)\ .$$

Hence (1.5.8) and (1.5.9) imply $\dot{v}_1 \in C(J,E_\gamma)$. By Theorem 1.4.1,

$$U_\gamma \in C\big(J_\Delta, \mathcal{L}_s(E_\gamma)\big)\ ,$$

which trivially gives $v_j \in C(J,E_\gamma)$, $j=1,2$. Thus Proposition 1.1.3 guarantees that $v_1 \in C^1(J,E_\gamma)$.

We use the notations of the proof of Lemma 1.3.1. Then it follows from (1.3.19) that

$$AU = -\partial_1 a - I\ , \tag{1.5.12}$$

where $I := I_1 + I_2 + I_3$. Thanks to (1.3.9), (1.3.14), (1.3.17), and (II.3.1.2),

$$I \in \mathfrak{K}(E_\gamma, 1-\gamma)\ . \tag{1.5.13}$$

Moreover, from Lemma II.4.2.1 we easily deduce by interpolation that

$$\partial_1 a \in \mathfrak{K}(E_{\gamma+\varepsilon}, E_\gamma, 1-\varepsilon)\ . \tag{1.5.14}$$

Thus, by (1.5.12)–(1.5.14), (II.3.1.2), and $E_{\gamma+\varepsilon} \hookrightarrow E_\gamma$,

$$AU \in \mathfrak{K}(E_{\gamma+\varepsilon}, E_\gamma, 1-\gamma\wedge\varepsilon)\ .$$

Consequently,

$$A(t)v_2(t) = \int_0^t AU(t,\tau)f_2(\tau)\,d\tau\ , \qquad t \in J\ ,$$

and $Av_2 \in C(J,E_\gamma)$. Hence $f_2 - Av_2 \in C(J,E_\gamma)$ and $\dot{v}_2(t) = f_2(t) - A(t)v_2(t)$ for $t \in \dot{J}$, together with Proposition 1.1.3, guarantee that $v_2 \in C^1(J,E_\gamma)$. Now (1.5.6) and the above considerations imply that $u \in C^1(\dot{J},E_\gamma)$ and that $u \in C^1(J,E_\gamma)$ if $x \in \operatorname{dom}\big(A_\gamma(0)\big)$. This and

$$Au = f - \dot{u} \in C(\dot{J},E_\gamma)$$

show that $u(t) \in \operatorname{dom}\big(A_\gamma(t)\big)$ for $t \in \dot{J}$, so that u is a solution of (1.5.4), which is strict if $x \in \operatorname{dom}\big(A_\gamma(0)\big)$. ■

Theorem 1.5.1, which is due to the author (cf. [Ama87, Section 6]), is an important *regularity theorem*. It implies that the solution u of the evolution equation (1.5.3), that a priori belongs to the class $C(\dot{J}, E_1) \cap C^1(\dot{J}, E_0)$, in fact satisfies $u \in C(\dot{J}, E_\gamma)$ and $A_\gamma u = f - \dot{u} \in C(\dot{J}, E_\gamma)$. Since, in general, E_γ and, consequently, $\mathrm{dom}\big(A_\gamma(t)\big)$ are smaller spaces than E_0 and E_1, respectively, we see that, given the hypotheses of Theorem 1.4.1, the solution u of (1.5.3) possesses better 'spacial regularity properties' than the ones we could deduce in Chapter II. This regularizing effect will be the basis for 'abstract bootstrapping arguments' for quasilinear parabolic evolution equations showing that suitable 'weak solutions' are already classical ones.

2 Constant Interpolation Spaces

In the preceding section we proved the existence of a parabolic evolution operator and a solvability theorem for parabolic Cauchy problems in the case where the domains of the generators $A(t)$ may depend on t. This was done by 'lifting' a 'constant domain case' to a suitable interpolation space. These results are important for their regularity aspects, but they depend on the fact that we can 'start' with a 'constant domain case'. It is the purpose of this section to study linear parabolic Cauchy problems in the case where the domains of $A(t)$ may vary with t but a suitable interpolation space is constant.

2.1 Semigroup and Convergence Estimates

Let E be a Banach space. Given $M > 0$, we denote by

$$\mathfrak{A}_M := \mathfrak{A}_M(E)$$

the set of all densely defined linear operators A in E such that $[\mathrm{Re}\, z \geq 0] \subset \rho(-A)$ and

$$(1 + |\lambda|)\, \|(\lambda + A)^{-1}\|_{\mathcal{L}(E)} \leq M\ , \qquad \mathrm{Re}\, \lambda \geq 0\ . \tag{2.1.1}$$

Note that $\mathfrak{A}_M \subset \mathcal{C}(E)$.

Let $M > 0$ be arbitrarily fixed. It follows from Corollary II.6.1.3 that there exist $\vartheta := \vartheta(M) \in (0, \pi/2)$ and $K := K(M)$ such that $\Sigma_\vartheta \subset \rho(-A)$ and

$$\|\lambda(\lambda + A)^{-1}\|_{\mathcal{L}(E)} \leq K\ , \qquad \lambda \in \Sigma_\vartheta\ , \quad A \in \mathfrak{A}_M\ . \tag{2.1.2}$$

Each $A \in \mathfrak{A}_M$ is the negative infinitesimal generator of a strongly continuous analytic semigroup $\{\, e^{-tA}\ ;\ t \geq 0 \,\}$ on E, and

$$A^k e^{-tA} = \frac{1}{2\pi i} \int_\Gamma (-\lambda)^k e^{t\lambda} (\lambda + A)^{-1}\, d\lambda\ , \qquad t > 0\ , \quad k \in \mathbb{N}\ , \tag{2.1.3}$$

Γ being a piece-wise smooth simple curve running in $\dot{\Sigma}_\vartheta$ from $\infty e^{-i(\varphi+\pi/2)}$ to $\infty e^{i(\varphi+\pi/2)}$ for some $\varphi \in (0, \vartheta]$. Note that $\mathfrak{A}_M \subset \mathcal{P}_M(E)$, so that the fractional powers A^z are well-defined for $A \in \mathfrak{A}_M$ and $z \in \mathbb{C}$.

The following lemma gives an extension of (2.1.3) to fractional exponents.

2.1.1 Lemma *Given* $\alpha \in \mathbb{R}^+$,

$$A^\alpha e^{-tA} = \frac{1}{2\pi i} \int_\Gamma (-\lambda)^\alpha e^{t\lambda} (\lambda + A)^{-1} \, d\lambda \,, \qquad t > 0 \,, \quad A \in \mathfrak{A}_M \,. \tag{2.1.4}$$

Proof Fix $t > 0$ and $k \in \mathbb{N}$ with $\alpha < k$, and define holomorphic functions φ_j on $\mathbb{C} \backslash (-\mathbb{R}^+)$ by $\varphi_1(z) := z^{\alpha-k}$ and $\varphi_2(z) := z^k e^{-tz}$, respectively. Given $\delta \in (0, k-\alpha)$,

$$|z|^\delta \, \varphi_j(z) \to 0 \qquad \text{as} \quad |z| \to \infty \text{ in } [\,|\arg z| \leq \pi/2 - \varepsilon]$$

for each $\varepsilon \in (0, \pi/2)$. Thus

$$\varphi_j(A) := \frac{1}{2\pi i} \int_\Gamma \varphi_j(-\lambda)(\lambda + A)^{-1} \, d\lambda \,, \qquad j = 1, 2 \,,$$

are well-defined in $\mathcal{L}(E)$ for $A \in \mathfrak{A}_M$, and the proof of Lemma III.4.6.1 shows that $\varphi_1(A)\varphi_2(A) = \varphi_1\varphi_2(A)$. Also note that $\varphi_1(A) = A^{\alpha-k}$ by Theorem III.4.6.5, and $\varphi_2(A) = A^k e^{-tA}$ by (2.1.3). Now the assertion follows since $A^{\alpha-k}A^k = A^\alpha$ by Theorem III.4.6.5. ■

2.1.2 Corollary *Suppose that* $\alpha \in \mathbb{R}^+$. *Then*

$$(t \mapsto A^\alpha e^{-tA}) \in C\big(\dot{\mathbb{R}}^+, \mathcal{L}(E)\big)$$

and

$$\|A^\alpha e^{-tA}\|_{\mathcal{L}(E)} \leq c(\alpha, M) t^{-\alpha} \,, \qquad t > 0 \,, \quad A \in \mathfrak{A}_M \,.$$

Proof This is an obvious consequence of (2.1.2), (2.1.4), and Lemma II.4.1.1. ■

Given $A \in \mathfrak{A}_M$, let $\{\, A_\varepsilon \,;\, \varepsilon > 0 \,\}$ be the Yosida approximation of A. Recall from Lemma II.6.1.2 that

$$|\lambda| \, \|(\lambda + A_\varepsilon)^{-1}\| \leq M + 1 \,, \qquad \operatorname{Re}\lambda \geq 0 \,, \quad \varepsilon > 0 \,, \quad A \in \mathfrak{A}_M \,. \tag{2.1.5}$$

From this, (2.1.2), and Corollary II.6.1.3 we deduce that there exist $\vartheta \in (0, \pi/2)$ and $c > 0$ such that $\Sigma_\vartheta \subset \rho(-A) \cap \rho(-A_\varepsilon)$ and

$$\|\lambda(\lambda + A)^{-1}\|_{\mathcal{L}(E)} + \|\lambda(\lambda + A_\varepsilon)^{-1}\|_{\mathcal{L}(E)} \leq c \tag{2.1.6}$$

for $\lambda \in \Sigma_\vartheta$, $\varepsilon > 0$, and $A \in \mathfrak{A}_M$. Using these facts we now prove some simple but important estimates and convergence results.

2.1.3 Lemma *Given $A \in \mathfrak{A}_M$ and $\lambda \in \Sigma_\vartheta$,*

$$(\lambda + A_\varepsilon)^{-1} \to (\lambda + A)^{-1} \qquad \text{in } \mathcal{L}(E)$$

as $\varepsilon \to 0$. The convergence is uniform with respect to $A \in \mathfrak{A}_M$ and $\lambda \in \Sigma_\vartheta$.

Proof Note that (cf. (II.6.1.12))

$$(\lambda + A_\varepsilon)^{-1} = \frac{1}{1+\varepsilon\lambda}\Big[\varepsilon + \frac{1}{1+\varepsilon\lambda}\Big(\frac{\lambda}{1+\varepsilon\lambda} + A\Big)^{-1}\Big] \tag{2.1.7}$$

for $\lambda \in \Sigma_\vartheta,\ \varepsilon > 0$, and $A \in \mathfrak{A}_M$. Hence

$$\begin{aligned}(\lambda + A_\varepsilon)^{-1} - (\lambda + A)^{-1} &= \frac{\varepsilon}{1+\varepsilon\lambda} + \Big(\frac{1}{(1+\varepsilon\lambda)^2} - 1\Big)\Big(\frac{\lambda}{1+\varepsilon\lambda} + A\Big)^{-1} \\ &\quad + \Big(\lambda - \frac{\lambda}{1+\varepsilon\lambda}\Big)\Big(\frac{\lambda}{1+\varepsilon\lambda} + A\Big)^{-1}(\lambda + A)^{-1}\end{aligned}$$

and, consequently, thanks to (2.1.6),

$$\|(\lambda + A_\varepsilon)^{-1} - (\lambda + A)^{-1}\|_{\mathcal{L}(E)} \le \varepsilon c\Big(1 + \frac{1}{|1+\varepsilon\lambda|}\Big) .$$

Since $|1+\varepsilon\lambda| \ge \cos\vartheta > 0$ for $\lambda \in \Sigma_\vartheta$, the assertion follows. ∎

By means of this lemma it is now easy to prove the following convergence result:

2.1.4 Lemma *Given $\alpha \in \mathbb{R}^+$ and $A \in \mathfrak{A}_M$,*

$$A_\varepsilon^\alpha e^{-tA_\varepsilon} \to A^\alpha e^{-tA} \qquad \text{in } \mathcal{L}(E)$$

as $\varepsilon \to 0$, locally uniformly with respect to $t \in \dot{\mathbb{R}}^+$.

Proof From (2.1.5) and Lemma 2.1.3 we infer the existence of $\varepsilon_0 > 0$ such that

$$A_\varepsilon \in \mathfrak{A}_{M+2} , \qquad 0 < \varepsilon < \varepsilon_0 . \tag{2.1.8}$$

Thus Lemma 2.1.1 implies

$$A_\varepsilon^\alpha e^{-tA_\varepsilon} = \frac{1}{2\pi i}\int_\Gamma (-\lambda)^\alpha e^{t\lambda}(\lambda + A_\varepsilon)^{-1}\, d\lambda , \qquad t > 0 , \quad \varepsilon > 0 .$$

Hence the assertion follows from Lemma 2.1.3 and the dominated convergence theorem. ∎

2.2 Assumptions and Consequences

Throughout the remainder of this section we fix a Banach space $E_0 := (E_0, \|\cdot\|_0)$, a perfect subinterval J of $\mathbb{R}^+$ containing 0, and a positive constant M.

Given $A\colon J \to \mathfrak{A}_M(E_0)$, we put

$$E_1\bigl(A(s)\bigr) := \bigl(\mathrm{dom}(A(s)), \|A(s)\cdot\|_0\bigr)\ , \qquad s \in J\ . \tag{2.2.1}$$

Note that $\bigl(E_0, E_1(A(s))\bigr)$ is a densely injected Banach couple and that

$$A(s) \in \mathcal{H}\bigl(E_1(A(s)), E_0\bigr)\ , \qquad s \in J\ . \tag{2.2.2}$$

We assume that

$$\left.\begin{array}{ll}\text{(i)} & \mathcal{A} \text{ is a nonempty set of maps } A\colon J \to \mathfrak{A}_M(E_0).\\ \text{(ii)} & \text{There exist constants } \theta \in (0,1) \text{ and } \kappa \geq 1,\\ & \text{an admissible interpolation functor } (\cdot,\cdot)_\theta,\\ & \text{and a Banach space } E_\theta := (E_\theta, \|\cdot\|_\theta) \text{ such that}\\ & \qquad E_\theta \doteq \bigl(E_0, E_1(A(s))\bigr)_\theta =: E_\theta\bigl(A(s)\bigr)\\ & \text{and}\\ & \qquad \kappa^{-1}\|x\|_\theta \leq \|x\|_{E_\theta(A(s))} \leq \kappa\|x\|_\theta\ , \qquad x \in E_\theta\ ,\\ & \text{for } s \in J \text{ and } A \in \mathcal{A}.\\ \text{(iii)} & \text{There exist } \rho \in (1-\theta, 1) \text{ and } \eta \geq 0 \text{ such that}\\ & A^{-1} \in C^\rho\bigl(J, \mathcal{L}(E_0, E_\theta)\bigr) \text{ and } [A^{-1}]_{\rho,J} \leq \eta \text{ for } A \in \mathcal{A}.\end{array}\right\} \tag{2.2.3}$$

Observe that (2.2.3(ii)) guarantees that the interpolation space $E_\theta\bigl(A(s)\bigr)$ between E_0 and $D\bigl(A(s)\bigr)$ is independent of $s \in J$ and $A \in \mathcal{A}$, except for (uniformly) equivalent norms.

Now we deduce some consequences of assumption (2.2.3). First we recall from (2.1.6) that there exists $\vartheta \in (0, \pi/2)$ such that

$$\Sigma_\vartheta \subset \rho\bigl(-A(s)\bigr) \cap \rho\bigl(-A_\varepsilon(s)\bigr)\ , \qquad s \in J\ , \quad \varepsilon > 0\ , \quad A \in \mathcal{A}\ ,$$

and estimate (2.1.6) is valid for $A(s)$, uniformly with respect to $s \in J$ and $A \in \mathcal{A}$.

2.2.1 Lemma *Given $A \in \mathcal{A}$,*

$$\bigl(\lambda + A(s)\bigr)^{-1} \in \mathcal{L}(E_0, E_\theta) \quad \textit{and} \quad A(s)\bigl(\lambda + A(s)\bigr)^{-1} \in \mathcal{L}(E_\theta, E_0) \tag{2.2.4}$$

for $\lambda \in \Sigma_\vartheta$ and $s \in J$. Moreover,

$$|\lambda|^{1-\theta}\,\bigl\|\bigl(\lambda + A(s)\bigr)^{-1}\bigr\|_{\mathcal{L}(E_0,E_\theta)} + |\lambda|^\theta\,\bigl\|A(s)\bigl(\lambda + A(s)\bigr)^{-1}\bigr\|_{\mathcal{L}(E_\theta,E_0)} \leq c \tag{2.2.5}$$

for $\lambda \in \Sigma_\vartheta$, $s \in J$, *and* $A \in \mathcal{A}$. *In addition,*

$$|\lambda|^\theta \left\| \left(\lambda + A(r)\right)^{-1} - \left(\lambda + A(s)\right)^{-1} \right\|_{\mathcal{L}(E_0)} \le c\eta\, |r-s|^\rho \tag{2.2.6}$$

for $\lambda \in \Sigma_\vartheta$, $r, s \in J$, *and* $A \in \mathcal{A}$.

Proof Since

$$A(s)\left(\lambda + A(s)\right)^{-1} = 1 - \lambda\left(\lambda + A(s)\right)^{-1} , \tag{2.2.7}$$

we deduce from (2.1.6) that

$$|\lambda| \left\| \left(\lambda + A(s)\right)^{-1} \right\|_{\mathcal{L}(E_0)} + \left\| \left(\lambda + A(s)\right)^{-1} \right\|_{\mathcal{L}(E_0, E_1(A(s)))} \le c \tag{2.2.8}$$

and

$$\left\| A(s)\left(\lambda + A(s)\right)^{-1} \right\|_{\mathcal{L}(E_0)} + |\lambda| \left\| A(s)\left(\lambda + A(s)\right)^{-1} \right\|_{\mathcal{L}(E_1(A(s)), E_0)} \le c \tag{2.2.9}$$

for $\lambda \in \Sigma_\vartheta$, $s \in J$, and $A \in \mathcal{A}$. Hence (2.2.4) and (2.2.5) follow by interpolation, thanks to (2.2.3(ii)).

Note that

$$\begin{aligned}
&\left(\lambda + A(r)\right)^{-1} - \left(\lambda + A(s)\right)^{-1} \\
&\quad = \left(\lambda + A(r)\right)^{-1} A(s)\left(\lambda + A(s)\right)^{-1} - A(r)\left(\lambda + A(r)\right)^{-1}\left(\lambda + A(s)\right)^{-1} \\
&\quad = A(r)\left(\lambda + A(r)\right)^{-1}\left[A^{-1}(r) - A^{-1}(s)\right] A(s)\left(\lambda + A(s)\right)^{-1} .
\end{aligned}$$

Thus (2.2.6) is a consequence of (2.2.5), assumption (2.2.3(iii)), and (2.2.9). ∎

Given $s \in J$ and $A \in \mathcal{A}$, the fractional powers $A^\alpha(s)$ of $A(s)$ are well-defined for $\alpha \in \mathbb{R}$, and

$$\left(E_0, E_1(A(s))\right)_{\alpha,1} \hookrightarrow D\left(A^\alpha(s)\right) \hookrightarrow \left(E_0, E_1(A(s))\right)^0_{\alpha,\infty} , \qquad 0 < \alpha < 1 ,$$

by (I.2.9.6). Hence (I.2.11.4), (I.2.5.2), and (2.2.3(ii)) imply $E_\theta \hookrightarrow D\left(A^\alpha(s)\right)$ for $\alpha < \theta$. This shows that $A^\alpha(s) \in \mathcal{L}(E_\theta, E_0)$. The following lemma gives more precise information on these operators.

2.2.2 Lemma *Suppose that* $0 < \alpha < \theta$. *Then*

$$(A \mapsto A^\alpha) \in B\left(\mathcal{A}, B(J, \mathcal{L}(E_\theta, E_0))\right) . \tag{2.2.10}$$

Moreover, given $A \in \mathcal{A}$,

$$A^\alpha \in C\left(J, \mathcal{L}_s(E_\theta, E_0)\right)$$

and

$$A^\alpha_\varepsilon(s) \to A^\alpha(s) \qquad \text{in } \mathcal{L}_s(E_\theta, E_0) \tag{2.2.11}$$

as $\varepsilon \to 0$, *uniformly with respect to* $s \in J$.

Proof By Lemma 2.2.1 and (2.2.9),

$$\begin{aligned}\big\|t^\alpha\big(t+A(s)\big)^{-2}A(s)x\big\|_0 &= t^\alpha\,\big\|\big(t+A(s)\big)^{-1}A(s)\big(t+A(s)\big)^{-1}x\big\|_0 \\ &\le ct^{\alpha-\theta-1}\,\|x\|_\theta\end{aligned}\tag{2.2.12}$$

for $x\in E_1\big(A(s)\big)$, $s\in J$, $t>0$, and $A\in\mathcal{A}$. By Theorem III.4.6.5(ii)

$$A^\alpha(s)x=\frac{\sin\alpha\pi}{\alpha\pi}\int_0^\infty t^\alpha\big(t+A(s)\big)^{-2}A(s)x\,dt\ ,\qquad x\in E_1\big(A(s)\big)\ .\tag{2.2.13}$$

Thus we infer from (2.2.12) and $\|A^{-1}(s)\|_{\mathcal{L}(E_0)}\le c$ that $A^\alpha(s)\in\mathcal{L}(E_\theta,E_0)$, that (2.2.13) is true for $x\in E_\theta$, and, consequently, that

$$\|A^\alpha(s)\|_{\mathcal{L}(E_\theta,E_0)}\le c\ ,\qquad s\in J\ ,\quad A\in\mathcal{A}\ .\tag{2.2.14}$$

This proves (2.2.10).

It follows from (2.2.7) that, given $A\in\mathcal{A}$,

$$A_\varepsilon(s)\big(t+A_\varepsilon(s)\big)^{-1}-A(s)\big(t+A(s)\big)^{-1}=t\big[\big(t+A(s)\big)^{-1}-\big(t+A_\varepsilon(s)\big)^{-1}\big]\ .$$

Thus, thanks to (2.1.6),

$$\begin{aligned}&\big\|\big(t+A_\varepsilon(s)\big)^{-2}A_\varepsilon(s)-\big(t+A_\varepsilon(s)\big)^{-1}A(s)\big(t+A(s)\big)^{-1}\big\|_{\mathcal{L}(E_0)}\\ &\qquad\le c\,\big\|\big(t+A(s)\big)^{-1}-\big(t+A_\varepsilon(s)\big)^{-1}\big\|_{\mathcal{L}(E_0)}\end{aligned}$$

for $s\in J$ and $t\ge 0$. Similarly, by (2.2.9),

$$\begin{aligned}&\big\|\big[\big(t+A_\varepsilon(s)\big)^{-1}-\big(t+A(s)\big)^{-1}\big]A(s)\big(t+A(s)\big)^{-1}\big\|_{\mathcal{L}(E_0)}\\ &\qquad\le c\,\big\|\big(t+A_\varepsilon(s)\big)^{-1}-\big(t+A(s)\big)^{-1}\big\|_{\mathcal{L}(E_0)}\end{aligned}$$

for $s\in J$ and $t\ge 0$. Hence we deduce from Lemma 2.1.3 that, given $x\in E_1\big(A(s)\big)$,

$$t^\alpha\big(t+A_\varepsilon(s)\big)^{-2}A_\varepsilon(s)x\to t^\alpha\big(t+A(s)\big)^{-2}A(s)x$$

as $\varepsilon\to 0$, uniformly with respect to $s\in J$ and t in bounded subintervals of $\mathbb{R}^+$. Since, thanks to (2.1.8), estimate (2.2.12) holds for A_ε also, uniformly with respect to $0<\varepsilon<\varepsilon_0$, we infer from (2.2.13) and the corresponding formula for $A_\varepsilon^\alpha(s)x$ that (2.2.11) is true.

By (2.1.7),

$$A_\varepsilon^{-1}=\varepsilon+A^{-1}\ .\tag{2.2.15}$$

From this, $E_\theta\hookrightarrow E_0$, and (2.2.3(iii)) it follows that $A_\varepsilon^{-1}\in C^\rho\big(J,\mathcal{L}(E_0)\big)$. Thus $A_\varepsilon\in C^\rho\big(J,\mathcal{L}(E_0)\big)$ by the analyticity of the inversion map $B\mapsto B^{-1}$ from $\mathcal{L}\mathrm{aut}(E_0)$

into itself. This implies $A^\alpha_\varepsilon \in C\big(J,\mathcal{L}(E_0)\big)$; hence $A^\alpha_\varepsilon \in C\big(J,\mathcal{L}(E_\theta,E_0)\big)$. Thus the uniformity of the convergence in (2.2.11) ensures $A^\alpha \in C\big(J,\mathcal{L}_s(E_\theta,E_0)\big)$. ■

2.2.3 Corollary *Suppose that $0<\alpha<\theta$. Then*

$$\big[(r,s)\mapsto A^\alpha(r)\big(A^{-1}(r)-A^{-1}(s)\big)\big] \in C\big(J\times J,\mathcal{L}(E_0)\big)$$

and

$$\big\|A^\alpha(r)\big(A^{-1}(r)-A^{-1}(s)\big)\big\|_{\mathcal{L}(E_0)} \le c\eta\,|r-s|^\rho\ , \qquad r,s\in J\ , \quad A\in\mathcal{A}\ .$$

Moreover, given $A\in\mathcal{A}$,

$$A^\alpha_\varepsilon(r)\big(A^{-1}_\varepsilon(r)-A^{-1}_\varepsilon(s)\big) \to A^\alpha(r)\big(A^{-1}(r)-A^{-1}(s)\big) \qquad \text{in } \mathcal{L}_s(E_0)$$

as $\varepsilon\to 0$, uniformly with respect to $r,s\in J$.

Proof These assertions are immediate consequences of (2.2.3(iii)), Lemma 2.2.2, and (2.2.15). ■

Next we derive continuity properties of $A^\alpha(s)e^{-tA(s)}$ as a function of $s\in J$ and $t\in\mathbb{R}^+$.

2.2.4 Lemma *Given $\alpha\in\mathbb{R}^+$,*

$$\|A^\alpha(r)e^{-tA(r)}-A^\alpha(s)e^{-tA(s)}\|_{\mathcal{L}(E_0)} \le c\eta t^{\theta-\alpha-1}\,|r-s|^\rho \tag{2.2.16}$$

for $r,s\in J,\ \ t>0$, and $A\in\mathcal{A}$. Moreover,

$$\big((r,t)\mapsto A^\alpha(r)e^{-tA(r)}\big) \in C\big(J\times\dot{\mathbb{R}}^+,\mathcal{L}(E_0)\big)\ .$$

Proof By Lemma 2.1.1

$$A^\alpha(r)e^{-tA(r)}-A^\alpha(s)e^{-tA(s)} = \frac{1}{2\pi i}\int_\Gamma (-\lambda)^\alpha e^{t\lambda}\big[\big(\lambda+A(r)\big)^{-1}-\big(\lambda+A(s)\big)^{-1}\big]\,d\lambda\ .$$

Hence the first assertion follows from Lemmas 2.2.1 and II.4.1.1.

Given $r,s\in J$ and $0<\tau<t<\infty$,

$$\begin{aligned}&A^\alpha(r)e^{-tA(r)}-A^\alpha(s)e^{-\tau A(s)}\\ &\quad = A^\alpha(r)e^{-tA(r)}-A^\alpha(s)e^{-tA(s)}+A^\alpha(s)[e^{-tA(s)}-e^{-\tau A(s)}]\ .\end{aligned} \tag{2.2.17}$$

Since

$$A^\alpha(s)[e^{-tA(s)}-e^{-\tau A(s)}] = -\int_\tau^t A^{\alpha+1}(s)e^{-\sigma A(s)}\,d\sigma\ , \tag{2.2.18}$$

we infer from Corollary 2.1.2 that

$$\|A^{\alpha}(s)[e^{-tA(s)} - e^{-\tau A(s)}]\|_{\mathcal{L}(E_0)} \leq c(\tau^{-\alpha} - t^{-\alpha}) \tag{2.2.19}$$

for $s \in J$ and $0 < \tau < t < \infty$. Now the second assertion is a consequence of (2.2.16)–(2.2.19). ■

Lastly, we prove the strong continuity of $(s,t) \mapsto e^{-tA(s)}$ on $J \times \mathbb{R}^+$.

2.2.5 Lemma *Given $A \in \mathcal{A}$,*

$$\big((s,t) \mapsto e^{-tA(s)}\big) \in \mathcal{L}\big(J \times \mathbb{R}^+, \mathcal{L}_s(E_0)\big) .$$

Proof Thanks to Lemma 2.2.4 we know that the assertion is true if we replace $J \times \mathbb{R}^+$ by $J \times \dot{\mathbb{R}}^+$. Thus it remains to show that, given $s_0 \in J$,

$$e^{-tA(s)} \to 1 \qquad \text{in } \mathcal{L}_s(E_0) \tag{2.2.20}$$

as $(s,t) \mapsto (s_0, 0)$. Fix $\alpha \in (0,\theta)$ and note that (2.2.18) and Lemma 2.2.2 imply

$$e^{-tA(s)} - e^{-\tau A(s)} = -\int_\tau^t A^{1-\alpha}(s) e^{-\sigma A(s)} A^{\alpha}(s)\, d\sigma \qquad \text{in } \mathcal{L}(E_\theta, E_0)$$

for $s \in J$ and $0 < \tau < t < \infty$. From this, Corollary 2.1.2, and the strong continuity of the semigroup $\{\, e^{-tA(s)} \ ; \ t \geq 0 \,\}$ on E_0 we infer that

$$\|e^{-tA(s)}x - x\|_0 \leq ct^{\alpha}\, \|x\|_\theta \ , \qquad x \in E_\theta \ , \quad s \in J \ , \quad t > 0 \ .$$

Now the uniform boundedness of the semigroup $\{\, e^{-tA(s)} \ ; \ t \geq 0 \,\}$, $s \in J$, on E_0 and the density of E_θ in E_0 imply (2.2.20). ■

2.3 Construction of Evolution Operators

Suppose, for the moment, that $A \in C^{\rho}\big(J, \mathcal{L}(E_0)\big)$ for some $\rho \in (0,1)$. Then Remark I.1.2.1(b) and Corollary II.4.4.2 imply the existence of a unique parabolic evolution operator U_A for A, and U_A solves the integral equations (II.2.2.2) and (II.2.2.3). Also assume that $\operatorname{type}\big(-A(s)\big) < 0$ for $s \in J$. Then

$$\begin{aligned} -\big[A(t) - A(s)\big] &= A(t)\big[A^{-1}(t) - A^{-1}(s)\big]A(s) \\ &= A^{1-\alpha}(t)A^{\alpha}(t)\big[A^{-1}(t) - A^{-1}(s)\big]A(s) \ . \end{aligned} \tag{2.3.1}$$

Thus, by applying $A^{1-\alpha}(s)$ from the right to (II.2.2.2) and putting

$$a_A(t,s) := A^{1-\alpha}(s) e^{-(t-s)A(s)}$$

and

$$k_A(t,s) := A^\alpha(t)\big[A^{-1}(t) - A^{-1}(s)\big]A^{2-\alpha}(s)e^{-(t-s)A(s)}$$

for $(t,s) \in J_\Delta$, we see that $V_A := U_A A^{1-\alpha}$ solves the linear Volterra equation

$$V = a_A + V \star k_A \ . \tag{2.3.2}$$

Similarly, let

$$b_A(t,s) := \int_s^t A^2(t)e^{-(t-\tau)A(t)}\big[A^{-1}(\tau) - A^{-1}(t)\big]A(\tau)e^{-(\tau-s)A(\tau)}\,d\tau$$

and

$$h_A(t,s) := -A^{2-\alpha}(t)e^{-(t-s)A(t)}A^\alpha(t)\big[A^{-1}(t) - A^{-1}(s)\big]$$

for $(t,s) \in J$. Then, by applying $A(t)$ from the left to (II.2.2.3) and using (2.3.1), it is easily verified that

$$W_A(t,s) := A(t)U_A(t,s) - A(t)e^{-(t-s)A(t)} \tag{2.3.3}$$

solves the Volterra equation

$$W = b_A + h_A \star W \ . \tag{2.3.4}$$

Now suppose that condition (2.2.3) is satisfied and fix α such that

$$1 - \rho < \alpha < \theta \ . \tag{2.3.5}$$

We shall prove that, given $A \in \mathcal{A}$, equations (2.3.2) and (2.3.4) have unique solutions V_A and W_A, respectively. Then we put

$$U_A := V_A A^{\alpha-1} \tag{2.3.6}$$

and show that W_A, indeed, equals the right-hand side of (2.3.3) and that U_A is a parabolic evolution operator for A.

It follows from Corollary 2.1.2 and Lemma 2.2.4 that

$$(A \mapsto a_A) \in B\big(\mathcal{A}, \mathfrak{K}_\infty(E_0, 1-\alpha)\big) \ . \tag{2.3.7}$$

Similarly, Corollaries 2.1.2 and 2.2.3 and Lemma 2.2.4 imply

$$(A \mapsto k_A) \in B\big(\mathcal{A}, \mathfrak{K}_\infty(E_0, 2-\alpha-\rho)\big) \tag{2.3.8}$$

and that the norm of this map is bounded by $c\eta$. Observe that $2-\alpha-\rho < 1$ by (2.3.5). Thus Theorem II.3.2.2 guarantees the existence of a unique solution

$$V_A \in \mathfrak{K}_\infty(E_0, 1-\alpha) \tag{2.3.9}$$

of (2.3.2), and

$$V_A = a_A + a_A \star w_A \tag{2.3.10}$$

with $w_A \in \mathfrak{K}(E_0, 2-\alpha-\rho)$. Note that (II.3.2.2) implies

$$w_A = k_A + w_A \star k_A = k_A + k_A \star w_A\ . \tag{2.3.11}$$

Let

$$k_A^0(t,s) := k_A A^{\alpha-1}(t,s) = A^\alpha(t)\big[A^{-1}(t) - A^{-1}(s)\big]A(s)e^{-(t-s)A(s)}$$

and

$$w_A^0 := w_A A^{\alpha-1}\ . \tag{2.3.12}$$

Then, thanks to (2.2.3) and Corollaries 2.1.2 and 2.2.3,

$$(A \mapsto k_A^0) \in B\big(\mathcal{A}, \mathfrak{K}_\infty(E_0, 1-\rho)\big)\ . \tag{2.3.13}$$

From (2.3.8) and Lemma II.3.2.1 we infer that

$$(t-s)^{2-\alpha-\rho}\, \|w_A(t,s)\|_{\mathcal{L}(E_0)} \le c(\varepsilon)e^{(\nu+\varepsilon)(t-s)} \tag{2.3.14}$$

for $(t,s) \in J_\Delta^*$, $\varepsilon > 0$, and $A \in \mathcal{A}$, where

$$\nu := c_0\eta^{1/(\alpha+\rho-1)}$$

for some positive constant c_0. Note that (2.3.11)–(2.3.13) imply

$$w_A^0 = k_A^0 + w_A \star k_A^0\ . \tag{2.3.15}$$

Hence we deduce from (2.3.13), (2.3.14), and the results of Subsection II.3.1 that

$$w_A^0 \in \mathfrak{K}(E_0, 1-\rho) \tag{2.3.16}$$

and

$$(t-s)^{1-\rho}\, \|w_A^0(t,s)\|_{\mathcal{L}(E_0)} \le c(\varepsilon)e^{(\nu+\varepsilon)(t-s)}\ , \qquad (t,s) \in J_\Delta^*\ , \quad A \in \mathcal{A}\ . \tag{2.3.17}$$

Put

$$a_A^0(t,s) := e^{-(t-s)A(s)} = a_A A^{\alpha-1}(t,s)\ , \qquad (t,s) \in J_\Delta\ . \tag{2.3.18}$$

We infer from (2.3.10), (2.3.12), and (2.3.16) that U_A, defined by (2.3.6), is given by

$$U_A = a_A^0 + a_A \star w_A^0\ . \tag{2.3.19}$$

Moreover, (2.3.7), (2.3.16), (2.3.17), and Subsection II.3.1 imply

$$a_A \star w_A^0 \in \mathfrak{K}(E_0, 1-\alpha-\rho) \hookrightarrow C(J_\Delta, E_0)$$

and

$$(t-s)^{1-\alpha-\rho}\,\|a_A \star w_A^0(t,s)\|_{\mathcal{L}(E_0)} \le c(\varepsilon)e^{(\nu+\varepsilon)(t-s)}$$

for $(t,s) \in J_\Delta$, $\varepsilon > 0$, and $A \in \mathcal{A}$. Consequently, thanks to (2.3.18), (2.3.19), Corollary 2.1.2, and Lemma 2.2.5, we see that

$$U_A \in \big(J_\Delta, \mathcal{L}_s(E_0)\big) \cap \mathfrak{K}(E_0, 0) \tag{2.3.20}$$

and

$$\|U_A(t,s)\|_{\mathcal{L}(E_0)} \le c(\varepsilon)e^{(\nu+\varepsilon)(t-s)}\ , \qquad (t,s) \in J_\Delta\ , \quad \varepsilon > 0\ , \quad A \in \mathcal{A}\ . \tag{2.3.21}$$

Next we consider the Volterra equation (2.3.4). For this we need the following technical result:

2.3.1 Lemma *Given $A \in \mathcal{A}$,*

$$b_A \in \mathfrak{K}(E_0, 2-\alpha-\rho)\ .$$

Moreover,

$$(t-s)^{2-\alpha-\rho}\,\|b_A(t,s)\|_{\mathcal{L}(E_0)} \le c(\varepsilon)e^{\varepsilon(t-s)}\ , \qquad (t,s) \in J_\Delta^*\ , \quad \varepsilon > 0\ , \quad A \in \mathcal{A}\ .$$

Proof Given $(t,s) \in J_\Delta^*$, we put $r := (s+t)/2$ and decompose b_A as

$$b_A = I_A + II_A + III_A + IV_A\ ,$$

where

$$I_A(t,s) := \int_r^t A^2(t)e^{-(t-\tau)A(t)}\big[A^{-1}(\tau) - A^{-1}(t)\big]A(\tau)e^{-(\tau-s)A(\tau)}\,d\tau\ ,$$

$$II_A(t,s) := \int_s^r A^2(t)e^{-(t-\tau)A(t)}\big[e^{-(\tau-s)A(\tau)} - e^{-(\tau-s)A(s)}\big]\,d\tau\ ,$$

$$III_A(t,s) := \int_s^r A(t)e^{-(t-\tau)A(t)}\big[A(s)e^{-(\tau-s)A(s)} - A(\tau)e^{-(\tau-s)A(\tau)}\big]\,d\tau\ ,$$

and

$$IV_A(t,s) := \int_s^r A^2(t)e^{-(t-\tau)A(t)}\big[e^{-(\tau-s)A(s)} - A^{-1}(t)A(s)e^{-(\tau-s)A(s)}\big]\,d\tau\ .$$

We rewrite I_A as

$$I_A(t,s) = -\int_r^t A^{2-\alpha}(t)e^{-(t-\tau)A(t)}A^\alpha(t)\big[A^{-1}(t) - A^{-1}(\tau)\big]A(\tau)e^{-(\tau-s)A(\tau)}\,d\tau\ .$$

Then we deduce from Corollaries 2.1.2 and 2.2.3 that

$$\begin{aligned}\|I_A(t,s)\|_{\mathcal{L}(E_0)} &\le c(r-s)^{-1}\int_r^t (t-\tau)^{\alpha+\rho-2}\,d\tau = c(r-s)^{-1}(t-r)^{\alpha+\rho-1}\\ &= c(t-s)^{\alpha+\rho-2}\end{aligned}$$

for $(t,s)\in J_\Delta^*$ and $A\in\mathcal{A}$. Corollary 2.1.2 and Lemma 2.2.4 imply

$$\|II_A(t,s)\|_{\mathcal{L}(E_0)} \le c(t-r)^{-2}\int_s^r (\tau-s)^{\theta+\rho-1}\,d\tau = c(t-s)^{\theta+\rho-2}$$

for $(t,s)\in J_\Delta^*$ and $A\in\mathcal{A}$. Similarly, we see that

$$\|III_A(t,s)\|_{\mathcal{L}(E_0)} \le c(t-r)^{-1}\int_s^r (\tau-s)^{\theta+\rho-2}\,d\tau = c(t-s)^{\theta+\rho-2}$$

for $(t,s)\in J_\Delta^*$ and $A\in\mathcal{A}$. Lastly, define IV_A^ε for $0<\varepsilon<r-s$ by replacing the lower limit s in the integral occurring in the definition of IV_A by $s+\varepsilon$. Then, by integrating the first summand of IV_A^ε by parts,

$$IV_A^\varepsilon(t,s) = A(t)e^{-(t-r)A(t)}\left[e^{-(r-s)A(s)} - e^{-(r-s-\varepsilon)A(t)}e^{-\varepsilon A(s)}\right].$$

Since $IV_A^\varepsilon(t,s)\to IV_A(t,s)$ in $\mathcal{L}_s(E_0)$ as $\varepsilon\to 0$, we see that

$$IV_A(t,s) = A(t)e^{-(t-r)A(t)}\left[e^{-(r-s)A(s)} - e^{-(r-s)A(t)}\right].$$

Hence, thanks to Corollary 2.1.2 and Lemma 2.2.4,

$$\|IV_A(t,s)\|_{\mathcal{L}(E_0)} \le c\,|t-s|^{\theta+\rho-2}$$

for $(t,s)\in J_\Delta^*$ and $A\in\mathcal{A}$. Now the assertion is obvious. ■

Finally, we are ready for the proof of the first main result of this section:

2.3.2 Theorem *Let assumption* (2.2.3) *be satisfied. Then, given $A\in\mathcal{A}$, there exists a unique parabolic evolution operator U_A for A. It possesses E_θ as regularity subspace. Moreover, the estimates*

$$\|U_A(t,s)\|_{\mathcal{L}(E_0)} + (t-s)\,\|A(t)U_A(t,s)\|_{\mathcal{L}(E_0)} \le c(\varepsilon)e^{(\nu+\varepsilon)(t-s)} \tag{2.3.22}$$

are valid for $(t,s)\in J_\Delta^*$, $\varepsilon>0$, and $A\in\mathcal{A}$, where

$$\nu := \nu(\alpha,\eta) := c_0\eta^{1/(\alpha+\rho-1)}$$

for some $\alpha\in(1-\rho,\theta)$ and c_0 is a positive constant.

Proof We shall show that the function U_A, defined by (2.3.6), has the desired properties.

First we note that Corollary 2.1.2 and Lemma 2.2.4 imply

$$h_A \in B\big(\mathcal{A}, \mathfrak{K}_\infty(E_0, 2-\alpha-\rho)\big) \; . \tag{2.3.23}$$

Hence, by Lemma 2.3.1 and Theorem II.3.2.2, the Volterra equation (2.3.4) possesses a unique solution $W_A \in \mathfrak{K}(E_0, 2-\alpha-\rho)$ and it is given by

$$W_A = b_A + r_A \star b_A \; , \tag{2.3.24}$$

where

$$r_A := \sum_{j=1}^{\infty} \underbrace{h_A \star \cdots \star h_A}_{j} \in \mathfrak{K}(E_0, 2-\alpha-\rho) \; .$$

It follows from (2.3.23) and Lemma II.3.2.1 that

$$(t-s)^{2-\alpha-\rho} \left\| r_A(t,s) \right\|_{\mathcal{L}(E_0)} \leq c(\varepsilon) e^{(\nu+\varepsilon)(t-s)} \tag{2.3.25}$$

for $(t,s) \in J_\Delta^*$, $\varepsilon > 0$, and $A \in \mathcal{A}$. This estimate, Lemma 2.3.1, formula (2.3.24), and the results of Section II.3.1 imply that $W_A \in \mathfrak{K}(E_0, 2-\alpha-\rho)$, and

$$(t-s)^{2-\alpha-\rho} \left\| W_A(t,s) \right\|_{\mathcal{L}(E_0)} \leq c(\varepsilon) e^{(\nu+\varepsilon)(t-s)} \tag{2.3.26}$$

for $(t,s) \in J_\Delta^*$, $\varepsilon > 0$, and $A \in \mathcal{A}$.

Now let $A \in \mathcal{A}$ be fixed, and let $\{\, A_\varepsilon \; ; \; \varepsilon > 0 \,\}$ be the Yosida approximation of $A \in \mathcal{A}$. Then $A_\varepsilon^{-1} \in C^\rho\big(J, \mathcal{L}(E_0)\big)$ by (2.2.15), and, consequently, $A_\varepsilon \in C^\rho\big(J, \mathcal{L}(E_0)\big)$ by (2.3.1). Hence all the considerations at the beginning of this subsection apply to A_ε. In particular (cf. (2.3.19)), the unique parabolic evolution operator U_ε of A_ε is given by

$$U_\varepsilon = a_\varepsilon^0 + a_\varepsilon \star w_\varepsilon^0 \; , \tag{2.3.27}$$

where we write a_ε^0 for $a_{A_\varepsilon}^0$, etc., for simplicity.

From Corollaries 2.1.2 and 2.2.3 we infer that

$$(\varepsilon \mapsto k_\varepsilon) \in B\big(\dot{\mathbb{R}}^+, \mathfrak{K}_\infty(E_0, 2-\alpha-\rho)\big) \; . \tag{2.3.28}$$

Moreover, given $(t,s) \in J_\Delta^*$, Lemma 2.1.4 and Corollary 2.2.3 imply

$$k_\varepsilon(t,s) \to k_A(t,s) \qquad \text{in } \mathcal{L}_s(E_0)$$

as $\varepsilon \to 0$. Now Lebesgue's dominated convergence theorem and an induction argument guarantee that, given $(t,s) \in J_\Delta^*$ and $j \geq 2$,

$$\underbrace{k_\varepsilon \star \cdots \star k_\varepsilon}_{j}(t,s) \to \underbrace{k_A \star \cdots \star k_A}_{j}(t,s) \qquad \text{in } \mathcal{L}_s(E_0)$$

as $\varepsilon \to 0$. Moreover, from (2.3.28) we deduce that $k_\varepsilon \star \cdots \star k_\varepsilon(t,s)$ satisfies an estimate of the form (II.3.2.1), independently of $\varepsilon > 0$. Thus, by applying the dominated convergence theorem (in $\ell_1(\mathbb{N}, \mathcal{L}(E_0))$) to the series (II.3.2.2), we infer that, given $(t,s) \in J_\Delta^*$,

$$w_\varepsilon(t,s) \to w_A(t,s) \qquad \text{in } \mathcal{L}_s(E_0) \tag{2.3.29}$$

as $\varepsilon \to 0$. Since $w_\varepsilon^0 = k_\varepsilon^0 + w_\varepsilon \star k_\varepsilon^0$ by (2.3.15), we see that $w_\varepsilon^0(t,s) \to w_A^0(t,s)$ in $\mathcal{L}_s(E_0)$ for $(t,s) \in J_\Delta^*$. Now Lemma 2.1.4 and (2.3.13) imply that, given $(t,s) \in J_\Delta^*$,

$$U_\varepsilon(t,s) \to U_A(t,s) \qquad \text{in } \mathcal{L}_s(E_0)\ . \tag{2.3.30}$$

Put $b_\varepsilon := b_{A_\varepsilon}$ and $h_\varepsilon := h_{A_\varepsilon}$, respectively, and let W_ε be the unique solution of the Volterra equation $W = b_\varepsilon + h_\varepsilon \star W$ for $\varepsilon > 0$. Then, by similar arguments as above, given $(t,s) \in J_\Delta^*$,

$$W_\varepsilon(t,s) \to W_A(t,s) \qquad \text{in } \mathcal{L}_s(E_0) \tag{2.3.31}$$

as $\varepsilon \to 0$. Since, by (2.3.3),

$$W_\varepsilon(t,s) = A_\varepsilon U_\varepsilon(t,s) - A_\varepsilon(t) e^{-(t-s)A_\varepsilon(t)}\ ,$$

it follows from (2.3.31) and Lemma 2.1.4 that, given $(t,s) \in J_\Delta^*$,

$$A_\varepsilon U_\varepsilon(t,s) = A(t)\big(1 + \varepsilon A(t)\big)^{-1} U_\varepsilon(t,s) \to W_A(t,s) + A(t) e^{-(t-s)A(t)} \tag{2.3.32}$$

in $\mathcal{L}_s(E_0)$ as $\varepsilon \to 0$. Note that, thanks to Lemma II.6.1.1 and (2.3.30),

$$\big(1 + \varepsilon A(t)\big)^{-1} U_\varepsilon(t,s) \to U_A(t,s) \qquad \text{in } \mathcal{L}_s(E_0)$$

as $\varepsilon \to 0$. Thus we infer from (2.3.32) and the closedness of $A(t)$ that

$$\operatorname{im}\big(U_A(t,s)\big) \subset \operatorname{dom}\big(A(t)\big)\ , \qquad (t,s) \in J_\Delta^*\ , \tag{2.3.33}$$

and that

$$AU_A(t,s) = W_A(t,s) + A(t) e^{-(t-s)A(t)}\ , \qquad (t,s) \in J_\Delta^*\ . \tag{2.3.34}$$

Hence Corollary 2.1.2 and (2.3.26) imply

$$AU_A \in \mathfrak{K}(E_0, 1) \tag{2.3.35}$$

and

$$(t-s)\, \|AU_A(t,s)\|_{\mathcal{L}(E_0)} \le c(\varepsilon) e^{(\nu+\varepsilon)(t-s)} \tag{2.3.36}$$

for $(t,s) \in J_\Delta^*$, $\varepsilon > 0$, and $A \in \mathcal{A}$.

Observe that

$$U_\varepsilon(t,s) - U_\varepsilon(\tau,s) = -\int_\tau^t A_\varepsilon U_\varepsilon(\sigma,s)\,d\sigma \qquad \text{in } \mathcal{L}(E_0)$$

for $0 \le s < \tau < t$ and $t \in J$. From this, (2.3.30), (2.3.32), and (2.3.36) we infer that

$$U_A(t,s) - U_A(\tau,s) = -\int_\tau^t AU_A(\sigma,s)\,d\sigma \qquad \text{in } \mathcal{L}_s(E_0)$$

for $0 \le s < \tau < t$ and $t \in J$. Hence

$$U_A(t+h,s) - U_A(t,s) + hAU_A(t,s) = \int_t^{t+h} \big[AU_A(t,s) - AU_A(\tau,s)\big]\,d\tau$$

for $(t,s) \in J_\Delta^*$ and $h > 0$ with $t + h \in J$, where the integral converges in $\mathcal{L}(E_0)$ thanks to (2.3.35). Now we deduce from Proposition 1.1.2 that

$$U_A(\cdot,s) \in C^1\big(J \cap (s,\infty), \mathcal{L}(E_0)\big)\ , \qquad s \in J\ , \tag{2.3.37}$$

and that

$$\partial_1 U_A(t,s) = -A(t)U_A(t,s)\ , \qquad (t,s) \in J_\Delta^*\ . \tag{2.3.38}$$

Since U_ε is the parabolic evolution operator for A_ε and since U_ε has the regularity subspace E_0, by Corollary II.4.4.2, we know that, given $x \in E_0$,

$$\begin{aligned} U_\varepsilon(t,s+h)x - U_\varepsilon(t,s)x &= \int_s^{s+h} U_\varepsilon(t,\tau)A_\varepsilon(\tau)x\,d\tau \\ &= \int_s^{s+h} U_\varepsilon(t,\tau)A_\varepsilon^{1-\alpha}(\tau)A_\varepsilon^\alpha(\tau)x\,d\tau \end{aligned} \tag{2.3.39}$$

for $(t,s) \in J_\Delta^*$ and $h > 0$ with $s + h < t$. Note that $U_\varepsilon A_\varepsilon^{1-\alpha} = V_\varepsilon$, where, thanks to (2.3.10),

$$V_\varepsilon = a_\varepsilon + a_\varepsilon \star w_\varepsilon\ , \qquad \varepsilon > 0\ .$$

From Lemma 2.1.4 and (2.3.29) we easily infer that, given $(t,s) \in J_\Delta$,

$$V_\varepsilon(t,s) \to V_A(t,s) \qquad \text{in } \mathcal{L}_s(E_0)$$

as $\varepsilon \to 0$. Thus (2.3.6), (2.3.39), and Lemma 2.2.2 imply

$$U_A(t,s+h)x - U_A(t,s)x = \int_s^{s+h} V_A(t,\tau)A^\alpha(\tau)x\,d\tau\ , \qquad x \in E_\theta\ ,$$

for $(t,s) \in J_\Delta^*$ and $h > 0$ with $s + h < t$. Since the integrand is a continuous function of τ, by (2.3.9) and (2.2.10), we see, by again invoking Proposition 1.1.2, that

$$U_A(t,\cdot) \in C^1\big([0,t), \mathcal{L}_s(E_\theta,E_0)\big)\ , \qquad t \in \dot{J}\ , \tag{2.3.40}$$

and that

$$\partial_2 U_A(t,s)x = V_A(t,s)A^\alpha(s)x\ , \qquad (t,s)\in J^*_\Delta\ , \quad x\in E_\theta\ . \tag{2.3.41}$$

If $x\in \mathrm{dom}\big(A(s)\big)$ then (2.3.6), (2.3.41), and Theorem III.4.6.5(iii) imply

$$V_A(t,s)A^\alpha(s)x = U_A(t,s)A(s)x\ .$$

Hence

$$U_A A\ . \tag{2.3.42}$$

Next, from (2.3.20), (2.3.35), and (2.3.36) we infer by interpolation that

$$U_A \in C\big(J^*_\Delta, \mathcal{L}(E_0,E_\theta)\big) \tag{2.3.43}$$

and that

$$(t-s)^\theta \, \|U_A(t,s)\|_{\mathcal{L}(E_0,E_\theta)} \le c(\varepsilon)e^{(\nu+\varepsilon)(t-s)} \tag{2.3.44}$$

for $(t,s)\in J^*_\Delta$, $\varepsilon>0$, and $A\in\mathcal{A}$.

Finally, note that (2.3.30) and the fact that U_ε is an evolution operator for $\varepsilon>0$ imply the validity of (II.2.1.3). Thus, thanks to (2.3.20), (2.3.21), (2.3.33), (2.3.35)–(2.3.38), and (2.3.40)–(2.3.43), the theorem has been completely proven. ■

Parabolic evolution equations with 'nonconstant domains' have been studied by several authors under a variety of different hypotheses. In several papers it has been assumed that an intermediate space between E_0 and $D\big(A(t)\big)$ is independent of t. The earliest investigations of this type are due to Kato [Kat61] and Sobolevskii [Sob61]. These authors assume that the domain of a suitable fractional power of $A(t)$ is constant. Such an assumption restricts the applicability of these results considerably since, in general — in particular, if $A(t)$ is induced by elliptic boundary value problems with coefficients having little regularity — it is rather difficult (or even not known how) to characterize these domains.

In [AcT86a] it is assumed that the interpolation spaces $\big(E_0, D(A(t))\big)_{\theta,\infty}$ are independent of t for some $\theta\in(0,1)$, except for equivalent norms. In that paper the authors do not construct an evolution operator but study the solvability of the Cauchy problem directly by means of suitable representation formulas (cf., however, [AcT86b], [Acq88]). This technique has already been used by the same authors in the case of constant domains in [AcT85]. In [AcT87] they succeeded to prove by means of their method the solvability of the Cauchy problem under hypotheses encompassing and generalizing most of the previously known results.

The theorems of [AcT87] have been further generalized by Yagi [Yag90], who weakened the assumption that $-A(t)$ should generate an analytic semigroup. In addition, he proposed a method of proof that is different from the one of Acquistapace and Terreni and closer to the one of Kato in [Kat61], also used in earlier papers by Yagi (e.g., [Yag88], [Yag89]). Our proof of Theorem 2.3.2 follows

[Yag90] inasmuch as we employ Volterra equations (2.3.2) and (2.3.4) and the representation of b_A in the proof of Lemma 2.3.1. It should be mentioned, however, that Theorem 2.3.2 seems to be new in this form and the given precision. It has the advantage that we do not make any assumption on the interpolation space E_θ, except that it is obtained by means of an admissible interpolation functor. This greatly facilitates its application to concrete problems.

It should be noted that Theorem 2.3.2 contains Theorem II.4.4.1 as a special case (except for slightly different exponential bounds). In fact, if assumption (II.4.2.1) is satisfied, it is easily verified that (2.3.1) implies (2.2.3) for any $\theta \in (1-\rho, 1)$. Thus the above considerations provide a new (and more complicated) proof of Theorem II.4.4.1.

Theorem 2.3.2 is closely related to Theorem 1.4.1. In fact — using the notations of the latter theorem — we see that E_1 is a constant *intermediate* space between E_γ and $D\big(A_\gamma(t)\big)$, that is, in general however, *not an interpolation* space. Thus, in general, Theorem 1.4.1 cannot be deduced from Theorem 2.3.2. Of course, conversely, we cannot obtain Theorem 2.3.2 from Theorem 1.4.1 since, usually, we do not have operators with constant domains to start with. In later chapters we shall see that each one of these theorems has its own range of applications.

The hypotheses in [AcT87] and [Yag90] are flexible enough to admit cases where no intermediate space between E_0 and $D\big(A(t)\big)$ is constant. In fact, those authors basically assume (2.2.3(i)) and

$$\big\|A(t)\big(\lambda+A(t)\big)^{-1}\big[A^{-1}(t)-A^{-1}(s)\big]\big\|_{\mathcal{L}(E_0)} \leq c\,|t-s|^{\rho}\,(1+|\lambda|)^{-\sigma} \tag{2.3.45}$$

for $s, t \in J$ and $\lambda \in \Sigma_\vartheta$, where $\rho, \sigma \in (0,1]$ satisfy $\sigma+\rho>1$. It is easily verified that (2.2.3(ii)) and (2.2.3(iii)) imply (2.3.45) with $\sigma=\theta$ (cf. (2.2.5)). On the other hand, if we do not assume that there exists a constant interpolation (or fractional power) space between E_0 and $D\big(A(t)\big)$, it can be shown that, given (2.2.3(i)), condition (2.3.45) is satisfied if A^{-1} is suitably continuously differentiable (cf. [AcT87, Section 7]). Moreover, given appropriate differentiability assumptions for A^{-1}, the existence of an evolution operator has first been shown by Kato and Tanabe [KaT62] (also cf. [Tan79], [Fat83] for expositions of these results, and [Yag76], [Yag77] for related investigations). It turns out, however, that differentiability assumptions for A^{-1} are too restrictive to be useful for dealing with *quasilinear* parabolic problems that — after all — are our main concern. For this reason we do not pursue this matter further.

2.4 Estimates for Evolution Operators

We begin by proving an additional continuity result for the parabolic evolution operators U_A.

2.4.1 Theorem *Let assumption* (2.2.3) *be satisfied. Then*

$$AU_A A^{-1} \in C\big(J_\Delta, \mathcal{L}_s(E_0)\big) \cap \mathfrak{K}(E_0, 0) , \qquad A \in \mathcal{A} ,$$

and, letting ν *have the same meaning as in Theorem* 2.3.2,

$$\|AU_A A^{-1}(t,s)\|_{\mathcal{L}(E_0)} \le c(\varepsilon) e^{(\nu+\varepsilon)(t-s)} \tag{2.4.1}$$

for $(t,s) \in J_\Delta$, $\varepsilon > 0$, *and* $A \in \mathcal{A}$.

Proof Thanks to (2.3.34),

$$AU_A A^{-1} = W_A A^{-1} + A\widehat{a}_A A^{-1} , \tag{2.4.2}$$

where $\widehat{a}_A(t,s) := e^{-(t-s)A(t)}$. Fixing $\alpha \in (1-\rho, \theta)$,

$$A\widehat{a}_A A^{-1}(t,s) = A^{1-\alpha}(t) e^{-(t-s)A(t)} A^{\alpha}(t)\big[A^{-1}(s) - A^{-1}(t)\big] + \widehat{a}_A(t,s) . \tag{2.4.3}$$

Hence Corollaries 2.1.2 and 2.2.3 imply

$$\|A\widehat{a}_A A^{-1}(t,s)\|_{\mathcal{L}(E_0)} \le c\big[(t-s)^{\alpha+\rho-1} + 1\big] , \qquad (t,s) \in J_\Delta , \quad A \in \mathcal{A} . \tag{2.4.4}$$

Moreover,

$$\begin{aligned} &A(t)\widehat{a}_A(t,s)A^{-1}(s) - A(\tau)\widehat{a}_A(\tau,\sigma)A^{-1}(\sigma) \\ &\quad = \big[A(t)e^{-(t-s)A(t)} - A(\tau)e^{-(\tau-\sigma)A(\tau)}\big]A^{-1}(s) \\ &\qquad + A(\tau)e^{-(\tau-\sigma)A(\tau)}\big[A^{-1}(s) - A^{-1}(\sigma)\big] , \end{aligned}$$

assumption (2.2.3(iii)), $E_\theta \hookrightarrow E_0$, and Lemma 2.2.4 guarantee that

$$A\widehat{a}_A A^{-1} \in C\big(J_\Delta^*, \mathcal{L}(E_0)\big) . \tag{2.4.5}$$

From (2.4.3), Corollaries 2.1.2 and 2.2.3, and Lemma 2.2.5 we see that, given $s_0 \in J$,

$$A\widehat{a}_A A^{-1}(t,s) \to 1 \qquad \text{in } \mathcal{L}_s(E_0)$$

as $(t,s) \to (s_0, s_0)$ in J_Δ. Thus

$$A\widehat{a}_A A^{-1} \in C\big(J_\Delta, \mathcal{L}_s(E_0)\big) . \tag{2.4.6}$$

Let

$$c_{A,j}(t,s) := A^{j-\alpha}(t) e^{-(t-s)A(t)} A^{\alpha}(t)\big[A^{-1}(s) - A^{-1}(t)\big]$$

for $j = 1, 2$, $(t,s) \in J_\Delta^*$, and $A \in \mathcal{A}$. Then

$$(A \mapsto c_{A,j}) \in B\big(\mathcal{A}, \mathfrak{K}_\infty(E_0, j-\alpha-\rho)\big) , \qquad j = 1, 2 , \tag{2.4.7}$$

by Corollaries 2.1.2 and 2.2.3. It is easily verified that

$$b_A A^{-1} = c_{A,2} \star c_{A,1} + c_{A,2} \star \widehat{a}_A , \qquad A \in \mathcal{A} .$$

Hence we deduce from (2.4.7), Corollary 2.1.2, and Section II.3.1 that

$$b_A A^{-1} \in \mathfrak{K}(E_0, 1-\alpha-\rho) \hookrightarrow C(J_\Delta, E_0)$$

and

$$(t-s)^{1-\alpha-\rho}\, \|b_A A^{-1}(t,s)\|_{\mathcal{L}(E_0)} \le c(\varepsilon) e^{\varepsilon(t-s)} , \qquad (t,s) \in J_\Delta^* , \quad A \in \mathcal{A} .$$

Thus (2.3.24) and (2.3.25) imply

$$W_A A^{-1} = b_A A^{-1} + (r_A \star b_A) A^{-1} = b_A A^{-1} + r_A \star (b_A A^{-1})$$

and, consequently,

$$W_A A^{-1} \in \mathfrak{K}(E_0, 1-\alpha-\rho) \hookrightarrow C\big(J_\Delta, \mathcal{L}(E_0)\big) \tag{2.4.8}$$

and

$$(t-s)^{1-\alpha-\rho}\, \|W_A A^{-1}(t,s)\|_{\mathcal{L}(E_0)} \le c(\varepsilon) e^{(\nu+\varepsilon)(t-s)} \tag{2.4.9}$$

for $(t,s) \in J_\Delta^*$, $\varepsilon > 0$, and $A \in \mathcal{A}$. Now the assertion follows from (2.4.2)–(2.4.6), (2.4.8), and (2.4.9). ■

2.4.2 Corollary *Let assumption* (2.2.3) *be satisfied. Given* $A \in \mathcal{A}$,

$$U_A \in \mathfrak{K}(E_0, E_\theta, \theta) \cap \mathfrak{K}(E_\theta, 0) \tag{2.4.10}$$

and

$$A U_A \in \mathfrak{K}(E_\theta, E_0, 1-\theta) . \tag{2.4.11}$$

Moreover,

$$\begin{aligned} \|U_A(t,s)\|_{\mathcal{L}(E_\theta)} &+ (t-s)^\theta\, \|U_A(t,s)\|_{\mathcal{L}(E_0,E_\theta)} \\ &+ (t-s)^{1-\theta}\, \|A U_A(t,s)\|_{\mathcal{L}(E_\theta,E_0)} \le c(\varepsilon) e^{(\nu+\varepsilon)(t-s)} \end{aligned} \tag{2.4.12}$$

for $(t,s) \in J_\Delta^*$, $\varepsilon > 0$, *and* $A \in \mathcal{A}$, *where* ν *has the same meaning as in Theorem* 2.3.2. *In addition,*

$$U_A(\cdot, s) \in C\big(J \cap [s,\infty), \mathcal{L}_s((E_0, E_\theta)_\xi)\big) , \qquad s \in J , \quad A \in \mathcal{A} , \tag{2.4.13}$$

for each admissible interpolation functor $(\cdot,\cdot)_\xi$.

Proof Observe that

$$\begin{aligned} \|U_A(t,s)\|_{\mathcal{L}(E_1(A(s)),E_1(A(t)))} &= \|A U_A(t,s)\|_{\mathcal{L}(E_1(A(s)),E_0)} \\ &= \|A U_A A^{-1}(t,s)\|_{\mathcal{L}(E_0)} \end{aligned} \tag{2.4.14}$$

and that

$$\|U_A(t,s)\|_{\mathcal{L}(E_0,E_1(A(t)))} = \|A(t) U_A(t,s)\|_{\mathcal{L}(E_0)}$$

for $(t,s) \in J_\Delta^*$ and $A \in \mathcal{A}$. Now we obtain (2.4.10)–(2.4.12) easily from Theorems 2.3.2 and 2.4.1 by interpolation (also cf. (2.3.43) and (2.3.44)).

To prove (2.4.13) we can assume that J is compact. Then, given $s \in J$ and $x \in E_\theta$,

$$\begin{aligned} &\|U_A(t,s)x - U_A(\tau,s)x\|_{(E_0,E_\theta)_\xi} \\ &\qquad \leq c\,\|U_A(t,s)x - U_A(\tau,s)x\|_\theta^\xi\,\|U_A(t,s)x - U_A(\tau,s)x\|_0^{1-\xi} \\ &\qquad \leq c\,\|x\|_\theta\,\|U_A(t,s)x - U_A(\tau,s)x\|_0^{1-\xi} \end{aligned}$$

for $t,\tau \in J \cap [s,\infty)$, as follows from (2.4.10). From (2.3.22) and (2.4.10) we deduce that $U_A \in \mathfrak{K}\big((E_0,E_\theta)_\xi,0\big)$. Now (2.4.13) is a consequence of $U_A \in C\big(J_\Delta,\mathcal{L}_s(E_0)\big)$ and the density of E_θ in $(E_0,E_\theta)_\xi$. ■

Lastly, we turn to the case of ordered Banach spaces and study the question of positivity for evolution operators.

2.4.3 Theorem *Let E_0 be an ordered Banach space and let assumption* (2.2.3) *be satisfied. Given $A \in \mathcal{A}$, suppose that $A(s)$ is resolvent positive for $s \in J$. Then U_A is positive.*

Proof Let $\{\, A_\varepsilon \;;\; \varepsilon > 0 \,\}$ be the Yosida approximation for A and let U_ε be the evolution operator for A_ε. Then, letting X be the positive cone of E_0, the proof of Theorem II.6.3.3 shows that U_ε is positive for $\varepsilon > 0$. Now the assertion follows from (2.3.30) and the closedness of the positive cone. ■

Recall from Theorem II.6.3.2 that $A(s)$ is resolvent positive iff the semigroup $\{\, e^{-tA(s)} \;;\; t \geq 0 \,\}$ is positive.

2.5 The Cauchy Problem

Let assumption (2.2.3) be satisfied. Let $(\cdot,\cdot)_\gamma$ be an admissible interpolation functor for some $\gamma \in (0,1)$ and put

$$F := (E_0,E_\theta)_\gamma \;.$$

Given $(x,A) \in E_0 \times \mathcal{A}$ and $f : J \to E_0$, consider the linear Cauchy problem

$$\dot{u} + A(t)u = f(t) \;, \quad t \in \dot{J} \;, \qquad u(0) = x \;. \tag{2.5.1$_{(x,A,f)}$}$$

The following solvability theorem is the second main result of this section.

2.5.1 Theorem *Given $(x,A) \in E_0 \times \mathcal{A}$ and*

$$f \in C^\sigma(J,E_0) + C(J,F) \tag{2.5.2}$$

for some $\sigma \in (0,1)$, the Cauchy problem $(2.5.1)_{(x,A,f)}$ has a unique solution

$$u \in C(J,E_0) \cap C^1(\dot{J},E_0) \;.$$

If $(\cdot,\cdot)_\xi$ *is an admissible interpolation functor and* $x \in (E_0, E_\theta)_\xi$ *then*

$$u \in C\big(J, (E_0, E_\theta)_\xi\big) ,$$

and u *is a strict solution if* $x \in \mathrm{dom}\big(A(0)\big)$.

Proof Thanks to Theorem 2.3.2 and Remark II.2.1.2(e) we have to show that the mild solution

$$u := U(\cdot,0)x + U \star f(\cdot,0) ,$$

where $U := U_A$, is a solution with the asserted regularity properties.

Put $u_0 := U(\cdot,0)x$ and

$$u_j := U \star f_j(\cdot,0) , \qquad j = 1,2 ,$$

where $f = f_1 + f_2$ with $f_1 \in C^\sigma(J, E_0)$ and $f_2 \in C(J, F)$. Then u_0 is a solution of $(2.5.1)_{(x,A,0)}$ by Theorem 2.3.2 and the properties of parabolic evolution operators. Theorem 2.4.1 implies that u_0 is strict if $x \in \mathrm{dom}\big(A(0)\big)$. If $x \in (E_0, E_\theta)_\xi$, it follows from (2.4.13) that $u_0 \in C\big(J, (E_0, E_\theta)_\xi\big)$.

Observe that u_j is a mild solution of $(2.5.1)_{(0,A,f_j)}$ for $j = 1, 2$. If we show that u_j is, in fact, a solution with the asserted continuity properties, the assertion follows from $u = u_0 + u_1 + u_2$.

(i) Let $\{ A_\varepsilon \; ; \; \varepsilon > 0 \}$ be the Yosida approximation for A, and let U_ε be the evolution operator for A_ε. Then $v_\varepsilon := U_\varepsilon \star f_1(\cdot,0)$ is a solution of $(2.5.1)_{(0,A_\varepsilon,f_1)}$. From (2.1.5), (2.3.21), and (2.3.30) it easily follows that, given $t \in J$,

$$v_\varepsilon(t) \to u_1(t) \qquad \text{in } E_0 \tag{2.5.3}$$

as $\varepsilon \to 0$. Moreover, (2.3.3) implies that

$$\begin{aligned} A_\varepsilon(t)v_\varepsilon(t) &= \int_0^t A_\varepsilon U_\varepsilon(t,\tau)\big[f_1(\tau) - f_1(t)\big]\,d\tau + \int_0^t A_\varepsilon U_\varepsilon(t,\tau) f_1(t)\,d\tau \\ &= \int_0^t A_\varepsilon U_\varepsilon(t,\tau)\big[f_1(\tau) - f_1(t)\big]\,d\tau + \int_0^t W_\varepsilon(t,\tau) f_1(t)\,d\tau \\ &\quad + [1 - e^{-tA_\varepsilon(t)}] f_1(t) \end{aligned} \tag{2.5.4}$$

for $t \in J$ and $\varepsilon > 0$. Observe that

$$A_\varepsilon U_\varepsilon(t,\tau) = W_\varepsilon(t,\tau) + A_\varepsilon(t) e^{-(t-\tau)A_\varepsilon(t)} , \qquad (t,\tau) \in J_\Delta^* .$$

Hence (2.3.31), (2.3.34), and Lemma 2.1.4 imply that, given $(t,\tau) \in J_\Delta^*$,

$$A_\varepsilon U_\varepsilon(t,\tau) \to AU(t,\tau) \qquad \text{in } \mathcal{L}_s(E_0)$$

as $\varepsilon \to 0$. By (2.1.5), (2.3.36), and $f_1 \in C^\sigma(J, E_0)$ we see that, given $t \in \dot{J}$,

$$\big\| A_\varepsilon U_\varepsilon(t,\tau)\big[f_1(\tau) - f_1(t)\big]\big\|_0 \le c(t-\tau)^{\sigma-1} , \qquad 0 \le \tau < t .$$

From this, (2.5.4), (2.3.31), and Lemma 2.1.4 we easily deduce that, given $t \in J$,

$$A_\varepsilon(t) v_\varepsilon(t) \to w(t) \qquad \text{in } E_0 \tag{2.5.5}$$

as $\varepsilon \to 0$, where

$$w(t) = \int_0^t AU(t,\tau)\big[f_1(\tau) - f_1(t)\big]\,d\tau + \int_0^t W(t,\tau) f_1(t)\,d\tau + [1 - e^{-tA(t)}] f_1(t) .$$

It is clear that $w \in C(J, E_0)$. Moreover, (2.5.3), (2.5.5), Lemma II.6.1.1, and the closedness of $A(t)$ guarantee that $u_1(t) \in \operatorname{dom}\big(A(t)\big)$ and $A(t)u_1(t) = w(t)$. From $(2.5.1)_{(0,A_\varepsilon,f_1)}$ we derive

$$v_\varepsilon(t) - v_\varepsilon(s) = \int_s^t \big[f_1(\tau) - A_\varepsilon(\tau) v_\varepsilon(\tau)\big]\,d\tau , \qquad (t,s) \in J_\Delta^* .$$

Thus, letting $\varepsilon \to 0$, it is easily verified that

$$u_1(t) - u_1(s) = \int_s^t \big[f_1(\tau) - w(\tau)\big]\,d\tau , \qquad (t,s) \in J_\Delta .$$

Hence $u_1 \in C^1(J, E_0)$ and

$$\dot{u}_1 = f_1 - w = f_1 - Au_1 ,$$

so that u_1 is a strict solution of $(2.5.1)_{(0,A,f_1)}$. Lastly, since $U \in \mathfrak{K}(E_0, E_\theta, \theta)$ by Corollary 2.4.2, it follows that

$$u_1 = U \star f_1(\cdot, 0) \in C(J, E_\theta) \hookrightarrow C\big(J, (E_0, E_\theta)_\xi\big) .$$

(ii) We can assume that J is compact. Then, by interpolating estimates (2.3.22) and (2.4.12),

$$(t-s)^{(1-\gamma)\theta} \|U(t,s)\|_{\mathcal{L}(F,E_\theta)} + (t-s)^{1-\gamma\theta} \|AU(t,s)\|_{\mathcal{L}(F,E_0)} \le c$$

for $(t,s) \in J_\Delta^*$. Hence

$$\|U(t,s)\|_{\mathcal{L}(F,E_\theta)} + \|AU(t,s)\|_{\mathcal{L}(F,E_0)} \le c(t-s)^{(\gamma-1)\theta} , \qquad (t,s) \in J_\Delta^* .$$

Now Theorems 1.2.1 and 1.2.2 guarantee that u_2 is a strict solution of the Cauchy problem $(2.5.1)_{(0,A,f_2)}$. Similarly as for u_1, we see that $u_2 \in C(J, E_\theta)$. This proves the theorem. ■

The fact that Hölder continuity of f implies the solvability of the Cauchy problem $(2.5.1)_{(x,A,f)}$ is well-known in the literature (e.g., [AcT87], [Yag89]). Results which are related to our assumption, namely, that f is continuous with values in the interpolation space F, can be found in [AcT87], where also singularities of f at $t = 0$ are admitted.

Of course, by using the estimates of Theorem 2.3.2 and Subsection 2.4, it is not difficult to give bounds for the solution of $(2.5.1)_{(x,A,f)}$ that are uniform with respect to $A \in \mathcal{A}$.

2.6 Abstract Boundary Value Problems

In this subsection we introduce a relatively simple class of hypotheses which imply assumption (2.2.3). These new conditions are motivated by the theory of elliptic boundary value problems that are discussed in Chapter VII. There it is shown how the present abstract hypotheses are met by suitable realizations of elliptic systems under rather general boundary conditions. In this subsection we illustrate the abstract theory by means of a simple second-order elliptic boundary value problem.

Let (E_0, E_1) be a densely injected Banach couple and let F be a Banach space. Given $\kappa \geq 1$, $\omega > 0$, and an increasing function $\beta \colon \dot{\mathbb{R}}^+ \to \dot{\mathbb{R}}^+$, we denote by

$$\mathfrak{B}(E_1, E_0, F, \kappa, \omega, \beta)$$

the set of all

$$(\mathcal{A}, \mathcal{B}) \in \mathcal{L}(E_1, E_0 \times F) \tag{2.6.1}$$

such that

$$(\omega + \mathcal{A}, \mathcal{B}) \in \mathcal{L}\mathrm{is}(E_1, E_0 \times F) \ , \tag{2.6.2}$$

that

$$|\lambda| \, \|x\|_0 + \|x\|_1 \leq \kappa \big(\|(\lambda + \mathcal{A})x\|_0 + \beta(|\lambda|) \, \|\mathcal{B}x\|_F \big) \ , \tag{2.6.3}$$

and that

$$\|\mathcal{A}x\|_0 + \|\mathcal{B}x\|_F \leq \kappa \, \|x\|_1 \tag{2.6.4}$$

for $x \in E_1$ and $\operatorname{Re} \lambda \geq \omega$. We also put

$$\mathfrak{B}(E_1, E_0, F, \kappa, \omega) := \bigcup_{\beta} \mathfrak{B}(E_1, E_0, F, \kappa, \omega, \beta)$$

and

$$\mathfrak{B}(E_1, E_0, F; \beta) := \bigcup_{\substack{\kappa \geq 1 \\ \omega > 0}} \mathfrak{B}(E_1, E_0, F, \kappa, \omega, \beta) \ ,$$

as well as

$$\mathfrak{B}(E_1, E_0, F) := \bigcup_{\beta} \mathfrak{B}(E_1, E_0, F; \beta) \ .$$

Observe that

$$\mathfrak{B}(E_1, E_0, F, \kappa, \omega, \beta) \subset \mathfrak{B}(E_1, E_0, F, \kappa, \omega) \subset \mathfrak{B}(E_1, E_0, F) \subset \mathcal{L}(E_1, E_0 \times F) \ ,$$

so that the Banach space $\mathcal{L}(E_1, E_0 \times F)$ induces the natural topologies on these subsets.

For definiteness we use ℓ_1-norms in product Banach spaces. If it is convenient, we also identify E_0 with $E_0 \times \{0\}$ etc.

2.6.1 Remarks **(a)** If $F = \{0\}$ (and, consequently, $\mathcal{B} = 0$) then

$$\mathfrak{B}(E_1, E_0, F, \kappa, \omega, \beta) = \mathcal{H}(E_1, E_0, \kappa, \omega) \ .$$

(b) Suppose that $(\mathcal{A}, \mathcal{B})$ satisfies (2.6.1), (2.6.2), and

$$|\lambda| \, \|x\|_0 \le \kappa\big(\|(\lambda + \mathcal{A})x\|_0 + \beta(|\lambda|) \, \|\mathcal{B}x\|_F\big) \ , \qquad x \in E_1 \ , \quad \operatorname{Re}\lambda \ge \omega \ .$$

Then

$$\|x\|_1 \le \mu(1 + 2\kappa) \, \|(\lambda + \mathcal{A})x\|_0 + \mu\big(1 + 2\kappa\beta(|\lambda|)\big) \, \|\mathcal{B}x\|_F$$

for $x \in E_1$ and $\operatorname{Re}\lambda \ge \omega$, where

$$\mu := \|(\omega + \mathcal{A}, \mathcal{B})^{-1}\|_{\mathcal{L}(E_0 \times F, E_1)} \ .$$

Proof This is an easy consequence of the identity

$$x = (\omega + \mathcal{A}, \mathcal{B})^{-1}(\lambda + \mathcal{A} + \omega - \lambda, \mathcal{B})x \ , \qquad x \in E_1 \ , \quad \lambda \in \mathbb{C} \ ,$$

and the fact that $|\lambda| \ge \omega$ for $\operatorname{Re}\lambda \ge \omega$. ■

(c) If $(\mathcal{A}, \mathcal{B}) \in \mathfrak{B}(E_1, E_0, F, \kappa, \omega, \beta)$ then

$$(\lambda + \mathcal{A}, \mathcal{B}) \in \mathcal{L}\mathrm{is}(E_1, E_0 \times F) \ , \qquad \operatorname{Re}\lambda \ge \omega \ .$$

Proof Given any $r > \omega$, it follows from (2.6.3), thanks to the monotonicity of β, that

$$\|(\lambda + \mathcal{A}, \mathcal{B})x\|_{E_0 \times F} \ge c \, \|x\|_1 \ , \qquad x \in E_1 \ , \quad \operatorname{Re}\lambda \ge \omega \ , \quad |\lambda| \le r \ ,$$

where $c := 1/\big(\kappa(1 \vee \beta(r))\big)$. Hence (2.6.2), Proposition I.1.1.1, and the arbitrariness of $r > \omega$ imply the assertion. ■

(d) If $(\mathcal{A}, \mathcal{B}) \in \mathfrak{B}(E_1, E_0, F, \kappa, \omega, \beta)$ then

$$\|(\lambda + \mathcal{A}, \mathcal{B})^{-1}\|_{\mathcal{L}(E_0 \times F, E_j)} \leq |\lambda|^{j-1} \kappa\big(1 \vee \beta(|\lambda|)\big)$$

for $\operatorname{Re}\lambda \geq \omega$ and $j = 0, 1$.

Proof This is a consequence of (2.6.3) and (c). ■

Suppose that $(\mathcal{A}, \mathcal{B}) \in \mathfrak{B}(E_1, E_0, F)$. Then

$$E_{1,\mathcal{B}} := \ker \mathcal{B}$$

is a closed linear subspace of E_1. We also put

$$A := A_{\mathcal{B}} := \mathcal{A} | E_{1,\mathcal{B}} \ . \tag{2.6.5}$$

It follows from Remark 2.6.1(c) that

$$\lambda + A \in \mathcal{L}\mathrm{is}(E_{1,\mathcal{B}}, E_0) \ , \qquad \operatorname{Re}\lambda \geq \omega \ . \tag{2.6.6}$$

Moreover,

$$A \in \mathcal{H}(E_{1,\mathcal{B}}, E_0, \kappa, \omega) \ , \tag{2.6.7}$$

provided $E_{1,\mathcal{B}}$ is dense in E_0. Lastly, we set

$$\|x\|_{E_{1,\mathcal{B}}} := \|(\omega + A)x\|_0 \ , \qquad x \in E_{1,\mathcal{B}} \ . \tag{2.6.8}$$

Then we infer from (2.6.3) and (2.6.4) that

$$\kappa_1^{-1} \|x\|_1 \leq \|x\|_{E_{1,\mathcal{B}}} \leq \kappa_1 \|x\|_1 \ , \qquad x \in E_{1,\mathcal{B}} \ , \tag{2.6.9}$$

where $\kappa_1 := \kappa + \omega \|i\|_{\mathcal{L}(E_1, E_0)}$ and $i \colon E_1 \hookrightarrow E_0$ is the injection map.

We also put

$$E^1_{1,\mathcal{B}} := (E_{1,\mathcal{B}}, \|\cdot\|_1)$$

and assume that

$$\left.\begin{array}{c}\text{there exist } \theta \in (0,1) \text{ and an admissible} \\ \text{interpolation functor } (\cdot,\cdot)_\theta \text{ such that} \\ (E_0, E^1_{1,\mathcal{B}})_\theta = (E_0, E_1)_\theta \ . \end{array}\right\} \tag{2.6.10}$$

The following lemma shows that this assumption implies that the interpolation spaces

$$E_{\theta,\mathcal{B}} := (E_0, E_{1,\mathcal{B}})_\theta \tag{2.6.11}$$

and

$$E_\theta := (E_\theta, \|\cdot\|_\theta) := (E_0, E_1)_\theta \tag{2.6.12}$$

coincide, except for equivalent norms, that is, $E_{\theta,\mathcal{B}}$ is independent of $\mathcal{B}$, except for equivalent norms.

2.6.2 Lemma *Given $\kappa \geq 1$ and $\omega > 0$, there exists a constant $\kappa_\theta \geq 1$ such that*

$$\kappa_\theta^{-1} \|x\|_\theta \leq \|x\|_{E_{\theta,\mathcal{B}}} \leq \kappa_\theta \|x\|_\theta , \qquad x \in E_\theta ,$$

for $(\mathcal{A}, \mathcal{B}) \in \mathfrak{B}(E_1, E_0, F, \kappa, \omega)$.

Proof From (2.6.9) we infer that

$$\mathrm{id} \in \mathcal{L}\mathrm{is}(E_{1,\mathcal{B}}, E_{1,\mathcal{B}}^1) \cap \mathcal{L}\mathrm{is}(E_0)$$

and that the norms of $\mathrm{id} \in \mathcal{L}(E_{1,\mathcal{B}}, E_{1,\mathcal{B}}^1)$ and $\mathrm{id} \in \mathcal{L}(E_{1,\mathcal{B}}^1, E_{1,\mathcal{B}})$ are bounded by κ_1. Now the assertion follows by interpolation from (2.6.10). ■

In order to prove the main result of this subsection we introduce the following assumptions:

$$\left.\begin{array}{ll}
\text{(i)} & (E_0, E_1) \text{ is a densely injected Banach couple, } F \text{ is a Banach} \\
& \text{space, and } J \text{ is a perfect subinterval of } \mathbb{R}^+ \text{ containing } 0. \\
\text{(ii)} & \kappa \geq 1,\ \ \omega, \eta_0 \in \dot{\mathbb{R}}^+,\ \ \rho \in (0,1), \text{ and} \\
& \qquad (\mathcal{A}, \mathcal{B}) \in C^\rho\big(J, \mathfrak{B}(E_1, E_0, F, \kappa, \omega)\big) \\
& \text{with} \\
& \qquad \big[(\mathcal{A}, \mathcal{B})\big]_{\rho,J} \leq \eta_0 . \\
\text{(iii)} & E_1(s) := E_{1,\mathcal{B}(s)} = \ker \mathcal{B}(s) \text{ is dense in } E_0 \text{ for } s \in J. \\
\text{(iv)} & 1 - \rho < \theta < 1 \text{ and} \\
& \qquad \big(E_0, (E_1(s), \|\cdot\|_1)\big)_\theta = (E_0, E_1)_\theta , \qquad s \in J .
\end{array}\right\} \tag{2.6.13}$$

Then we put

$$A(s) := \mathcal{A}(s) | E_1(s) , \qquad s \in J , \tag{2.6.14}$$

and

$$A \colon J \to \mathcal{C}(E_0) , \quad s \mapsto A(s) \tag{2.6.15}$$

and consider the linear Cauchy problem

$$\dot{u} + A(t)u = f(t) , \quad t \in \dot{J} , \qquad u(0) = x , \tag{2.6.16}$$

where $x \in E_0$ and $f : J \to E_0$. Note that this means that we study the evolution equation

$$\dot{u} + \mathcal{A}(t)u = f(t) , \quad t \in \dot{J} , \qquad u(0) = x , \tag{2.6.17}$$

subject to the side condition

$$\mathcal{B}(t)u(t) = 0 , \qquad t \in \dot{J} . \tag{2.6.18}$$

Since in practical applications the side condition is most often induced by suitable boundary conditions, $(\mathcal{A}, \mathcal{B})$ is said to be an abstract **boundary value problem** (BVP) whenever $(\mathcal{A}, \mathcal{B}) \in \mathfrak{B}(E_1, E_0, F)$. Then (2.6.17), (2.6.18) is an abstract **initial-boundary value problem** (IBVP) with homogeneous boundary conditions.

It may be worthwhile to illustrate assumption (2.6.13) by means of a prototype problem already now, although we have to rely on results that will be proven in later chapters only. Thus in the following example we use, without further explanations, facts from the theory of elliptic BVPs and the theory of function spaces and refer the reader to Chapters VI and VII for notations, definitions, and proofs.

2.6.3 Example: Second Order Elliptic BVPs We denote by X a bounded domain in $\mathbb{R}^n$ of class C^2 and fix $\delta \in C(\partial X, \{0,1\})$. Then we put

$$\Gamma_j := \delta^{-1}(j) = \{\, x \in \partial\Omega \ ;\ \delta(x) = j \,\} , \qquad j = 0, 1 .$$

Note that Γ_j is a union of components of ∂X, hence an open and closed submanifold of ∂X. We fix $p \in (1, \infty)$ and observe that

$$\bigl(L_p(X), W_p^2(X)\bigr)$$

is a densely injected Banach couple. We also put

$$\partial W_p^2(X) := W_p^{2-1/p}(\Gamma_0) \oplus W_p^{1-1/p}(\Gamma_1) ,$$

with obvious modifications if Γ_0 [resp. Γ_1] is empty, in which case Γ_1 [resp. Γ_0] equals ∂X.

Set

$$\mathbb{BVP}(X) := C(\overline{X})^{n^2} \times L_\infty(X)^n \times L_\infty(X) \times C^1(\Gamma_1)^n \times C^1(\Gamma_1) \times C^2(\Gamma_0) .$$

Given

$$\bigl((a_{jk}), (a_j), a_0, (b_j), b_0, c_0\bigr) \in \mathbb{BVP}(X) ,$$

we define a second order (scalar) differential operator $\mathcal{A}$ and a first order (scalar) boundary operator $\mathcal{B}$ on X by

$$\mathcal{A}u := a_{jk}\partial_j\partial_k u + a_j\partial_j u + a_0 u$$

and

$$\mathcal{B}u := \delta(b_j\gamma_\partial\partial_j u + b_0\gamma_\partial u) + (1-\delta)c_0\gamma_\partial u ,$$

respectively, where γ_∂ is the trace operator for ∂X. Here and in the following, we use the summation convention, j and k running from 1 to n. Then $(\mathcal{A}, \mathcal{B})$ is a **linear BVP on** X of order at most two. Observe that $\mathcal{B}$ reduces to the Dirichlet boundary operator on Γ_0, whereas it is a Neumann type boundary operator on Γ_1.

We topologize the set of these BVPs by identifying them with the Banach space $\mathbb{BVP}(X)$ by means of the identification

$$(\mathcal{A},\mathcal{B}) \quad\longleftrightarrow\quad \big((a_{jk}),(a_j),a_0,(b_j),b_0,c_0\big)\ .$$

We denote by $\nu := (\nu^1,\ldots,\nu^n)$ the outer unit normal vector field on ∂X. Given positive constants $\underline{\alpha}$ and M, we write

$$\mathbb{EBVP}(X,\underline{\alpha},M)$$

for the subset of $\mathbb{BVP}(X)$ consisting of all regularly elliptic BVPs on X of class $(\underline{\alpha},M)$. By definition, $(\mathcal{A},\mathcal{B})\in\mathbb{BVP}(X)$ is a **regularly elliptic BVP of class** $(\underline{\alpha},M)$ if

$$a_{jk}=a_{kj}\ ,\quad 1\le j,k\le n\ ,\quad \text{and}\quad a_{jk}(x)\xi^j\xi^k\ge\underline{\alpha}\,|\xi|^2\ ,\quad x\in\overline{X}\ ,\quad \xi\in\mathbb{R}^n\ ,$$

if

$$b_j(x)\nu^j(x)\ge\underline{\alpha}\ ,\quad x\in\Gamma_1\ ,\quad\text{and}\quad c_0(x)\ge\underline{\alpha}\ ,\quad x\in\Gamma_0\ ,$$

and if

$$\|(\mathcal{A},\mathcal{B})\|_{\mathbb{BVP}(X)}\le M\ .$$

It follows from the results proven in Chapter VII that, given positive constants $\underline{\alpha}$ and M, there exist $\kappa\ge 1$ and $\omega>0$ such that

$$\mathbb{EBVP}(X,\underline{\alpha},M)\hookrightarrow\mathfrak{B}\big(W_p^2(X),L_p(X),\partial W_p^2(X),\kappa,\omega,\beta\big)\ ,\tag{2.6.19}$$

where

$$\beta(t):=\begin{cases} t^{(1-1/p)/2}\ , & 0<t<1\ ,\\ t^{(2-1/p)/2}\ , & 1\le t<\infty\ .\end{cases}$$

Now we put

$$(E_0,E_1):=\big(L_p(X),W^2_{p,(1-\delta)\gamma_\partial}(X)\big)\ ,\tag{2.6.20}$$

where

$$W^2_{p,(1-\delta)\gamma_\partial}(X):=\big\{u\in W_p^2(X)\ ;\ (1-\delta)\gamma_\partial u=0\big\}$$

is a closed linear subspace of $W_p^2(X)$. Then (E_0,E_1) is a densely injected Banach couple. We also put

$$F:=W_p^{1-1/p}(\Gamma_1)\ .$$

Then it follows from (2.6.19) that, given $\underline{\alpha},M\in\dot{\mathbb{R}}^+$, there exist $\kappa\ge1$ and $\omega>0$ such that

$$\mathbb{EBVP}(X,\underline{\alpha},M)\hookrightarrow\mathfrak{B}(E_1,E_0,F,\kappa,\omega,\beta)$$

(by a slight abuse of notation). Hence assumption (2.6.13(ii)) is satisfied if there exist $\rho\in(0,1)$ and $\eta\in\mathbb{R}^+$ such that

$$(\mathcal{A},\mathcal{B})\in C^\rho\big(J,\mathbb{EBVP}(X,\underline{\alpha},M)\big)\tag{2.6.21}$$

with

$$\big[(\mathcal{A},\mathcal{B})\big]_{C^\rho(J,\mathbb{BVP}(X))} \le \eta \, .$$

Note that

$$\mathcal{D}(X) \hookrightarrow W^2_{p,\mathcal{B}}(X) := \big\{\, u \in W^2_p(X) \ ; \ \mathcal{B}u = 0 \,\big\}$$

and that

$$E_{1,\mathcal{B}} = W^2_{p,\mathcal{B}}(X) \ , \qquad (\mathcal{A},\mathcal{B}) \in \mathbb{BVP}(X) \, . \tag{2.6.22}$$

Hence $E_{1,\mathcal{B}}$ is dense in E_0 for $(\mathcal{A},\mathcal{B}) \in \mathbb{BVP}(X)$. Lastly, it follows from the results of Chapter VI that

$$\big(L_p(X), W^2_{p,\mathcal{B}}(X)\big)_\theta = (L_p, W^2_{p,(1-\delta)\gamma_\partial})_\theta \ , \qquad 0 < 2\theta < 1 + 1/p \ , \quad 2\theta \ne 1/p \ ,$$

provided $(\cdot,\cdot)_\theta := (\cdot,\cdot)_{\theta,p}$ if $\theta \ne 1/2$, and $(\cdot,\cdot)_{1/2} := [\cdot,\cdot]_{1/2}$. Thus we see from (2.6.20) and (2.6.22) that assumption (2.6.13(iv)) is satisfied if (2.6.21) is true with $2\rho > 1 - 1/p$ and if we fix $2\theta \in (2-2\rho, 1+1/p)$ with $2\theta \ne 1/p$. ■

Now it is easy to prove the main result of this subsection.

2.6.4 Theorem *Let assumption* (2.6.13) *be satisfied. Then there exists a unique parabolic evolution operator* $U := U_A$ *for* A. *It has* E_θ *as regularity subspace and satisfies*

$$(t-s)^{1-j}\, \|U(t,s)\|_{\mathcal{L}(E_0,E_j)} \le c(\varepsilon) e^{(\omega+\nu+\varepsilon)(t-s)} \tag{2.6.23}$$

for $(t,s) \in J^*_\Delta$, $\varepsilon > 0$, *and* $j = 0,1$, *where* $\nu := c_0 \eta_0^{1/(\alpha+\rho-1)}$ *for suitable constants* c_0 *and* $\alpha \in (1-\rho, \theta)$. *Given an increasing* $\beta : \dot{\mathbb{R}}^+ \to \dot{\mathbb{R}}^+$, *estimate* (2.6.23) *is uniform with respect to*

$$(\mathcal{A},\mathcal{B}) \in C^\rho\big(J, \mathfrak{B}(E_1,E_0,F,\kappa,\omega,\beta)\big) \quad \textit{with} \quad \big[(\mathcal{A},\mathcal{B})\big]_{\rho,J} \le \eta_0 \, .$$

If

$$f \in C^\sigma(J,E_0) + C(J,E_\beta)$$

for some $\sigma \in (0,1)$ *and* $\beta \in (0,\theta)$, *the Cauchy problem* (2.6.16) *has a unique solution*

$$u \in C(J,E_0) \cap C^1(\dot{J},E_0) \, .$$

It is strict if $x \in E_1(0)$, *and* $u \in C(J,E_\theta)$ *if* $x \in E_\theta$.

Proof It follows from assumption (2.6.13), in particular from (2.6.3) and Remark 2.6.1(c), that

$$\big\|\big(\lambda + A(t)\big)^{-1}\big\|_{\mathcal{L}(E_0,E_j)} \le \kappa\, |\lambda|^{j-1} \ , \qquad \operatorname{Re}\lambda \ge \omega \ , \quad t \in J \ , \quad j = 0,1 \, . \tag{2.6.24}$$

Thus, since

$$M := \sup_{\operatorname{Re}\lambda \ge 0} \frac{1+|\lambda|}{|\lambda+\omega|} < \infty \ ,$$

we infer that $\omega + A\colon J \to \mathfrak{A}_{\kappa M}(E_0)$. Hence assumption (2.2.3(i)) is satisfied for the set

$$\big\{\, \omega + A\ ;\ (\mathcal{A},\mathcal{B}) \in C^\rho\big(J, \mathfrak{B}(E_1,E_0,F,\kappa,\omega)\big)\,\big\}\ . \tag{2.6.25}$$

Lemma 2.6.2 implies assumption (2.2.3(ii)).

Observe that, given $(\mathcal{A},\mathcal{B}), (\widetilde{\mathcal{A}},\widetilde{\mathcal{B}}) \in \mathfrak{B}(E_1,E_0,F)$, the 'generalized resolvent equation'

$$(\lambda + A)^{-1} - (\lambda + \widetilde{A})^{-1} = (\lambda + \mathcal{A},\mathcal{B})^{-1}\big[(\widetilde{\mathcal{A}},\widetilde{\mathcal{B}}) - (\mathcal{A},\mathcal{B})\big](\lambda + \widetilde{A})^{-1} \tag{2.6.26}$$

is valid for $\operatorname{Re}\lambda \geq \omega$. From this, from (2.6.24), and from Remark 2.6.1(d) we deduce by interpolation that

$$\begin{aligned}&\big\|\big(\omega + A(s)\big)^{-1} - \big(\omega + A(t)\big)^{-1}\big\|_{\mathcal{L}(E_0,E_\theta)}\\ &\qquad \leq \kappa^2\omega^{\theta-1}\big(1 \vee \beta(\omega)\big)\, \|(\mathcal{A},\mathcal{B})(s) - (\mathcal{A},\mathcal{B})(t)\|_{\mathcal{L}(E_1,E_0\times F)}\end{aligned}$$

for $s,t \in J$. Thus, letting $\eta := \kappa^2\omega^{\theta-1}\big(1 \vee \beta(\omega)\big)\eta_0$, assumption (2.2.3(iii)) is also satisfied. Now the assertions, but the last one, follow from Theorem 2.3.2, Remark II.2.1.2(d), estimates (2.6.9), and Theorem 2.5.1, since the almost reiteration property of Remark I.2.11.2(a) guarantees that

$$f \in C^\sigma(J,E_0) + C(J,F)\ ,$$

where $F := (E_0,E_\theta)_\gamma$, provided $0 < \gamma < \beta/\theta$.

To derive the last assertion, the proof of Theorem 2.5.1 shows that it suffices to verify that $U(\cdot,s) \in C\big(J \cap [s,\infty), \mathcal{L}_s(E_\theta)\big)$ for $s \in J$. For this we can assume that J is compact. Then, given $x \in E_1(s)$,

$$\|U(t,s)x - U(\tau,s)x\|_\theta \leq c\,\|U(t,s)x - U(\tau,s)x\|_1^\theta\, \|U(t,s)x - U(\tau,s)x\|_0^{1-\theta}$$

for $t,\tau \in J \cap [s,\infty)$. Thus assumption (2.6.13(iv)), (2.6.9), formula (2.4.14), and Theorem 2.4.1 imply

$$\|U(t,s)x - U(\tau,s)x\|_\theta \leq c\,\|x\|_1^\theta\, \|U(t,s)x - U(\tau,s)x\|_0^{1-\theta}\ ,$$

which shows that $U(\cdot,s)x \in C\big(J \cap [s,\infty), E_\theta\big)$ for $x \in E_1(s)$. Now the assertion follows from (2.4.10) and the density of $E_1(s)$ in E_θ. ■

We close this section by an illustration of this abstract theorem in a prototype situation building on Example 2.6.3.

2.6.5 Example: Second Order Parabolic BVPs We use the notations of Example 2.6.3 and suppose that there exist constants $\underline{\alpha}, M \in \dot{\mathbb{R}}^+$ and $2\rho \in (1-1/p, 2)$ such that

$$(\mathcal{A},\mathcal{B}) \in C^\rho\big(J, \mathbb{EBVP}(X,\underline{\alpha},M)\big)\ .$$

Then, given $u^0 \in L_p(X)$ and

$$f \in C^\sigma\big(J, L_p(X)\big) + C\big(J, W_p^\sigma(X)\big)$$

for some $\sigma \in (0, 1/p)$, the parabolic IBVP

$$\begin{aligned} \partial_t u + \mathcal{A}(t)u &= f(t) && \text{in} \quad X \times \dot{J} \ , \\ \mathcal{B}(t)u &= 0 && \text{on } \partial X \times \dot{J} \ , \\ u(\cdot, 0) &= u^0 && \text{on} \quad X \ , \end{aligned} \tag{2.6.27}$$

has a unique solution

$$u \in C\big(J, L_p(X)\big) \cap C^1\big(\dot{J}, L_p(X)\big) \ .$$

If $u^0 \in W_p^s(X)$ for some $s \in (0, 1 + 1/p) \backslash \{1/p\}$ and

$$(1 - \delta)\gamma_\partial u^0 = 0 \qquad \text{if} \quad 1/p < s < 1 + 1/p \ ,$$

then

$$u \in C\big(J, W_p^s(X)\big) \ .$$

If $u^0 \in W^2_{p,\mathcal{B}(0)}(X)$ then

$$u \in C^1\big(J, L_p(X)\big) \quad \text{and} \quad \big(t \mapsto \mathcal{A}(t)u(t)\big) \in C\big(J, L_p(X)\big) \ ,$$

that is, u is a strict solution of (2.6.27).

This follows from Example 2.6.3, Theorem 2.6.4, and the fact, proven in Chapter VI, that

$$\big(L_p(X), W^2_{p,(1-\delta)\gamma_\partial}(X)\big)_{\theta,p} = W_p^{2\theta}(X)$$

for $0 < 2\theta < 1/p$. ■

3 Maximal Regularity

In this section we are interested in situations in which it is not assumed that an interpolation space between E_0 and $D\big(A(t)\big)$ is independent of t. Moreover, we want to impose mild regularity assumptions for the t-dependence only. For this we choose a particular setting that is an abstract version of parabolic initial-boundary value problems and employ the concept of maximal regularity of Chapter III in this case of 'nonconstant domains'. However, we consider in this chapter only the case where $A(t)$ is a suitably small perturbation of some constant operator $A_0 \in \mathcal{H}\big(D(A_0), E_0\big)$. In order to avoid excessive lists of hypotheses we relegate the case of 'strongly varying' functions $t \mapsto A(t)$ to Chapter VIII, where we will study more specific settings.

3.1 Abstract Initial Boundary Value Problems

Throughout the following, J denotes a perfect subinterval of $\mathbb{R}^+$ containing 0. We assume that

$$\left.\begin{array}{l}(E_0, E_1) \text{ is a densely injected Banach couple} \\ \text{and } F \text{ is a Banach space.}\end{array}\right\} \tag{3.1.1}$$

We also suppose that $\mathbb{E}_j := \mathbb{E}_j(J)$ and $\mathbb{F} := \mathbb{F}(J)$ are Banach spaces such that

$$\left.\begin{array}{l}\mathbb{E}_1 \hookrightarrow L_{1,\mathrm{loc}}(J, E_1) \cap W^1_{1,\mathrm{loc}}(J, E_0), \quad \mathbb{E}_0 \hookrightarrow L_{1,\mathrm{loc}}(J, E_0), \\ \text{and } \mathbb{F} \hookrightarrow L_{1,\mathrm{loc}}(J, F).\end{array}\right\} \tag{3.1.2}$$

Recall from Subsection III.1.4 that the trace map

$$\gamma := \gamma_0 : \mathbb{E}_1 \to E_0 \ , \quad u \mapsto u(0)$$

is well-defined, and that $\gamma\mathbb{E}_1$ is the trace space of $\mathbb{E}_1$.

Next we assume that

$$(\mathcal{A}, \mathcal{B}) \in \mathcal{L}(\mathbb{E}_1, \mathbb{E}_0 \times \mathbb{F}) \ .$$

Then, given $(f, g, u^0) \in \mathbb{E}_0 \times \mathbb{F} \times \gamma\mathbb{E}_1$, we consider the abstract IBVP

$$\begin{aligned} \partial_t u + \mathcal{A}(t)u &= f(t) \ , \\ \mathcal{B}(t)u &= g(t) \ , \\ u(0) &= u^0 \ . \end{aligned} \qquad t \in \dot{J} \ , \tag{3.1.3}$$

This means, of course, that we consider the (nonautonomous) initial value problem

$$\dot{u} + \mathcal{A}(t)u = f(t) \ , \quad t \in \dot{J} \ , \qquad u(0) = u^0 \ ,$$

subject to the (nonautonomous) side condition

$$\mathcal{B}(t)u(t) = g(t) \ , \qquad t \in \dot{J} \ . \tag{3.1.4}$$

Suppose that (3.1.3) has a solution

$$u \in C(J, E_1) \tag{3.1.5}$$

and that

$$\big(t \mapsto (\mathcal{B}, g)(t)\big) \in C\big(J, \mathcal{L}(E_1, F) \times F\big) \ . \tag{3.1.6}$$

Then it follows from (3.1.3) that the 'compatibility condition'

$$\mathcal{B}(0)u^0 = g(0)$$

has to be satisfied. In general, however, u may not possess the regularity (3.1.5), so that $u(0) = u^0$ may not belong to E_1 but to the superspace $\gamma\mathbb{E}_1$. Then $\mathcal{B}(0)u^0$ may not be defined. However, there may exist a linear operator

$$\beta\mathcal{B}(0) \in \mathcal{L}(\gamma\mathbb{E}_1, F_0) ,$$

suitably derived from $\mathcal{B}(0)$, where F_0 is a Banach space with $F \hookrightarrow F_0$, such that the modified compatibility condition

$$\beta\mathcal{B}(0)u^0 = \rho g$$

is true, where ρg may be a 'partial trace' of g at $t = 0$. This is illustrated by the following example where we use again results that will be proven in later chapters.

3.1.1 Example: Second Order Parabolic IBVPs We denote by X a bounded subdomain of $\mathbb{R}^n$ of class C^2 and use the notations and conventions of Example 2.6.3. We fix $p \in (1, \infty)$, define a densely injected Banach couple by

$$(E_0, E_1) := \big(L_p(X), W_p^2(X)\big) ,$$

and put

$$(\mathbb{E}_0, \mathbb{E}_1) := \big(L_p(J, E_0), L_p(J, E_1) \cap W_p^2(J, E_0)\big) .$$

It follows that

$$\gamma\mathbb{E}_1 = W_p^{2(1-1/p)}(X) .$$

We choose $\delta \in C\big(\partial X, \{0, 1\}\big)$ and recall that $\Gamma_j := \delta^{-1}(j)$ for $j = 0, 1$. Then we let

$$F_j := W_p^{2-j-1/p}(\Gamma_j) , \qquad j = 0, 1 ,$$

and put

$$F := F_0 \oplus F_1 =: \partial W_p^2(X) .$$

We also put

$$\mathbb{F}_j := L_p(J, F) \cap W_p^{(2-j-1/p)/2}\big(J, L_p(\Gamma_j)\big) , \qquad j = 0, 1 ,$$

and

$$\mathbb{F} := \mathbb{F}_0 \oplus \mathbb{F}_1 .$$

Furthermore,

$$\mathbb{BO}(\Gamma_0) := C^2(\Gamma_0) , \quad \mathbb{BO}(\Gamma_1) := C^1(\Gamma_1)^n \times C^1(\Gamma_1)$$

and

$$\mathbb{BO}(\partial X) := \mathbb{BO}(\Gamma_0) \oplus \mathbb{BO}(\Gamma_1) .$$

Then, given $c_0 \in \mathbb{BO}(\Gamma_0)$ and $\big((b_j), b_0\big) \in \mathbb{BO}(\Gamma_1)$, we define boundary operators

$$\mathcal{B}_j \in \mathcal{L}(E_1, F_j) , \qquad j = 0, 1 ,$$

by

$$\mathcal{B}_0 := (1-\delta)c_0\gamma_\partial \ , \quad \mathcal{B}_1 := \delta(b_j\gamma_\partial\partial_j + b_0\gamma_\partial) \ ,$$

respectively. Then

$$\mathcal{B} := \mathcal{B}_0 + \mathcal{B}_1 \in \mathcal{L}(E_1, F) \ .$$

We identify the set of all boundary operators of type $\mathcal{B}_j$ with $\mathbb{BO}(\Gamma_j)$ by means of the identifications

$$\mathcal{B}_0 \longleftrightarrow c_0 \ , \quad \mathcal{B}_1 \longleftrightarrow \big((b_j), b_0\big) \ .$$

Now we assume that

$$\big(t \mapsto (\mathcal{A},\mathcal{B})(t)\big) \in BUC\big(J, \mathbb{BVP}(X)\big)$$

such that

$$\big(t \mapsto \mathcal{B}_j(t)\big) \in BUC^{\rho_j}\big(J, \mathbb{BO}(\Gamma_j)\big) \ , \qquad j = 0, 1 \ ,$$

with $2\rho_j > 1 + j - 1/p$. It follows that

$$(\mathcal{A},\mathcal{B}) \in \mathcal{L}(\mathbb{E}_1, \mathbb{E}_0 \times \mathbb{F}) \quad \text{and} \quad \big(t \mapsto \mathcal{B}(t)\big) \in C\big(J, \mathcal{L}(E_1, L_p(\partial X))\big) \ .$$

Next we define

$$\beta \in \mathcal{L}\big(\mathbb{BO}(\partial X), \mathcal{L}(\gamma\mathbb{E}_1, F_0)\big)$$

by

$$\beta\mathcal{B} := \begin{cases} \mathcal{B} & \text{if} \quad 3 < p < \infty \ , \\ (1-\delta)c_0\gamma_\partial & \text{if } 3/2 < p < 3 \ , \\ 0 & \text{if} \quad 1 < p < 3/2 \ , \end{cases}$$

and

$$\rho \in \mathcal{L}\big(\mathbb{F}, L_p(\partial X)\big)$$

by

$$\rho := \begin{cases} \gamma & \text{if} \quad 3 < p < \infty \ , \\ \gamma(1-\delta) & \text{if } 3/2 < p < 3 \ , \\ 0 & \text{if} \quad 1 < p < 3/2 \ , \end{cases}$$

where γ denotes again the trace at $t = 0$, that is, $\gamma v := v(0)$ for $v \in \mathbb{F}$, whenever this makes sense. Then the **compatibility condition**

$$\beta\mathcal{B}(0)\gamma u = \rho\mathcal{B}u \ , \qquad u \in \mathbb{E}_1 \ ,$$

is valid. Note that

$$\mathbb{F}(\mathcal{B}) := \big\{ (g, u^0) \in \mathbb{F} \times \gamma\mathbb{E}_1 \ ; \ \beta\mathcal{B}(0)u^0 = \rho g \big\}$$

is a closed linear subspace of $\mathbb{F} \times \gamma\mathbb{E}_1$. It is shown in Chapter VI that there exists $\mathcal{R} \in \mathcal{L}\big(\mathbb{F}(\mathcal{B}), \mathbb{E}_1\big)$ satisfying

$$(\mathcal{B}, \gamma)\mathcal{R} = 1_{\mathbb{F}(\mathcal{B})} \ ,$$

provided there exists $\underline{\alpha} > 0$ such that $c_0(x) \geq \underline{\alpha}$ for $x \in \Gamma_0$ and $b_j(x)\nu^j(x) \geq \underline{\alpha}$ for $x \in \Gamma_1$. Thus, in this case, $(\mathcal{B}, \gamma)$ is a retraction of $\mathbb{E}_1$ onto $\mathbb{F}(\mathcal{B})$. ■

3.2 Isomorphism Theorems

We presuppose assumptions (3.1.1) and (3.1.2) and assume that

$$(\mathcal{A}, \mathcal{B}) \in \mathcal{L}(E_1, E_0 \times F) . \tag{3.2.1}$$

Denoting the constant map $\big(t \mapsto (\mathcal{A}, \mathcal{B})\big)$ again by $(\mathcal{A}, \mathcal{B})$, we also suppose that

$$(\partial + \mathcal{A}, \mathcal{B}) \in \mathcal{L}(\mathbb{E}_1, \mathbb{E}_0 \times \mathbb{F}) . \tag{3.2.2}$$

Next we introduce an abstract 'compatibility condition' by assuming that

$$\left.\begin{aligned} &\mathbb{F}(\mathcal{B}) \text{ is a Banach space such that} \\ &\qquad \mathbb{F}(\mathcal{B}) \hookrightarrow \mathbb{F} \times \gamma\mathbb{E}_1 \\ &\text{and } (\mathcal{B}, \gamma) \text{ is a continuous retraction from } \mathbb{E}_1 \text{ onto } \mathbb{F}(\mathcal{B}). \end{aligned}\right\} \tag{3.2.3}$$

We denote by $E_{1,\mathcal{B}}$ and $\mathbb{E}_{1,\mathcal{B}}$ the kernels of $\mathcal{B} \in \mathcal{L}(E_1, F)$ and $\mathcal{B} \in \mathcal{L}(\mathbb{E}_1, \mathbb{F})$, respectively, and put

$$A := \mathcal{A} | E_{1,\mathcal{B}} .$$

Then we assume that

$$\left.\begin{aligned} &(E_0, E_{1,\mathcal{B}}) \text{ is a densely injected Banach couple,} \\ &\qquad A \in \mathcal{H}(E_{1,\mathcal{B}}, E_0) , \\ &\text{and } (\mathbb{E}_0, \mathbb{E}_{1,\mathcal{B}}) \text{ is a pair of maximal regularity for } A. \end{aligned}\right\} \tag{3.2.4}$$

Recalling definition (III.1.5.7) of K_A, we can easily prove the following isomorphism result:

3.2.1 Theorem *Let assumptions* (3.1.1), (3.1.2), *and* (3.2.1)–(3.2.4) *be satisfied. Then*

$$\big(\partial + \mathcal{A}, (\mathcal{B}, \gamma)\big) \in \mathcal{L}\mathrm{is}\big(\mathbb{E}_1, \mathbb{E}_0 \times \mathbb{F}(\mathcal{B})\big)$$

and

$$\big(\partial + \mathcal{A}, (\mathcal{B}, \gamma)\big)^{-1} = \big[K_A, \big(1 - K_A(\partial + \mathcal{A})\big)(\mathcal{B}, \gamma)^c\big] , \tag{3.2.5}$$

where $(\mathcal{B}, \gamma)^c$ *is a coretraction for* $(\mathcal{B}, \gamma)$.

Proof It follows from (3.2.2) and (3.2.3) that

$$\big(\partial + \mathcal{A}, (\mathcal{B}, \gamma)\big) \in \mathcal{L}\big(\mathbb{E}_1, \mathbb{E}_0 \times \mathbb{F}(\mathcal{B})\big) . \tag{3.2.6}$$

Moreover, (3.2.4) implies

$$(\partial + \mathcal{A}, \gamma) \in \mathcal{L}\mathrm{is}(\mathbb{E}_{1,\mathcal{B}}, \mathbb{E}_0 \times \gamma\mathbb{E}_1) \tag{3.2.7}$$

and, thanks to Corollary III.1.6.2,

$$[(\partial + A)_\gamma]^{-1} = K_A \;, \tag{3.2.8}$$

where $(\partial + A)_\gamma := (\partial + A, \gamma) | \ker(\gamma)$. From (3.2.7) we infer that $\big(\partial + \mathcal{A}, (\mathcal{B}, \gamma)\big)$ is injective. Given $\big(f, (g, u^0)\big) \in \mathbb{E}_0 \times \mathbb{F}(\mathcal{B})$, let

$$u := K_A f + \big[1 - K_A(\partial + \mathcal{A})\big](\mathcal{B}, \gamma)^c (g, u^0) \;.$$

Then $u \in \mathbb{E}_1$ and, thanks to (3.2.8),

$$(\partial + \mathcal{A})u = (\partial + A)_\gamma K_A f + \big[(\partial + \mathcal{A}) - (\partial + A)_\gamma K_A (\partial + \mathcal{A})\big](\mathcal{B}, \gamma)^c (g, u^0) = f \;.$$

Moreover, $(\mathcal{B}, \gamma) K_A = 0$ implies

$$(\mathcal{B}, \gamma) u = (\mathcal{B}, \gamma)(\mathcal{B}, \gamma)^c (g, u^0) = (g, u^0) \;.$$

Hence the map (3.2.6) is surjective and its inverse is given by (3.2.5). The latter representation (or Banach's homomorphism theorem) also implies the continuity of the inverse. ∎

For later purposes we close this section by a qualitative version of a simple time-dependent perturbation result.

3.2.2 Proposition *Let assumptions* (3.1.1), (3.1.2), *and* (3.2.1)–(3.2.4) *be satisfied. Also suppose that*

$$\mathcal{A}, \widetilde{\mathcal{A}} \in \mathcal{L}(\mathbb{E}_1, \mathbb{E}_0) \quad \text{and} \quad (\widetilde{\mathcal{B}}, \gamma) \in \mathcal{L}\big(\mathbb{E}_1, \mathbb{F}(\mathcal{B})\big) \tag{3.2.9}$$

and that

$$\begin{aligned} &\big\|\big(\widetilde{\mathcal{A}}, (\widetilde{\mathcal{B}}, \gamma)\big) - \big(\mathcal{A}, (\mathcal{B}, \gamma)\big)\big\|_{\mathcal{L}(\mathbb{E}_1, \mathbb{E}_0 \times \mathbb{F}(\mathcal{B}))} \\ &\qquad \le \lambda \big\| \big(\partial + \mathcal{A}, (\mathcal{B}, \gamma)\big)^{-1} \big\|^{-1}_{\mathcal{L}(\mathbb{E}_0 \times \mathbb{F}(\mathcal{B}), \mathbb{E}_1)} \end{aligned} \tag{3.2.10}$$

for some $\lambda \in (0,1)$. *Then*

$$\big(\partial + \widetilde{\mathcal{A}}, (\widetilde{\mathcal{B}}, \gamma)\big) \in \mathcal{L}\mathrm{is}\big(\mathbb{E}_1, \mathbb{E}_0 \times \mathbb{F}(\mathcal{B})\big)$$

and

$$\begin{aligned} &\big\| \big(\partial + \widetilde{\mathcal{A}}, (\widetilde{\mathcal{B}}, \gamma)\big)^{-1} \big\|_{\mathcal{L}(\mathbb{E}_0 \times \mathbb{F}(\mathcal{B}), \mathbb{E}_1)} \\ &\qquad \le (1 - \lambda)^{-1} \big\| \big(\partial + \mathcal{A}, (\mathcal{B}, \gamma)\big)^{-1} \big\|_{\mathcal{L}(\mathbb{E}_0 \times \mathbb{F}(\mathcal{B}), \mathbb{E}_1)} \;. \end{aligned}$$

Proof From $\mathcal{A} \in \mathcal{L}(\mathbb{E}_1, \mathbb{E}_0)$ and (3.2.2) we infer $\partial \in \mathcal{L}(\mathbb{E}_1, \mathbb{E}_0)$. Hence, thanks to (3.2.9),

$$\widetilde{\mathcal{C}} := \big(\partial + \widetilde{\mathcal{A}}, (\widetilde{\mathcal{B}}, \gamma)\big) \in \mathcal{L}\big(\mathbb{E}_1, \mathbb{E}_0 \times \mathbb{F}(\mathcal{B})\big) \;.$$

Put $\mathcal{C} := \bigl(\partial + \mathcal{A}, (\mathcal{B}, \gamma)\bigr)$ and note that $\mathcal{C} \in \mathcal{L}\mathrm{is}\bigl(\mathbb{E}_1, \mathbb{E}_0 \times \mathbb{F}(\mathcal{B})\bigr)$ by Theorem 3.2.1. Since, by (3.2.10),

$$\|\widetilde{\mathcal{C}} - \mathcal{C}\|_{\mathcal{L}(\mathbb{E}_1, \mathbb{E}_0 \times \mathbb{F}(\mathcal{B}))} \, \|\mathcal{C}^{-1}\|_{\mathcal{L}(\mathbb{E}_0 \times \mathbb{F}(\mathcal{B}), \mathbb{E}_1)} \leq \lambda \ ,$$

the relation $\widetilde{\mathcal{C}} = \bigl[1 + (\widetilde{\mathcal{C}} - \mathcal{C})\mathcal{C}^{-1}\bigr]\mathcal{C}$ implies the assertion. ■

Given the assumptions of Proposition 3.2.2, it follows that the IBVP

$$\begin{aligned} \partial_t u + \widetilde{\mathcal{A}}(t) u &= f(t) \ , \\ \widetilde{\mathcal{B}}(t) u &= g(t) \ , \\ u(0) &= u^0 \end{aligned} \qquad t \in \dot{J} \ , \tag{3.2.11}$$

has for each $(f, g, u^0) \in \mathbb{E}_0 \times \mathbb{F} \times \gamma\mathbb{E}_1$ a unique solution $u \in \mathbb{E}_1$, provided g and u^0 satisfy the compatibility condition $(g, u^0) \in \mathbb{F}(\mathcal{B})$. This result is the basis for proving, in Chapter VIII, solvability results for (3.2.11) without the smallness assumption (3.2.10).

Chapter V

Scales of Banach Spaces

So far we have studied linear parabolic evolution equations in a fixed Banach space. However, we have already seen, in connection with considerations of higher regularity and problems with variable domains, that it is useful and necessary to consider induced equations in interpolation spaces as well.

The theory of linear and nonlinear partial differential equations relies to a large extent on the concept of weak solutions. As a rule, weak solutions belong to some space (of distributions) that is larger than the space in which the problem has originally been set up. These larger spaces are often characterized by duality arguments and the corresponding extended, or 'extrapolated', operators are implicitly given via suitable bilinear forms (Dirichlet forms, for example).

In this chapter we develop an abstract framework for these situations. It allows us to handle rather easily evolution equations in continuous scales of Banach spaces. These scales are — in most practical cases — constructed with the help of the generators of semigroups involved and are very well adapted to the given problems. On the one hand, the scales are being used as precise measures for the regularity of the solutions. On the other hand, the 'extrapolated' parts of the scales provide the appropriate setting for an abstract theory of weak solutions. In concrete applications, some of which are given in the second volume of this treatise, it turns out that these extrapolation spaces often allow the reduction of problems with variable domains to fixed domain problems. This is of particular importance for the study of quasilinear equations.

Later chapters show that the general abstract theory of this chapter is very flexible and applies to a great variety of concrete situations. In particular, the theory of interpolation-extrapolation scales provides the right framework for 'continuous bootstrapping' arguments that can be used to prove that solutions of the 'extrapolated weak equations' are, in fact, solutions of the original problem. It also allows for a priori estimates in 'weak norms' that can be used in proving global existence results. For this we refer to the third volume. Lastly, it has the advantage — and this is most important for applications to systems of equations coming from physics, for example — that it does not depend on coercivity conditions.

In Section 1 we develop the general theory of Banach scales. Section 2 is devoted to the investigation of the behavior of semigroups and evolution equations in Banach scales, particularly adapted to a given problem.

1 Banach Scales

In this section we study, in some detail, the concept of Banach scales. These scales differ from the usual concept of 'scales of Banach spaces', that are simply one-parameter families of suitably injected Banach spaces, by an additional structure, expressed in terms of linear operators. In particular, it is shown how Banach scales can be constructed starting with one Banach space E and a closed and densely defined linear operator in it, having a nonempty resolvent set. It is most important that we are able to construct superspaces of E and that we can characterize these superspaces by a simple duality theorem. These 'extrapolated' spaces and the techniques presented in this section constitute the abstract basis for our approach to quasilinear parabolic problems in a 'weak setting'.

1.1 General Concepts

Before introducing the concept of Banach scales we prove a simple technical result concerning maximal restrictions of linear operators. Recall that B_F is the F-realization of the linear operator B in G if $F \hookrightarrow G$.

1.1.1 Lemma *Let F and G be Banach spaces with $F \hookrightarrow G$ and let $B \in \mathcal{C}(G)$. Suppose that $(\mu + B)^{-1}(F) \subset F$ for some $\mu \in \rho(-B)$. Then $\mu \in \rho(-B_F)$ and*

$$(\mu + B_F)^{-1} = (\mu + B)^{-1}|F \; . \tag{1.1.1}$$

If also $\mathrm{dom}(B) \subset F$ *then* $\rho(B) = \rho(B_F)$ *and* (1.1.1) *holds for each* $\mu \in \rho(-B)$.

Proof Since $B_F \in \mathcal{C}(F)$, the first assertion is obvious. Suppose that $\mathrm{dom}(B) \subset F$. Then $(\lambda + B)^{-1}(G) \subset F$ for each $\lambda \in \rho(-B)$. Hence $\rho(B) \subset \rho(B_F)$ and

$$(\lambda + B_F)^{-1} = (\lambda + B)^{-1}|F \; , \qquad \lambda \in \rho(-B) \; .$$

Given $\lambda \in \rho(-B_F)$ and $y \in G$, put

$$x := (\mu + B)^{-1}y + (\mu - \lambda)(\lambda + B_F)^{-1}(\mu + B)^{-1}y \; .$$

Then $x \in \mathrm{dom}(B)$ and $(\lambda + B)x = y$. Hence $\lambda + B$ is surjective. If $x \in \mathrm{dom}(B)$ and $(\lambda + B)x = 0$, $\mathrm{dom}(B) \subset F$ implies $x \in \mathrm{dom}(B_F)$. Consequently, $(\lambda + B_F)x = 0$, so that $x = 0$. Thus $\lambda + B$ is bijective. Now $\lambda \in \rho(-B)$ follows from the closed graph theorem. ■

It is easily seen that $\operatorname{dom}(B) \subset F$ iff $(\mu + B)^{-1}(G) \subset F$ for some $\mu \in \rho(-B)$. Moreover, by iterating the resolvent equation it is easy to generalize the preceding lemma to the case $(\mu + B)^{-k}(G) \subset F$ for some $k \in \dot{\mathbb{N}}$ (cf. [Are94, Proposition 1.1]).

Denote by A any one of the sets $\mathbb{N}$, $\mathbb{Z}$, $\mathbb{R}^+$, or $\mathbb{R}$. Let, for each $\alpha \in \mathsf{A}$, be given a Banach space $E_\alpha := (E_\alpha, \|\cdot\|_\alpha)$ and an operator $A_\alpha \in \mathcal{L}\mathrm{is}(E_{\alpha+1}, E_\alpha)$ such that, for each pair $\alpha, \beta \in \mathsf{A}$ with $\alpha > \beta$,

$$E_\alpha \hookrightarrow E_\beta$$

and the diagram

$$\begin{array}{ccccccc} \cdots\cdots & \hookrightarrow & E_{\alpha+1} & \hookrightarrow & E_{\beta+1} & \hookrightarrow & \cdots\cdots \\ & & A_\alpha \big\downarrow \cong & & \cong \big\downarrow A_\beta & & \\ \cdots\cdots & \hookrightarrow & E_\alpha & \hookrightarrow & E_\beta & \hookrightarrow & \cdots\cdots \end{array} \tag{1.1.2}$$

is commutative. Then

$$\big[(E_\alpha, A_\alpha) \ ; \ \alpha \in \mathsf{A}\big]$$

is a **Banach scale** over the index set A. It is said to be **one-sided** if $\mathsf{A} \in \{\mathbb{N}, \mathbb{R}^+\}$ and **two-sided** otherwise. It is **discrete** if $\mathsf{A} \in \{\mathbb{N}, \mathbb{Z}\}$ and **continuous** otherwise. Of course, the diagram (1.1.2) ends on the right at $\beta = 0$ if the scale is one-sided. The Banach scale $\big[(E_\alpha, A_\alpha) \ ; \ \alpha \in \mathsf{A}\big]$ is **densely injected** if $E_\alpha \overset{d}{\hookrightarrow} E_\beta$ for $\alpha > \beta$.

Let $\big[(E_\alpha, A_\alpha) \ ; \ \alpha \in \mathsf{A}\big]$ be a Banach scale over some $\mathsf{A} \in \{\mathbb{N}, \mathbb{Z}, \mathbb{R}^+, \mathbb{R}\}$. Then we have the following easy but useful consequences of the above definitions.

1.1.2 Remarks **(a)** $A_\alpha \in \mathcal{C}(E_\alpha)$ for $\alpha \in \mathsf{A}$.

Proof Since A_α is a toplinear isomorphism, the graph norm of A_α is an equivalent norm for $E_{\alpha+1}$. Hence the assertion follows from Lemma I.1.1.2. ■

(b) $A_\beta \supset A_\alpha$ for $\alpha > \beta$. Thus A_α, considered as a linear operator in E_α, is closable in E_β and its closure equals the restriction of A_β to

$$E_{\alpha+1,\beta+1} := \text{ closure of } E_{\alpha+1} \text{ in } E_{\beta+1} \ ,$$

considered as a linear operator in E_β.

Proof This is a consequence of (a) and (1.1.2). ■

(c) A_α is the E_α-realization of A_β if $\alpha > \beta$.

(d) If the scale is densely injected, the operators A_α are completely determined by A_0 (or by A_{α_0} for any fixed $\alpha_0 \in \mathsf{A}$).

Proof This follows from (b) and (c). ■

(e) $\rho(A_\alpha) = \rho(A_\beta)$ for $\alpha, \beta \in \mathsf{A}$.

Proof This is a consequence of Lemma 1.1.1 and (a)–(c), and $\rho(A_\alpha) \neq \emptyset$. ■

(f) $(A_\alpha)^j \in \mathcal{L}\mathrm{is}(E_{\alpha+j}, E_\alpha)$ for $\alpha \in \mathsf{A}$ and $j \in \dot{\mathbb{N}}$, where A_α is considered as a linear operator in E_α with domain $E_{\alpha+1}$.

(g) If the scale is discrete, it is densely injected iff $E_{\alpha+1} \overset{d}{\hookrightarrow} E_\alpha$ for $\alpha \in \mathsf{A}$. ■

We put

$$E_\infty := \bigcap_{\alpha \in \mathsf{A}} E_\alpha \ ,$$

equipped with its natural projective limit toplogy, that is, the coarsest locally convex topology for which all the natural injections $j_\alpha : E_\infty \to E_\alpha, \quad x \mapsto x$ are continuous. We denote by A_∞ the E_∞-realization of A_0.

1.1.3 Proposition *E_∞ is a Fréchet space such that $E_\infty \hookrightarrow E_\alpha$ for $\alpha \in \mathsf{A}$, and $\{\, \|\cdot\|_j \ ; \ j \in \mathbb{N} \,\}$ is a generating family of (semi)norms for E_∞. Moreover,*

$$A_\infty \in \mathcal{L}\mathrm{aut}(E_\infty)$$

and A_∞ is the E_∞-realization of A_α for each $\alpha \in \mathsf{A}$.

Proof Since $\mathbb{N}$ is cofinal in A, it is clear that the family $\{\, \|\cdot\|_j \ ; \ j \in \mathbb{N} \,\}$ generates the topology. Thus E_∞ has a countable neighborhood basis of zero and is obviously a Hausdorff space. Let (x_k) be a Cauchy sequence in E_∞. Then it is a Cauchy sequence in E_j for each $j \in \mathbb{N}$. Now the completeness of E_j and an obvious diagonal sequence argument show that (x_k) converges in E_∞. Thus E_∞ is a Fréchet space, and $E_\infty \hookrightarrow E_\alpha$ for each $\alpha \in \mathsf{A}$ by definition of the topology of E_∞. The last part of the assertion is an easy consequence of $A_\alpha \in \mathcal{L}\mathrm{is}(E_{\alpha+1}, E_\alpha)$ for each $\alpha \in \mathsf{A}$, and of Remark 1.1.2(c). ■

1.1.4 Proposition *A one-sided discrete Banach scale is densely injected iff E_∞ is dense in E_0.*

Proof Suppose that E_∞ is dense in E_0. Let $k \in \mathbb{N}, \ x \in E_k$, and $\varepsilon > 0$ be given. By Remark 1.1.2(f) we know that $(A_0)^k \in \mathcal{L}\mathrm{is}(E_k, E_0)$. Since E_∞ is dense in E_0, there exists $y \in E_\infty$ such that

$$\|y - (A_0)^k x\|_0 \le \varepsilon / \|(A_0)^{-k}\|_{\mathcal{L}(E_0,E_k)} \ .$$

Proposition 1.1.3 implies $z := (A_0)^{-k} y \in E_\infty$. Hence

$$\|z - x\|_k = \big\|(A_0)^{-k}\big(y - (A_0)^k x\big)\big\|_k \le \|(A_0)^{-k}\|_{\mathcal{L}(E_0,E_k)}\, \|y - (A_0)^k x\|_0 < \varepsilon \ .$$

This shows that E_∞ is dense in E_k. Hence $E_\infty \subset E_{k+1} \hookrightarrow E_k$ guarantees that $E_{k+1} \overset{d}{\hookrightarrow} E_k$. Thus the scale is densely injected by Remark 1.1.2(g).

Conversely, suppose that the scale is densely injected. Define a new norm $|\cdot|_k$ on E_k by

$$|x|_k := \sum_{j=0}^{k} \|x\|_j \ , \qquad x \in E_k \ .$$

Then, denoting by $i_k \in \mathcal{L}(E_{k+1}, E_k)$ the natural injection,

$$\|x\|_k \le |x|_k \le \Big(1 + \sum_{j=0}^{k-1} \prod_{\ell=j}^{k-1} \|i_\ell\|_{\mathcal{L}(E_{\ell+1},E_\ell)}\Big)\, \|x\|_k \ , \qquad x \in E_k \ .$$

Hence $|\cdot|_k$ is an equivalent norm for E_k, and

$$|x|_k \le |x|_j \ , \qquad x \in E_j \ , \quad k \le j \ . \tag{1.1.3}$$

Now let $x_0 \in E_0$ and $\varepsilon > 0$ be given. Since $E_{k+1} \overset{d}{\hookrightarrow} E_k$ for $k \in \mathbb{N}$, we find inductively $x_k \in E_k$ such that $|x_{k+1} - x_k|_k \le \varepsilon 2^{-k-1}$ for $k \in \mathbb{N}$. Hence, given $\ell > m \ge k$, we infer from (1.1.3) that

$$|x_\ell - x_m|_k \le \sum_{j=m}^{\ell-1} |x_{j+1} - x_j|_k \le \sum_{j=m}^{\ell-1} |x_{j+1} - x_j|_j \le \varepsilon \sum_{j=m}^{\infty} 2^{-j-1} = \varepsilon 2^{-m} \ .$$

Hence $(x_j)_{j \ge k}$ is a Cauchy sequence in E_k for each $k \in \mathbb{N}$. Thus, given any $k \in \mathbb{N}$, the sequence $(x_j)_{j\ge k}$ converges in E_k to some y_k. Since $E_\ell \hookrightarrow E_k$ for $\ell > k$, it follows that $y_\ell = y_k$ for $0 \le k < \ell < \infty$. This shows that $y := y_0 \in E_\infty$ and $\|y - x_0\|_0 = |y - x_0|_0 \le \varepsilon$. Thus E_∞ is dense in E_0. ■

The proof of the sufficiency part of Proposition 1.1.4 follows [Est84, Corollary 2.2]. There a corresponding result, a so-called abstract Mittag-Leffler theorem, is proven for a projective limit of complete metric spaces (also cf. [Bou53]). For an application of the abstract Mittag-Leffler theorem to a situation closely related to our case we refer to [ArEK94, Proposition 6.2].

1.1.5 Corollary *Let* $\big[(E_\alpha, A_\alpha)\ ;\ \alpha \in \mathsf{A}\big]$ *be a densely injected Banach scale. Then* $E_\infty \overset{d}{\hookrightarrow} E_\alpha$ *for* $\alpha \in \mathsf{A}$.

Proof Let $k \in \mathbb{N}$ with $k > \alpha$ be given. Then $E_\infty \hookrightarrow E_k \overset{d}{\hookrightarrow} E_\alpha$. By applying Proposition 1.1.4 to the densely injected discrete scale $\big[(E_{k+j}, A_{k+j})\ ;\ j \in \mathbb{N}\big]$ it follows that $E_\infty \overset{d}{\hookrightarrow} E_k$. This proves the assertion. ∎

A Banach scale is said to be **compactly injected** if $E_\alpha \lhook\joinrel\hookrightarrow E_\beta$ for $\alpha > \beta$.

1.1.6 Proposition *Let $\big[(E_\alpha, A_\alpha)\ ;\ \alpha \in \mathsf{A}\big]$ be a compactly injected Banach scale. Then E_∞ is a Montel space and $E_\infty \lhook\joinrel\hookrightarrow E_\alpha$ for $\alpha \in \mathsf{A}$.*

Proof Since E_∞ is a Fréchet space, it is barreled and metrizable. Hence it is a Montel space if bounded subsets are relatively sequentially compact. Let $B \subset E_\infty$ be bounded. Then B is bounded in E_j for each $j \in \mathbb{N}$. Since $E_{j+1} \lhook\joinrel\hookrightarrow E_j$, it follows that B is relatively sequentially compact in E_j for each $j \in \mathbb{N}$. Now an obvious diagonal sequence argument shows that B is relatively sequentially compact in E_∞. Hence the assertion follows from the factorization $E_\infty \hookrightarrow E_{\alpha+1} \lhook\joinrel\hookrightarrow E_\alpha$ for $\alpha \in \mathsf{A}$. ∎

1.1.7 Proposition *Let $\big[(E_\alpha, A_\alpha)\ ;\ \alpha \in \mathsf{A}\big]$ be a discrete Banach scale. Then it is compactly injected iff $(A_0)^{-1} \in \mathcal{K}(E_0)$.*

Proof Denote by $i_0 \in \mathcal{L}(E_1, E_0)$ the natural injection of E_1 into E_0. Then we deduce from the factorization

$$\begin{array}{ccc} E_0 & \xrightarrow[\cong]{A_0^{-1}} & E_1 \\ & {\scriptstyle A_0^{-1}}\searrow \quad \swarrow{\scriptstyle i_0} & \\ & E_0 & \end{array}$$

that $E_1 \lhook\joinrel\hookrightarrow E_0$ iff $(A_0)^{-1} \in \mathcal{K}(E_0)$. Thus we have to show that $E_1 \lhook\joinrel\hookrightarrow E_0$ guarantees $E_\alpha \lhook\joinrel\hookrightarrow E_\beta$ for $\alpha > \beta$. Since $\mathsf{A} \subset \mathbb{Z}$, this follows if we show that $E_{j+1} \lhook\joinrel\hookrightarrow E_j$ for $j \in \mathsf{A}$.

If $j \geq 1$, the commutativity of the diagram (see Remarks 1.1.2(b) and (f))

$$\begin{array}{ccc} E_{j+1} & \hookrightarrow & E_j \\ {\scriptstyle (A_1)^j}\big\downarrow\cong & & \cong\big\uparrow{\scriptstyle (A_0)^{-j}} \\ E_1 & \lhook\joinrel\hookrightarrow & E_0 \end{array}$$

implies the desired result. If $j \leq -1$, we infer the assertion from the diagram which is obtained from the preceding one by replacing A_1 by A_{j+1} and A_0 by A_j, respectively. ∎

Two Banach scales $\big[(E_\alpha, A_\alpha)\ ;\ \alpha \in \mathsf{A}\big]$ and $\big[(F_\alpha, B_\alpha)\ ;\ \alpha \in \mathsf{A}\big]$ over the same index set A are **isomorphic** if there exist $I_\alpha \in \mathcal{L}\mathrm{is}(E_\alpha, F_\alpha),\ \alpha \in \mathsf{A}$, such that, given $\alpha, \beta \in \mathsf{A}$ with $\alpha > \beta$, the diagram

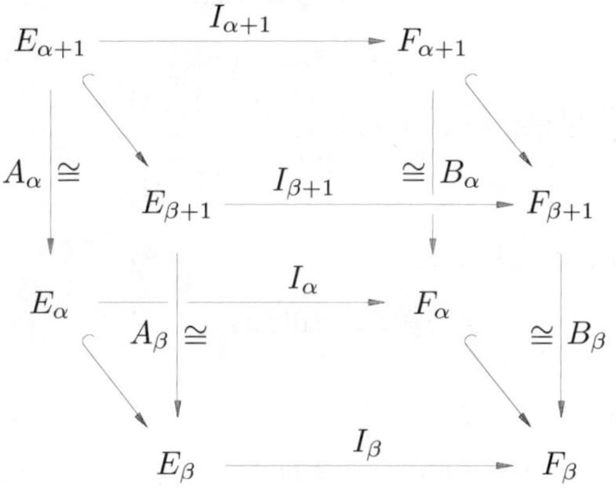

(1.1.4)

is commutative. If each I_α is also isometric, the two scales are **isometrically isomorphic**. Two Banach scales are **equivalent** if they are isomorphic and if $E_\alpha \doteq F_\alpha$ for $\alpha \in \mathsf{A}$.

A Banach scale is said to be **reflexive** if each E_α is a reflexive Banach space. It is a **Hilbert scale** if each E_α is a Hilbert space. Lastly, a Banach scale is said to **satisfy interpolation inequalities** if

$$\|x\|_\beta \le c(\alpha,\beta,\gamma)\, \|x\|_\gamma^{(\alpha-\beta)/(\alpha-\gamma)}\, \|x\|_\alpha^{(\beta-\gamma)/(\alpha-\gamma)}\ , \qquad x \in E_\alpha\ ,$$

provided $\alpha, \beta, \gamma \in \mathsf{A}$ with $\gamma < \beta < \alpha$.

The concepts and results of this subsection are new, except for Lemma 1.1.1, whose proof follows [Are94]. A different notion of Banach scales has been introduced and studied by Kreĭn [Kre60] and Kreĭn and Petunin [KrP66] (also see [KrPS82, Chapter III]).

1.2 Power Scales

In this subsection we exhibit a simple general technique for constructing one-sided Banach scales. For this we assume that

$$E \text{ is a Banach space, } A \in \mathcal{C}(E), \text{ and } 0 \in \rho(A)\ . \tag{1.2.1}$$

Then it is known that

$$A^j \in \mathcal{C}(E)\ , \qquad j \in \mathbb{N}\ , \tag{1.2.2}$$

where, of course, $A^0 := 1_E$ (e.g., [HiP57, Theorem 2.16.4]).

We put

$$E_j := (E_j, \|\cdot\|_j) := E_j(A) := \big(\mathrm{dom}(A^j), \|A^j \cdot\|\big) , \qquad j \in \mathbb{N} , \tag{1.2.3}$$

so that $E_0 = E$. Since $0 \in \rho(A^j)$, it follows that $\|\cdot\|_j$ is equivalent to the graph norm of A^j, that is,

$$E_j \doteq D(A^j) . \tag{1.2.4}$$

Hence E_j is a Banach space by (1.2.2), and

$$\|x\|_k \leq \|A^{k-j}\|_{\mathcal{L}(E)} \|x\|_j , \qquad x \in E_j , \quad j > k ,$$

shows that $E_j \hookrightarrow E_k$ if $j > k$. Thus

$$A_j := E_j\text{-realization of } A \tag{1.2.5}$$

is well-defined.

1.2.1 Theorem *Let assumption* (1.2.1) *be satisfied and define E_j and A_j for $j \in \mathbb{N}$ by* (1.2.3) *and* (1.2.5), *respectively. Then*

$$\big[(E_j, A_j) \; ; \; j \in \mathbb{N}\big]$$

is a Banach scale, the **power scale** *generated by (E, A). It is densely injected iff A is densely defined, and it is compactly injected iff $A^{-1} \in \mathcal{K}(E)$.*

Proof The first assertion is easily verified. If A is densely defined, that is, $E_1 \overset{d}{\hookrightarrow} E$, it follows from $A^{-j} \in \mathcal{L}\mathrm{is}(E, E_j) \cap \mathcal{L}\mathrm{is}(E_1, E_{j+1})$ that $E_{j+1} \overset{d}{\hookrightarrow} E_j$ for $j \geq 1$. Now the density assertion is a consequence of Remark 1.1.2(g). The last part of the assertion follows from Proposition 1.1.7. ■

1.2.2 Remarks (a) Every one-sided discrete Banach scale is essentially a power scale. More precisely: if $\big[(E_j, A_j) \; ; \; j \in \mathbb{N}\big]$ is a Banach scale then it is isomorphic to the power scale generated by (E_0, A_0).

Proof This is a consequence of (1.1.2) and Remarks 1.1.2(a) and (b). ■

(b) If no confusion seems likely, we simply say: a power scale is generated by A, instead of: by (E, A).

(c) Suppose that $B \in \mathcal{C}(E)$ with $0 \in \rho(B)$. Then $\big[\big(E_j(B), B_j\big) \; ; \; j \in \mathbb{N}\big]$, the power scale generated by (E, B), is well-defined. Suppose that

$$\mathrm{dom}(B^k) = E_k , \qquad 1 \leq k \leq n ,$$

for some $n \in \dot{\mathbb{N}}$, where $E_k := E_k(A)$. Then $E_k \doteq E_k(B)$ and

$$\|A^k B^{-k}\|_{\mathcal{L}(E)}^{-1} \|x\|_k \leq \|x\|_{E_k(B)} \leq \|B^k A^{-k}\|_{\mathcal{L}(E)} \|x\|_k , \qquad x \in E_k ,$$

for $1 \leq k \leq n$.

Proof It follows from (1.2.2) and the closed graph theorem that $A^k B^{-k}$ and $B^k A^{-k}$ belong to $\mathcal{L}(E)$. Hence the assertion is obvious. ■

(d) Given $\lambda \in \rho(-A)$, we have $E_n(\lambda + A) \doteq E_n$ and

$$\left(1+|\lambda|\,\|(\lambda+A)^{-1}\|\right)^{-n}\|x\|_n \le \|x\|_{E_n(\lambda+A)} \le (1+|\lambda|\,\|A^{-1}\|)^n\,\|x\|_n\ , \qquad x \in E_n\ ,$$

for $n \in \mathbb{N}$.

Proof This follows from $A^n(\lambda+A)^{-n} = \left[A(\lambda+A)^{-1}\right]^n = \left[1-\lambda(\lambda+A)^{-1}\right]^n$ and $(\lambda+A)^n A^{-n} = \left[(\lambda+A)A^{-1}\right]^n = [\lambda A^{-1}+1]^n$, respectively, and from (c). ■

(e) Fix $m \in \dot{\mathbb{N}}$ and put $(F_j, B_j) := (E_{j+m}, A_{j+m})$ for $j \in \mathbb{N}$. Then $\left[(F_j, B_j)\,;\, j \in \mathbb{N}\right]$ is the power scale generated by (E_m, A_m). It is isometrically isomorphic to the power scale generated by (E, A).

Proof The first assertion is obvious from Remarks 1.1.2(c) and (e). Then, putting $I_\alpha := (A_\alpha)^{-m}$ for $\alpha \in \mathbb{N}$, Remark 1.1.2(f) implies that $I_\alpha \in \mathcal{L}\mathrm{is}(E_\alpha, F_\alpha)$ and that the diagram (1.1.4) is commutative. It is easily verified that I_α is an isometric isomorphism. ■

Now we turn to the situation that A is of positive type, that is, $A \in \mathcal{P}(E)$. Recall that this means that $A \in \mathcal{P}_K(E)$ for some $K \ge 1$ which, in turn, means that $A \in \mathcal{C}(E)$ with dense domain, $\mathbb{R}^+ \subset \rho(-A)$, and $(1+s)\,\|(s+A)^{-1}\| \le K$ for $s \in \mathbb{R}^+$. Thus, in particular, the power scale generated by A is well-defined. We also recall the definition of the classes $\mathcal{P}(E; K, \vartheta)$ of Subsection III.4.7.

For later purposes we prepare the following simple technical lemma, including operators having bounded imaginary powers. Of course, $\left[(E_j, A_j)\ ;\ j \in \mathbb{N}\right]$ is the power scale generated by (E, A).

1.2.3 Lemma (i) *Suppose that $A \in \mathcal{P}(E; K, \vartheta)$ for some $K \ge 1$ and $\vartheta \in [0, \pi)$. Then $A_j \in \mathcal{P}(E_j; K, \vartheta)$ for $j \in \mathbb{N}$.*

(ii) *If $A \in \mathcal{BIP}(E; M, \theta)$ for some $M \ge 1$ and $\theta \ge 0$ then $A_j \in \mathcal{BIP}(E_j; M, \theta)$ for $j \in \mathbb{N}$.*

Proof (i) From Lemma III.4.9.1(i) and Remarks 1.1.2 we infer that

$$\|(\lambda + A_j)^{-1}x\|_j = \|(\lambda + A)^{-1}A^j x\| \le \|(\lambda + A)^{-1}\|\,\|x\|_j$$

for $j \in \mathbb{N}$, $x \in E_j$, and $\lambda \in \rho(-A)$. This implies the assertion.

(ii) From (i) and Theorem III.4.6.5 it follows that $(A_j)^z$ is well-defined for $j \in \mathbb{N}$ and $z \in \mathbb{C}$. If $-1 < \mathrm{Re}\, z < 1$ and $x \in E_{j+1}$ then $(A_j)^z x$ is given by replacing in Theorem III.4.6.5(ii) the operator A by A_j everywhere. Now we infer from Remarks 1.1.2 that

$$(A_j)^z x = A^z x\ , \qquad j \in \mathbb{N}\ , \quad x \in E_{j+1}\ , \quad -1 < \mathrm{Re}\, z < 1\ .$$

Thus Lemma III.4.9.2(ii) implies

$$\|(A_j)^z x\|_j = \|A^z A^j x\| \le \|A^z\| \, \|x\|_j \ , \qquad j \in \mathbb{N} \ , \quad x \in E_{j+1} \ , \quad \operatorname{Re} z = 0 \ .$$

Now the assertion follows from the density of E_{j+1} in E_j. ■

Suppose that $A \in \mathcal{P}(E)$. Then the fractional powers, A^α, of A are well-defined for $\alpha \in \mathbb{R}$, and we put

$$E_\alpha := (E_\alpha, \|\cdot\|_\alpha) := E_\alpha(A) := \big(\operatorname{dom}(A^\alpha), \|A^\alpha \cdot\|\big) \ , \qquad \alpha \in \mathbb{R}^+ \ . \tag{1.2.6}$$

Observe that this is consistent with (1.2.3) if $\alpha \in \mathbb{N}$.

1.2.4 Theorem *Let E be a Banach space and let A be a linear operator of positive type in E. Define E_α by* (1.2.6) *and let A_α be the E_α-realization of A for $\alpha \in \mathbb{R}^+$. Then*

$$\big[(E_\alpha, A_\alpha) \ ; \ \alpha \in \mathbb{R}^+\big]$$

is a densely injected Banach scale, the one-sided **fractional power scale** *generated by (E, A). It satisfies interpolation inequalities. Moreover, it is compactly injected iff $A^{-1} \in \mathcal{K}(E)$.*

Proof The fact that the fractional power scale is a densely injected Banach scale is an easy consequence of Theorem III.4.6.5(v) and (vi).

Suppose that $0 < \beta < \alpha$ and that $m \in \mathbb{N}$ satisfies $m \le \alpha < m + 1$. Then, thanks to Proposition III.4.6.3,

$$A^{-\beta} = \frac{\sin \pi\beta}{\pi} \frac{m!}{(1-\beta)(2-\beta)\cdots(m-\beta)} \Big[\int_0^t s^{m-\beta} A^\alpha (s+A)^{-m-1} A^{-\alpha} \, ds + \int_t^\infty s^{m-\beta} (s+A)^{-m-1} \, ds \Big]$$

for $0 < t < \infty$. Hence

$$\|A^{-\beta} x\| \le c(m) \big[\varphi(t) \, \|A^{-\alpha} x\| + \psi(t) \, \|x\| \big] \ , \qquad x \in E \ , \tag{1.2.7}$$

where

$$\varphi(t) := \int_0^t s^{m-\beta} \, \|A^{\alpha-m}(s+A)^{-1}\| \, \|A(s+A)^{-1}\|^m \, ds \tag{1.2.8}$$

and

$$\psi(t) := \int_t^\infty s^{m-\beta} \, \|(s+A)^{-1}\|^{m+1} \, ds$$

for $0 < t < \infty$. From (III.4.6.1) it follows that

$$\|(s+A)^{-1}\|^{m+1} \le K^{m+1} (1+s)^{-m-1} \ , \qquad 0 < s < \infty \ .$$

Thus

$$\psi(t) \leq K^{m+1} \int_t^\infty s^{-1-\beta}\,ds = \beta^{-1} K^{m+1} t^{-\beta}\;, \qquad 0 < t < \infty\;. \tag{1.2.9}$$

Since $A(s+A)^{-1} = 1 - s(s+A)^{-1}$, we see from (III.4.6.1) that

$$\big\| \big[A(s+A)^{-1}\big] \big\|^m \leq (1+K)^m\;, \qquad 0 < s < \infty\;. \tag{1.2.10}$$

Moreover, putting $z := m+1-\alpha$, we infer from Theorem III.4.6.5 and the resolvent equation that

$$\begin{aligned} A^{\alpha-m}(s+A)^{-1} &= A^{-z}A(s+A)^{-1} = \frac{1}{2\pi i}\int_\Gamma (-\lambda)^{-z}(\lambda+A)^{-1}A(s+A)^{-1}\,d\lambda \\ &= \frac{1}{2\pi i}\Big[\int_\Gamma \frac{(-\lambda)^{-z}}{s-\lambda} A(\lambda+A)^{-1}\,d\lambda - \int_\Gamma \frac{(-\lambda)^{-z}}{s-\lambda}\,d\lambda\,A(s+A)^{-1}\Big]\;. \end{aligned}$$

Thanks to Cauchy's theorem and the choice of Γ, the last integral is zero. In the second to the last integral we can replace Γ by the curve consisting of the two half-lines $\{\, re^{\pm i\varphi}\;;\; r \geq 0 \,\}$ for a suitable $\varphi \in (0,\pi/2)$ and running from $\infty e^{-i\varphi}$ to $\infty e^{i\varphi}$. Then using (1.2.10) with $m=1$, we obtain the estimate

$$\begin{aligned} \|A^{\alpha-m}(s+A)^{-1}\| &\leq \frac{1+K}{2\pi}\int_0^\infty r^{-z}\big[\,|s-re^{i\varphi}|^{-1} + |s-re^{-i\varphi}|^{-1}\big]\,dr \\ &= s^{-z}\frac{1+K}{2\pi}\int_0^\infty \tau^{-z}\big[\,|1-\tau e^{i\varphi}|^{-1} + |1-\tau e^{-i\varphi}|^{-1}\big]\,d\tau = cs^{\alpha-m-1}\;. \end{aligned}$$

Consequently, (1.2.8) and (1.2.10) imply

$$\varphi(t) \leq c\int_0^t s^{\alpha-\beta-1}\,ds = \frac{c}{\alpha-\beta}t^{\alpha-\beta}\;, \qquad 0 < t < \infty\;.$$

Now (1.2.7) gives

$$\|A^{-\beta}x\| \leq c\Big[\frac{t^{\alpha-\beta}}{\alpha-\beta}\,\|A^{-\alpha}x\| + \frac{t^{-\beta}}{\beta}\,\|x\|\Big]\;, \qquad 0 < t < \infty\;.$$

By minimizing the right-hand side with respect to t, it follows that

$$\|A^{-\beta}x\| \leq c\,\|x\|^{1-\beta/\alpha}\,\|A^{-\alpha}x\|^{\beta/\alpha}\;, \qquad x \in E\;.$$

We apply this inequality to $A^\alpha x$ for $x \in E_\alpha$. Then Theorem III.4.6.5(iii) implies

$$\|A^{\alpha-\beta}x\| \leq c\,\|x\|^{\beta/\alpha}\,\|A^{\alpha}x\|^{(\alpha-\beta)/\alpha}\;, \qquad x \in E_\alpha\;.$$

Thus, putting $\beta' := \alpha-\beta$ and omitting the dash afterwards, we infer from the latter estimate that, given $0 < \beta < \alpha < \infty$,

$$\|A^{\beta}x\| \leq c\,\|x\|^{1-\beta/\alpha}\,\|A^{\alpha}x\|^{\beta/\alpha}\;, \qquad x \in E_\alpha\;. \tag{1.2.11}$$

Suppose that $-\infty < \gamma < \beta < \alpha < \infty$. Then (1.2.11) implies

$$\|A^{\beta-\gamma}x\| \le c\,\|x\|^{(\alpha-\beta)/(\alpha-\gamma)}\,\|A^{\alpha-\gamma}x\|^{(\beta-\gamma)/(\alpha-\gamma)}\ , \qquad x \in E_{\alpha-\gamma}\ .$$

By replacing in this estimate x by $A^\gamma x$ and using Theorem III.4.6.5(i) and (iii) we see that

$$\|A^\beta x\| \le c(\alpha,\beta,\gamma)\,\|A^\gamma x\|^{(\alpha-\beta)/(\alpha-\gamma)}\,\|A^\alpha x\|^{(\beta-\gamma)/(\alpha-\gamma)} \tag{1.2.12}$$

for $x \in \mathrm{dom}(A^\alpha)$, provided $-\infty < \gamma < \beta < \alpha < \infty$. This proves, in particular, that the fractional power scale satisfies interpolation inequalities.

If the fractional power scale is compactly injected, it follows from Proposition 1.1.7 that $A^{-1} \in \mathcal{K}(E)$. Thus assume $A^{-1} \in \mathcal{K}(E)$. Then $A^{-k} \in \mathcal{K}(E)$ for $k \in \dot{\mathbb{N}}$. Suppose that $\alpha > 0$ and fix $k \in \mathbb{N}$ with $k > \alpha$. Then (1.2.12) implies

$$\|A^{-\alpha}x\| \le c\,\|A^{-k}x\|^{\alpha/k}\,\|x\|^{1-\alpha/k}\ , \qquad x \in E\ . \tag{1.2.13}$$

Let (x_j) be a bounded sequence in E. Since $A^{-k} \in \mathcal{K}(E)$, there exists a subsequence, again denoted by (x_j), such that $(A^{-k}x_j)$ is a Cauchy sequence in E. Then we deduce from (1.2.13) that $(A^{-\alpha}x_j)$ is a Cauchy sequence in E, which proves that $A^{-\alpha} \in \mathcal{K}(E)$. Thus

$$E_\alpha \overset{d}{\hookrightarrow} E_0\ , \qquad 0 < \alpha < \infty\ . \tag{1.2.14}$$

Suppose that $0 < \beta < \alpha < \infty$. Let (x_j) be a bounded sequence in E_α. Then, thanks to (1.2.14), there exists a subsequence, again denoted by (x_j), such that (x_j) is a Cauchy sequence in E_0. Hence

$$\|x_j - x_k\|_\beta \le c\,\|x_j - x_k\|^{1-\beta/\alpha}\,\|x_j - x_k\|_\alpha^{\beta/\alpha}$$

implies that (x_j) is a Cauchy sequence in E_β. This proves that $E_\alpha \hookrightarrow E_\beta$ which, together with (1.2.14), guarantees that the fractional power scale is compactly injected. ■

1.2.5 Corollary *Let A be an operator of positive type in some Banach space E. Then $A^{-1} \in \mathcal{K}(E)$ iff $A^{-\alpha} \in \mathcal{K}(E)$ for some $\alpha > 0$.*

We close this subsection with the following useful technical observation.

1.2.6 Proposition *Suppose that $A \in \mathcal{P}(E)$ and let $\big[(E_\alpha, A_\alpha)\ ;\ \alpha \in \mathbb{R}^+\big]$ be the fractional power scale generated by (E, A). Fix $\beta \in (0,1]$ and $\gamma \in \mathbb{R}^+$ and put $(F_\alpha, B_\alpha) := (E_{\alpha\beta+\gamma}, A_{\alpha\beta+\gamma})$ for $\alpha \in \mathbb{R}^+$. Then $\big[(F_\alpha, B_\alpha)\ ;\ \alpha \in \mathbb{R}^+\big]$ is the fractional power scale generated by $\big(E_\gamma, (A_\gamma)^\beta\big)$. If $\beta = 1$, it is isometrically isomorphic to the fractional power scale generated by (E, A).*

Proof By an obvious modification of the proof of Lemma 1.2.3(i) we infer that $A_\gamma \in \mathcal{P}(E_\gamma)$. Hence $(A_\gamma)^\beta \in \mathcal{P}(E_\gamma)$ by Proposition III.4.6.10, so that the fractional power scale generated by $\big(E_\gamma, (A_\gamma)^\beta\big)$ is well-defined. Theorem III.4.6.13 and Remark 1.1.2(b) imply $\big[(A_\gamma)^\beta\big]^\alpha x = A^{\alpha\beta} x$ for $\alpha \geq 0$ and $x \in E_{\alpha\beta+\gamma} = F_\alpha$. Thus

$$\begin{aligned} \|B^\alpha x\|_F = \big\|\big[(A_\gamma)^\beta\big]^\alpha x\big\|_\gamma &= \|A^\gamma A^{\alpha\beta} x\| \\ &= \|A^{\alpha\beta+\gamma} x\| = \|x\|_{\alpha\beta+\gamma} = \|x\|_{F_\alpha} \end{aligned} \tag{1.2.15}$$

for $\alpha \in \mathbb{R}^+$ and $x \in E_{\alpha\beta+\gamma}$. This proves the first assertion. If $\beta = 1$, it is now clear that

$$I_\alpha := (A_\alpha)^{-\gamma} \in \mathcal{L}\mathrm{is}(E_\alpha, F_\alpha)\ , \qquad \alpha \in \mathbb{R}^+\ ,$$

these isomorphisms are isometric, and the diagrams (1.1.4) are commutative. ■

The principal results of this subsection are well-known, at least implicitly and if $-A$ generates a strongly continuous analytic semigroup (e.g., [Sob66], [Fri69, Section II.14], [Hen81, Section 1.4], [Paz83, Section 2.6]). Under this assumption, to the best of our knowledge, the first proof of Theorem 1.2.4 is in [Ama78, Lemma 2.1]. The above derivation of the **moment inequality** (1.2.12) follows [Kre72, Section I.5.3].

1.3 Extrapolation Spaces

Let F be a normed vector space. Recall that a **completion** $(\widetilde{F}, \widetilde{\jmath})$ of F consists of a Banach space $\widetilde{F}$ and an isometric linear isomorphism $\widetilde{\jmath}$ from F onto a dense linear subspace of $\widetilde{F}$.

A particular completion can be obtained by taking for $\widetilde{F}$ the Banach space whose elements are equivalence classes $\widetilde{(x_j)}$ of Cauchy sequences in F, where two Cauchy sequences (x_j) and (y_j) are said to be equivalent if $\|x_j - y_j\| \to 0$. This space is given the natural linear structure, and its norm is defined by

$$\|\widetilde{(x_j)}\|_{\widetilde{F}} := \lim_{j\to\infty} \|x_j\|\ .$$

Then $\widetilde{\jmath}(x) := \widetilde{(x)}$, where (x) is the constant sequence $(x, x, \ldots)$ for $x \in F$ (e.g., [Yos65, Section I.10]).

Let $(\widetilde{F}_1, \widetilde{\jmath}_1)$ be another completion of F. Then there exists an isometric $h \in \mathcal{L}\mathrm{is}(\widetilde{F}, \widetilde{F}_1)$ such that the diagram

$$\begin{array}{ccc} & F & \\ {\scriptstyle \widetilde{\jmath}}\swarrow & & \searrow{\scriptstyle \widetilde{\jmath}_1} \\ \widetilde{F} & \xrightarrow[\cong]{\;h\;} & \widetilde{F}_1 \end{array}$$

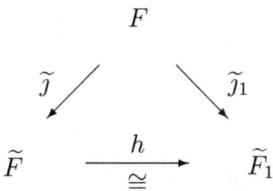

is commutative. In this sense completions are unique, except for isometric linear isomorphisms. If no confusion seems likely, we fix a particular completion $\widetilde{F} := (\widetilde{F}, \widetilde{\jmath})$, **the** completion of F, and identify F with the subspace $\widetilde{\jmath}(F)$ of $\widetilde{F}$.

Throughout the remainder of this subsection we presuppose that $E := (E, \|\cdot\|)$ is a Banach space and $A \in \mathcal{C}(E)$ with $0 \in \rho(A)$. Then $\big[(E_j, A_j)\ ;\ j \in \mathbb{N}\big]$ is the discrete power scale generated by (E, A).

Observe that $\|A^{-1}\cdot\|$ is a norm on E which is weaker than $\|\cdot\|$. Thus the completion

$$(E_{-1}, \|\cdot\|_{-1}) := E_{-1}(A) := (E, \|A^{-1}\cdot\|)^{\sim} \tag{1.3.1}$$

of the normed linear space $(E, \|A^{-1}\cdot\|)$ is a well-defined Banach space and

$$E \overset{d}{\hookrightarrow} E_{-1}\ . \tag{1.3.2}$$

It is said to be the **extrapolation space of** E **generated by** A. Note that

$$\|Ax\|_{-1} = \|x\|\ , \qquad x \in E_1\ , \tag{1.3.3}$$

and that

$$\|A^{-1}y\| = \|y\|_{-1}\ , \qquad y \in E\ . \tag{1.3.4}$$

Thus let

$$E_{1,0} := \text{closure of } E_1 \text{ in } E\ . \tag{1.3.5}$$

Then it follows from (1.3.3) and (1.3.4) that there exists a unique continuous extension A_{-1} of A such that A_{-1} is an isometric isomorphism from $E_{1,0}$ onto E_{-1} and $(A_{-1})^{-1} \supset A^{-1}$. In the following, A_{-1} is said to be the E_{-1}**-realization** of A.

1.3.1 Proposition (i) $A_{-1} \in \mathcal{L}\mathrm{is}(E_{1,0}, E_{-1}) \cap \mathcal{C}(E_{-1})$ *and this isomorphism is isometric.*

(ii) A *is closable in* E_{-1} *and its closure equals* A_{-1}.

(iii) $\rho(A_{-1}) = \rho(A)$.

Proof (i) Since $A_{-1} \in \mathcal{L}\mathrm{is}(E_{1,0}, E_{-1})$, the graph norm of A_{-1} is an equivalent norm on $E_{1,0}$. Thus $A_{-1} \in \mathcal{C}(E_{-1})$ by Lemma I.1.1.2. (ii) is now obvious. Lastly, (iii) is an easy consequence of (i), (ii), and Lemma 1.1.1. ■

Since $A_{-1} \in \mathcal{C}(E_{-1})$ and $0 \in \rho(A_{-1})$, it follows that

$$\begin{aligned} &E_{-2} := E_{-1}(A_{-1}),\ \text{the extrapolation space of } E_{-1} \text{ generated by } A_{-1},\\ &\text{and } A_{-2},\ \text{the } E_{-2}\text{-realization of } A_{-1}, \end{aligned} \tag{1.3.6}$$

are well-defined. It follows that

$$E \overset{d}{\hookrightarrow} E_{-1} \overset{d}{\hookrightarrow} E_{-2}\ , \tag{1.3.7}$$

that

$$A_{-2} \in \mathcal{L}\mathrm{is}(E_{-1}, E_{-2}) \cap \mathcal{C}(E_{-2}) \ , \tag{1.3.8}$$

that this isomorphism is isometric, and that A_{-1} is closable in E_{-2} and A_{-2} is its closure.

However, relations (1.3.7) and (1.3.8) have to be interpreted with care. To be more precise, let (E_{-1}, j_{-1}) and (E_{-2}, j_{-2}) be the completions of $(E, \|A^{-1}\cdot\|)$ and $(E_{-1}, \|(A_{-1})^{-1}\cdot\|)$, respectively. Then

$$E \xrightarrow{j_{-1}} E_{-1} \xrightarrow{j_{-2}} E_{-2}$$

and j_{-1} and j_{-2} are linear isometries onto dense linear subspaces of E_{-1} and E_{-2}, respectively. Thus if we identify, as above, E_{-1} with its image $j_{-2}(E_{-1})$ in E_{-2}, we also identify $j_{-1}(E)$ with $j_{-2}\big(j_{-1}(E)\big)$. Therefore, by constructing E_{-2} we have to replace the already constructed extrapolation space E_{-1} by an isomorphic image so that (1.3.7) is true.

1.3.2 Theorem *Suppose that $A \in \mathcal{C}(E)$ is densely defined with $0 \in \rho(A)$ and fix $m \in \dot{\mathbb{N}}$. Define E_{-k} and A_{-k} for $1 \le k \le m$ inductively by*

$$E_{-k} := (E_{-k}, \|\cdot\|_{-k}) := \big(E_{-k+1}, \|(A_{-k+1})^{-1}\cdot\|\big)^{\sim}$$

and

$$A_{-k} := \textit{closure of } A_{-k+1} \textit{ in } E_{-k} \ .$$

Put

$$(F_j, B_j) := (E_{j-m}, A_{j-m}) \ , \qquad j \in \mathbb{N} \ .$$

Then $\big[(F_j, B_j) \ ; \ j \in \mathbb{N}\big]$ is the power scale generated by (E_{-m}, A_{-m}). It is isometrically isomorphic to the power scale generated by (E, A).

Proof The above considerations imply that the pairs (E_{-k}, A_{-k}) are well-defined for $1 \le k \le m$. Since $E_{1,0} = E_0 = E$, thanks to the fact that A is densely defined, it is an easy consequence of (1.3.3)–(1.3.5) that $\big[(F_j, B_j) \ ; \ j \in \mathbb{N}\big]$ is indeed the power scale generated by (E_{-m}, A_{-m}). The last part of the theorem follows now from Remark 1.2.2(e). ■

Suppose that A is densely defined and $m \in \dot{\mathbb{N}}$. Then $\big[(E_{j-m}, A_{j-m}) \ ; \ j \in \mathbb{N}\big]$ is said to be the **extrapolated power scale of order** m, **generated by** (E, A), and we also denote it by

$$\big[(E_j, A_j) \ ; \ j \in \mathbb{Z} \cap [-m, \infty)\big] \ .$$

Thanks to Theorem 1.2.1, it is a densely injected, one-sided, discrete Banach scale.

1.3.3 Remark Given $k \in \mathbb{Z} \cap [-m, \infty)$, it is clear that $\big[(E_{j+k}, A_{j+k}) \ ; \ j \in \mathbb{N}\big]$ is the power scale generated by (E_k, A_k) and that $\big[(E_j, A_j) \ ; \ j \in \mathbb{Z} \cap [-m, \infty)\big]$ is the extrapolated power scale of order $m + k$, generated by (E_k, A_k). ■

1.3.4 Proposition *Suppose that A is densely defined and $m \in \dot{\mathbb{N}}$. Then*

$$\big[(E_j, A_j)\ ;\ j \in \mathbb{Z} \cap [-m, \infty)\big]$$

is compactly injected iff $(A_k)^{-1} \in \mathcal{K}(E_k)$ for some $k \in \mathbb{Z}$ with $k \geq -m$. If this is the case, $(A_k)^{-1} \in \mathcal{K}(E_k)$ for each $k \in \mathbb{Z}$ with $k \geq -m$.

Proof If the scale

$$\big[(E_j, A_j)\ ;\ j \in \mathbb{Z} \cap [-m, \infty)\big] \tag{1.3.9}$$

is compactly injected, it follows that

$$\big[(E_{j+k}, A_{j+k})\ ;\ j \in \mathbb{N}\big] \tag{1.3.10}$$

is compactly injected for each $k \in \mathbb{Z}$ with $k \geq -m$. Hence $(A_k)^{-1} \in \mathcal{K}(E_k)$ by Proposition 1.1.7. Conversely, suppose that $(A_k)^{-1} \in \mathcal{K}(E_k)$ for some $k \in \mathbb{Z}$ with $k \geq -m$. Then the scale (1.3.10) is compactly injected by Proposition 1.1.7. Due to Theorem 1.3.2 and Remark 1.3.3 the full scale (1.3.9) is isomorphic to (1.3.10). Hence it is compactly injected as well. ■

It is tempting to put

$$E_{-\infty} := \bigcup_{j \in \mathbb{Z}} E_j\ .$$

Note, however, that this is not meaningful, in general, since we have to construct the negative spaces inductively and, by constructing E_{-m-1} from E_{-m} (and A_{-m}), we have to replace the spaces $E_{-1}, \ldots, E_{-m}$ already obtained by isomorphic images. On the other hand, if we do not use the above identifications but work with the continuous injections $j_{-k} : E_{-k+1} \to E_{-k}$ provided by the completion process, we can construct a 'superspace' $E_{-\infty}$ as the inductive limit of the spaces E_{-j}, $j \in \dot{\mathbb{N}}$, with respect to these injections (e.g., [Sch71, Section II.6.3]). Since these constructions are rather cumbersome and we do not need them, we refrain from giving details. Moreover, in Subsection 1.4 we give a simple sufficient condition guaranteeing the existence of a universal superspace $E_{-\infty}$ such that $E_j \hookrightarrow E_{-\infty}$ for each $j \in \mathbb{Z}$.

Suppose that $B \in \mathcal{C}(E)$ is densely defined with $0 \in \rho(B)$. Then the extrapolated power scale $\big[(E_j(B), B_j)\ ;\ j \in \mathbb{Z} \cap [-m, \infty)\big]$ of order m, generated by (E, B), is well-defined. In the following proposition we give a sufficient condition for $E_{-k}(B) \doteq E_{-k} := E_{-k}(A)$ for $1 \leq k \leq m$. Recall that the dual of a closed and densely defined linear operator is well-defined and that it has the same resolvent set as its predual. Hence the following assumptions are meaningful.

1.3.5 Proposition *Suppose that A and B are closed and densely defined in E with $0 \in \rho(A) \cap \rho(B)$, and that*

$$\operatorname{dom}\big((A')^k\big) = \operatorname{dom}\big((B')^k\big)\ , \qquad 1 \leq k \leq m\ . \tag{1.3.11}$$

Then $E_{-k}(B) \doteq E_{-k}$ *and*

$$\|(B')^k(A')^{-k}\|^{-1}_{\mathcal{L}(E')}\,\|x\|_{-k} \le \|x\|_{E_{-k}(B)} \le \|(A')^k(B')^{-k}\|_{\mathcal{L}(E')}\,\|x\|_{-k}$$

for $x \in E_{-k}$ *and* $1 \le k \le m$.

Proof Since $(A')^k$ and $(B')^k$ are closed in E' by (1.2.2), it follows from (1.3.11) and the closed graph theorem that $(B')^k(A')^{-k}$ and $(A')^k(B')^{-k}$ belong to $\mathcal{L}(E')$. Hence we deduce from

$$\left\langle (B')^k(A')^{-k}x', x\right\rangle = \langle x', A^{-k}B^k x\rangle\ , \qquad x \in E_k(B)\ , \quad x' \in E'\ ,$$

and the density of $E_k(B)$ in E that $A^{-k}B^k$ has an extension $(A^{-k}B^k)^- \in \mathcal{L}(E)$ and that

$$\|(A^{-k}B^k)^-\|_{\mathcal{L}(E)} = \|(B')^k(A')^{-k}\|_{\mathcal{L}(E')}\ .$$

Thus, given $x \in E$,

$$\|x\|_{-k} = \|A^{-k}x\| = \|A^{-k}B^kB^{-k}x\| \le \|(B')^k(A')^{-k}\|_{\mathcal{L}(E')}\,\|x\|_{E_{-k}(B)}\ .$$

Similarly, there exists a unique extension $(B^{-k}A^k)^- \in \mathcal{L}(E)$ of $B^{-k}A^k$ satisfying

$$\|(B^{-k}A^k)^-\|_{\mathcal{L}(E)} = \|(A')^k(B')^{-k}\|_{\mathcal{L}(E')}\ .$$

Thus, given $x \in E$,

$$\|x\|_{E_{-k}(B)} = \|B^{-k}x\| = \|B^{-k}A^kA^{-k}x\| \le \|(A')^k(B')^{-k}\|_{\mathcal{L}(E')}\,\|x\|_{-k}\ .$$

Now the assertion follows from the density of E in E_{-k}. ■

1.3.6 Corollary *Suppose that* $A \in \mathcal{C}(E)$ *is densely defined with* $0 \in \rho(A)$. *Given* $m \in \dot{\mathbb{N}}$ *and* $\lambda \in \rho(-A)$, *it follows that* $E_{-k}(\lambda + A) \doteq E_{-k} := E_{-k}(A)$ *with*

$$(1 + |\lambda|\,\|A^{-1}\|)^{-k}\,\|x\|_{-k} \le \|x\|_{E_{-k}(\lambda+A)} \le \left(1 + |\lambda|\,\|(\lambda + A)^{-1}\|\right)^k\,\|x\|_{-k}$$

for $x \in E_{-k}$ *and* $1 \le k \le m$.

Proof This follows from

$$(A')^k\left[(\lambda + A)'\right]^{-k} = \left(\left[A(\lambda + A)^{-1}\right]^k\right)' = \left(\left[1 - \lambda(\lambda + A)^{-1}\right]^k\right)'$$

and

$$\left[(\lambda + A)'\right]^k(A')^{-k} = \left(\left[(\lambda + A)A^{-1}\right]^k\right)' = \left[(\lambda A^{-1} + 1)^k\right]'$$

and Proposition 1.3.5. ■

Next we turn again to the case that A is of positive type. First we prove the following simple extension of Lemma 1.2.3.

1.3.7 Lemma (i) *Suppose that $A \in \mathcal{P}(E; K, \vartheta)$ for some $K \geq 1$ and $\vartheta \in [0, \pi)$. Then $A_j \in \mathcal{P}(E_j; K, \vartheta)$ for $j \in \mathbb{Z} \cap [-m, \infty)$.*

(ii) *If $A \in \mathcal{BIP}(E; M, \theta)$ for some $M \geq 1$ and $\theta \geq 0$ then $A_j \in \mathcal{BIP}(E_j; M, \theta)$ for $j \in \mathbb{Z} \cap [-m, \infty)$.*

Proof Thanks to Theorem 1.3.2 and Lemma 1.2.3 it suffices to prove the assertions for $j = -m$.

(i) Since

$$\|(\lambda + A_{-m})^{-1} x\|_{-m} = \|(\lambda + A)^{-1} A^{-m} x\| \leq \|(\lambda + A)^{-1}\| \, \|x\|_{-m}$$

for $x \in E$ and $\lambda \in \rho(-A_{-m}) = \rho(-A)$, the first assertion follows using the density of E in E_{-m}.

(ii) From (i) and Theorem III.4.6.5 we infer that $(A_{-m})^z$ is well-defined for $z \in \mathbb{C}$. Similarly as in the proof of Lemma 1.2.3 we see that $(A_{-m})^z x = A^z x$ for $x \in E_1$ and $-1 < \operatorname{Re} z < 1$. Thus

$$\|(A_{-m})^z x\|_{-m} = \|A^{-m} A^z x\| \leq \|A^z\| \, \|x\|_{-m}$$

for $x \in E_1$ and $\operatorname{Re} z = 0$, and the desired conclusion follows from the density of E_1 in E_{-m}. ■

Suppose that $A \in \mathcal{P}(E)$ and $m \in \dot{\mathbb{N}}$. It follows from Lemma 1.3.7 by induction that $A_{-m} \in \mathcal{P}(E)$. Thus, letting $(F, B) := (E_{-m}, A_{-m})$, Theorem 1.2.4 guarantees that the one-sided fractional power scale $\big[(F_\alpha, B_\alpha)\ ;\ \alpha \in \mathbb{R}^+\big]$ is well-defined. In the following, we put

$$(E_\alpha, A_\alpha) := (F_{\alpha+m}, B_{\alpha+m}) \ , \qquad -m \leq \alpha < \infty \ , \tag{1.3.12}$$

and call

$$\big[(E_\alpha, A_\alpha)\ ;\ \alpha \in [-m, \infty)\big]$$

the **extrapolated fractional power scale of order** m, generated by (E, A). Note that (1.3.12) is consistent with the earlier definitions of (E_α, A_α) for $\alpha \in \mathbb{R}^+$ and $\alpha \in \{-1, -2, \ldots, -m\}$.

1.3.8 Theorem *Suppose that $A \in \mathcal{P}(E)$ and $m \in \dot{\mathbb{N}}$. Then the extrapolated fractional power scale of order m, generated by (E, A), is a densely injected Banach scale satisfying interpolation inequalities. It is isometrically isomorphic to the fractional power scale*

$$\big[(E_\alpha, A_\alpha)\ ;\ \alpha \in \mathbb{R}^+\big] \ ,$$

generated by (E, A). It is compactly injected iff $A^{-1} \in \mathcal{K}(E)$. If $0 < \alpha \leq m$ then $E_{-\alpha}$ is a completion of $(E, \|A^{-\alpha} \cdot\|)$ and $A_{-\alpha}$ is the closure of A in $E_{-\alpha}$.

Proof The first assertion is an immediate consequence of Theorem 1.2.4. The second one follows from Proposition 1.2.6. Now the assertion concerning compact injections is a consequence of Proposition 1.3.4 and Theorem 1.2.4.

Suppose that $0 < \alpha < m$. Then, thanks to Theorem III.4.6.5(iii),

$$\|A^{-\alpha}x\| = \|A^{-m}A^{-\alpha+m}x\| = \|B^{-\alpha+m}x\|_{-m} = \|x\|_{-\alpha}\ , \qquad x \in E_{m-\alpha}\ .$$

Since $E_{m-\alpha} \overset{d}{\hookrightarrow} E$, it follows that $\|A^{-\alpha}x\| = \|x\|_{-\alpha}$ for $x \in E$. Thus we deduce from $E \overset{d}{\hookrightarrow} E_{-\alpha}$ that $E_{-\alpha}$ is a completion of $(E, \|A^{-\alpha}\cdot\|)$. Since

$$D(A) = E_1 \overset{d}{\hookrightarrow} E_{-\alpha+1} = D(A_{-\alpha})$$

and $A \subset A_{-\alpha}$, we see that A is closable in $E_{-\alpha}$ and its closure equals $A_{-\alpha}$. ■

1.3.9 Corollary *Suppose that $A \in \mathcal{P}(E)$. Let $\big[(E_\alpha, A_\alpha)\ ;\ \alpha \in [-m,\infty)\big]$ be the extrapolated fractional power scale of order m, generated by (E, A). Then $E_\alpha \cong E_\beta$ for $-m \le \alpha < \beta < \infty$, and $(A_\alpha)^{\alpha-\beta}$ is an isometric isomorphism from E_α onto E_β.*

In order to have a unified language we call one-sided [fractional] power scales **extrapolated [fractional] power scales of order zero**.

The above construction of E_{-1} and A_{-1} occurs perhaps for the first time in [Nag83] (also cf. [Wal86]). Without being aware of this, it has been reintroduced in [Ama87] (that paper had already been submitted in 1984; also cf. [Ama86]) and, again independently, in [Liu89]. Around the same time Da Prato and Grisvard [DaPG84] proposed a different construction of an extrapolation space $\widetilde{E}_{-1}$ and an extension $\widetilde{A}_{-1}$ of A. In [Ama88c, Remark 6.2] it has been shown that the extrapolation pairs (E_{-1}, A_{-1}) and $(\widetilde{E}_{-1}, \widetilde{A}_{-1})$ are isomorphic (cf. [DaP84] for a corresponding statement; also see [vN92]).

In [Nag83] and [Wal86] only extrapolated discrete scales have been studied. Extrapolated fractional power scales have been investigated in [Ama87] and in [Liu89]. The latter author also introduces at each point α of the scale inductive limit spaces $\bigcup_{\beta>\alpha} E_\beta$ and projective limit spaces $\bigcap_{\beta<\alpha} E_\alpha$. All the papers cited above have been motivated by the theory of semigroups and, as a rule, it is assumed that $A \in \mathcal{G}(E)$.

Lastly, it should be mentioned that more recently there has been developed an extrapolation theory by B. Jawerth and M. Milman (e.g., [JM91]) that is of an entirely different nature than the one considered here (also see [Mil94]).

1.4 Dual Scales

Throughout this subsection we suppose that E is a reflexive Banach space and $A \in \mathcal{C}(E)$ with dense domain and $0 \in \rho(A)$, unless explicitly stated otherwise. Recall that $A' \in \mathcal{C}(E')$ is densely defined and $\rho(A') = \rho(A)$. In order to avoid confusion with too many dashes we put

$$E^\sharp := E' \ , \quad A^\sharp := A' \ .$$

Then, given $m \in \dot{\mathbb{N}}$, the extrapolated power scales of order m

$$\big[(E_j, A_j) \ ; \ j \in \mathbb{Z} \cap [-m, \infty)\big] \tag{1.4.1}$$

and

$$\big[(E_j^\sharp, A_j^\sharp) \ ; \ j \in \mathbb{Z} \cap [-m, \infty)\big] \ , \tag{1.4.2}$$

generated by (E, A) and $(E^\sharp, A^\sharp)$, respectively, are well-defined. The scale (1.4.2) is said to be the **dual scale** of the Banach scale (1.4.1).

Now we establish an important duality theorem for which we need some preparation. Here and hereafter we denote by $\langle \cdot, \cdot \rangle := \langle \cdot, \cdot \rangle_E$ the duality pairing between E' and E.

1.4.1 Lemma *Given $k \in \mathbb{N}$ with $1 \le k \le m$, put*

$$\langle x^\sharp, x \rangle_k := \big\langle (A_{-k}^\sharp)^{-k} x^\sharp, A^k x \big\rangle \ , \qquad x \in E_k \ , \quad x^\sharp \in E_{-k}^\sharp \ .$$

Then

$$\langle \cdot, \cdot \rangle_k : E_{-k}^\sharp \times E_k \to \mathbb{K}$$

is a separating continuous bilinear form of norm 1.

Proof We infer $A^k \in \mathcal{L}\mathrm{is}(E_k, E)$ as well as $(A_{-k}^\sharp)^{-k} \in \mathcal{L}\mathrm{is}(E_{-k}^\sharp, E^\sharp)$ from Remark 1.1.2(f). It is clear that these isomorphisms are isometric. Now the assertion is obvious. ■

The following simple observation is rather useful and deserves to be singled out.

1.4.2 Lemma *Let X and Y be vector spaces and let B and C be linear operators with domains in X and ranges in Y such that $B \supset C$. If B is injective and C is surjective then $B = C$.*

Proof Suppose that $x \in \mathrm{dom}(B)$. Since C is surjective, there exists $y \in \mathrm{dom}(C)$ such that $Cy = Bx$. Thus $B \supset C$ implies $Bx = By$, that is, $x - y \in \ker(B) = \{0\}$. Hence $x = y \in \mathrm{dom}(C)$ so that $\mathrm{dom}(B) \subset \mathrm{dom}(C)$, which proves the assertion. ■

After these preparations we can prove the following duality result. Here and in similar situations $(A_k)'$ denotes the dual of $A_k \in \mathcal{C}(E_k)$, that is, as a linear (in general: unbounded) operator in E_k and not as a bounded linear map from E_{k+1} to E_k.

1.4.3 Proposition *Suppose that $k \in \mathbb{N}$ with $1 \le k \le m$. Then*

$$(E_k)' = E^\sharp_{-k} \tag{1.4.3}$$

and

$$(A_k)' = A^\sharp_{-k} \tag{1.4.4}$$

with respect to the duality pairing $\langle \cdot, \cdot \rangle_k$.

Proof It follows from Lemma 1.4.1 that $x^\sharp \mapsto \langle x^\sharp, \cdot \rangle_k$ defines a linear isometry from $E^\sharp_{-k}$ into $(E_k)'$.

Given $f \in (E_k)'$, we deduce from $A^k \in \mathcal{L}\mathrm{is}(E_k, E)$ the existence of $y^\sharp \in E^\sharp$ such that the diagram

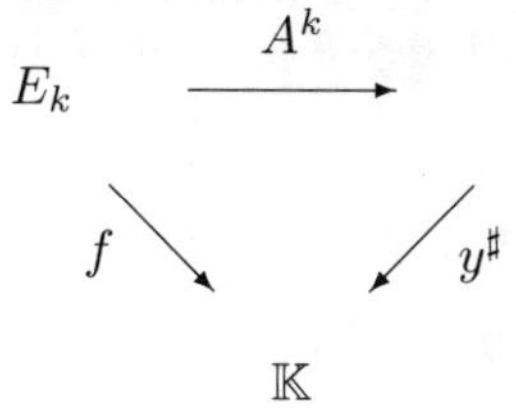

is commutative, that is,

$$f(x) = \langle y^\sharp, A^k x \rangle , \qquad x \in E_k . \tag{1.4.5}$$

Since $(A^\sharp_{-k})^k \in \mathcal{L}\mathrm{is}(E^\sharp, E^\sharp_{-k})$, the element $x^\sharp := (A^\sharp_{-k})^k y^\sharp \in E^\sharp_{-k}$ is well-defined, and (1.4.5) implies $f(x) = \langle x^\sharp, x \rangle_k$ for $x \in E_k$. Hence $(E_k)' = E^\sharp_{-k}$.

Since $\mathrm{dom}(A_k) = E_{k+1}$ is dense in E_k, the dual $(A_k)'$ of $A_k \in \mathcal{C}(E_k)$ with respect to the duality pairing $\langle \cdot, \cdot \rangle_k$ is well-defined, and $(A_k)' \in \mathcal{C}(E^\sharp_{-k})$ by (1.4.3). As A_k is the E_k-realization of A and $A^\sharp_{-k} \supset A^\sharp$, it follows that

$$\begin{aligned} \langle x^\sharp, A_k x \rangle_k &= \big\langle (A^\sharp_{-k})^{-k} x^\sharp, A A^k x \big\rangle = \big\langle A^\sharp (A^\sharp_{-k})^{-k} x^\sharp, A^k x \big\rangle \\ &= \big\langle (A^\sharp_{-k})^{-k} A^\sharp_{-k} x^\sharp, A^k x \big\rangle = \big\langle A^\sharp_{-k} x^\sharp, x \big\rangle_k \end{aligned}$$

for $x^\sharp \in E^\sharp_{-k+1}$ and $x \in E_{k+1}$. Hence $(A_k)' \supset A^\sharp_{-k}$. Thanks to $\rho(A_k) = \rho\big((A_k)'\big)$, Remark 1.1.2(e), and Theorem 1.3.2, $0 \in \rho\big((A_k)'\big) \cap \rho(A^\sharp_{-k})$. Thus $(A_k)' = A^\sharp_{-k}$ by Lemma 1.4.2. ■

1.4.4 Corollary *If $0 \le \ell < k \le m$,*

$$\langle x^\sharp, x \rangle_k = \langle x^\sharp, x \rangle_\ell , \qquad x \in E_k , \quad x^\sharp \in E^\sharp_{-\ell} .$$

Proof Suppose that $x^\sharp \in E^\sharp$ and $x \in E_k$. Then, thanks to $A^\sharp_{-k} \supset A^\sharp_{-\ell} \supset A^\sharp$ and $A \supset A_\ell$,

$$\begin{aligned}\langle x^\sharp, x\rangle_k &= \langle (A^\sharp)^{-k}x^\sharp, A^k x\rangle = \langle (A^\sharp)^{\ell-k}(A^\sharp)^{-\ell}x^\sharp, A^{k-\ell}A^\ell x\rangle \\ &= \langle (A^{\ell-k})'(A^\sharp)^{-\ell}x^\sharp, A^{k-\ell}A^\ell x\rangle = \langle (A^\sharp)^{-\ell}x^\sharp, A^\ell x\rangle = \langle x^\sharp, x\rangle_\ell \ .\end{aligned}$$

Now the assertion follows from $E^\sharp \overset{d}{\hookrightarrow} E^\sharp_{-\ell}$. ■

It is a consequence of Corollary 1.4.4 that

$$\langle x^\sharp, x\rangle_k = \langle x^\sharp, x\rangle \ , \qquad x \in E_k \ , \quad x^\sharp \in E^\sharp \ , \qquad 0 < k \le m \ .$$

Since $E^\sharp$ is dense in $E^\sharp_{-k}$, we see that $\langle\cdot,\cdot\rangle_k$ is uniquely determined by $\langle\cdot,\cdot\rangle$, that is, $\langle\cdot,\cdot\rangle_k$ is **naturally induced by** $\langle\cdot,\cdot\rangle$. Hereafter, in abuse of notation, we simply write $\langle\cdot,\cdot\rangle$ for $\langle\cdot,\cdot\rangle_k$, $k \in \mathbb{N}$ with $0 \le k \le m$, if no confusion seems possible.

1.4.5 Remarks **(a)** It should be observed that everything said above remains valid if E is not reflexive but A' is densely defined.

(b) If E is not reflexive and A' is not densely defined, it is still true that $A' \in \mathcal{C}(E')$ and $0 \in \rho(A')$. Thus, letting $E^\sharp := E'$ and $A^\sharp := A'$, the dual power scale $\big[(E^\sharp_j, A^\sharp_j)\ ;\ j \in \mathbb{N}\big]$ is still well-defined. ■

As an easy consequence of Proposition 1.4.3 we prove now the following fundamental theorem whose corollary contains a most important characterization of the extrapolation spaces E_{-k} for $1 \le k \le m$.

1.4.6 Theorem *E_k and $E^\sharp_k$ are reflexive for $k \in \mathbb{Z} \cap [-m, \infty)$. Moreover, with respect to the duality pairing $\langle\cdot,\cdot\rangle$,*

$$(E_k)' = E^\sharp_{-k} \ , \quad (A_k)' = A^\sharp_{-k} \ , \qquad -m \le k \le m \ . \tag{1.4.6}$$

Proof Since E_k [resp. $E^\sharp_k$] is toplinearly isomorphic to E [resp. $E^\sharp$] and since E, hence $E^\sharp$, is reflexive, the first assertion follows. From Proposition 1.4.3 we know the validity of (1.4.6) for $0 \le k \le m$. Thus

$$E_k = (E^\sharp_{-k})' \ , \quad A_k = (A^\sharp_{-k})' \ , \qquad 0 \le k \le m \ . \tag{1.4.7}$$

Put $(F_k, B_k) := (E^\sharp_k, A^\sharp_k)$ for $k \in \mathbb{Z} \cap [-m, \infty)$. Then, using (1.4.7) for $k = 0$, we infer that the dual scale of $\big[(F_k, B_k)\ ;\ k \in \mathbb{Z} \cap [-m, \infty)\big]$ equals the scale (1.4.1). Thus, by applying (1.4.7) to F_k and B_k, we see that

$$E^\sharp_k = (E_{-k})' \ , \quad A^\sharp_k = (A_{-k})' \ , \qquad 0 \le k \le m \ ,$$

with respect to $\langle\cdot,\cdot\rangle$. This proves the assertion. ■

1.4.7 Corollary *If* $1 \le k \le m$ *then* $E_{-k} = (E_k^\sharp)'$ *and* $A_{-k} = (A_k^\sharp)'$ *with respect to the duality pairing* $\langle \cdot, \cdot \rangle$.

On the basis of Theorem 1.4.6 we can now extend our extrapolation scales to get two-sided scales. For this we need some preparation.

Let X and Y be *LCSs* such that $i \colon X \overset{d}{\hookrightarrow} Y$. Then $i' \in \mathcal{L}(Y', X')$ and i' is injective by the density of X in Y. Observe that

$$i'(y') = y' \circ i = y' | X \in X'_Y \ , \qquad y' \in Y' \ ,$$

where

$$X'_Y := \{ x' \in X' \ ; \ x' \text{ is continuous with respect to the } Y\text{-topology of } X \} \ .$$

Conversely, given $x' \in X'_Y$, the density of X in Y implies the existence of a unique $y' \in Y'$ with $y' \supset x'$. Thus $i'(Y') = X'_Y$. In the following, **we identify** Y' **and** X'_Y **by means of the injection** i'.

Recall that a *LCS* X is **semireflexive** if the **canonical injection**

$$\kappa \colon X \to X'' \ , \quad x \mapsto \langle \cdot, x \rangle_X$$

is surjective. (It is reflexive if κ is a toplinear isomorphism.)

1.4.8 Proposition *Let* X *and* Y *be LCSs. Then*

$$X \overset{d}{\hookrightarrow} Y \quad \textit{implies} \quad Y' \hookrightarrow X'$$

and

$$\langle y', x \rangle_X = \langle y', x \rangle_Y \ , \qquad y' \in Y' \ , \quad x \in X \ .$$

If X *is semireflexive,*

$$X \overset{d}{\hookrightarrow} Y \quad \textit{implies} \quad Y' \overset{d}{\hookrightarrow} X' \ .$$

Proof The first assertion follows from the considerations above. Thus suppose $X \overset{d}{\hookrightarrow} Y$ and X is semireflexive. Let $x'' \in X''$ satisfy $\langle x'', y' \rangle_{X'} = 0$ for $y' \in Y'$. Then, putting $x := \kappa^{-1}(x'') \in X$,

$$\langle y', x \rangle_X = \langle y', x \rangle_Y = 0 \ , \qquad y' \in Y' \ ,$$

so that $x = 0$, hence $x'' = 0$. Now the density of Y' in X' follows from the Hahn-Banach theorem. ■

Recall from Proposition 1.1.4 that

$$E_\infty \overset{d}{\hookrightarrow} E_j \overset{d}{\hookrightarrow} E_k \overset{d}{\hookrightarrow} E\ , \qquad 0 < k < j < \infty\ ,$$

and that E_∞ is the projective limit of the sequence of Banach spaces (E_j) with respect to the continuous injections $E_\infty \hookrightarrow E_j$. Since E_j is reflexive for $j \in \mathbb{N}$, it follows that E_∞ is semireflexive (e.g., [Sch71, Section IV.5.8]). Hence Proposition 1.4.8 implies

$$E' \overset{d}{\hookrightarrow} (E_k)' \overset{d}{\hookrightarrow} (E_j)' \overset{d}{\hookrightarrow} (E_\infty)'$$

and

$$\langle x', x\rangle_{E_\infty} = \langle x', x\rangle_{E_j} = \langle x', x\rangle_{E_k} = \langle x', x\rangle\ , \qquad x' \in E'\ , \quad x \in E_\infty\ ,$$

for $0 < k < j < \infty$. By density, the E_j'-E_j-duality pairing is for each $j \in \bar{\mathbb{N}}$ completely determined by the E'-E-duality pairing $\langle \cdot, \cdot \rangle$. Thus $\langle \cdot, \cdot \rangle_{E_k} = \langle \cdot, \cdot \rangle_k$ for $0 \le k \le m$, and, letting

$$E^\sharp_{-\infty} := (E_\infty)'\ ,$$

it follows from Theorem 1.4.6 that

$$E^\sharp \overset{d}{\hookrightarrow} E^\sharp_{-j} \overset{d}{\hookrightarrow} E^\sharp_{-k} \overset{d}{\hookrightarrow} E^\sharp_{-\infty}\ , \qquad 0 \le k < j \le m\ . \tag{1.4.8}$$

Note that this is true for every $m \in \dot{\mathbb{N}}$. Furthermore, if we replace in (1.4.8) the integer m by some larger integer $\widetilde{m}$, we obtain the extrapolation scale

$$\big[(E^\sharp_j, A^\sharp_j)\ ;\ j \in \mathbb{Z} \cap [-\widetilde{m}, \infty)\big]$$

of order $\widetilde{m}$ without changing the previously constructed extrapolation scale of order $-m$, generated by $A^\sharp$. Thus, proceeding inductively, we can now indeed construct the two-sided power scale generated by $(E^\sharp, A^\sharp)$,

$$\big[(E^\sharp_k, A^\sharp_k)\ ;\ k \in \mathbb{Z}\big]\ .$$

By replacing in the above arguments the power scale $\big[(E_j, A_j)\ ;\ j \in \mathbb{N}\big]$ by the scale $\big[(E^\sharp_j, A^\sharp_j)\ ;\ k \in \mathbb{N}\big]$ and using Corollary 1.4.7 we obtain the **two-sided power scale** generated by (E, A),

$$\big[(E_k, A_k)\ ;\ k \in \mathbb{Z}\big]\ .$$

In the following theorem we collect these observations:

1.4.9 Theorem *Let E be a reflexive Banach space and suppose that $A \in \mathcal{C}(E)$ with dense domain and $0 \in \rho(A)$. Then the two-sided power scales*

$$\big[(E_k, A_k)\ ;\ k \in \mathbb{Z}\big] \quad \textit{and} \quad \big[(E^\sharp_k, A^\sharp_k)\ ;\ k \in \mathbb{Z}\big]\ ,$$

generated by (E, A) *and* $(E^\sharp, A^\sharp)$, *where* $E^\sharp := E'$ *and* $A^\sharp := A'$, *respectively, are well-defined densely injected reflexive Banach scales. Moreover,*

$$(E_k)' = E^\sharp_{-k} , \quad (A_k)' = A^\sharp_{-k} , \qquad k \in \mathbb{Z} ,$$

with respect to the duality pairing naturally induced by the E'-E *pairing* $\langle \cdot, \cdot \rangle$.

1.4.10 Remark Since, by definition, $E_{-\infty}$ is the dual of the projective limit space $E^\sharp_\infty = \lim_{\leftarrow} E^\sharp_j$, which is a Fréchet space, and since $E^\sharp_\infty \overset{d}{\hookrightarrow} E^\sharp_j$ for $j \in \mathbb{N}$ by Proposition 1.1.4, it follows that $E_{-\infty}$ is the inductive limit of the Banach spaces E_j, $j \in \mathbb{Z}$, with respect to the natural injections $E_j \overset{d}{\hookrightarrow} E_k$ for $j > k$, that is, $E_{-\infty} = \lim_{\rightarrow} E_j$ (cf. [Sch71, Sections IV.3.4, IV.4.4, and Corollary 2 to Theorem IV.4.3]. Thus

$$E_\infty = \lim_{\leftarrow} E_j \overset{d}{\hookrightarrow} E_j \overset{d}{\hookrightarrow} E_k \overset{d}{\hookrightarrow} E_{-\infty} = \lim_{\rightarrow} E_j , \qquad j > k .$$

Moreover, Proposition 1.1.3 implies

$$A_{-\infty} := (A^\sharp_\infty)' \in \mathcal{L}\mathrm{aut}(E_{-\infty})$$

and it is clear that A_j is the E_j-realization of $A_{-\infty}$ for $j \in \mathbb{Z}$. Similarly,

$$E^\sharp_\infty = \lim_{\leftarrow} E^\sharp_j \overset{d}{\hookrightarrow} E^\sharp_j \overset{d}{\hookrightarrow} E^\sharp_k \overset{d}{\hookrightarrow} E^\sharp_{-\infty} = \lim_{\rightarrow} E^\sharp_j , \qquad j > k ,$$

and

$$A^\sharp_{-\infty} := (A_\infty)' \in \mathcal{L}\mathrm{aut}(E^\sharp_{-\infty}) .$$

If $A^{-1} \in \mathcal{K}(E)$ then $(A^\sharp)^{-1} \in \mathcal{K}(E^\sharp)$ by Schauder's theorem. Therefore E_∞ and $E^\sharp_\infty$ are Montel spaces, hence reflexive, by Proposition 1.1.6. Thus $E_{-\infty}$ and $E^\sharp_{-\infty}$ are Montel spaces, consequently also reflexive (e.g., [Sch71, Section IV.5.9]). ■

Lastly, we extend the above results to the case of reflexive fractional power scales provided A is of positive type. For this and for later use we prepare the following technical observation:

1.4.11 Lemma *Let* E *be a reflexive Banach space and let* $A \in \mathcal{P}(E)$. *Then*

$$(A^z)' = (A')^z , \qquad \mathrm{Re}\, z \neq 0 . \tag{1.4.9}$$

If $A \in \mathcal{BIP}(E; M, \theta)$ *for some* $M \geq 1$ *and* $\theta \geq 0$ *then* $A' \in \mathcal{BIP}(E'; M, \theta)$ *and* (1.4.9) *holds for* $\mathrm{Re}\, z = 0$ *as well.*

Proof If $\mathrm{Re}\, z < 0$, relation (1.4.9) is an easy consequence of the integral representation (III.4.6.22). Then we obtain (1.4.9) for $\mathrm{Re}\, z > 0$ from (III.4.6.12). Thus suppose that $A \in \mathcal{BIP}(E; M, \theta)$. Then $\|A^z\| \leq KMe^{\theta|\mathrm{Im}\, z|}$ for $-1 < \mathrm{Re}\, z < 0$ and a suitable constant $K \geq 1$ by Lemma III.4.7.4(ii). Hence we obtain from (1.4.9) that $\|(A')^z\| \leq KMe^{\theta|\operatorname{Im} z|}$ for $-1 < \operatorname{Re} z < 0$ which, thanks to Lemma III.4.7.4(i), entails $A' \in \mathcal{BIP}(E'; KM, \theta)$. By (1.4.9),

$$\big\langle (A')^z x', x \big\rangle = \langle x, A^z x \rangle \ , \qquad x' \in E' \ , \quad x \in E \ , \quad \operatorname{Re} z < 0 \ .$$

Thus, letting z tend to it for some $t \in \mathbb{R}$, we infer from Theorem III.4.7.1 that

$$\big\langle (A')^{it} x', x \big\rangle = \langle x, A^{it} x \rangle \ , \qquad x' \in E' \ , \quad x \in E \ , \quad t \in \mathbb{R} \ ,$$

which implies that (1.4.9) is also true for $\operatorname{Re} z = 0$. Lastly,

$$\|(A')^{it}\| = \|(A^{it})'\| = \|A^{it}\| \leq Me^{\theta|t|} \ , \qquad t \in \mathbb{R} \ ,$$

shows that $A' \in \mathcal{BIP}(E'; M, \theta)$. ■

Now it is easy to prove the following theorem on fractional power scales. For simplicity we restrict ourselves to the reflexive case, the most important one for applications.

1.4.12 Theorem *Let E be a reflexive Banach space and suppose that $A \in \mathcal{P}(E)$. Then the two-sided fractional power scale generated by (E, A),*

$$\big[(E_\alpha, A_\alpha) \ ; \ \alpha \in \mathbb{R}\big] \ , \tag{1.4.10}$$

and its dual scale,

$$\big[(E_\alpha^\sharp, A_\alpha^\sharp) \ ; \ \alpha \in \mathbb{R}\big] \ , \tag{1.4.11}$$

generated by $(E^\sharp, A^\sharp)$, are well-defined densely injected reflexive Banach scales satisfying interpolation inequalities. Moreover,

$$(E_\alpha)' = E_{-\alpha}^\sharp \ , \quad (A_\alpha)' = A_{-\alpha}^\sharp \ , \qquad \alpha \in \mathbb{R} \ , \tag{1.4.12}$$

with respect to the duality pairing naturally induced by $\langle \cdot, \cdot \rangle$.

Proof Theorem 1.3.8 and the considerations leading to Theorem 1.4.9 imply that the scales (1.4.10) and (1.4.11) are well-defined and satisfy interpolation inequalities. Given $\alpha > 0$, put

$$\langle x^\sharp, x \rangle_\alpha := \big\langle (A_{-\alpha}^\sharp)^{-\alpha} x^\sharp, A^\alpha x \big\rangle \ , \qquad x \in E_\alpha \ , \quad x^\sharp \in E_{-\alpha} \ .$$

Note that this definition coincides with the one of Lemma 1.4.1 if $\alpha \in \dot{\mathbb{N}}$. From Theorem 1.3.8 we also infer that

$$\|(A_{-\alpha}^\sharp)^{-\alpha} x^\sharp\| = \|x^\sharp\|_{-\alpha} \ , \qquad x^\sharp \in E_{-\alpha}^\sharp \ .$$

Hence

$$\langle \cdot, \cdot \rangle_\alpha : E_{-\alpha}^\sharp \times E_\alpha \to \mathbb{K}$$

is a separating continuous bilinear form of norm 1. From this and Lemma 1.4.11 we infer, by replacing in the proof of Proposition 1.4.3 everywhere k by α, that (1.4.12) is true for $\alpha > 0$ with respect to the duality pairing $\langle \cdot, \cdot \rangle_\alpha$.

Suppose that $0 < \beta < \alpha$. Then, by replacing k by α and ℓ by β in the proof of Corollary 1.4.4 and by using (1.4.9), we see that

$$\langle x^\sharp, x\rangle_\alpha = \langle x^\sharp, x\rangle_\beta \ , \qquad x^\sharp \in E^\sharp_{-\beta} \ , \quad x \in E_\alpha \ .$$

This shows that $\langle\cdot,\cdot\rangle_\alpha$ coincides with the duality pairing naturally induced by $\langle\cdot,\cdot\rangle$.

Lastly, since $E_\alpha = \big(\mathrm{dom}(A^\alpha), \|A^\alpha\cdot\|\big)$ and $E_{-\alpha}$ is a completion of $(E, \|A^{-\alpha}\cdot\|)$ for $\alpha > 0$, it follows that $A^\alpha \in \mathcal{L}\mathrm{is}(E_\alpha, E)$ for $\alpha \in \mathbb{R}$. Hence E_α and, consequently, $E^\sharp_\alpha$ also, are reflexive for $\alpha \in \mathbb{R}$. Now the proof of Theorem 1.4.6 applies to give (1.4.12) for each $\alpha \in \mathbb{R}$. ■

The preceding duality theorem for reflexive fractional power scales has first been proven in [Ama87] under the assumption that $A \in \mathcal{G}(E)$. It has then been rediscovered in [Liu89]. To be precise, the results in [Ama87] and [Liu89] are only valid for extrapolated fractional power scales of order m for some $m \in \mathbb{N}$, since in neither paper — and also not in the papers by [Nag83] and [Wal86], where two-sided discrete power scales are being studied — the existence of universal superspaces $E_{-\infty}$ and $E^\sharp_{-\infty}$ has been established. The existence of these spaces, hence the fact that the two-sided [fractional] power scales are well-defined indeed, is proven here for the first time.

1.5 Interpolation-Extrapolation Scales

Let E be a Banach space and suppose that $A \in \mathcal{C}(E)$ is densely defined with $0 \in \rho(A)$. Then, given $m \in \mathbb{N}$, the extrapolated discrete power scale

$$\big[(E_j, A_j) \ ; \ j \in \mathbb{Z} \cap [-m, \infty)\big]$$

of order m, generated by (E, A), is a well-defined densely injected Banach scale. We specify for each $\theta \in (0,1)$ an admissible interpolation functor $(\cdot,\cdot)_\theta$ and put

$$E_\alpha := (E_j, E_{j+1})_{\alpha - j} \ , \quad A_\alpha := E_\alpha\text{-realization of } A_j \tag{1.5.1}$$

for $j < \alpha < j+1$ and $j \in \mathbb{Z} \cap [-m, \infty)$.

1.5.1 Theorem *$\big[(E_\alpha, A_\alpha) \ ; \ \alpha \in [-m, \infty)\big]$ is a densely injected Banach scale, the* **interpolation-extrapolation scale of order** m, **generated by** (E, A) **and** $(\cdot,\cdot)_\theta$, *$0 < \theta < 1$. It is compactly injected iff $A^{-1} \in \mathcal{K}(E)$. If this is the case,*

$$(A_\alpha)^{-1} \in \mathcal{K}(E_\alpha) \ , \qquad \alpha \in [-m, \infty) \ .$$

Proof Thanks to the admissibility of the interpolation functors $(\cdot,\cdot)_\theta$ we obtain the first assertion by interpolation from Theorems 1.2.1 and 1.3.2 and Remark 1.1.2(c). The second assertion is an easy consequence of Proposition 1.3.4

and Theorem I.2.11.1. The last assertion is implied by the commutativity of the diagram

$$\begin{array}{ccc} E_\alpha & \xrightarrow[\cong]{(A_\alpha)^{-1}} & E_{\alpha+1} \\ & {\scriptstyle (A_\alpha)^{-1}}\searrow \quad \swarrow{\scriptstyle i_\alpha} & \\ & E_\alpha & \end{array}$$

where $i_\alpha : E_{\alpha+1} \hookrightarrow E_\alpha$ is the natural injection. ∎

If we assume that A is of positive type, we can exploit the relations between the domains of fractional powers and suitable real interpolation spaces to obtain further information.

1.5.2 Theorem *Suppose that $A \in \mathcal{P}(E)$. Then the interpolation-extrapolation scale of order m, generated by (E, A) and $(\cdot,\cdot)_\theta$, $0 < \theta < 1$, satisfies interpolation inequalities.*

Proof Suppose that $-m \leq k \leq j < \ell < \infty$, where k, j, and ℓ are integers. Then, letting $F := E_k$ and $B := A_k$, it follows from Theorem 1.3.2 that

$$E_i \doteq D(B^{i-k}) \ , \qquad i \in \{k, j, \ell\} \ .$$

Hence (I.2.9.6) implies

$$(E_k, E_\ell)_{(j-k)/(\ell-k),1} \overset{d}{\hookrightarrow} E_j \overset{d}{\hookrightarrow} (E_k, E_\ell)^0_{(j-k)/(\ell-k),\infty} \ , \tag{1.5.2}$$

with the obvious interpretation if $j = k$. Moreover, we infer from (I.2.11.4) that

$$(E_j, E_{j+1})_{\alpha-j,1} \overset{d}{\hookrightarrow} E_\alpha \overset{d}{\hookrightarrow} (E_j, E_{j+1})^0_{\alpha-j,\infty} \ , \qquad j < \alpha < j+1 \ .$$

From (1.5.2) and the reiteration property (I.2.8.2) for the real method we deduce that

$$(E_j, E_{j+1})_{\alpha-j,q} \doteq (E_k, E_\ell)_{(\alpha-k)/(\ell-k),q} \ , \qquad 1 \leq q \leq \infty \ ,$$

provided $j < \alpha < j+1$. Thus it follows that

$$(E_k, E_\ell)_{(\alpha-k)/(\ell-k),1} \overset{d}{\hookrightarrow} E_\alpha \overset{d}{\hookrightarrow} (E_k, E_\ell)^0_{(\alpha-k)/(\ell-k),\infty} \tag{1.5.3}$$

for $k < \alpha < \ell$.

Finally, suppose that $-m \leq \gamma < \beta < \alpha < \infty$ and fix $k, \ell \in \mathbb{Z} \cap [-m, \infty)$ such that $k < \gamma < \beta < \alpha < \ell$ if $\gamma > -m$, and put $k := \gamma$ otherwise. Then (1.5.3) and the reiteration property (I.2.8.2) imply

$$(E_\gamma, E_\alpha)_{\eta,1} \hookrightarrow (E_k, E_\ell)_{(\beta-k)/(\ell-k),1} \hookrightarrow E_\beta \ , \tag{1.5.4}$$

where $\eta := (\beta-\gamma)/(\alpha-\gamma)$. Hence

$$\|x\|_\beta \le c\,\|x\|_{(E_\gamma,E_\alpha)_{\eta,1}} \le c\,\|x\|_\gamma^{1-\eta}\,\|x\|_\alpha^\eta \ , \qquad x \in E_\alpha \ ,$$

by Proposition I.2.2.1 and Example I.2.4.1. This proves the assertion. ■

It should be noted that, thanks to Theorem I.2.11.1, interpolation inequalities are always satisfied if $j \le \gamma < \beta < \alpha \le j+1$ for some $j \in \mathbb{Z}$, without the assumption that A be of positive type. The latter assumption has only been used to interpolate 'across integers'.

Now it is easy to show that the interpolation-extrapolation scale enjoys an almost reiteration property similar to the one of Remark I.2.11.2(a). Note that the latter remark is not applicable since, in general, $E_\alpha \ne (E_j, E_k)_{(\alpha-j)/(k-j)}$ if $j < \alpha < k$ with $j,k \in \mathbb{Z}$ and $k-j \ge 2$.

1.5.3 Theorem *Suppose that $A \in \mathcal{P}(E)$. Then the interpolation-extrapolation scale of order m, generated by (E,A) and $(\cdot,\cdot)_\theta$, $0<\theta<1$, possesses the following* **almost reiteration property**: *if $-m \le \alpha < \beta < \infty$ and $0 < \eta_- < \eta < \eta_+ < 1$ then*

$$(E_\alpha, E_\beta)_{\eta_+} \overset{d}{\hookrightarrow} E_{(1-\eta)\alpha+\eta\beta} \overset{d}{\hookrightarrow} (E_\alpha, E_\beta)_{\eta_-} \ .$$

Proof From (I.2.11.4), (I.2.5.2), and (1.5.4) we easily deduce that

$$(E_\alpha, E_\beta)_{\eta_+} \hookrightarrow (E_\alpha, E_\beta)_{\eta,1} \hookrightarrow E_{(1-\eta)\alpha+\eta\beta} \ .$$

Similarly, we see that

$$E_{(1-\eta)\alpha+\eta\beta} \hookrightarrow (E_\alpha, E_\beta)_{\eta,\infty} \hookrightarrow (E_\alpha, E_\beta)_{\eta_-} \ .$$

Since the density assertion is obvious, the theorem follows. ■

Suppose that $(\cdot,\cdot)_\theta$ is one of the real interpolation functors $(\cdot,\cdot)_{\theta,q}$, $1 \le q < \infty$, or $(\cdot,\cdot)_\theta = (\cdot,\cdot)_{\theta,\infty}^0$ *for each* $\theta \in (0,1)$, or $(\cdot,\cdot)_\theta = [\cdot,\cdot]_\theta$ *for each* $\theta \in (0,1)$. Then it follows from Remark I.2.11.2(b) that the reiteration property

$$(E_\alpha, E_\beta)_\eta \doteq E_{(1-\eta)\alpha+\eta\beta} \ , \qquad 0 < \eta < 1 \ , \tag{1.5.5}$$

holds, provided $j \le \alpha < \beta \le j+1$ for some $j \in \mathbb{Z}$. In general, however, this is not true if we interpolate 'across integers'. For example, $E_j \ne (E_{j-1}, E_{j+1})_{1/2}$, in general (see Chapter VI). In the following, we give a sufficient condition for the reiteration property (1.5.5) to hold unrestrictedly. Recall that $[\cdot,\cdot]_\theta$ denotes the complex interpolation functor of exponent θ.

1.5.4 Theorem *Suppose that $A \in \mathcal{BIP}(E)$. Then the interpolation-extrapolation scale $\big[(E_\alpha, A_\alpha)\ ;\ \alpha \in [-m,\infty)\big]$, generated by (E,A) and $[\cdot,\cdot]_\theta$, $0<\theta<1$, is equivalent to the fractional power scale of order m, generated by (E,A), that is,*

$$E_\alpha \doteq \begin{cases} \big(\mathrm{dom}(A^\alpha), \|A^\alpha\cdot\|\big) & 0 \le \alpha < \infty\ , \\ (E, \|A^\alpha\cdot\|)^\sim & -m \le \alpha < 0\ . \end{cases}$$

It possesses the **reiteration property**

$$[E_\alpha, E_\beta]_\eta \doteq E_{(1-\eta)\alpha+\eta\beta}\ , \qquad -m \le \alpha < \beta < \infty\ , \quad 0<\eta<1\ .$$

Moreover, $(A_\alpha)^{\alpha-\beta} \in \mathcal{L}\mathrm{is}(E_\alpha, E_\beta)$ for $-m \le \alpha < \beta < \infty$.

Proof Fix $\alpha, \beta \in [-m,\infty)$. By extending the scale, if necessary, we can assume that $-m<\alpha<\beta<m$. Put $F := E_{-m}$ and $B := A_{-m}$. Then $B \in \mathcal{BIP}(F)$ by Lemma 1.3.7, and $E_{i-m} \doteq D(B^i)$ for $i \in \mathbb{N}$ by Theorem 1.3.2. Hence we infer from (I.2.9.8) that

$$[E_{-m}, E_m]_{i/(2m)} \doteq E_{i-m}\ , \qquad i \in \mathbb{N} \cap [0,2m]\ . \tag{1.5.6}$$

Moreover, if $-m \le j \le \gamma < j+1 \le m$ with $j \in \mathbb{Z}$, we see from $E_\gamma = [E_j, E_{j+1}]_{\gamma-j}$, from (1.5.6), and the reiteration property (I.2.8.4) that

$$\begin{aligned} E_\gamma &= \big[[E_{-m}, E_m]_{(j+m)/(2m)}, [E_{-m}, E_m]_{(j+1+m)/(2m)}\big]_{\gamma-j} \\ &\doteq [E_{-m}, E_m]_{(\gamma+m)/(2m)}\ . \end{aligned}$$

Thus, again by reiteration,

$$\begin{aligned} [E_\alpha, E_\beta]_\eta &\doteq \big[[E_{-m}, E_m]_{(\alpha+m)/(2m)}, [E_{-m}, E_m]_{(\beta+m)/(2m)}\big]_\eta \\ &\doteq [E_{-m}, E_m]_{[(1-\eta)\alpha+\eta\beta+m]/(2m)} \doteq E_{(1-\eta)\alpha+\eta\beta}\ . \end{aligned}$$

Note that

$$\begin{aligned} E_\alpha \doteq [E_{-m}, E_m]_{(\alpha+m)/(2m)} &\doteq \big[F, D(B^{2m})\big]_{(\alpha+m)/(2m)} \\ &\doteq D(B^{\alpha+m}) = D\big((A_{-m})^{\alpha+m}\big) \end{aligned}$$

by (I.2.9.9) and thanks to the fact that $A_{-m} \in \mathcal{BIP}(E_{-m})$ by Lemma 1.3.7. Now the assertions follow from Theorem 1.3.8 and Corollary 1.3.9. ∎

The following technical proposition extends Lemma 1.3.7 and is of independent interest.

1.5.5 Proposition *Let $A \in \mathcal{P}(E)$ and let $\big[(E_\alpha, A_\alpha)\ ;\ \alpha \in [-m,\infty)\big]$ be the interpolation-extrapolation scale generated by (E, A) and $(\cdot,\cdot)_\theta,\ 0<\theta<1$, where each $(\cdot,\cdot)_\theta$ is exact.*

(i) *If $A \in \mathcal{P}(E; K, \vartheta)$ then $A_\alpha \in \mathcal{P}(E_\alpha; K, \vartheta)$.*

(ii) *If $A \in \mathcal{BIP}(E; M, \sigma)$ then $A_\alpha \in \mathcal{BIP}(E_\alpha; M, \sigma)$.*

Proof This follows from Lemma 1.3.7 by interpolation. ∎

Of course, if we drop the assumption that the interpolation functors be exact, Proposition 1.5.5 remains valid if we replace in the conclusions K and M by $c(\alpha)K$ and $c(\alpha)M$, respectively, for $\alpha \in \mathbb{R}\backslash\mathbb{Z}$, where $c(\alpha)$ is a suitable constant determined by the interpolation functor.

Suppose that E is reflexive and $A \in \mathcal{P}(E)$. Then the two-sided interpolation-extrapolation scale $\big[(E_\alpha, A_\alpha)\ ;\ \alpha \in \mathbb{R}\big]$ generated by (E, A) and $(\cdot,\cdot)_\theta,\ 0<\theta<1$, is well-defined. Fix $\gamma \in \mathbb{R}\backslash\mathbb{Z}$ and put $(F_j, B_j) := (E_{j+\gamma}, A_{j+\gamma})$ for $j \in \mathbb{Z}$. Since $B_0 \in \mathcal{P}(F_0)$ by Proposition 1.5.5, it is obvious that $\big[(F_j, B_j)\ ;\ j \in \mathbb{Z}\big]$ is the discrete power scale generated by $(F, B) := (E_\gamma, A_\gamma)$. (Note that in this case we do not need to restrict ourselves to extrapolation scales of finite order since the spaces F_{-j} are already given for each $j \in \mathbb{N}$ and since F_{-j} is obviously a completion of $(F, \|B^{-j}\cdot\|)$ for $j \in \dot{\mathbb{N}}$.) It should be observed that this scale need not be reflexive.

Determine for each $\theta \in (0,1)$ an admissible interpolation functor $\{\cdot,\cdot\}_\theta$. Then the interpolation-extrapolation scale $\big[(F_\alpha, B_\alpha)\ ;\ \alpha \in \mathbb{R}\big]$, generated by (F, B) and $\{\cdot,\cdot\}_\theta,\ 0<\theta<1$, is well-defined.

1.5.6 Remark In general, $(F_\alpha, B_\alpha) \neq (E_{\alpha+\gamma}, A_{\alpha+\gamma})$, as will be seen in the next chapter. Suppose, however, that $(\cdot,\cdot)_\theta = \{\cdot,\cdot\}_\theta = [\cdot,\cdot]_\theta$ for $0<\theta<1$ and let $A \in \mathcal{BIP}(E)$. Then it is a consequence of Theorem 1.5.4 and Lemma 1.5.5 that $(F_\alpha, B_\alpha) = (E_{\alpha+\gamma}, A_{\alpha+\gamma})$ for $\alpha \in \mathbb{R}$, except for equivalent norms. ∎

Now we give a sufficient condition for the scale $\big[(F_\alpha, B_\alpha)\ ;\ \alpha \in \mathbb{R}\big]$ to possess the reiteration property. To allow for a simple formulation, we put $(\cdot,\cdot)^0_{\theta,q} := (\cdot,\cdot)_{\theta,q}$ for $0<\theta<1$ and $1 \le q < \infty$.

1.5.7 Theorem *Suppose that E is reflexive, $A \in \mathcal{P}(E)$, and $\big[(E_\alpha, A_\alpha)\ ;\ \alpha \in \mathbb{R}\big]$ is the interpolation-extrapolation scale generated by (E, A) and $(\cdot,\cdot)_\theta,\ 0<\theta<1$. Fix $\gamma \in \mathbb{R}\backslash\mathbb{Z}$ and denote by $\big[(F_\alpha, B_\alpha)\ ;\ \alpha \in \mathbb{R}\big]$ the interpolation-extrapolation scale generated by $(F, B) := (E_\gamma, A_\gamma)$ and $\{\cdot,\cdot\}_\theta,\ 0<\theta<1$. Suppose there exists $q \in [1,\infty]$ such that $\{\cdot,\cdot\}_\theta = (\cdot,\cdot)^0_{\theta,q}$ for $0<\theta<1$ and $(\cdot,\cdot)_{\gamma-j} = (\cdot,\cdot)^0_{\gamma-j,q}$, where $j \in \mathbb{Z}$ satisfies $j<\gamma<j+1$. Then*

$$\{F_\alpha, F_\beta\}_\eta \doteq F_{(1-\eta)\alpha+\eta\beta}\ , \qquad -\infty<\alpha<\beta<\infty\ , \quad 0<\eta<1\ .$$

Proof Fix $k, \ell \in \mathbb{Z}$ such that $k \le \alpha < \beta < \ell$. Then it follows from (1.5.2) (applied to the F-scale) that

$$(F_k, F_\ell)_{(j-k)/(\ell-k),1} \hookrightarrow F_j \hookrightarrow (F_k, F_\ell)^0_{(j-k)/(\ell-k),\infty} \qquad (1.5.7)$$

for $j \in \mathbb{Z}$ and $k < j < \ell$. If $k \le j < \xi < j+1 \le \ell$ then $F_\xi = \{F_j, F_{j+1}\}_{\xi - j}$ by definition. Thus we obtain from (1.5.7) and the reiteration property (I.2.8.2) that

$$F_\xi \doteq \{F_k, F_\ell\}_{(\xi-k)/(\ell-k)} , \qquad \xi \in (k,\ell)\backslash\mathbb{Z} . \qquad (1.5.8)$$

Now we deduce from (1.5.7) and (1.5.8), thanks to (I.2.5.2) and (I.2.8.2), that

$$\{F_\alpha, F_\beta\}_\eta \doteq \{F_k, F_\ell\}_{[(1-\eta)\alpha+\eta\beta-k]/(\ell-k)} , \qquad 0 < \eta < 1 . \qquad (1.5.9)$$

Hence (1.5.8) and (1.5.9) guarantee that $\{F_\alpha, F_\beta\}_\eta \doteq F_{(1-\eta)\alpha+\eta\beta}$ for $\eta \in (0,1)$ with $(1-\eta)\alpha + \eta\beta \notin \mathbb{Z}$.

Thus suppose that $m := (1-\eta)\alpha + \eta\beta \in \mathbb{Z}$. There are $i \in \mathbb{Z}$ and $\sigma \in (0,1)$ such that $\gamma = i + \sigma$. Hence $F_j = E_{j+\gamma} = E_{i+j+\sigma} = \{E_{i+j}, E_{i+j+1}\}_\sigma$ for $j \in \mathbb{Z}$, since $\{\cdot,\cdot\}_\sigma = (\cdot,\cdot)_\sigma$ by assumption. So we deduce from (1.5.2) by reiteration that

$$F_n = \{E_{i+n}, E_{i+n+1}\}_\sigma \doteq \{E_{i+k}, E_{i+\ell+1}\}_{(n-k+\sigma)/(\ell+1-k)} \qquad (1.5.10)$$

for $n \in \mathbb{Z} \cap [k,\ell]$. By means of this and by using (I.2.5.2) and (I.2.8.2) we find that

$$\begin{aligned} &\{F_k, F_\ell\}_{(m-k)/(\ell-k)} \\ &\doteq \big\{\{E_{i+k}, E_{i+\ell+1}\}_{\sigma/(\ell+1-k)}, \{E_{i+k}, E_{i+\ell+1}\}_{(\ell-k+\sigma)/(\ell+1-k)}\big\}_{(m-k)/(\ell-k)} \\ &\doteq \{E_{i+k}, E_{i+\ell+1}\}_{(m-k+\sigma)/(\ell+1-k)} \doteq F_m , \end{aligned} \qquad (1.5.11)$$

which, together with (1.5.9), proves the theorem. ■

1.5.8 Corollary *Suppose that $\xi \in \mathbb{R}$ with $\gamma + \xi \notin \mathbb{Z}$. Also suppose $(\cdot,\cdot)_\eta = \{\cdot,\cdot\}_\eta$ for $\eta := \gamma + \xi - [\gamma + \xi]$. Then $F_\xi \doteq E_{\gamma+\xi}$.*

Proof Fix $k, \ell \in \mathbb{Z}$ with $k < \xi < \ell$. Then (1.5.8) and (1.5.10) imply

$$F_\xi \doteq \{E_{i+k}, E_{i+\ell+1}\}_{(\xi-k+\sigma)/(\ell+1-k)} \qquad (1.5.12)$$

by the same arguments that led to (1.5.11). Note that

$$i + \xi + \sigma = \xi + \gamma = \eta + [\gamma + \xi] \notin \mathbb{Z} .$$

Hence, by replacing in (1.5.8) the spaces F_k and F_ℓ by E_{i+k} and $E_{i+\ell+1}$, respectively, and using $(\cdot,\cdot)_\eta = \{\cdot,\cdot\}_\eta$ we find that

$$\{E_{i+k}, E_{i+\ell+1}\}_{(\xi-k+\sigma)/(\ell+1-k)} \doteq E_{i+\xi+\sigma} = E_{\gamma+\xi} .$$

Now the assertion follows from (1.5.12). ■

The above results allow a considerable strengthening of the almost reiteration property of Theorem 1.5.3, provided we restrict the interpolation functors suitably.

1.5.9 Theorem *Suppose that E is reflexive, $A \in \mathcal{P}(E)$, and $q \in [1,\infty]$, and put $(\cdot,\cdot)_\theta := (\cdot,\cdot)^0_{\theta,q}$ for $0<\theta<1$. Let $\big[(E_\alpha, A_\alpha)\ ;\ \alpha \in \mathbb{R}\big]$ be the interpolation-extrapolation scale generated by (E,A) and $(\cdot,\cdot)_\theta$, $0<\theta<1$. Put*

$$E^\bullet_\alpha := \begin{cases} E_\alpha & \text{if } \alpha \in \mathbb{R}\backslash\mathbb{Z}\ , \\ (E_{\alpha-1/2}, E_{\alpha+1/2})_{1/2} & \text{if } \alpha \in \mathbb{Z}\ . \end{cases} \tag{1.5.13}$$

Then $(E^\bullet_\alpha, E^\bullet_\beta)_\eta \doteq E^\bullet_{(1-\eta)\alpha+\eta\beta}$ for $-\infty<\alpha<\beta<\infty$ and $0<\eta<1$.

Proof This follows from Theorem 1.5.7 and Corollary 1.5.8 by letting $\gamma := 1/2$. ■

Of course, we put

$$A^\bullet_\alpha := \begin{cases} A_\alpha & \text{if } \alpha \in \mathbb{R}\backslash\mathbb{Z}\ , \\ E^\bullet_\alpha\text{-realization of } A_{\alpha-1/2} & \text{if } \alpha \in \mathbb{Z}\ . \end{cases} \tag{1.5.14}$$

Then $(E^\bullet_\alpha, A^\bullet_\alpha) = (F_{\alpha-1/2}, B_{\alpha-1/2})$ for $\alpha \in \mathbb{R}$, where $\big[(F_\alpha, B_\alpha)\ ;\ \alpha\in\mathbb{R}\big]$ is the interpolation-extrapolation scale generated by $(F,B) := (E_{1/2}, A_{1/2})$ and $(\cdot,\cdot)_\theta$, $0<\theta<1$. The following theorem shows that this scale is, in fact, a fractional power scale, provided we restrict the admitted interpolation functors slightly.

1.5.10 Theorem *Suppose that E is reflexive, $A \in \mathcal{P}(E,\vartheta)$ for some $\vartheta \in [0,\pi)$, that $1 \le q < \infty$, and $(\cdot,\cdot)_\theta = (\cdot,\cdot)_{\theta,q}$ for $1\le q<\infty$. Let $\big[(E_\alpha, A_\alpha)\ ;\ \alpha\in\mathbb{R}\big]$ be the interpolation-extrapolation scale generated by (E,A) and $(\cdot,\cdot)_\theta$, $0<\theta<1$, and define $(E^\bullet_\alpha, A^\bullet_\alpha)$ for $\alpha\in\mathbb{R}$ by (1.5.13) and (1.5.14). Then $A^\bullet_\alpha \in \mathcal{BIP}(E^\bullet_\alpha;\pi-\vartheta)$. Moreover, $\big[(E^\bullet_\alpha, A^\bullet_\alpha)\ ;\ \alpha\in\mathbb{R}\big]$ is equivalent to the fractional power scale generated by $(E^\bullet_0, A^\bullet_0)$, and $(A^\bullet_\alpha)^{\alpha-\beta} \in \mathcal{L}\mathrm{is}(E^\bullet_\alpha, E^\bullet_\beta)$ for $-\infty<\alpha<\beta<\infty$.*

Proof Theorem III.4.7.5 guarantees that $A_{1/2} \in \mathcal{BIP}(E_{1/2};\pi-\vartheta)$. Thus, letting $(F,B) := (E_{1/2}, A_{1/2})$ and denoting by $\big[(F_\alpha, B_\alpha)\ ;\ \alpha\in\mathbb{R}\big]$ the interpolation-extrapolation scale generated by (F,B) and $(\cdot,\cdot)_\theta$, $0<\theta<1$, it follows from Proposition 1.5.5 that $B_\alpha \in \mathcal{BIP}(F_\alpha;\pi-\vartheta)$ for $\alpha\in\mathbb{R}$. Consequently, using the fact that $(F_\alpha, B_\alpha) \doteq (E^\bullet_{\alpha+1/2}, A^\bullet_{\alpha+1/2})$, we get $A^\bullet_\alpha \in \mathcal{BIP}(E^\bullet_\alpha;\pi-\vartheta)$ for $\alpha\in\mathbb{R}$.

Let $\big[(G_\alpha, C_\alpha)\ ;\ \alpha\in\mathbb{R}\big]$ be the fractional power scale generated by (F,B). Then Theorem 1.5.4 implies that it is equivalent to the interpolation-extrapolation scale generated by (F,B) and $[\cdot,\cdot]_\theta$, $0<\theta<1$. Let $\alpha\in\mathbb{R}$ be given and let $k\in\mathbb{Z}$ satisfy $k-1<\alpha<k+1$. Then, since $G_j = F_j$ for $j\in\mathbb{Z}$, we deduce from the reiteration properties of Theorems 1.5.4 and 1.5.7 and from (I.2.8.6)

$$\begin{aligned} G_\alpha &\doteq [G_{k-1}, G_{k+1}]_{(\alpha-k+1)/2} = [F_{k-1}, F_{k+1}]_{(\alpha-k+1)/2} \\ &\doteq \big[(F_{k-2}, F_{k+2})_{1/4}, (F_{k-2}, F_{k+2})_{3/4}\big]_{(\alpha-k+1)/2} \\ &\doteq (F_{k-2}, F_{k+2})_{(\alpha-k+2)/4} \doteq F_\alpha\ . \end{aligned}$$

This proves that the interpolation-extrapolation scale $\big[(F_\alpha, B_\alpha)\ ;\ \alpha \in \mathbb{R}\big]$ is equivalent to the fractional power scale generated by $(E^\bullet_{1/2}, A^\bullet_{1/2})$. Now the assertion follows from Proposition 1.2.6 and Corollary 1.3.9. ■

1.5.11 Remark The assumption that E be reflexive has only been used to be able to deal with two-sided scales which simplifies the presentation. It is clear that everything proven above remains valid — with obvious adaptations — if we do not presuppose the reflexivity of E but consider extrapolation scales of finite order only. ■

Now we suppose that

$$\left.\begin{aligned} &E \text{ is a reflexive Banach space and} \\ &(\cdot,\cdot)_\theta \in \big\{[\cdot,\cdot]_\theta, (\cdot,\cdot)_{\theta,q}, (\cdot,\cdot)^0_{\theta,\infty}\ ;\ 1<q<\infty\big\},\quad 0<\theta<1. \end{aligned}\right\} \tag{1.5.15}$$

Then we define the **dual interpolation functor** $(\cdot,\cdot)^\sharp_\theta$ by

$$(\cdot,\cdot)^\sharp_\theta := \begin{cases} [\cdot,\cdot]_\theta & \text{if } (\cdot,\cdot)_\theta = [\cdot,\cdot]_\theta\ , \\ (\cdot,\cdot)_{\theta,1} & \text{if } (\cdot,\cdot)_\theta = (\cdot,\cdot)^0_{\theta,\infty}\ , \\ (\cdot,\cdot)_{\theta,q'} & \text{if } (\cdot,\cdot)_\theta = (\cdot,\cdot)_{\theta,q}\ , \quad 1<q<\infty\ . \end{cases}$$

Note that $(\cdot,\cdot)^\sharp_\theta$ is an admissible interpolation functor for $0<\theta<1$. Thus the **dual interpolation-extrapolation scale**

$$\big[(E^\sharp_\alpha, A^\sharp_\alpha)\ ;\ \alpha \in \mathbb{R}\big]\ ,$$

that is, the interpolation-extrapolation scale generated by $(E^\sharp, A^\sharp)$ and $(\cdot,\cdot)^\sharp_\theta$, $0<\theta<1$, is a well-defined densely injected Banach scale. The next theorem extends Theorem 1.4.9 to interpolation-extrapolation-scales.

1.5.12 Theorem *Let condition* (1.5.15) *be satisfied and suppose that* $A \in \mathcal{C}(E)$ *is densely defined with* $0 \in \rho(A)$. *Then* $(E_\alpha)' \doteq E^\sharp_{-\alpha}$ *and* $(A_\alpha)' \doteq A^\sharp_{-\alpha}$ *for* $\alpha \in \mathbb{R}$ *with respect to the duality pairing naturally induced by the* E'-E-*duality pairing* $\langle\cdot,\cdot\rangle$. *If*

$$(\cdot,\cdot)_\theta \in \big\{[\cdot,\cdot]_\theta, (\cdot,\cdot)_{\theta,q}\ ;\ 1<q<\infty\big\}\ , \qquad 0<\theta<1\ ,$$

then E_α *is reflexive for* $\alpha \in \mathbb{R}$.

Proof Thanks to Theorem 1.4.9 it remains to consider the case $\alpha \in \mathbb{R}\backslash\mathbb{Z}$. Thus $j<\alpha<j+1$ for some $j \in \mathbb{Z}$. Then it follows from (I.2.6.1)–(I.2.6.3), (I.2.5.4), and Theorem 1.4.9 that

$$\begin{aligned} (E_\alpha)' = (E_j, E_{j+1})'_{\alpha-j} &\doteq (E'_j, E'_{j+1})^\sharp_{\alpha-j} = (E^\sharp_{-j}, E^\sharp_{-j-1})^\sharp_{\alpha-j} \\ &= (E^\sharp_{-j-1}, E^\sharp_{-j})^\sharp_{1-\alpha+j} = E^\sharp_{-\alpha}\ . \end{aligned}$$

Suppose that $x \in E_{\alpha+1} \hookrightarrow E_{j+1}$ and $x^\sharp \in E^\sharp_{-j+1} \hookrightarrow E^\sharp_{-\alpha+1}$. Then, using $A_j \supset A_\alpha$ and $A^\sharp_{-\alpha} \supset A^\sharp_{-j}$ together with Theorem 1.4.9, it follows that

$$\langle x^\sharp, A_\alpha x\rangle = \langle x^\sharp, A_j x\rangle = \langle A^\sharp_{-j} x^\sharp, x\rangle = \langle A^\sharp_{-\alpha} x^\sharp, x\rangle \ .$$

Since $A^\sharp_{-\alpha} \in \mathcal{L}(E^\sharp_{-\alpha+1}, E^\sharp_{-\alpha})$ and $E^\sharp_{-j+1}$ is dense in $E^\sharp_{-\alpha+1}$, we deduce that

$$\langle x^\sharp, A_\alpha x\rangle = \langle A^\sharp_{-\alpha} x^\sharp, x\rangle \ , \qquad x^\sharp \in E^\sharp_{-\alpha+1} \ , \quad x \in E_{\alpha+1} \ .$$

Hence $(A_\alpha)' \supset A^\sharp_{-\alpha}$. Now $0 \in \rho\big((A_\alpha)'\big) = \rho(A_\alpha)$ and $A^\sharp_{-\alpha} \in \mathcal{L}\mathrm{is}(E^\sharp_{-\alpha+1}, E^\sharp_{-\alpha})$, together with Lemma 1.4.2, guarantee that $(A_\alpha)' = A^\sharp_{-\alpha}$.

By reflexivity, $(E^\sharp)^\sharp = E$ and $(A^\sharp)^\sharp = A$. If $(\cdot,\cdot)_\theta \notin \big\{(\cdot,\cdot)_{\theta,1}, (\cdot,\cdot)^0_{\theta,\infty}\big\}$ for $0<\theta<1$ then $\big((\cdot,\cdot)^\sharp_\theta\big)^\sharp = (\cdot,\cdot)_\theta$. Hence $\big[(E_\alpha, A_\alpha) \ ; \ \alpha \in \mathbb{R}\big]$ is the dual scale of the scale $\big[(E^\sharp_\alpha, A^\sharp_\alpha) \ ; \ \alpha \in \mathbb{R}\big]$ generated by $(E^\sharp, A^\sharp)$ and $(\cdot,\cdot)^\sharp_\theta$, $0<\theta<1$. Now, by what has already been shown,

$$\big((E_\alpha)'\big)' \doteq (E^\sharp_{-\alpha})' = (E^\sharp_{-\alpha})^\sharp = E_\alpha \ , \qquad \alpha \in \mathbb{R} \ .$$

This implies the asserted reflexivity. ■

1.5.13 Corollary *Suppose that $A \in \mathcal{BIP}(E)$ and let $\big[(E_\alpha, A_\alpha) \ ; \ \alpha \in \mathbb{R}\big]$ be the interpolation-extrapolation scale generated by (E, A) and $[\cdot,\cdot]_\theta$, $0<\theta<1$. Then $\big[(E_\alpha, A_\alpha) \ ; \ \alpha \in \mathbb{R}\big]$ is equivalent to the fractional power scale generated by (E, A), and its dual scale $\big[(E^\sharp_\alpha, A^\sharp_\alpha) \ ; \ \alpha \in \mathbb{R}\big]$ is equivalent to the fractional power scale generated by (E', A').*

Proof This is a consequence of Theorem 1.5.4, Lemma 1.4.11, and the foregone theorem. ■

Let $n \in \dot{\mathbb{N}}$ and suppose that $E_1, \ldots, E_n$ and F are *LCSs* over the same field. Then

$$\mathcal{L}(E_1, \ldots, E_n; F)$$

is the vector space of all continuous n-linear maps from $E_1 \times \cdots \times E_n$ to F, and

$$\mathcal{L}^n(E, F) := \mathcal{L}(\underbrace{E, \ldots, E}_{n}; F) \ . \tag{1.5.16}$$

Note that $\mathcal{L}^1(E, F) = \mathcal{L}(E, F)$. If $E_1, \ldots, E_n$ and F are normed vector spaces,

$$\|T\| := \sup\big\{ \|T(x_1, \ldots, x_n)\|_F \ ; \ \|x_j\|_{E_j} \le 1, \ 1 \le j \le n \big\} \tag{1.5.17}$$

defines a norm on $\mathcal{L}(E_1, \ldots, E_n; F)$, and the latter space is a Banach space if F is one.

Let $b \in \mathcal{L}(E_1, E_2; \mathbb{K})$ be given. Then there exist unique

$$B_1 \in \mathcal{L}\big(E_1, (E_2)'\big) \quad \text{and} \quad B_2 \in \mathcal{L}\big(E_2, (E_1)'\big) \tag{1.5.18}$$

satisfying

$$b(x_1, x_2) = \langle B_1 x_1, x_2 \rangle_{E_2} = \langle B_2 x_2, x_1 \rangle_{E_1} \ , \qquad (x_1, x_2) \in E_1 \times E_2 \ , \tag{1.5.19}$$

the **form operators** induced by b. Note that

$$\|B_1\| = \|B_2\| = \|b\| \tag{1.5.20}$$

if E_1 and E_2 are normed vector spaces.

The following observation is almost trivial. However, in concrete situations it plays an important rôle since it allows to identify extrapolated operators $A_{-\alpha}$ for certain $\alpha > 0$ explicitly.

1.5.14 Proposition *Let condition* (1.5.15) *be satisfied. Suppose that* $0 \leq \alpha \leq 1$ *and that* $b \in \mathcal{L}(E^\sharp_{1-\alpha}, E_\alpha; \mathbb{K})$, *and let* B_j, $j = 1, 2$, *be the form operators induced by* b. *If*

$$b(x^\sharp, x) = \langle x^\sharp, Ax \rangle \ , \qquad (x^\sharp, x) \in E^\sharp_1 \times E_1 \ , \tag{1.5.21}$$

then $B_1 = A^\sharp_{-\alpha}$ *and* $B_2 = A_{\alpha-1}$.

Proof From (1.5.18), (1.5.19), and Theorem 1.5.12 we infer $B_1 \in \mathcal{L}(E^\sharp_{1-\alpha}, E^\sharp_{-\alpha})$ and $B_2 \in \mathcal{L}(E_\alpha, E_{\alpha-1})$, and (1.5.21) implies $B_2 \supset A$. Since $A_{\alpha-1} \in \mathcal{L}(E_\alpha, E_{\alpha-1})$ and $A_{\alpha-1} \supset A$, we obtain $A_{\alpha-1} = B_2$ by the density of E_1 in E_α. A similar argument shows that $A^\sharp_{-\alpha} = B_1$. ■

Lastly, we consider the case that E is a Hilbert space and A is self-adjoint and positive definite, that is, $A = A^* \geq a$ for some $a > 0$.

1.5.15 Theorem *Let* $E := \big(E, (\cdot|\cdot)\big)$ *be a Hilbert space and let* $A = A^* \geq a$ *for some* $a > 0$. *Then the fractional power scale* $\big[(E_\alpha, A_\alpha) \ ; \ \alpha \in \mathbb{R}\big]$ *generated by* (E, A) *is a Hilbert scale. More precisely,* $E_\alpha = \big(E_\alpha, (\cdot|\cdot)_\alpha\big)$, *where*

$$(x|y)_\alpha = (A^\alpha x | A^\alpha y) \ , \qquad x, y \in E_{(\alpha+1)\vee 1} \ ,$$

and $A_\alpha = (A_\alpha)^* \geq a$ *for* $\alpha \in \mathbb{R}$. *It is equivalent to the interpolation-extrapolation scale generated by* (E, A) *and* $[\cdot, \cdot]_\theta$, $0 < \theta < 1$, *which, in turn, is equivalent to the interpolation-extrapolation scale generated by* (E, A) *and* $(\cdot, \cdot)_{\theta,2}$, $0 < \theta < 1$.

Proof Thanks to Theorem III.4.6.7 and Lemma 1.4.11, the first assertion is easily verified. The second one follows from Theorem 1.5.4 and Example III.4.7.3(a). The last assertion is a consequence of (I.2.9.11). ■

1.5.16 Remarks **(a)** Let $E := \bigl(E, (\cdot|\cdot)_E\bigr)$ be a Hilbert space. The **Riesz duality mapping** $\vartheta := \vartheta_E$ is the conjugate linear map from E to E' defined by

$$\langle \vartheta x, y\rangle_E = (y|x)_E \ , \qquad x, y \in E \ . \tag{1.5.22}$$

Thanks to the Riesz representation theorem, ϑ is a well-defined bijective isometry. Defining the **dual inner product** $(\cdot|\cdot)_{E'}$ by

$$(x'|y')_{E'} := (\vartheta^{-1}y', \vartheta^{-1}x')_E \ , \qquad x', y' \in E' \ , \tag{1.5.23}$$

it follows that $\bigl(E', (\cdot|\cdot)_{E'}\bigr)$ is a Hilbert space as well.

Let $\bigl(F, (\cdot|\cdot)_F\bigr)$ be a second Hilbert space and let $B \in \mathcal{C}(E, F)$ be densely defined. Then the dual $B' \in \mathcal{C}(F', E')$ and the adjoint $B^* \in \mathcal{C}(F, E)$ are well-defined, and it is easily verified that the diagram

$$\begin{array}{ccccc} F & \supset & \operatorname{dom}(B^*) & \xrightarrow{\ B^*\ } & E \\ \vartheta_F \big\downarrow \cong & & \vartheta_F \big\downarrow \cong & & \cong \big\downarrow \vartheta_E \\ F' & \supset & \operatorname{dom}(B') & \xrightarrow{\ B'\ } & E' \end{array} \tag{1.5.24}$$

is commutative, where, in abuse of notation, we use $\cong$ to denote conjugate-linear bijections. Thus

$$B' = \vartheta_E B^* \vartheta_F^{-1} \ . \tag{1.5.25}$$

Given $x \in \operatorname{dom}(B)$ and $y' \in \operatorname{dom}(B')$, it follows from (1.5.22) and (1.5.23) that

$$\begin{aligned} \langle y', Bx\rangle_F &= \langle B'y', x\rangle_E = (x|\vartheta_E^{-1}B'y')_E = (B'y'|\vartheta_E x)_{E'} \\ &= \bigl(y'|(B')^*\vartheta_E x\bigr)_{F'} = \bigl(\vartheta_F^{-1}(B')^*\vartheta_E x|\vartheta_F^{-1}y'\bigr)_F \\ &= \bigl\langle y', \vartheta_F^{-1}(B')^*\vartheta_E x\bigr\rangle \ . \end{aligned}$$

Consequently, by the analogue to (1.5.24),

$$(B')^* = \vartheta_F B \vartheta_E^{-1} \ . \tag{1.5.26}$$

Now suppose that $E = F$ and that $B = B^*$. Then we infer from (1.5.25) and (1.5.26) that $B' \in \mathcal{C}(E')$ is self-adjoint also. Moreover, if $B = B^* \geq b$ for some $b \in \mathbb{R}$, it follows from (1.5.25) and (1.5.23) that

$$(B'x'|x')_{E'} = (\vartheta B \vartheta^{-1}x'|x')_{E'} = (B\vartheta^{-1}x'|\vartheta^{-1}x')_E \geq b\,\|\vartheta^{-1}x'\|_E^2 = b\,\|x'\|_{E'}^2$$

for $x' \in \operatorname{dom}(B')$, which shows that $B' \geq b$.

(b) Let (X, μ) be a σ-finite positive measure space and let $E := L_2(X, \mu; \mathbb{K}^N)$ for some $N \in \dot{\mathbb{N}}$. Then

$$\langle u, v \rangle = \int_X \langle u, v \rangle \, d\mu \quad \text{and} \quad (u|v) = \langle u, \bar{v} \rangle \ , \qquad u, v \in E \ ,$$

where

$$\langle \xi, \eta \rangle := \sum_{j=1}^{N} \xi^j \eta^j \ , \qquad \xi := (\xi^1, \ldots, \xi^N), \eta := (\eta^1, \ldots, \eta^N) \in \mathbb{K}^N \ .$$

In this case the duality mapping is given by complex conjugation, that is, $\vartheta u = \bar{u}$ for $u \in E$.

(c) Let $E := \big(E, (\cdot\,|\,\cdot)\big)$ be a Hilbert space and let $A \in \mathcal{C}(E)$ be positive definite and self-adjoint. Then it follows from (a) and Theorems 1.5.12 and 1.5.15 that the scale $\big[(E_\alpha^\sharp, A_\alpha^\sharp) \ ; \ \alpha \in \mathbb{R}\big]$, dual to the fractional power scale $\big[(E_\alpha, A_\alpha) \ ; \ \alpha \in \mathbb{R}\big]$ generated by (E, A), is a Hilbert scale. ■

Most of the theorems of this subsection are generalizations and refinements of earlier results of the author (cf. [Ama86], [Ama87], [Ama88b], and [Ama88c]). For Theorem 1.5.4 we also refer to [Tri78, Section 1.5.3]. Theorem 1.5.7 is due to Simonett [Sim92]. Theorem 1.5.10 is new. In [DaPG84] Da Prato and Grisvard introduced the extrapolation spaces $(E_{-1}, E_0)_{\theta,\infty}^0$, $0 < \theta < 1$.

2 Evolution Equations in Banach Scales

Having established a powerful theory of interpolation-extrapolation scales, we now study the behavior of semigroups and evolution equations in these scales. We show that, roughly and imprecisely speaking, we can 'shift' a given evolution equation 'along a scale' to obtain a continuum of related problems. Of particular importance is the fact that we can develop a good duality theory. This leads to interpretations of solutions to 'extrapolated' Cauchy problems as naturally defined weak solutions. We also investigate the case where the pair (E, A), generating the scale, consists of an ordered Banach space and a resolvent positive operator of positive type. In the last subsection, for completeness we briefly discuss some implications of our general results to evolution equations that are not of parabolic type and lie, in principle, outside the range of our interest.

2.1 Semigroups in Interpolation-Extrapolation Scales

Interpolation-extrapolation scales are most useful for getting precise information on regularity and smoothing properties of linear operators. In this subsection we study, in particular, the behavior of semigroups in interpolation-extrapolation scales. These investigations are the abstract basis for the 'weak theory' of linear and quasilinear parabolic systems to be developed in later chapters.

The following simple lemma is of general interest and is given in a form that is more general than presently needed.

2.1.1 Lemma *Let E and F be Fréchet spaces and suppose that there exists a LCS G such that $E \hookrightarrow G$ and $F \hookrightarrow G$. If E is a vector subspace of F then $E \hookrightarrow F$.*

Proof Let $i \colon E \to F$ be the natural injection. Let (x_j) be a sequence in E such that $x_j \to x$ in E and $i(x_j) \to y$ in F. Since $E \hookrightarrow G$ and $F \hookrightarrow G$, we see that $x_j \to x$ in G and $x_j = i(x_j) \to y$ in G. Hence $x = y$, that is, $i(x) = y$, so that the graph of i is closed in $E \times F$. Now the assertion follows from the closed graph theorem. ■

2.1.2 Lemma *Let F and G be Banach spaces with $F \hookrightarrow G$. Suppose that $B \in \mathcal{G}(G)$ with $\operatorname{dom}(B) \overset{d}{\subset} F$, that the semigroup $\{\, e^{-tB} \ ; \ t \geq 0 \,\}$ leaves F invariant, and that $\|e^{-tB}\|_{\mathcal{L}(F)} \leq c$ for $0 \leq t \leq 1$. Then $B_F \in \mathcal{G}(F)$ and $e^{-tB_F} = e^{-tB} | F$ for $t \geq 0$.*

Proof Fix $\omega \in \rho(-B)$ and let $G_1 := \operatorname{dom}(B)$, endowed with the norm $\|(\omega + B)\cdot\|$. Denote by B_1 the G_1-realization of B. Then $(\omega + B)e^{-tB} \supset e^{-tB}(\omega + B)$, as follows from Lemma III.4.9.1. This implies the commutativity of the diagram

$$
\begin{array}{ccc}
G & \xrightarrow{\quad e^{-tB} \quad} & G \\
{\scriptstyle \omega + B}\Big\uparrow{\scriptstyle \cong} & & {\scriptstyle \cong}\Big\uparrow{\scriptstyle \omega + B} \\
G_1 & \xrightarrow{\quad U_1(t) := e^{-tB} | G_1 \quad} & G_1
\end{array}
$$

From this we see that $\big\{\, U_1(t) \ ; \ t \geq 0 \,\big\}$ is a strongly continuous semigroup on G_1.

Let $U_F(t) := e^{-tB} | F$ for $t \geq 0$. Then $\{\, U_F(t) \ ; \ t \geq 0 \,\}$ is a semigroup on F and $U_F(t) \supset U_1(t)$ for $t \geq 0$. Since $G_1 \subset F$ and $G_1 \hookrightarrow G$, it follows from Lemma 2.1.1 that $G_1 \hookrightarrow F$. Thus, given $x \in G_1$, we see that $U_F(t)x = U_1(t)x \to x$ in G_1, hence in F, as $t \to 0$. Since $\|U_F(t)\|_{\mathcal{L}(F)} \leq c$ for $0 \leq t \leq 1$ and since G_1 is dense in F by assumption, it follows that $\{\, U_F(t) \ ; \ t \geq 0 \,\}$ is a strongly continuous semigroup on F. Let A be the negative infinitesimal generator of this semigroup. Given $\lambda > \operatorname{type}(-A) \vee \operatorname{type}(-B)$, we infer from the semigroup representation of the resolvent (cf. (II.6.3.3)) that

$$
(\lambda + A)^{-1} x = \int_0^\infty e^{-\lambda t} U_F(t) x \, dt = \int_0^\infty e^{-\lambda t} e^{-tB} x \, dt = (\lambda + B)^{-1} x \ , \qquad x \in F \ .
$$

Thus, thanks to Lemma 1.1.1, $(\lambda + A)^{-1} = (\lambda + B_F)^{-1}$, which implies $A = B_F$. ■

Now we suppose that

$$\left.\begin{array}{l} E \text{ is a Banach space,} \\ B \in \mathcal{C}(E) \text{ and is densely defined with } 0 \in \rho(B), \quad m \in \mathbb{N}, \\ \text{and } (\cdot,\cdot)_\theta, \quad 0<\theta<1, \text{ are exact admissible interpolation functors.} \end{array}\right\} \tag{2.1.1}$$

We denote by

$$\big[(E_\alpha, B_\alpha)\ ;\ \alpha \in [-m,\infty)\big] \tag{2.1.2}$$

the interpolation-extrapolation scale of order m, generated by (E,B) and $(\cdot,\cdot)_\theta$, $0<\theta<1$, and by

$$\big[(E_j^\sharp, B_j^\sharp)\ ;\ j \in \mathbb{N}\big] \tag{2.1.3}$$

the dual power scale generated by $(E^\sharp, B^\sharp) := (E', B')$ (recall Remark 1.4.5(b)).

Given any densely defined $A \in \mathcal{C}(E)$, we put $A^\sharp := A'$ and denote by A_α the E_α-realization of A for $\alpha \in \mathbb{R}$, which is well-defined. Moreover, if A is closable in $E_{-\alpha}$ for some $\alpha \in (0,m]$, we put

$$A_{-\alpha} := \text{closure of } A \text{ in } E_{-\alpha} \tag{2.1.4}$$

and call $A_{-\alpha}$ the **$E_{-\alpha}$-realization** of A. Note that $A_\alpha \in \mathcal{C}(E_\alpha)$, whenever it is well-defined.

Suppose that $n \in \mathbb{N}$ and $A \in \mathcal{C}(E)$ satisfy $\rho(A) \neq \emptyset$ and

$$\operatorname{dom}(A^k) = E_k\ , \qquad 0 \le k \le n\ , \quad k \in \mathbb{N}\ . \tag{2.1.5}$$

Then, given $\mu \in \rho(-A)$ and using the fact that $\operatorname{dom}\big((\mu+A)^k\big) = \operatorname{dom}(A^k)$ for $k \in \mathbb{N}$ (cf. [HiP57, Theorem 2.16.4]), it follows from the closed graph theorem that $(\mu+A)^k B^{-k}$ and $B^k(\mu+A)^{-k}$ belong to $\mathcal{L}(E)$ for $0 \le k \le n$, or, equivalently, that

$$(\mu+A)^k \in \mathcal{L}(E_k, E)\ , \quad (\mu+A)^{-k} \in \mathcal{L}(E, E_k) \tag{2.1.6}$$

for $0 \le k \le n$. Similarly, if

$$\operatorname{dom}\big((A^\sharp)^k\big) = E_k^\sharp\ , \qquad 0 \le k \le m\ , \tag{2.1.7}$$

it follows from $\rho(A) = \rho(A^\sharp)$ that

$$(\mu+A^\sharp)^k \in \mathcal{L}(E_k^\sharp, E^\sharp)\ , \quad (\mu+A^\sharp)^{-k} \in \mathcal{L}(E^\sharp, E_k^\sharp) \tag{2.1.8}$$

for $0 \le k \le m$. Thus

$$\left.\begin{array}{r} N(A,\mu,m,n) := \text{maximum of the norms of all the operators} \\ \text{occurring in (2.1.6) and (2.1.8)} \end{array}\right\} \tag{2.1.9}$$

is well-defined if (2.1.5) and (2.1.7) are satisfied.

The following fundamental theorem is the main result of this subsection. In fact, it is one of the main results in the theory of interpolation-extrapolation scales — at least for our purposes. It shows that, if $A \in \mathcal{G}(E)$ and satisfies (2.1.5) and (2.1.7) then A_α exists and belongs to $\mathcal{G}(E_\alpha)$ for $-m \le \alpha \le n$, and the semigroups $\{ e^{-tA_\alpha} \; ; \; t \ge 0 \}$ are obtained naturally from $\{ e^{-tA} \; ; \; t \ge 0 \}$ by restriction and extension, respectively (cf. diagram (2.1.12)). Moreover, if $-A$ generates an analytic semigroup, the same is true for $-A_\alpha$. In addition, the theorem contains useful quantitative information. This theorem is the abstract basis for weak formulations of rather general initial-boundary value problems for a wide variety of concrete differential equations. It is also the foundation of abstract 'bootstrapping' arguments leading to rather precise regularity results. This will be clear from the results developed in the subsequent sections (of this and the following volume).

2.1.3 Theorem *Let assumption* (2.1.1) *be satisfied, let* $\big[(E_\alpha, B_\alpha) \; ; \; \alpha \in [-m, \infty) \big]$ *be the interpolation-extrapolation scale of order* m*, generated by* (E, B) *and* $(\cdot,\cdot)_\theta$*,* $0 < \theta < 1$*, and let* $\big[(E_j^\sharp, B_j^\sharp) \; ; \; j \in \mathbb{N} \big]$ *be its dual power scale.*

Suppose that $A \in \mathcal{G}(E)$ *and there exists* $n \in \mathbb{N}$ *such that*

$$\mathrm{dom}(A^k) = E_k \; , \qquad 0 \le k \le n \; , \tag{2.1.10}$$

and

$$\mathrm{dom}\big((A^\sharp)^k \big) = E_k^\sharp \; , \qquad 0 \le k \le m \; . \tag{2.1.11}$$

Then $A_\alpha \in \mathcal{G}(E_\alpha)$ *and* $\rho(A_\alpha) = \rho(A)$ *for* $-m \le \alpha \le n$*. Moreover, the diagrams*

$$\begin{array}{ccc} E_\alpha & \xrightarrow{\;\; e^{-tA_\alpha} \;\;} & E_\alpha \\ \Big\downarrow & & \Big\downarrow \\ E_\beta & \xrightarrow{\;\; e^{-tA_\beta} \;\;} & E_\beta \end{array} \tag{2.1.12}$$

are commutative for $-m \le \beta \le \alpha \le n$ *and* $t \ge 0$*.*

More precisely, suppose that $A \in \mathcal{G}(E, M, \omega)$ *for some* $M \ge 1$ *and* $\omega \in \mathbb{R}$*, fix* $\mu \in \rho(-A)$*, and put* $N := N(A, \mu, m, n)$*. Then*

$$A_\alpha \in \mathcal{G}(E_\alpha, N^2 M, \omega) \; , \qquad -m \le \alpha \le n \; , \tag{2.1.13}$$

and, if $n \ge 1$*,*

$$\lambda + A_\alpha \in \mathcal{L}\mathrm{is}(E_{\alpha+1}, E_\alpha) \; , \qquad \lambda \in \rho(-A) \; , \quad -m \le \alpha \le n - 1 \; . \tag{2.1.14}$$

Moreover,

$$\|A_\alpha\|_{\mathcal{L}(E_{\alpha+1}, E_\alpha)} \le N^2 \, \|A\|_{\mathcal{L}(E_1, E)} \tag{2.1.15}$$

and

$$\|(\lambda + A_\alpha)^{-1}\|_{\mathcal{L}(E_\alpha, E_{\alpha+j})} \le N^2 \, \|(\lambda + A)^{-1}\|_{\mathcal{L}(E, E_j)} \tag{2.1.16}$$

for $\lambda \in \rho(-A)$*,* $j = 0, 1$*, and* $-m \le \alpha \le n - 1$*.*

If, in addition, $A \in \mathcal{H}(E_1, E_0)$, *so that* $n \geq 1$, *then* $A_\alpha \in \mathcal{H}(E_{\alpha+1}, E_\alpha)$ *for* $-m \leq \alpha \leq n-1$. *Indeed, if* $A \in \mathcal{H}(E_1, E_0, \kappa, \omega_0)$ *for some* $\kappa \geq 1$ *and* $\omega_0 > 0$ *then*

$$A_\alpha \in \mathcal{H}(E_{\alpha+1}, E_\alpha, \kappa_0, \omega_0) \ , \qquad -m \leq \alpha \leq n-1 \ ,$$

where $\kappa_0 := N^2(1 + \kappa + \kappa N^2 \, \|A\|_{\mathcal{L}(E_1, E_0)})$. *Given* $\sigma > \mathrm{type}(-A)$,

$$\|e^{-tA_\beta}\|_{\mathcal{L}(E_\beta, E_\alpha)} \leq c t^{\beta-\alpha} e^{\sigma t} \ , \qquad t > 0 \ , \tag{2.1.17}$$

for $-m \leq \beta \leq \alpha \leq n$, *where* c *depends on* $\alpha - \beta$.

Proof (i) Suppose that $A \in \mathcal{G}(E, M, \omega)$ and $\mu \in \rho(-A)$. First we consider the case $B := B(\mu) := \mu + A$. Then (2.1.10) and (2.1.11) are satisfied for each $n \in \mathbb{N}$ and $N := N(A, \mu, m, n) = 1$, thanks to the fact that $B = \mu + A$ restricts or extends, respectively, to an isometric isomorphism from E_{k+1} to E_k for $k \in \mathbb{Z}$.

Lemma III.4.9.1 implies

$$B^j e^{-tA} \supset e^{-tA} B^j \ , \qquad t \geq 0 \ , \quad j \in \mathbb{Z} \ . \tag{2.1.18}$$

Hence, in particular, E_1 is invariant under $\{\, e^{-tA} \ ; \ t \geq 0 \,\}$ and

$$\|e^{-tA} x\|_1 = \|e^{-tA} Bx\| \leq \|e^{-tA}\|_{\mathcal{L}(E)} \, \|x\|_1 \ , \qquad t \geq 0 \ , \quad x \in E_1 \ .$$

Now we deduce from Lemma 2.1.2 that $A_1 \in \mathcal{G}(E_1, M, \omega)$ with $e^{-tA_1} = e^{-tA} | E_1$, and Lemma 1.1.1 gives $\rho(A_1) = \rho(A)$. By Lemma III.4.9.1 we also see that

$$\|(\lambda + A_1)^{-1} x\|_{1+k} = \|(\lambda + A)^{-1} Bx\|_k \leq \|(\lambda + A)^{-1}\|_{\mathcal{L}(E, E_k)} \, \|x\|_1$$

for $\lambda \in \rho(-A)$, $k \in \{0, 1\}$, and $x \in E_1$, which proves (2.1.16) for $\alpha = 1$.

It is clear that A_1 and B_1 are resolvent commuting. Thus, by applying the above arguments inductively, we obtain the corresponding assertions for each $\alpha \in \mathbb{N}$.

Letting $j = -1$ in (2.1.18), we infer that e^{-tA} extends to an element $U_{-1}(t)$ of $\mathcal{L}(E_{-1})$. Since $B_{-1} \in \mathcal{L}\mathrm{is}(E, E_{-1})$, it follows that

$$U_{-1}(t) = B_{-1} e^{-tA} (B_{-1})^{-1} \ , \qquad t \geq 0 \ . \tag{2.1.19}$$

Hence $\big\{\, U_{-1}(t) \ ; \ t \geq 0 \,\big\}$ is a strongly continuous semigroup on E_{-1} satisfying

$$\|U_{-1}(t)\|_{\mathcal{L}(E_{-1})} \leq \|e^{-tA}\|_{\mathcal{L}(E)} \ , \qquad t \geq 0 \ .$$

Let C be its negative infinitesimal generator. Then we deduce from (2.1.19) that $C = B_{-1} A B^{-1}$ with $\mathrm{dom}(C) = E$. Lemma III.4.9.1 implies $C \supset A$. Hence A is closable in E_{-1} and $C \supset A_{-1}$. Since $e^{-tA_1} \subset e^{-tA} \subset U_{-1}(t)$, we infer that E_1 is invariant under $\{\, U_{-1}(t) \ ; \ t \geq 0 \,\}$. Thus, thanks to the density of E_1 in E_{-1}, the core theorem of semigroup theory guarantees that E_1 is a core for C, that is, C is the closure of its restriction to E_1 (e.g., [Dav80, Theorem 1.9]). Consequently,

$C = A_{-1}$. This proves that A_{-1} is well-defined, that $A_{-1} \in \mathcal{G}(E_{-1}, M, \omega)$, and $e^{-tA} \subset e^{-tA_{-1}}$. Lemma 1.1.1 implies $\rho(A_{-1}) = \rho(A)$, and (2.1.16) for $\alpha = -1$ follows easily from Lemma III.4.9.1 and the definition of E_{-1}. It is also clear that A_{-1} and B_{-1} are resolvent commuting. Hence we can extend the results already proven to each $\alpha \in \mathbb{Z} \cap [-m, \infty)$.

Now let $j < \alpha < j+1$ for some $j \in \mathbb{Z} \cap [-m, \infty)$. Then we deduce by interpolation from what is already proven that E_α is invariant under $\{ e^{-tA_j} \ ; \ t \geq 0 \}$ and that

$$\|e^{-tA_j}\|_{\mathcal{L}(E_\alpha)} \leq \|e^{-tA}\|_{\mathcal{L}(E)} \leq Me^{\omega t} , \qquad t \geq 0 .$$

Hence Lemma 2.1.2 guarantees that $C_\alpha := (A_j)_{E_\alpha} \in \mathcal{G}(E_\alpha, M, \omega)$ and that e^{-tC_α} is the restriction of e^{-tA_j} to E_α for $t \geq 0$. If $j \in \mathbb{N}$, this gives $C_\alpha = A_\alpha$. If $j < 0$, we infer that E_1 is invariant under $\{ e^{-tC_\alpha} \ ; \ t \geq 0 \}$ and $e^{-tC_\alpha} | E_1 = e^{-tA_1}$ for $t \geq 0$. Thus, similarly as above, the core theorem guarantees that A_α is well-defined and that $A_\alpha = C_\alpha$. The remaining assertions of the theorem, pertaining to the case $A \in \mathcal{G}(E, M, \omega)$, except estimates (2.1.15), are now obvious.

Note that Lemma III.4.9.1 implies $AB^{-j} \supset B^{-j}A$ for $j \in \mathbb{N}$. Hence, given $j \in \mathbb{N}$ and $x \in E_{j+1}$,

$$\begin{aligned} \|A_j x\|_j &= \|AB^{-j}B^j x\|_j = \|B^{-j}AB^j x\|_j = \|AB^j x\| \\ &\leq \|A\|_{\mathcal{L}(E_1,E_0)} \, \|B^j x\|_1 = \|A\|_{\mathcal{L}(E_1,E_0)} \, \|x\|_{j+1} \ . \end{aligned}$$

If $j \in \mathbb{N} \cap [0, m]$,

$$\|A_{-j} x\|_{-j} = \|B^{-j}Ax\| = \|AB^{-j}x\| \leq \|A\|_{\mathcal{L}(E_1,E_0)} \, \|x\|_{1-j}$$

for $x \in E_1$. Thus by density we obtain the validity of the last estimate for each $x \in E_{1-j}$. Now interpolation proves (2.1.15).

Suppose that $A \in \mathcal{H}(E_1, E_0, \kappa, \omega_0)$. Then (I.1.2.3) and (2.1.16) entail

$$\|(\lambda + A_\alpha)^{-1}\|_{\mathcal{L}(E_\alpha, E_{\alpha+1})} \leq \|(\lambda + A)^{-1}\|_{\mathcal{L}(E_0, E_1)} \leq \kappa$$

for $\operatorname{Re}\lambda \geq \omega_0$ and $\alpha \in [-m, \infty)$. Thus $A_\alpha \in \mathcal{H}(E_{\alpha+1}, E_\alpha, \kappa_0, \omega_0)$ follows from Remark I.1.2.1(a) and (2.1.15).

Fix $\sigma > \operatorname{type}(-A)$. Corollary I.1.4.3 guarantees the existence of $\vartheta \in (0, \pi/2)$ and $M_0 \geq 1$ such that $\sigma + \Sigma_\vartheta \subset \rho(-A)$ and

$$\|\sigma + A\|_{\mathcal{L}(E_1,E_0)} + (1 + |\lambda|)^{1-k} \, \|(\lambda + \sigma + A)^{-1}\|_{\mathcal{L}(E_0, E_k)} \leq M_0$$

for $\lambda \in \Sigma_\vartheta$ and $k = 0, 1$. Hence $\rho(A_\alpha) = \rho(A)$ and (2.1.15) and (2.1.16), applied to $\sigma + A$ and $\mu - \sigma$ instead of A and μ, imply

$$\sigma + \Sigma_\vartheta \subset \rho(-A_\alpha) \tag{2.1.20}$$

and

$$\|\sigma + A_\alpha\|_{\mathcal{L}(E_{\alpha+1},E_\alpha)} + (1 + |\lambda|)^{1-k} \, \|(\lambda + \sigma + A_\alpha)^{-1}\|_{\mathcal{L}(E_\alpha, E_{\alpha+k})} \leq M_0 \tag{2.1.21}$$

for $\lambda \in \Sigma_\vartheta$, $k = 0, 1$, and $-m \le \alpha < \infty$. From this and Lemma II.4.2.1 we infer, in particular, that, given $\ell \in \mathbb{N}$,

$$\|e^{-(\sigma+A_j)t}\|_{\mathcal{L}(E_j)} + t^{\ell+1} \|(\sigma + A_j)^\ell e^{-(\sigma+A_j)t}\|_{\mathcal{L}(E_j,E_{j+1})} \le c(\ell) \tag{2.1.22}$$

for $j \in \mathbb{Z} \cap [-m, \infty)$ and $t \ge 0$. Since (2.1.16) implies

$$\|(\sigma + A_j)^{-\ell}\|_{\mathcal{L}(E_j,E_{j+\ell})} \le \big(\|(\sigma + A)^{-1}\|_{\mathcal{L}(E,E_1)}\big)^\ell ,$$

estimate (2.1.17) follows from (2.1.22), provided $\alpha, \beta \in \mathbb{Z} \cap [-m, \infty)$.

We denote by $\big[(E_{\alpha,q}, B_{\alpha,q})\ ;\ \alpha \in [-m, \infty)\big]$ the interpolation-extrapolation scales generated by (E, B) and $(\cdot,\cdot)^0_{\theta,q}$, $0 < \theta < 1$, where $1 \le q \le \infty$. Then we deduce by interpolation from what has just been proven that

$$t^{\alpha-j} \|e^{-tA_j}\|_{\mathcal{L}(E_j,E_{\alpha,q})} + t^{1+j-\alpha} \|e^{-tA_j}\|_{\mathcal{L}(E_{\alpha,q},E_{j+1})} + \|e^{-tA_j}\|_{\mathcal{L}(E_{\alpha,q})} \le ce^{\sigma t} \tag{2.1.23}$$

for $t > 0$, $j \in \mathbb{Z} \cap [-m, \infty)$, $\alpha \in (j, j + 1)$, and $1 \le q \le \infty$. Hence, by interpolating again and using the reiteration property (I.2.8.2), we see that

$$\|e^{-tA_j}\|_{\mathcal{L}(E_{\beta,\infty},E_{\alpha,1})} \le ct^{\beta-\alpha}e^{\sigma t} , \qquad t \ge 0 , \tag{2.1.24}$$

for $j \in \mathbb{Z} \cap [-m, \infty)$ and $j < \beta < \alpha < j + 1$. Now we infer from (2.1.23), (2.1.24), and (I.2.11.4) the validity of (2.1.17), provided $j \le \beta \le \alpha \le j + 1$ for some $j \in \mathbb{Z}$ with $j \ge -m$.

Lastly, suppose that $-m \le \beta < \alpha < \infty$ and $(\alpha, \beta) \cap \mathbb{Z} \ne \emptyset$. Then there are integers j and k such that $-m \le j \le \beta < j + 1 \le k \le \alpha < k + 1$. Letting $s := t/3$, the semigroup property implies

$$\|e^{-tA_\beta}\|_{\mathcal{L}(E_\beta,E_\alpha)} \le \|e^{-sA_k}\|_{\mathcal{L}(E_k,E_\alpha)} \|e^{-sA_{j+1}}\|_{\mathcal{L}(E_{j+1},E_k)} \|e^{-sA_\beta}\|_{\mathcal{L}(E_\beta,E_{j+1})}$$

for $t > 0$, and (2.1.17) follows.

(ii) Now we consider the general case. We denote by

$$\big[(E_\alpha(\mu), B_\alpha(\mu))\ ;\ \alpha \in [-m, \infty)\big]$$

the interpolation-extrapolation scale of order m, generated by $\big(E, B(\mu)\big)$ and $(\cdot,\cdot)_\theta$, $0 < \theta < 1$. It follows from (2.1.10) that $\mathrm{dom}\big(B^k(\mu)\big) = E_k$ for $0 \le k \le n$, where $B^k(\mu) := \big[B(\mu)\big]^k$. Hence Remark 1.1.2(f) implies $E_k \doteq E_k(\mu)$ and

$$\|B^k B^{-k}(\mu)\|^{-1}_{\mathcal{L}(E)} \|x\|_k \le \|x\|_{E_k(\mu)} \le \|B^k(\mu)B^{-k}\|_{\mathcal{L}(E)} \|x\|_k$$

for $x \in E_k$ and $0 \le k \le n$. Similarly, $\mathrm{dom}\big([(B(\mu))']^k\big) = \mathrm{dom}\big([B']^k\big)$ for $0 \le k \le m$ follows from (2.1.11). Consequently, by Proposition 1.3.5, $E_{-k} \doteq E_{-k}(\mu)$ and

$$\|\big(B'(\mu)\big)^k (B')^{-k}\|^{-1}_{\mathcal{L}(E')} \|x\|_{-k} \le \|x\|_{E_{-k}(\mu)} \le \|(B')^k \big(B'(\mu)\big)^{-k}\|_{\mathcal{L}(E')} \|x\|_{-k}$$

for $x \in E_{-k}$ and $0 \le k \le m$, where $B'(\mu) := \big(B(\mu)\big)'$. Since B^k [resp. $(B')^k$] is an isometric isomorphism from E_k onto E [resp. $E_k^\sharp$ onto $E^\sharp$], we see that

$$\|B^k B^{-k}(\mu)\|_{\mathcal{L}(E)} = \|(\mu + A)^{-k}\|_{\mathcal{L}(E,E_k)}$$

and

$$\|B^k(\mu) B^{-k}\|_{\mathcal{L}(E)} = \|(\mu + A)^{k}\|_{\mathcal{L}(E_k,E)}$$

for $0 \le k \le n$, and

$$\big\|\big(B'(\mu)\big)^j (B')^{-j}\big\|_{\mathcal{L}(E')} = \|(\mu + A^\sharp)^j\|_{\mathcal{L}(E_j^\sharp,E^\sharp)}$$

and

$$\big\|(B')^j \big(B'(\mu)\big)^{-j}\big\|_{\mathcal{L}(E')} = \|(\mu + A^\sharp)^{-j}\|_{\mathcal{L}(E^\sharp,E_j^\sharp)}$$

for $0 \le j \le m$. Thus it follows that $E_k \doteq E_k(\mu)$ and

$$N^{-1}\,\|x\|_k \le \|x\|_{E_k(\mu)} \le N\,\|x\|_k \ , \qquad x \in E_k \ ,$$

for $-m \le k \le n$. From this we obtain by interpolation that

$$E_\alpha \doteq E_\alpha(\mu) \ , \qquad -m \le \alpha \le n \ , \tag{2.1.25}$$

and

$$N^{-1}\,\|x\|_\alpha \le \|x\|_{E_\alpha(\mu)} \le N\,\|x\|_\alpha \ , \qquad x \in E_\alpha \ , \quad -m \le \alpha \le n \ . \tag{2.1.26}$$

Now the assertion is an obvious consequence of (2.1.25), (2.1.26), and part (i) of this proof. ■

2.1.4 Corollary *Suppose that $A \in \mathcal{G}(E)$ and $B = \mu + A$ for some $\mu \in \rho(-A)$. Then* (2.1.10) *is true for every $n \in \mathbb{N}$ and* (2.1.11) *is also satisfied. Hence the assertions of Theorem* 2.1.3 *hold for $-m \le \beta \le \alpha < \infty$. Moreover, $N = 1$ in this case.*

2.1.5 Remarks **(a)** Let $A \in \mathcal{G}(E)$. Given $\mu \in \rho(-A)$, denote by

$$\big[\big(E_\alpha(\mu), B_\alpha(\mu)\big) \ ; \ \alpha \in [-m, \infty)\big]$$

the interpolation-extrapolation scale generated by $\mu + A$ and $(\cdot,\cdot)_\theta$, $0 < \theta < 1$. Then

$$E_\alpha(\mu) \doteq E_\alpha(\nu) \ , \qquad \mu, \nu \in \rho(-A) \ .$$

More precisely,

$$\|x\|_{E_\alpha(\nu)} \le \big(1 + |\mu - \nu|\,\|(\mu + A)^{-1}\|_{\mathcal{L}(E)}\big)^\alpha\,\|x\|_{E_\alpha(\mu)} \ , \qquad \alpha \ge 0 \ ,$$

and

$$\|x\|_{E_{-\alpha}(\nu)} \le \big(1 + |\mu - \nu|\,\|(\nu + A)^{-1}\|_{\mathcal{L}(E)}\big)^\alpha\,\|x\|_{E_{-\alpha}(\mu)} \ , \qquad 0 < \alpha \le m \ ,$$

for $\mu, \nu \in \rho(-A)$.

Proof This follows from Remark 1.2.2(d) and Corollary 1.3.6 by interpolation. ■

(b) It is obvious that Theorem 2.1.3 remains valid if we drop the assumption that the interpolation functors be exact, provided N is replaced by $c(\alpha)N$, where $c\colon \mathbb{R} \to \dot{\mathbb{R}}^+$ is a 1-periodic function satisfying $c(k) = 1$ for $k \in \mathbb{Z}$ and depending on the interpolation functors $(\cdot,\cdot)_\theta,\ 0 < \theta < 1$, only. ■

Theorem 2.1.3 gives extensions and sharpenings of earlier results due to the author (cf. [Ama86], [Ama87]).

2.2 Parabolic Evolution Equations in Banach Scales

Let $E := (E, \|\cdot\|)$ be a Banach space, $m \in \mathbb{N}$, and $(\cdot,\cdot)_\theta$ exact admissible interpolation functors for $0 < \theta < 1$.

Denote by $\mathfrak{A}$ a subset of $\mathcal{G}(E, M, \omega)$ for some $M \geq 1$ and $\omega \in \mathbb{R}$ such that $0 \in \rho(A)$ for each $A \in \mathfrak{A}$. Then, given $A \in \mathfrak{A}$, the interpolation-extrapolation scale $\bigl[(E_\alpha(A), A)\ ;\ \alpha \in [-m, \infty)\bigr]$ generated by (E, A) and $(\cdot,\cdot)_\theta,\ 0 < \theta < 1$, is well-defined. It follows from Corollary 2.1.4 that

$$A_\alpha \in \mathcal{G}(E_\alpha, M, \omega)\ , \tag{2.2.1}$$

that

$$\|A_\alpha\|_{\mathcal{L}(E_{\alpha+1}(A), E_\alpha(A))} \leq \|A\|_{\mathcal{L}(E_1(A), E)} \tag{2.2.2}$$

and

$$\|(\lambda + A_\alpha)^{-1}\|_{\mathcal{L}(E_\alpha(A), E_{\alpha+j}(A))} \leq \|(\lambda + A)^{-1}\|_{\mathcal{L}(E, E_j(A))}\ , \qquad j = 0, 1\ , \tag{2.2.3}$$

for $\lambda \in \rho(-A) = \rho(-A_\alpha)$ and $-m \leq \alpha < \infty$, and that the diagrams

$$\begin{array}{ccc} E_\alpha(A) & \xrightarrow{\ e^{-tA_\alpha}\ } & E_\alpha(A) \\ \big\downarrow & & \big\downarrow \\ E_\beta(A) & \xrightarrow{\ e^{-tA_\beta}\ } & E_\beta(A) \end{array} \tag{2.2.4}$$

are commutative for $-m \leq \beta < \alpha < \infty$ and $t \geq 0$. Moreoever, if $-A$ generates an analytic semigroup, this is true for $-A_\alpha$ and, given $\sigma > \operatorname{type}(-A)$,

$$\|e^{-tA_\beta}\|_{\mathcal{L}(E_\beta(A), E_\alpha(A))} \leq c t^{\beta-\alpha} e^{\sigma t}\ , \qquad t > 0\ , \tag{2.2.5}$$

for $-m \leq \beta < \alpha < \infty$, where c is a function of $\alpha - \beta$.

Let J be a perfect subinterval of $\mathbb{R}^+$ containing 0 and suppose that

$$A\colon J \to \mathfrak{A}\ .$$

Put

$$E_\alpha(t) := E_\alpha\bigl(A(t)\bigr)\ , \quad t \in J\ , \qquad E_\alpha := E_\alpha(0)\ , \quad -m \le \alpha < \infty\ ,$$

and denote the norm of E_α by $\|\cdot\|_\alpha$. In general, the Banach spaces $E_\alpha(s)$ and $E_\alpha(t)$ are distinct, even as vector spaces, if $\alpha \neq 0$ and $s \neq t$. Suppose, however, that there exists $\alpha \in (0,1)$ such that $E_{\alpha-j}(t) \doteq E_{\alpha-j}$ for $j \in \{0,1\}$ and $t \in J$. Then the Cauchy problem

$$\dot u + A(t)u = f(t)\ , \quad t \in \dot J\ , \qquad u(0) = x \tag{2.2.6}$$

in E and the **extrapolated Cauchy problem** (**of order** $\alpha - 1$)

$$\dot u + A_{\alpha-1}(t)u = f(t)\ , \quad t \in \dot J\ , \qquad u(0) = x \qquad (2.2.7)_{\alpha-1}$$

in $E_{\alpha-1}$ are well-defined, provided, of course, $f : J \to E$ and $x \in E$ in case (2.2.6) and $f : J \to E_{\alpha-1}$ and $x \in E_{\alpha-1}$ in case $(2.2.7)_{\alpha-1}$. Whereas the 'original' Cauchy problem (2.2.6) can be highly irregular, due to the fact that the domains $E_1(t)$ of $A(t)$ vary with $t \in J$, the operators $A_{\alpha-1}(t)$ of the extrapolated Cauchy problem $(2.2.7)_{\alpha-1}$ have constant domain, namely E_α. Thus we can apply the general results developed in Chapters II and III to the extrapolated problem $(2.2.7)_{\alpha-1}$ and use the results on variable domain problems exposed in Chapter IV to get solvability and regularity results for the original Cauchy problem (2.2.6). This is done in the following fundamental existence and regularity theorem, where we use the notations introduced in Subsection IV.2.1. Also recall that $(\cdot,\cdot)^0_{\theta,q} := (\cdot,\cdot)_{\theta,q}$ if $0 < \theta < 1$ and $1 \le q < \infty$.

2.2.1 Theorem *Suppose* $A\colon J \to \mathfrak{A}_M(E)$ *for some* $M > 0$, $0 < \alpha < \beta < \gamma < 1$, *and*

$$E_\xi(t) \doteq E_\xi\ , \qquad \xi \in \{\alpha, \alpha-1, \beta-1, \gamma-1\}\ , \quad t \in J\ .$$

Also suppose that there exists $q \in [1,\infty]$ *such that*

$$(\cdot,\cdot)_\beta = (\cdot,\cdot)^0_{\beta,q}\ , \quad \textit{that} \quad \beta - \alpha < \rho < 1\ , \tag{2.2.8}$$

and that $A_{\alpha-1} \in C^\rho\bigl(J, \mathcal{L}(E_\alpha, E_{\alpha-1})\bigr)$ *and*

$$f \in C(J, E_{\gamma-1})\ . \tag{2.2.9}$$

Then, given $x \in E_{\alpha-1}$, *the extrapolated parabolic Cauchy problem*

$$\dot u + A_{\alpha-1}(t)u = f(t)\ , \quad t \in \dot J\ , \qquad u(0) = x \qquad (2.2.10)_{\alpha-1}$$

possesses a unique solution $u(\cdot,x)$ *and*

$$u(\cdot,x) \in C(J, E_{\alpha-1}) \cap C(\dot J, E_\alpha) \cap C^1(\dot J, E_{\beta-1})\ . \tag{2.2.11}$$

It is also a solution of the extrapolated evolution equation

$$\dot{v} + A_{\beta-1}(t)v = f(t)\ , \quad t \in \dot{J}\ , \tag{2.2.12$_{\beta-1}$}$$

in $E_{\beta-1}$*. If there exists a constant* $\kappa \geq 1$ *such that*

$$\kappa^{-1}\left\|x\right\|_\alpha \leq \|x\|_{E_\alpha(t)} \leq \kappa\left\|x\right\|_\alpha\ , \qquad x \in E_\alpha\ , \quad t \in J\ , \tag{2.2.13}$$

if $\rho > 1-\alpha$*, and if*

$$f \in C^\sigma(J,E) + C\big(J,(E,E_\alpha)_\sigma\big) \tag{2.2.14}$$

for some $\sigma \in (0,1)$ *then* $u(\cdot,x) \in C(J,E) \cap C^1(\dot{J},E)$ *and it is a solution of the 'original' evolution equation*

$$\dot{u} + A(t)u = f(t)\ , \qquad t \in \dot{J}\ ,$$

in E*, which is strict if* $x \in E_1$*.*

Proof Put $(F_1,F_0) := (E_\alpha, E_{\alpha-1})$ and $B := A_{\alpha-1}$. Since $A(t) \in \mathcal{H}\big(E_1(t),E\big)$, it follows from $B \in C^\rho\big(J,\mathcal{L}(F_1,F_0)\big)$ and Theorem 2.1.3 that

$$B \in C^\rho\big(J,\mathcal{H}(F_1,F_0)\big)\ .$$

Fix μ and ν such that $\beta-\alpha < \mu < \nu < \gamma-\alpha$ and $\mu < \rho$. Put $\{\cdot,\cdot\}_\theta := (\cdot,\cdot)^0_{\theta,q}$, $0<\theta<1$, and let $F_\theta := \{F_0,F_1\}_\theta$. Then, by the almost reiteration Theorem 1.5.3,

$$E_{\gamma-1} \hookrightarrow F_\nu \hookrightarrow F_\mu \hookrightarrow E_{\beta-1}\ . \tag{2.2.15}$$

Consequently, $f \in C(J,F_\nu)$ by (2.2.9). Since $x \in F_0$ and $\rho > \mu$, Theorem II.1.2.2 guarantees that the extrapolated Cauchy problem $(2.2.10)_{\alpha-1}$ possesses a unique solution $u(\cdot,x)$ and

$$u(\cdot,x) \in C(J,F_0) \cap C(\dot{J},F_1) \cap C^1(\dot{J},F_\mu)\ .$$

Hence the validity of (2.2.11) follows from (2.2.15). By Theorem II.1.2.2 we also know that $u(\cdot,x)$ is a solution of the parabolic equation

$$\dot{v} + B_\mu(t)v = f(t)\ , \qquad t \in \dot{J}\ , \tag{2.2.16}$$

in F_μ, where B_μ is the F_μ-realization of B. Given $t \in J$, let $\big[\big(F_\zeta(t),B_\zeta(t)\big)\ ;\ \zeta \in \mathbb{R}^+\big]$ be the interpolation-extrapolation scale of order zero generated by $\big(F,B(t)\big)$ and the interpolation functors $\{\cdot,\cdot\}_\theta$, $0<\theta<1$. Then we infer from (2.2.8), Corollary 1.5.8, and Remark 1.5.11 (letting $\xi := \beta-\alpha+1$ since γ of Corollary 1.5.8 corresponds to $\alpha-1$ in our situation) that $F_{\beta-\alpha+1}(t) \doteq E_\beta(t)$ for $t \in J$. Consequently, $B_{\beta-\alpha}(t) = A_{\beta-1}(t)$ for $t \in J$, and since $B_\mu(t) \subset B_{\beta-\alpha}(t)$, we deduce from (2.2.16) that $u(\cdot,x)$ is a solution of $(2.2.12)_{\beta-1}$.

Now let assumptions (2.2.13) through (2.2.14) be satisfied. Since it suffices to prove the corresponding assertion for each compact subinterval of J, we can assume that J is compact. As the inversion map

$$\mathcal{L}\mathrm{is}(F_1, F_0) \to \mathcal{L}\mathrm{is}(F_0, F_1)\ , \quad C \mapsto C^{-1}$$

is analytic, our assumptions imply that $(A_{\alpha-1})^{-1} \in BUC^{\rho}\big(J, \mathcal{L}(E_{\alpha-1}, E_\alpha)\big)$. Thus, using $A_{\alpha-1} \supset A$ and $E \hookrightarrow E_{\alpha-1}$,

$$A^{-1} \in BUC^{\rho}\big(J, \mathcal{L}(E, E_\alpha)\big)\ ,$$

where $\rho > 1 - \alpha$. Now the last part of the assertion follows from Theorem IV.2.5.1, the fact that $A_{\alpha-1} \supset A$, and by uniqueness. ■

2.3 Duality

In this subsection we derive important and useful results about semigroups in reflexive Banach spaces for which the duality Theorem 1.5.12 is valid. In particular, we show that the dual operators $-A^\sharp_\alpha$ generate strongly continuous semigroups as well that can be obtained naturally by dualizing. We restrict the class of admissible Banach spaces and interpolation functors by assuming that

$$\left.\begin{aligned} &E \text{ is a reflexive Banach space and} \\ &(\cdot,\cdot)_\theta \in \big\{\,[\cdot,\cdot]_\theta, (\cdot,\cdot)_{\theta,q}, (\cdot,\cdot)^0_{\theta,\infty}\ ;\ 1 < q < \infty\big\},\quad 0 < \theta < 1. \end{aligned}\right\} \tag{2.3.1}$$

We also suppose that

$$A \in \mathcal{G}(E, M, \omega) \text{ for some } M \geq 1 \text{ and } \omega \in \mathbb{R}\ . \tag{2.3.2}$$

We fix $\mu > \mathrm{type}(-A)$ and put $B := \mu + A$, so that $B \in \mathcal{P}(E)$. Then we denote by

$$\big[(E_\alpha, B_\alpha)\ ;\ \alpha \in \mathbb{R}\big]$$

the interpolation-extrapolation scale generated by (E, B) and $(\cdot,\cdot)_\theta,\ \ 0 < \theta < 1$, and by

$$\big[(E^\sharp_\alpha, B^\sharp_\alpha)\ ;\ \alpha \in \mathbb{R}\big]$$

its dual scale. It follows from Theorem 1.4.12 that these scales are well-defined. For each $\alpha \in \mathbb{R}$ we denote by A_α the E_α-realization of A and by $A^\sharp_\alpha$ the $E^\sharp_\alpha$-realiztion of $A^\sharp = A'$.

In the following theorem we collect some of the basic properties of the operators $A^\sharp_\alpha$. In particular, we show that $-A^\sharp_\alpha$ generates a strongly continuous semigroup on $E^\sharp_\alpha$.

2.3.1 Theorem *Let* (2.3.1) *and* (2.3.2) *be satisfied. Then* $A^\sharp_\alpha \in \mathcal{G}(E^\sharp_\alpha, M, \omega)$ *and* $\rho(A^\sharp_\alpha) = \rho(A^\sharp) = \rho(A)$ *for* $\alpha \in \mathbb{R}$. *The diagrams*

$$\begin{array}{ccc} E^\sharp_\alpha & \xrightarrow{\;e^{-tA^\sharp_\alpha}\;} & E^\sharp_\alpha \\ \big\downarrow & & \big\downarrow \\ E^\sharp_\beta & \xrightarrow{\;e^{-tA^\sharp_\beta}\;} & E^\sharp_\beta \end{array}$$

are commutative for $-\infty < \beta \leq \alpha < \infty$ *and* $t \geq 0$, *and*

$$\lambda + A^\sharp_\alpha \in \mathcal{L}\mathrm{is}(E^\sharp_{\alpha+1}, E^\sharp_\alpha) \ , \qquad \lambda \in \rho(A) \ , \quad \alpha \in \mathbb{R} \ .$$

If $A \in \mathcal{H}(E_1, E_0, \kappa, \omega_0)$ *for some* $\kappa \geq 1$ *and* $\omega_0 \geq 0$ *then* $A^\sharp_\alpha \in \mathcal{H}(E^\sharp_{\alpha+1}, E^\sharp_\alpha, \kappa_0, \omega_0)$, *where* $\kappa_0 := 1 + \kappa + \kappa \left\| A^\sharp \right\|_{\mathcal{L}(E^\sharp_1, E^\sharp)}$.

Proof Recall from semigroup theory that $A^\sharp \in \mathcal{G}(E^\sharp, M, \omega)$, thanks to (2.3.1) and (2.3.2). From [HiP57, Theorem 2.16.4] we infer that $\mathrm{dom}(A^k) = E_k$ as well as $\mathrm{dom}\big((A^\sharp)^k\big) = E^\sharp_k$ for $k \in \mathbb{N}$. The reflexivity of E implies $(A^\sharp)^\sharp = A$ and $(B^\sharp)^\sharp = B$. Thus

$$\mathrm{dom}\big(\big[(A^\sharp)^\sharp\big]^k\big) = \mathrm{dom}(A^k) = E_k = \mathrm{dom}(B^k) = \mathrm{dom}\big(\big[(B^\sharp)^\sharp\big]^k\big) = (E^\sharp)^\sharp_k$$

for $k \in \mathbb{N}$. This shows that the assumptions of Theorem 2.1.3 are satisfied for $A^\sharp \in \mathcal{G}(E^\sharp, M, \omega)$. Now that theorem implies the assertions. ■

In the following theorem we prove important duality results. For this we recall from Theorem 1.5.12 that

$$E^\sharp_{-\alpha} = (E_\alpha)' \ , \quad B^\sharp_{-\alpha} = (B_\alpha)' \ , \qquad \alpha \in \mathbb{R} \ , \tag{2.3.3}$$

with respect to the duality pairings naturally induced by the E'-E-duality pairing $\langle \cdot, \cdot \rangle$. We use these duality pairings without further mention.

2.3.2 Theorem *Let assumptions* (2.3.1) *and* (2.3.2) *be satisfied. Then*

$$A^\sharp_\alpha = (A_{-\alpha})' \ , \qquad \alpha \in \mathbb{R} \ , \tag{2.3.4}$$

as unbounded operators, and

$$e^{-tA^\sharp_\alpha} = (e^{-tA_{-\alpha}})' \ , \qquad t \geq 0 \ , \quad \alpha \in \mathbb{R} \ , \tag{2.3.5}$$

that is, $\{\, e^{-tA^\sharp_\alpha} \ ; \ t \geq 0 \,\}$ *is the dual semigroup to* $\{\, e^{-tA_{-\alpha}} \ ; \ t \geq 0 \,\}$. *Furthermore,*

$$\langle A^\sharp_\alpha x^\sharp, x \rangle = \langle x^\sharp, A_{-\alpha-1} x \rangle \ , \qquad (x^\sharp, x) \in E^\sharp_{\alpha+1} \times E_{-\alpha} \ , \tag{2.3.6}$$

for $\alpha \in \mathbb{R}$, *that is,* $A^\sharp_\alpha \in \mathcal{L}(E^\sharp_{\alpha+1}, E^\sharp_\alpha)$ *is the dual of the* bounded *linear operator* $A_{-\alpha-1} \in \mathcal{L}(E_{-\alpha}, E_{-\alpha-1})$.

Proof Given $\alpha \in \mathbb{R}$ and $\lambda \in \mathbb{K}$, the definition of the E_α-realization implies

$$(\lambda + A)_\alpha x = (\lambda + A)x = \lambda x + Ax = \lambda x + A_\alpha x = (\lambda + A_\alpha)x \;, \qquad x \in E_{(\alpha+1)\vee 1} \;.$$

Thus we infer from Remark 1.1.2(b) and the density of $E_{(\alpha+1)\vee 1}$ in $E_{\alpha+1}$ that

$$(\lambda + A)_\alpha = \lambda + A_\alpha \;, \qquad \alpha \in \mathbb{R} \;, \quad \lambda \in \mathbb{K} \;. \tag{2.3.7}$$

Similarly,

$$(\lambda + A^\sharp)_\alpha = \lambda + A^\sharp_\alpha \;, \qquad \alpha \in \mathbb{R} \;, \quad \lambda \in \mathbb{K} \;. \tag{2.3.8}$$

From this and (2.3.3) we deduce that

$$\mu + A^\sharp_\alpha = (\mu + A^\sharp)_\alpha = B^\sharp_\alpha = (B_{-\alpha})' = \bigl[(\mu + A)_{-\alpha}\bigr]' = (\mu + A_{-\alpha})' = \mu + (A_{-\alpha})'$$

for $\alpha \in \mathbb{R}$, which proves (2.3.4). Assertion (2.3.5) is now a well-known consequence of (2.3.4).

From (2.3.4), $\operatorname{dom}(A^\sharp_\alpha) = E^\sharp_{\alpha+1}$, and $\operatorname{dom}(A_{-\alpha}) = E_{1-\alpha}$ it follows that

$$\langle A^\sharp_\alpha x^\sharp, x\rangle = \langle x^\sharp, A_{-\alpha} x\rangle \;, \qquad (x^\sharp, x) \in E^\sharp_{\alpha+1} \times E_{1-\alpha} \;. \tag{2.3.9}$$

Since $A_{-\alpha-1} \supset A_{-\alpha}$ and $A_{-\alpha-1} \in \mathcal{L}(E_{-\alpha}, E_{-\alpha-1})$, we obtain (2.3.6) by continuity from (2.3.9), the density of $E_{1-\alpha}$ in $E_{-\alpha}$, the fact that $A^\sharp_\alpha x^\sharp \in E^\sharp_\alpha$ for $x^\sharp \in E^\sharp_{\alpha+1}$, and from the first part of (2.3.3). ∎

The following technical proposition gives useful estimates for the norms of the operators $A^\sharp_\alpha$ and $(\lambda + A^\sharp_\alpha)^{-1}$.

2.3.3 Proposition *Let assumptions* (2.3.1) *and* (2.3.2) *be satisfied. Then*

$$\|A^\sharp_\alpha\|_{\mathcal{L}(E^\sharp_{\alpha+1}, E^\sharp_\alpha)} \le \|A\|_{\mathcal{L}(E_1, E)} \wedge \|A^\sharp\|_{\mathcal{L}(E^\sharp_1, E^\sharp)}$$

and

$$\|(\lambda + A^\sharp_\alpha)^{-1}\|_{\mathcal{L}(E^\sharp_\alpha, E^\sharp_{\alpha+1})} \le \|(\lambda + A)^{-1}\|_{\mathcal{L}(E, E_1)} \wedge \|(\lambda + A^\sharp)^{-1}\|_{\mathcal{L}(E^\sharp, E^\sharp_1)}$$

for $\lambda \in \rho(-A)$ *and* $\alpha \in \mathbb{R}$.

Proof Thanks to the proof of Theorem 2.3.1 we obtain from Theorem 2.1.3 that

$$\|A^\sharp_\alpha\|_{\mathcal{L}(E^\sharp_{\alpha+1}, E^\sharp_\alpha)} \le \|A^\sharp\|_{\mathcal{L}(E^\sharp_1, E^\sharp)} \tag{2.3.10}$$

and

$$\|(\lambda + A^\sharp_\alpha)^{-1}\|_{\mathcal{L}(E^\sharp_\alpha, E^\sharp_{\alpha+1})} \le \|(\lambda + A^\sharp)^{-1}\|_{\mathcal{L}(E^\sharp, E^\sharp_1)} \tag{2.3.11}$$

for $\lambda \in \rho(-A)$ and $\alpha \in \mathbb{R}$.

From (2.3.3) and (2.3.6) it follows that

$$\|A^\sharp_\alpha\|_{\mathcal{L}(E^\sharp_{\alpha+1},E^\sharp_\alpha)} = \|A_{-\alpha-1}\|_{\mathcal{L}(E_{-\alpha},E_{-\alpha-1})}\ , \qquad \alpha\in\mathbb{R}\ .$$

Thus we infer from (2.1.15) that

$$\|A^\sharp_\alpha\|_{\mathcal{L}(E^\sharp_{\alpha+1},E^\sharp_\alpha)} \le \|A\|_{\mathcal{L}(E_1,E_0)}\ , \qquad \alpha\in\mathbb{R}\ ,$$

which, together with (2.3.10), proves the first assertion.

Recall from Theorems 2.1.3 and 2.3.1 that $\lambda + A_{-\alpha-1}\in\mathcal{L}\mathrm{is}(E_{-\alpha},E_{-\alpha-1})$ and $\lambda + A^\sharp_\alpha\in\mathcal{L}\mathrm{is}(E^\sharp_{\alpha+1},E^\sharp_\alpha)$ for $\lambda\in\rho(-A)$. Hence we deduce from (2.3.6) that

$$\langle (\lambda + A^\sharp_\alpha)^{-1}y^\sharp, y\rangle = \langle y^\sharp, (\lambda + A_{-\alpha-1})^{-1}y\rangle\ , \qquad (y^\sharp,y)\in E^\sharp_\alpha\times E_{-\alpha-1}\ .$$

This implies

$$\|(\lambda + A^\sharp_\alpha)^{-1}\|_{\mathcal{L}(E^\sharp_\alpha,E^\sharp_{\alpha+1})} = \|(\lambda + A_{-\alpha-1})^{-1}\|_{\mathcal{L}(E_{-\alpha-1},E_{-\alpha})}\ .$$

Now the second assertion is a consequence of (2.1.16) and (2.3.11). ■

2.4 Approximation Theorems

At this stage we make a digression and prove some useful approximation theorems that are of general interest.

Let X be a nonempty set, V a vector space, and $\Phi(X)$ a vector subspace of $\mathbb{K}^X$. We define a bilinear map

$$\Phi(X)\times V\to V^X\ , \quad (\varphi,v)\mapsto\varphi\otimes v$$

by $\varphi\otimes v(x) := \varphi(x)v$. Then $\varphi\otimes v$ is the **tensor product** of φ and v and

$$\Phi(X)\otimes V := \mathrm{span}\{\,\varphi\otimes v\ ;\ \varphi\in\Phi(X),\ v\in V\,\}\ ,$$

the **tensor product** of $\Phi(X)$ and V, is a well-defined linear subspace of V^X. (Recall that in the proof of Theorem III.4.3.6 we have already used the tensor product $L_r(X,\mu)\otimes E$.)

Throughout the remainder of this subsection X denotes a nonempty open subset of $\mathbb{R}^n$ and $E := (E,|\cdot|)$ a Banach space.

2.4.1 Proposition *$\mathcal{D}(X)\otimes E$ is sequentially dense in $\mathcal{D}(X,E)$.*

Proof Let $u\in\mathcal{D}(X,E)$ and an open set K with $\mathrm{supp}(u)\subset\subset K\subset\subset X$ be given. The family $\{\,\mathbb{B}(e,\varepsilon)\ ;\ e\in E\,\}$ is, for each $\varepsilon>0$, an open covering of the compact set $u(X)=u(\overline{K})$ in E. Thus there are finite subcoverings $\{\,\mathbb{B}(e_j,\varepsilon)\ ;\ 1\le j\le m(\varepsilon)\,\}$.

Fix $\varepsilon > 0$ and put $U_0 := \operatorname{supp}(u)^c$ and $U_j := u^{-1}\big(\mathbb{B}(e_j,\varepsilon)\big) \cap K$ for $1 \le j \le m$, where $m := m(\varepsilon)$. Then $\big\{ U_j \; ; \; 0 \le j \le m \big\}$ is an open covering of X. Choose a smooth partition of unity $\{ \psi_j \; ; \; 0 \le j \le m \}$ on X subordinate to $\big\{ U_j \; ; \; 0 \le j \le m \big\}$ and, letting $e_0 := 0$, put $v := \sum_{j=0}^m \psi_j \otimes e_j$. Then $v \in \mathcal{D}(X) \otimes E$ with $\operatorname{supp}(v) \subset \overline{K}$ and

$$|u(x) - v(x)| \le \sum_{j=0}^{m} \psi_j(x)\, |u(x) - e_j| \le \varepsilon \; , \qquad x \in X \; .$$

This shows that we can find a sequence (v_k) in $\mathcal{D}(X) \otimes E$ such that $\operatorname{supp}(v_k) \subset K$ and $v_k \to u$ in $BUC(\mathbb{R}^n, E)$.

Let $\{ \varphi_\varepsilon \; ; \; \varepsilon > 0 \}$ be a mollifier. Then $\varphi_{1/j} * v_k \in \mathcal{D}(X) \otimes E$ provided $j \ge j_0$, where j_0 is independent of k, thanks to the fact that the support of $\varphi_{1/j} * v_k$ is contained in the $(1/j)$-neighborhood of K. Now we infer from (III.4.2.10) and (III.4.2.19) that $\varphi_{1/j} * v_k \to \varphi_{1/j} * u$ in $\mathcal{E}(\mathbb{R}^n, E)$ as $k \to \infty$. By reapplying the last argument and using (III.4.2.23) we find that $\varphi_{1/j} * u \to u$ in $\mathcal{E}(\mathbb{R}^n, E)$. Since the supports of $\varphi_{1/j} * v$ are contained in a fixed compact subset of X, standard properties of LF-spaces guarantee that the convergence of the above sequences takes place in $\mathcal{D}(X, E)$. ■

2.4.2 Corollary *Suppose that (E_0, E_1) is a densely injected Banach couple. Then $\mathcal{D}(X, E_1) \overset{d}{\hookrightarrow} \mathcal{D}(X, E_0)$. In fact, $\mathcal{D}(X, E_1)$ is sequentially dense in $\mathcal{D}(X, E_0)$.*

Proof The first assertion is trivial. It is obvious that $\mathcal{D}(X) \otimes E_1$ is sequentially dense in $\mathcal{D}(X) \otimes E_0$ with respect to the topology induced by $\mathcal{D}(X, E_0)$. Hence the sequential density of $\mathcal{D}(X) \otimes E_0$ in $\mathcal{D}(X, E_0)$ implies the assertion. ■

Next we turn to Sobolev spaces and prove the following important approximation result.

2.4.3 Theorem *Suppose that (E_0, E_1) is a densely injected Banach couple. Then $\mathcal{D}(\mathbb{R}^n, E_1) \overset{d}{\hookrightarrow} W_p^m(\mathbb{R}^n, E_0)$ for $1 \le p < \infty$ and $m \in \mathbb{N}$.*

Proof Thanks to Corollary 2.4.2 we can assume that $E_0 = E_1$. Then, letting $\mathcal{D} := \mathcal{D}(\mathbb{R}^n, E_0)$ and $W_p^m := W_p^m(\mathbb{R}^n, E_0)$, it follows from

$$\|u\|_{m,p} \le c(K) p_{m,K}(u) \; , \qquad u \in \mathcal{D} \; , \quad \operatorname{supp}(u) \subset K \subset\subset \mathbb{R}^n \; ,$$

that $\mathcal{D} \hookrightarrow W_p^m$.

To prove the asserted density, let $u \in W_p^m$ be given. Let $\psi \in C^\infty\big(\mathbb{R}^n, [0,1]\big)$ satisfy $\psi | \mathbb{B}^n = 1$ and $\operatorname{supp}(\psi) \subset 2\mathbb{B}^n$. Recall that σ_α denotes dilation and put $\psi_j := \sigma_{1/j} \psi$ for $j \in \dot{\mathbb{N}}$. Then

$$\partial^\alpha(\psi_j u) = \sum_{\beta \le \alpha} \tbinom{\alpha}{\beta} j^{-|\beta|} \sigma_{1/j}(\partial^\beta \psi) \partial^{\alpha-\beta} u \; , \qquad |\alpha| \le m \; ,$$

by Leibniz' rule. Observe that $\|\sigma_{1/j}(\partial^\beta\psi)\|_\infty \le c$ and $\operatorname{supp}\bigl(\sigma_{1/j}(\partial^\beta\psi)\bigr) \subset 2j\mathbb{B}^n$ for $j \in \dot{\mathbb{N}}$ and $|\beta| \le m$. Hence $\psi_j u \in W_p^m$ and

$$\|\partial^\alpha(\psi_j u - u)\|_p \le 2\,\|u\|_{m,p,(j\mathbb{B}^n)^c} + cj^{-1}\,\|u\|_{m,p}$$

for $j \in \dot{\mathbb{N}}$ and $|\alpha| \le m$. Thus, thanks to Lebesgue's theorem, $\psi_j u \to u$ in W_p^m as $j \to \infty$. Hence we can assume that u has compact support. Let $\{\,\varphi_\varepsilon\ ;\ \varepsilon > 0\,\}$ be a mollifier. Then $\varphi_\varepsilon * u \in \mathcal{D}$ and $\partial^\alpha(\varphi_\varepsilon * u) = \varphi_\varepsilon * \partial^\alpha u$ for $|\alpha| \le m$ by (III.4.2.13) and (III.4.2.10), respectively. Thus (III.4.2.23) implies $\varphi_\varepsilon * u \to u$ in W_p^m as $\varepsilon \to 0$, which proves the assertion. ■

Given $u \in \mathcal{D}'(\mathbb{R}^n, E)$, define its **restriction**, $r_X u$, to X by

$$r_X u(\varphi) := u(\varphi)\ , \qquad \varphi \in \mathcal{D}(X)\ .$$

Then $r_X u \in \mathcal{D}'(X, E)$, and $r_X := (u \mapsto r_X u)$, the **restriction operator** for X, is linear and continuous:

$$r_X \in \mathcal{L}\bigl(\mathcal{D}'(\mathbb{R}^n, E), \mathcal{D}'(X, E)\bigr)\ . \tag{2.4.1}$$

Observe that r_X is independent of E.

It is obvious that r_X commutes with differentiation,

$$r_X \partial^\alpha = \partial^\alpha r_X\ , \qquad \alpha \in \mathbb{N}^n\ , \tag{2.4.2}$$

and that

$$r_X u = u|X\ , \qquad u \in C(\mathbb{R}^n, E)\ , \tag{2.4.3}$$

where $u|X$ is the usual point-wise restriction. It is also clear that

$$r_X \in \mathcal{L}\bigl(L_p(\mathbb{R}^n, E), L_p(X, E)\bigr)\ , \qquad 1 \le p \le \infty\ , \tag{2.4.4}$$

with norm at most one.

It follows from (2.4.2)–(2.4.4) that

$$r_X \in \mathcal{L}\bigl(W_p^m(\mathbb{R}^n, E), W_p^m(X, E)\bigr)\ , \qquad 1 \le p \le \infty\ , \quad m \in \mathbb{N}\ , \tag{2.4.5}$$

and that r_X has norm at most one.

The set X is said to possess the **extension property** if there exists a map

$$e_X : \bigcup_{1 \le p < \infty} L_p(X, E) \to \bigcup_{1 \le p < \infty} L_p(\mathbb{R}^n, E)\ ,$$

an **extension operator** for X, such that

$$e_X \in \mathcal{L}\bigl(W_p^m(X, E), W_p^m(\mathbb{R}^n, E)\bigr)\ , \qquad 1 \le p < \infty\ , \quad m \in \mathbb{N}\ ,$$

and

$$r_X e_X = \mathrm{id}\ ,$$

independently of E.

2.4.4 Remark Let X have the extension property. Then r_X is a retraction from $W_p^m(\mathbb{R}^n, E)$ onto $W_p^m(X, E)$ for $m \in \mathbb{N}$ and $1 \le p < \infty$, and e_X is a corresponding coretraction. ■

Following Stein [Ste70a, Section VI.3], X is said to be a **special Lipschitz domain** if it is obtained by rotation from a set of the form

$$\{ (y, t) \subset \mathbb{R}^{n-1} \times \mathbb{R} \; ; \; t > \varphi(y) \} ,$$

where $\varphi \in BUC^{1\text{-}}(\mathbb{R}^{n-1})$. Then $\|\varphi\|_{C^{1\text{-}}}$ is the **bound** of X. Moreover, X has a **minimally smooth boundary** if there exist $\varepsilon > 0$, $N \in \dot{\mathbb{N}}$, $M \ge 0$, and a sequence $U_0, U_1, \ldots$ of open subsets of $\mathbb{R}^n$ such that:

(i) if $x \in \partial X$ then $\mathbb{B}(x, \varepsilon) \subset U_j$ for some j;

(ii) no point of $\mathbb{R}^n$ is contained in more than N of the U_js;

(iii) for each j there exists a special Lipschitz domain X_j, whose bound does not exceed M, such that $U_j \cap X = U_j \cap X_j$.

Of course, $\mathbb{R}^n$ and each half-space in $\mathbb{R}^n$ have minimally smooth boundaries. More interestingly, X has a minimally smooth boundary if $\overline{X}$ is an n-dimensional $C^{1\text{-}}$-submanifold of $\mathbb{R}^n$ with compact boundary. For further examples of open sets with minimally smooth boundaries we refer to [Fra79].

2.4.5 Theorem *If X has a minimally smooth boundary, it has the extension property.*

Proof See [Ste70a, Section VI.3] (also cf. [EdE87, Section V.4.4]) and observe that these proofs extend trivially to the E-valued case. ■

Now it is easy to prove the following useful approximation theorem for Sobolev spaces on X.

2.4.6 Theorem *Let (E_0, E_1) be a densely injected Banach couple and let X have a minimally smooth boundary. Then*

$$\mathcal{D}(\overline{X}, E_1) := r_X \mathcal{D}(\mathbb{R}^n, E_1)$$

is dense in $W_p^m(X, E_0)$ for $1 \le p < \infty$ and $m \in \mathbb{N}$.

Proof Given $u \in W_p^m(X, E_0)$, we deduce from Theorems 2.4.5 and 2.4.3 the existence of a sequence $u_j \in \mathcal{D}(\mathbb{R}^n, E_1)$ converging in $W_p^m(\mathbb{R}^n, E_0)$ towards $e_X u$. Hence, thanks to (2.4.5), we see that $r_X u_j$ converges in $W_p^m(X, E_0)$ towards u. ■

As a first application of the above approximation results we prove a useful *product rule*. For this we recall from Proposition 1.4.8 that, given a densely injected Banach couple (E_0, E_1) such that E_1 is reflexive if $E_0 \ne E_1$, (E_1', E_0') is a densely injected Banach couple as well. Moreover, the E_0'-E_0-duality pairing determines the E_1'-E_1-duality pairing so that we can denote either one by $\langle\cdot,\cdot\rangle$ without fearing confusion.

2.4.7 Proposition *Suppose that X has a minimally smooth boundary. Let (E_0, E_1) be a densely injected Banach couple such that E_1 is reflexive if $E_0 \neq E_1$. Also suppose that $p \in [1,\infty]$,*

$$u \in L_p(X, E_1) \cap W_p^1(X, E_0) ,$$

and

$$v \in L_{p'}(X, E_0') \cap W_{p'}^1(X, E_1') .$$

Then $\langle v, u\rangle \in W_1^1(X)$ and

$$\partial_j \langle v, u\rangle = \langle \partial_j v, u\rangle + \langle v, \partial_j u\rangle , \qquad j = 1, \ldots, n . \tag{2.4.6}$$

Proof First note that (2.4.6) holds point-wise if $u \in \mathcal{D}(\overline{X}, E_1)$ and $v \in \mathcal{D}(\overline{X}, E_0')$. Next, Hölder's inequality implies that, given any Banach space F,

$$\big[(w', w) \mapsto \langle w', w\rangle\big] \in \mathcal{L}\big(L_{p'}(X, F'), L_p(X, F); L_1(X)\big) . \tag{2.4.7}$$

Using this, with F equal to E_0 or E_1 at the appropriate places, the assertion follows from Theorem 2.4.6, provided $1 < p < \infty$, thanks to the fact that (E_0, E_1) and (E_1', E_0') are densely injected Banach couples. If $p = 1$, we see from (2.4.7) that $\langle v, u\rangle$ and the right-hand side of (2.4.6) belong to $L_1(X)$. Let $\varphi \in \mathcal{D}(X)$ be given and fix an open set K such that $\operatorname{supp}(\varphi) \subset\subset K \subset\subset X$. Suppose that $u \in \mathcal{D}(\overline{X}, E_1)$. Then (writing again u for $r_K u$ etc.), $u \in L_2(K, E_1) \cap W_2^1(K, E_0)$ and $v \in L_2(K, E_0') \cap W_2^1(K, E_1')$. Hence

$$-\int_X \langle u, v\rangle \partial_j \varphi \, dx = \int_X \big[\langle \partial_j u, v\rangle + \langle u, \partial_j v\rangle\big] \varphi \, dx$$

by what has just been shown. Thus $\langle u, v\rangle \in W_1^1(X)$ and (2.4.6) is true, provided $u \in \mathcal{D}(\overline{X}, E_1)$ and $v \in L_\infty(X, E_0') \cap W_\infty^1(X, E_1')$. By applying again Theorem 2.4.6 and (2.4.7), we see that the assertion is true if $p = 1$. Finally, if $p = \infty$, we obtain the stated result by interchanging the rôles of u and v in the last step of the proof. ■

2.5 Final Value Problems

In this short subsection we collect some (almost trivial) results about final value problems. These problems come up naturally in connection with weak solutions of evolution equations in the next subsection. They are also of importance in the theory of optimal control for linear and quasilinear parabolic systems.

We fix $T \in \mathbb{R}^+$ and put $J := [0, T]$. We let E and F be Banach spaces with $F \hookrightarrow E$, assume that

$$B : J \to \mathcal{C}(E) \quad \text{with} \quad \operatorname{dom}\bigl(B(t)\bigr) \subset F , \qquad t \in J , \tag{2.5.1}$$

and that

$$(x, g) \in E \times L_{1,\mathrm{loc}}(J, E) .$$

Then we consider the final value problem

$$-\dot{v} + B(t)v = g(t) , \qquad 0 \leq t < T , \quad v(T) = x . \tag{2.5.2}$$

Of course, a **solution** v of (2.5.2) is a function $v \in C^1\bigl([0, T), E\bigr) \cap C(J, E)$ such that $v(t) \in \operatorname{dom}\bigl(B(t)\bigr)$ for $t \in [0, T)$ and (2.5.2) is point-wise satisfied. Thus a solution of (2.5.2) is defined by an obvious modification of the definition of a solution to an initial value problem. Clearly, a solution v of (2.5.2) is **strict** if $v \in C^1(J, E)$ and $v(T) \in \operatorname{dom}\bigl(B(T)\bigr)$. Similarly, the concepts of a $W^1_{p,\mathrm{loc}}$**-solution** and a W^1_p**-solution**, $1 \leq p \leq \infty$, of (2.5.2) are clear.

With (2.5.2) we associate the initial value problem

$$\dot{u} + A(t)u = f(t) , \qquad 0 < t \leq T , \quad u(0) = x , \tag{2.5.3}$$

where $A : J \to \mathcal{C}(E)$ and $f \in L_{1,\mathrm{loc}}(J, E)$ are defined by

$$A(t) := B(T - t) , \quad f(t) := g(T - t) , \qquad t \in J , \tag{2.5.4}$$

respectively.

2.5.1 Lemma *Put*

$$u(t) := v(T - t) , \qquad 0 \leq t \leq T . \tag{2.5.5}$$

Then v is a [strict] solution (resp. a $W^1_{p,\mathrm{loc}}$-solution for some $p \in [1, \infty]$) of the final value problem (2.5.2) *iff u is a [strict] solution (resp. a $W^1_{p,\mathrm{loc}}$-solution) of the initial value problem* (2.5.3).

Proof Obvious. ■

A function $V : J_\Delta \to \mathcal{L}(E)$ is said to be a **parabolic evolution operator for the final value problem** (2.5.2) (with regularity subspace F) if $U : J_\Delta \to \mathcal{L}(E)$, defined by

$$U(t, s) := V(T - s, T - t) , \qquad (t, s) \in J_\Delta , \tag{2.5.6}$$

is a parabolic evolution operator for A, defined by (2.5.4) (with regularity subspace F). Using (2.5.6), it is trivial to deduce from properties (II.2.1.2)–(II.2.1.9) the properties of V. In particular,

$$V \in C\big(J_\Delta, \mathcal{L}_s(E)\big) \cap C\big(J_\Delta^*, \mathcal{L}(E,F)\big) \tag{2.5.7}$$

and

$$V(t,t) = 1 \ , \quad V(t,s) = V(\tau,s)V(t,\tau) \ , \qquad 0 \le s \le \tau \le t \le T \ . \tag{2.5.8}$$

Moreover,

$$V(t,\cdot) \in C^1\big([0,t), \mathcal{L}(E)\big) \ , \qquad t \in (0,T] \ , \tag{2.5.9}$$

and

$$\partial_2 V(t,s) = B(s)V(t,s) \ , \qquad 0 \le s < t \le T \ . \tag{2.5.10}$$

Note that, thanks to Remark II.2.1.2(c), there exists at most one parabolic evolution operator V for the final value problem (2.5.2).

2.5.2 Proposition *Suppose that V is a parabolic evolution operator for the final value problem* (2.5.2). *If v is a solution of* (2.5.2),

$$v(t) = V(T,t)x + \int_t^T V(\tau,t)g(\tau)\,d\tau \ , \qquad 0 \le t \le T \ . \tag{2.5.11}$$

If $g = 0$ then $v := V(T,\cdot)x$ is a solution of (2.5.2).

Proof This follows by means of the transformations (2.5.5) and (2.5.6) from the corresponding results for initial value problems. ■

Of course, formula (2.5.11) is meaningful whenever $g \in L_{1,\mathrm{loc}}(J,E)$. Hence the function v defined by (2.5.11) is said to be a **mild solution of the final value problem** (2.5.2). Furthermore, (2.5.11) is the **variation-of-constants formula for the final value problem** (2.5.2).

2.5.3 Theorem *Suppose that (E_0,E_1) is a densely injected Banach couple and*

$$B \in C^\rho\big(J, \mathcal{H}(E_1,E_0)\big)$$

for some $\rho \in (0,1)$. Then there exists a unique parabolic evolution operator V for the final value problem (2.5.2), *and it possesses E_1 as regularity subspace. If $f \in C^\rho(J,E_0)$ then* (2.5.2) *has a unique solution, which is strict if $x \in E_1$.*

Proof Thanks to Lemma 2.5.1, the assertions are obvious consequences of Corollary II.4.4.2 and Theorem II.1.2.1, respectively. ■

We leave it to the reader to formulate further existence theorems for (2.5.2) based upon Theorem II.1.2.2 or the existence results of Chapter III.

2.6 Weak Solutions and Duality

In this subsection we generalize the concept of solutions of Cauchy problems. These generalizations are motivated by the notions of weak solutions in the theory of partial differential equations. Thus they are of particular importance in applications of the abstract theory to linear and quasilinear parabolic systems given in later chapters.

Assume that

$$\left.\begin{aligned}&E \text{ is a reflexive Banach space and}\\&(\cdot,\cdot)_\theta \in \{\,[\cdot,\cdot]_\theta, (\cdot,\cdot)_{\theta,q}, (\cdot,\cdot)^0_{\theta,\infty}\ ;\ 1<q<\infty\},\quad 0<\theta<1.\end{aligned}\right\}\tag{2.6.1}$$

Denote by J a perfect subinterval of $\mathbb{R}^+$ containing 0, and suppose that

$$A\colon J \to \mathcal{G}(E) \quad \text{and} \quad \bigcap_{t\in J} \rho\big(A(t)\big) \neq \emptyset\,. \tag{2.6.2}$$

Fix $\mu \in \bigcap_{t\in J} \rho\big(-A(t)\big)$, and denote by

$$\big[\big(E_\alpha(t), (\mu + A(t))_\alpha\big)\ ;\ \alpha \in \mathbb{R}\big]\,, \qquad t \in J\,,$$

the interpolation-extrapolation scales generated by $\big(E, \mu + A(t)\big)$ and the interpolation functors $(\cdot,\cdot)_\theta,\ 0<\theta<1$, where the latter are independent of $t \in J$. Let

$$\big[\big(E^\sharp_\alpha(t), (\mu + A(t))^\sharp_\alpha\big)\ ;\ \alpha \in \mathbb{R}\big]\,, \qquad t \in J\,,$$

be the respective dual scales. Put

$$E_\alpha := (E_\alpha, \|\cdot\|_\alpha) := E_\alpha(0)\,, \quad E^\sharp_\alpha := (E^\sharp_\alpha, \|\cdot\|^\sharp_\alpha) := E^\sharp_\alpha(0)\,, \qquad \alpha \in \mathbb{R}\,. \tag{2.6.3}$$

Moreover, let $A_\alpha(t)$ and $A^\sharp_\alpha(t)$ be the $E_\alpha(t)$- and $E^\sharp_\alpha(t)$-realizations of A and $A^\sharp$, respectively, for $\alpha \in \mathbb{R}$ and $t \in J$. Lastly, assume that

$$\left.\begin{aligned}&\text{there exists } \beta \in [0,1] \text{ such that}\\&E_\gamma \doteq E_\gamma(t),\quad t\in J,\quad \gamma \in \{\beta, \beta-1\}.\end{aligned}\right\}\tag{2.6.4}$$

Note that Theorem 1.5.12 implies

$$E^\sharp_\gamma \doteq E^\sharp_\gamma(t)\,, \qquad t \in J\,, \quad \gamma \in \{1-\beta, -\beta\}\,. \tag{2.6.5}$$

In the following proposition we establish a simple but important '*generalized Green's formula*'.

2.6.1 Proposition *Let assumptions* (2.6.1), (2.6.2), *and* (2.6.4) *be satisfied and suppose that*

$$A_{\beta-1} \in \mathcal{L}_{\infty,\mathrm{loc}}\big(J, \mathcal{L}(E_\beta, E_{\beta-1})\big) . \tag{2.6.6}$$

Then, given $p \in [1,\infty]$,

$$u \in L_{p,\mathrm{loc}}(J, E_\beta) \cap W^1_{p,\mathrm{loc}}(J, E_{\beta-1}) , \tag{2.6.7}$$

and

$$u^\sharp \in L_{p',\mathrm{loc}}(J, E^\sharp_{1-\beta}) \cap W^1_{p',\mathrm{loc}}(J, E^\sharp_{-\beta}) , \tag{2.6.8}$$

it follows that

$$\int_s^t \langle u^\sharp, (\partial + A_{\beta-1})u\rangle \, d\tau + \langle u^\sharp, u\rangle(s) = \int_s^t \langle(-\partial + A^\sharp_{-\beta})u^\sharp, u\rangle \, d\tau + \langle u^\sharp, u\rangle(t)$$

for $s, t \in J$ *with* $s < t$.

Proof From (2.3.6) we know that

$$\big\langle x^\sharp, A_{\beta-1}(\tau)x\big\rangle = \big\langle A^\sharp_{-\beta}(\tau)x^\sharp, x\big\rangle , \qquad (x^\sharp, x) \in E^\sharp_{1-\beta} \times E_\beta , \quad \tau \in J . \tag{2.6.9}$$

Thus (2.6.6)–(2.6.8) and Hölder's inequality imply

$$\langle u^\sharp, A_{\beta-1}u\rangle = \langle A^\sharp_{-\beta}u^\sharp, u\rangle \in L_{1,\mathrm{loc}}(J) . \tag{2.6.10}$$

Now the assertion is an easy consequence of Proposition 2.4.7, Theorem 1.5.12, and Theorem III.1.2.2. ∎

The generalized Green's formula suggests the following definition of 'weak solutions' of linear Cauchy problems. Let the hypotheses of Proposition 2.6.1 be satisfied. Suppose that

$$p \in [1,\infty] , \quad (x, f) \in E_{\beta-1} \times L_{p,\mathrm{loc}}(J, E_{\beta-1}) . \tag{2.6.11}$$

Then u is said to be a **weak $L_{p,\mathrm{loc}}(J, E_\beta)$-solution** of the Cauchy problem

$$\dot u + A(t)u = f(t) , \quad t \in \dot J , \qquad u(0) = x , \tag{2.6.12}$$

provided $u \in L_{p,\mathrm{loc}}(J, E_\beta)$ and

$$\int_J \langle(-\partial + A^\sharp_{-\beta})u^\sharp, u\rangle \, dt = \int_J \langle u^\sharp, f\rangle \, dt + \big\langle u^\sharp(0), x\big\rangle$$

for each $u^\sharp \in L_{p'}(J, E^\sharp_{1-\beta}) \cap W^1_{p'}(J, E^\sharp_{-\beta})$ having compact support in $J\backslash\{\sup(J)\}$. Of course, if u is a weak $L_{p,\mathrm{loc}}(J, E_\beta)$-solution of (2.6.12) and $u \in L_p(J, E_\beta)$ then u is a **weak $L_p(J, E_\beta)$-solution**.

For the next proposition we recall that $W^1_{p,\mathrm{loc}}$-solutions have been defined in Subsection III.1.3.

2.6.2 Proposition *Let assumptions* (2.6.1), (2.6.2), *and* (2.6.4) *be satisfied and suppose that* $A_{\beta-1}$ *and* (x, f) *satisfy* (2.6.6) *and* (2.6.11), *respectively. If* u *is a* $W^1_{p,\mathrm{loc}}$*-solution of the extrapolated Cauchy problem*

$$\dot{u} + A_{\beta-1}(t)u = f(t)\ , \quad t \in J\ , \qquad u(0) = x \tag{2.6.13$_{\beta-1}$}$$

then u *is a weak* $L_{p,\mathrm{loc}}(J, E_\beta)$*-solution of* (2.6.12).

Proof This is an immediate consequence of the above definition and Proposition 2.6.1. ■

Let $T \in \dot{J}$ be fixed and supppose that

$$(x^\sharp, f^\sharp) \in E^\sharp_{-\beta} \times L_{1,\mathrm{loc}}\big((0,T), E^\sharp_{-\beta}\big)\ . \tag{2.6.14}$$

Then the generalized Green's formula suggests the study of the **dual final value problem** of (2.6.12) defined by

$$-\dot{u}^\sharp + A^\sharp(t)u^\sharp = f^\sharp(t)\ , \quad 0 \le t < T\ , \qquad u^\sharp(T) = x^\sharp\ . \tag{2.6.15^T}$$

With $(2.6.15)^T$ we also associate the **extrapolated** dual final value problem:

$$-\dot{u}^\sharp + A^\sharp_{-\beta}(t)u^\sharp = f^\sharp(t)\ , \quad 0 \le t < T\ , \qquad u^\sharp(T) = x^\sharp\ . \tag{2.6.16$^T_{-\beta}$}$$

It is now easy to give a general sufficient condition for weak $L_{p,\mathrm{loc}}(J, E_\beta)$-solutions of (2.6.12) to be unique.

2.6.3 Proposition *Let assumptions* (2.6.1), (2.6.2), *and* (2.6.4) *be satisfied and suppose that* $A_{\beta-1}$ *and* (x, f) *satisfy* (2.6.6) *and* (2.6.11), *respectively. Then problem* (2.6.12) *has at most one weak* $L_{p,\mathrm{loc}}(J, E_\beta)$*-solution if, given* $T \in \mathring{J}$ *and* $f^\sharp \in \mathcal{D}(0,T) \otimes E^\sharp_{1-\beta}$, *the extrapolated dual final value problem* $(2.6.16)^T_{-\beta}$ *with* $x^\sharp = 0$ *has a* $W^1_{p'}$*-solution.*

Proof Suppose that $u \in L_{p,\mathrm{loc}}(J, E_\beta)$ satisfies

$$\int_J \langle (-\partial + A^\sharp_{-\beta})v^\sharp, u\rangle\, d\tau = 0 \tag{2.6.17}$$

for each

$$v^\sharp \in L_{p'}(J, E^\sharp_{1-\beta}) \cap W^1_{p'}(J, E^\sharp_{-\beta}) \tag{2.6.18}$$

having compact support in $J \backslash \{\sup(J)\}$. Then we have to show that $u = 0$. Given $T \in \mathring{J}$ and $f^\sharp \in \mathcal{D}(0,T) \otimes E^\sharp_{1-\beta}$, let $u^\sharp$ be a $W^1_{p'}$-solution of $(2.6.16)^T_{-\beta}$ for $x^\sharp = 0$.

Define $v^\sharp$ by extending $u^\sharp$ by zero over $J \cap (T,\infty)$. Then it follows from Theorem III.1.2.2 that $v^\sharp$ satisfies (2.6.18) and has compact support in $J\backslash\{\sup(J)\}$. Thus, thanks to (2.6.17),

$$\int_J \langle f^\sharp, u\rangle \, d\tau = 0 , \qquad f^\sharp \in \mathcal{D}(0,T) \otimes E^\sharp_{1-\beta} . \tag{2.6.19}$$

Hence, using $E^\sharp_{1-\beta} \hookrightarrow E^\sharp_{-\beta}$,

$$\Bigl\langle x^\sharp, \int_J \varphi u \, dt \Bigr\rangle = \int_J \langle \varphi \otimes x^\sharp, u\rangle \, dt = 0 , \qquad x^\sharp \in E^\sharp_{1-\beta} , \quad \varphi \in \mathcal{D}(\mathring{J}) .$$

Since $E^\sharp_{1-\beta}$ is dense in $E^\sharp_{-\beta}$, the above relation is valid for each $x^\sharp \in E^\sharp_{-\beta}$, which shows that

$$\int_J \varphi u \, dt = 0 , \qquad \varphi \in \mathcal{D}(\mathring{J}) .$$

From this we easily infer, by invoking (III.4.2.27), for example, that $u = 0$. ■

So far we have required neither restrictive regularity assumptions nor a parabolicity hypothesis. If we impose conditions of this type, we obtain the following uniqueness and existence result for weak $L_{p,\mathrm{loc}}(J, E_\beta)$-solutions of (2.6.12) and solutions of the extrapolated Cauchy problem $(2.6.13)_{\beta-1}$, respectively.

2.6.4 Theorem *Let assumptions* (2.6.1), (2.6.2), *and* (2.6.4) *be satisfied. Suppose that $-A(t)$ generates an analytic semigroup on E for $t \in J$, and*

$$A_{\beta-1} \in C^\rho\bigl(J, \mathcal{L}(E_\beta, E_{\beta-1})\bigr) \tag{2.6.20}$$

for some $\rho \in (0,1)$. Then, given $p \in [1,\infty]$ and

$$(x, f) \in E_{\beta-1} \times L_{p,\mathrm{loc}}(J, E_{\beta-1}) ,$$

the Cauchy problem (2.6.12) *has at most one weak $L_{p,\mathrm{loc}}(J, E_\beta)$-solution. If there exists $\varepsilon \in (0,1)$ such that*

$$x \in E_\beta , \quad f \in C(J, E_{\beta-1+\varepsilon}) + C^\varepsilon(J, E_{\beta-1}) , \tag{2.6.21}$$

the extrapolated Cauchy problem $(2.6.13)_{\beta-1}$ has a unique strict solution, and (2.6.12) *has a unique weak $L_{p,\mathrm{loc}}(J, E_\beta)$-solution.*

Proof Since $A(t) \in \mathcal{H}\bigl(E_1(t), E\bigr)$ for $t \in J$ by assumption, Theorem 2.3.1 implies

$$A^\sharp_{-\beta}(t) \in \mathcal{H}(E^\sharp_{1-\beta}, E^\sharp_{-\beta}) , \qquad t \in J . \tag{2.6.22}$$

From (2.3.6) we deduce that

$$\|A^\sharp_{-\beta}(s) - A^\sharp_{-\beta}(t)\|_{\mathcal{L}(E^\sharp_{1-\beta}, E^\sharp_{-\beta})} = \|A_{\beta-1}(s) - A_{\beta-1}(t)\|_{\mathcal{L}(E_\beta, E_{\beta-1})} , \qquad s,t \in J .$$

Consequently, thanks to (2.6.20) and (2.6.22),

$$A^\sharp_{-\beta} \in C^\rho\big(J, \mathcal{H}(E^\sharp_{1-\beta}, E^\sharp_{-\beta})\big) . \tag{2.6.23}$$

Thus, given $T \in \mathring{J}$ and $f^\sharp \in \mathcal{D}(0,T) \otimes E^\sharp_{1-\beta} \subset C^\rho(J, E^\sharp_{-\beta})$, it follows from Theorem 2.5.3 that the final value problem $(2.6.16)^T_{-\beta}$ has a solution on $[0,T]$, hence a $W^1_{p'}$-solution. Now the first assertion is a consequence of Proposition 2.6.3. From $A(t) \in \mathcal{H}\big(E_1(t), E\big)$, Corollary 2.1.4, and (2.6.20) we infer that

$$A_{\beta-1} \in C^\rho\big(J, \mathcal{H}(E_\beta, E_{\beta-1})\big) . \tag{2.6.24}$$

Put $(F_0, F_1) := (E_{\beta-1}, E_\beta)$ and $F_\theta := (F_0, F_1)_\theta$ for $0 < \theta < 1$. Fix $\sigma \in (0, \varepsilon)$. The almost reiteration property of Theorem 1.5.3 implies $E_{\beta-1+\varepsilon} \hookrightarrow F_\sigma$. Hence (2.6.21) gives

$$f \in C(J, F_\sigma) + C^\sigma(J, F_0) .$$

Choose a decomposition $f = f_0 + f_1$ with $f_0 \in C^\sigma(J, F_0)$ and $f_1 \in C(J, F_\sigma)$ and, letting $B := A_{\beta-1}$, consider the Cauchy problems

$$\dot{v} + Bv = f_j(t) , \quad t \in \dot{J} , \qquad v(0) = x_j , \quad j = 0, 1 ,$$

where $x_0 := x$ and $x_1 := 0$. It follows from Theorems II.1.2.1 and II.1.2.2, respectively, that these problems have unique strict solutions u_0 and u_1, respectively. Hence $u := u_0 + u_1$ is a strict solution of the extrapolated Cauchy problem $(2.6.13)_{\beta-1}$; thus a $W^1_{p,\mathrm{loc}}$-solution of $(2.6.13)_{\beta-1}$. Consequently, thanks to Proposition 2.6.2, it is a weak $L_{p,\mathrm{loc}}(J, E_\beta)$-solution of (2.6.12). Therefore it is unique by the first part of the assertion. ■

We close this subsection by identifying the dual of the evolution operator $U_{\beta-1}$, where the latter is considered as an element of $\mathcal{L}(E_\beta, E_{\beta-1})$. This identification is particularly useful in problems of control theory.

2.6.5 Proposition *Let assumptions* (2.6.1), (2.6.2), *and* (2.6.4) *be satisfied. Suppose that $-A(t)$ generates an analytic semigroup on E for $t \in J$, and*

$$A_{\beta-1} \in C^\rho\big(J, \mathcal{L}(E_\beta, E_{\beta-1})\big) .$$

Let $U_{\beta-1}$ be the parabolic evolution operator of $A_{\beta-1}$ and, given $T \in \mathring{J}$, let $V^\sharp_{-\beta}$ be the evolution operator of the extrapolated dual final value problem $(2.6.16)^T_{-\beta}$. *Then*

$$\big\langle x^\sharp, U_{\beta-1}(t,s)x \big\rangle = \big\langle V^\sharp_{-\beta}(t,s)x^\sharp, x \big\rangle , \qquad (x^\sharp, x) \in E^\sharp_{1-\beta} \times E_\beta , \quad 0 \le s \le t \le T ,$$

that is, the dual operator in $\mathcal{L}(E^\sharp_{1-\beta}, E^\sharp_{-\beta})$ of $U(t,s) \in \mathcal{L}(E_\beta, E_{\beta-1})$ equals $V(t,s)$ for $0 \le s \le t \le T$.

Proof Let $s,t\in J$ with $s<t$ be fixed. Since the assumptions of Theorem 2.6.4 are satisfied, (2.6.23) and (2.6.24) are true. Hence Theorem 2.5.3 and Corollary II.4.4.2 guarantee the existence and uniqueness of the evolution operators $V^\sharp_{-\beta}$ of $(2.6.16)^T_{-\beta}$ and $U_{\beta-1}$ of $A_{\beta-1}$, respectively. Given $(x^\sharp,x)\in E^\sharp_{1-\beta}\times E_\beta$, put $u(\tau):=U_{\beta-1}(\tau,s)x$ and $u^\sharp(\tau):=V^\sharp_{-\beta}(t,\tau)x^\sharp$ for $s\le\tau\le t$. Then Theorem II.1.2.1, Remark II.2.1.2(a), and Theorem 2.5.3 imply

$$(u^\sharp,u)\in C\big([s,t],E^\sharp_{1-\beta}\times E_\beta\big)\cap C^1\big([s,t],E^\sharp_{-\beta}\times E_{\beta-1}\big)$$

and

$$(-\partial+A^\sharp_{-\beta})u^\sharp=0\ ,\quad (\partial+A_{\beta-1})u=0 \qquad \text{on } [s,t]\ .$$

Thus u and $u^\sharp$ satisfy (2.6.7) and (2.6.8), respectively, on the interval $[s,t]$, with $u(s)=x$ and $u(t)=x^\sharp$. Hence it follows from the generalized Green's formula of Proposition 2.6.1 that

$$\big\langle x^\sharp,U_{\beta-1}(t,s)x\big\rangle=\langle u^\sharp,u\rangle(t)=\langle u^\sharp,u\rangle(s)=\big\langle V^\sharp_{-\beta}(t,s)x^\sharp,x\big\rangle\ .$$

This proves the assertion. ∎

2.7 Positivity

Let $E:=(E,P)$ be an *OBS* and suppose that A is of positive type, that is, $A\in\mathcal{P}(E)$. Fix $m\in\mathbb{N}$ and, for each $\theta\in(0,1)$, an admissible interpolation functor $(\cdot,\cdot)_\theta$. Denote by $\big[(E_\alpha,A_\alpha)\ ;\ \alpha\in[-m,\infty)\big]$ the interpolation-extrapolation scale generated by (E,A) and $(\cdot,\cdot)_\theta$, $0<\theta<1$. Lastly, assume that A is resolvent positive.

Given $\alpha\in[-m,\infty)$, put

$$P_\alpha:=\left\{\begin{array}{ll} i_\alpha^{-1}(P)\ , & \alpha\ge 0\ ,\\ \mathrm{cl}_{E_\alpha}(P)\ , & \alpha<0\ ,\end{array}\right. \tag{2.7.1}$$

where $i_\alpha: E_\alpha\hookrightarrow E$ is the canonical injection. Then P_α is a closed convex cone in E_α, which is proper if $\alpha\ge 0$. Thus P_α induces a preorder in E_α, the **natural preorder**. Henceforth we put

$$E_\alpha=(E_\alpha,P_\alpha)\ ,\qquad -m\le\alpha<\infty\ ,$$

that is, we endow each E_α with its natural preorder, which we simply denote by $\le$ without fearing confusion.

2.7.1 Proposition *Suppose that $A\in\mathcal{P}(E)$ and is resolvent positive. Then E_α is an OBS for $\alpha\ge 0$ with respect to the natural order induced by P_α. Given $-m\le\beta\le\alpha<\infty$, the injection $E_\alpha\overset{d}{\hookrightarrow}E_\beta$ is positive, A_α is resolvent positive, and P_α is dense in P_β.*

Proof The first three assertions are trivial. Observe that $-m \le \beta \le \alpha < \infty$ implies $P_\alpha \subset P_\beta$. Thus $P_\alpha \overset{d}{\subset} P_\beta$ if $\alpha \le 0$ since, by definition, $P = P_0 \overset{d}{\subset} P_\beta$. Hence it remains to show that P_α is dense in P_β if $\beta \ge 0$. Given $\beta \ge 0$ and $x \in P_\beta$, note that $(1+\varepsilon A_\beta)^{-1}x \in P_{\beta+1}$. Since A is of positive type, Proposition 1.5.5 and the remarks following it imply $A_\beta \in \mathcal{P}(E_\beta)$. Hence it follows from Lemma II.6.1.1 that $(1+\varepsilon A_\beta)^{-1}x \to x$ in E_β as $\varepsilon \to 0$. This shows that $P_{\beta+1} \overset{d}{\subset} P_\beta$ for $\beta \ge 0$. Thus $P_{\beta+k} \overset{d}{\subset} P_\beta$ for $\beta \ge 0$ and $k \in \mathbb{N}$. Consequently, given $\alpha \ge \beta \ge 0$, it follows from $P_{\beta+k} \subset P_\alpha \subset P_\beta$, where $k \in \dot{\mathbb{N}}$ satisfies $\beta + k > \alpha$, that $P_\alpha \overset{d}{\subset} P_\beta$. ■

Now we assume that

$$\left.\begin{array}{l} E \text{ is a reflexive Banach space ordered by a total cone } P, \text{ and} \\ (\cdot,\cdot)_\theta \in \big\{ [\cdot,\cdot]_\theta, (\cdot,\cdot)_{\theta,q}, (\cdot,\cdot)_\theta^0 \; ; \; 1<q<\infty \big\}, \quad 0<\theta<1. \end{array}\right\} \qquad (2.7.2)$$

In addition, we retain the assumption that

$$A \in \mathcal{P}(E) \text{ and is resolvent positive.} \qquad (2.7.3)$$

Then the interpolation-extrapolation scale $\big[(E_\alpha, A_\alpha)\; ; \; \alpha \in \mathbb{R}\big]$, generated by (E, A) and $(\cdot,\cdot)_\theta$, $0<\theta<1$, and its dual scale $\big[(E_\alpha^\sharp, A_\alpha^\sharp)\; ; \; \alpha \in \mathbb{R}\big]$ are well-defined. We denote by

$$P^\sharp \text{ the dual cone of } E^\sharp.$$

Then $E^\sharp = (E^\sharp, P^\sharp)$ is an *OBS* with the so-defined **natural (dual) order**. Thus, by replacing P in (2.7.1) by $P^\sharp$, the cones $P_\alpha^\sharp$ of $E_\alpha^\sharp$ are well-defined and induce the natural preorder of $E_\alpha^\sharp$ for $\alpha \in \mathbb{R}$. Also note that $A^\sharp \in \mathcal{P}(E^\sharp)$ and

$$\big\langle (1+\varepsilon A^\sharp)^{-1} x^\sharp, x \big\rangle = \big\langle x^\sharp, (1+\varepsilon A)^{-1} x \big\rangle \ge 0 \, , \qquad (x^\sharp, x) \in P^\sharp \times P \, , \quad \varepsilon > 0 \, ,$$

which shows that $A^\sharp$ is resolvent positive as well. Hence it is a consequence of Proposition 2.7.1 that $P_\alpha^\sharp$ is dense in $P_\beta^\sharp$ if $-\infty < \beta < \alpha < \infty$.

The next *duality theorem*, the main result of this subsection, shows that $P_\alpha^\sharp$ is the dual cone of $P_{-\alpha}$ for $\alpha \in \mathbb{R}$.

2.7.2 Theorem *Let assumptions* (2.7.2) *and* (2.7.3) *be satisfied and let E_α and $E_\alpha^\sharp$ be given their natural preorders, induced by the positive cones P_α and $P_\alpha^\sharp$, respectively. Then E_α and $E_\alpha^\sharp$ are OBSs for $\alpha \ge 0$, the injections $E_\alpha \hookrightarrow E_\beta$ and $E_\alpha^\sharp \hookrightarrow E_\beta^\sharp$ are positive for $-\infty < \beta < \alpha < \infty$, the operators A_α and $A_\alpha^\sharp$ are resolvent positive, and P_α [resp. $P_\alpha^\sharp$] is dense in P_β [resp. $P_\beta^\sharp$] for $-\infty < \beta < \alpha < \infty$. Moreover,*

$$P_\alpha^\sharp = (P_{-\alpha})' \, , \qquad \alpha \in \mathbb{R} \, , \qquad (2.7.4)$$

that is, given $\alpha \in \mathbb{R}$ *and* $x^\sharp \in E_\alpha^\sharp$,

$$x^\sharp \in P_\alpha^\sharp \quad \text{iff} \quad \langle x^\sharp, x\rangle \geq 0 , \qquad x \in P_{-\alpha} .$$

Proof Only (2.7.4) remains to be proven. Thus assume first that $\alpha > 0$. Given $x^\sharp \in P_\alpha^\sharp = P^\sharp \cap E_\alpha^\sharp$, it follows that $\langle x^\sharp, x\rangle \geq 0$ for $x \in P$. Hence the density of P in $P_{-\alpha}$ and Theorem 1.5.12 imply $\langle x^\sharp, x\rangle \geq 0$ for $x \in P_{-\alpha}$, that is,

$$P_\alpha^\sharp \subset (P_{-\alpha})' . \tag{2.7.5}$$

Now let $\alpha < 0$. Given $x^\sharp \in P_\alpha^\sharp$, there exists a sequence $(x_j^\sharp)$ in $P^\sharp$ converging in $E_\alpha^\sharp$ towards $x^\sharp$. Hence $\langle x_j^\sharp, x\rangle \to \langle x^\sharp, x\rangle$ as $j \to \infty$ for $x \in P_{-\alpha} \subset P$, and $\langle x_j^\sharp, x\rangle \geq 0$ implies that (2.7.5) is true in this case as well.

To prove the inclusion reverse to (2.7.5), let $x^\sharp \in E_\alpha^\sharp$ satisfy $\langle x^\sharp, x\rangle \geq 0$ for $x \in P_{-\alpha}$. If $\alpha > 0$, it follows that $\langle x^\sharp, x\rangle \geq 0$ for $x \in P$ so that $x^\sharp \in P^\sharp$. Consequently, $x^\sharp \in P_\alpha^\sharp = P^\sharp \cap E_\alpha$, which shows that $(P_{-\alpha})' \subset P_\alpha^\sharp$ in this case. Thus let $\alpha < 0$ and fix $k \in \dot{\mathbb{N}}$ with $\alpha + k > 0$. Then, thanks to Theorem 1.5.12 and Remark 1.1.2(f),

$$\big\langle (1+\varepsilon A_\alpha^\sharp)^{-k} x^\sharp, x\big\rangle = \big\langle x^\sharp, (1+\varepsilon A_{-\alpha})^{-k} x\big\rangle \geq 0 , \qquad x \in P_{-\alpha} , \tag{2.7.6}$$

due to the resolvent positivity of $A_{-\alpha}$. Since $(1+\varepsilon A_\alpha^\sharp)^{-k} x^\sharp \in E_{\alpha+k}^\sharp \subset E^\sharp$, we deduce from (2.7.6) and the density of $P_{-\alpha}$ in P that $(1+\varepsilon A_\alpha^\sharp)^{-k} x^\sharp \in P^\sharp$. Finally, we infer from Lemma II.6.1.1 that $(1+\varepsilon A_\alpha^\sharp)^{-k} x^\sharp \to x^\sharp$ in $E_\alpha^\sharp$. This proves that $(P_{-\alpha})' \subset P_\alpha^\sharp$ in this case as well. ■

2.7.3 Corollary *Suppose that assumptions* (2.7.2) *and* (2.7.3) *are satisfied and let* $\alpha > 0$. *Then* $E_{-\alpha}^\sharp$ *is an OBS iff* P_α *is total in* E_α. *If this is the case,* P_β *is total in* E_β, *and* $E_{-\beta}^\sharp$ *is an OBS for* $0 \leq \beta \leq \alpha$.

Proof Recall that $(P_\alpha)'$ is proper iff P_α is total. Clearly, if P_α is total in E_α then $P_\alpha \subset P_\beta$ and $E_\alpha \overset{d}{\hookrightarrow} E_\beta$ imply that P_β is total in E_β for $0 \leq \beta < \alpha$. Thus the assertions follow from (2.7.4). ■

It is clear how to combine the results of this subsection with Theorems 2.1.3 and 2.3.2 and the assertions of Subsection II.6.4 to deduce information on positive semigroups and evolution operators in Banach scales. We leave details to the reader.

2.8 General Evolution Equations

Although this treatise is concerned with parabolic equations, large parts of this chapter require only the assumption that $A \in \mathcal{G}(E)$. For this reason we discuss briefly some implications of the theory of interpolation-extrapolation scales to the case of evolution equations

$$\dot{u} + Au = f(t) , \quad t \in J , \qquad u(0) = x , \tag{2.8.1}$$

where it is not required that $-A$ generates an analytic semigroup.

Thus we assume E is a Banach space and J is a perfect subinterval of $\mathbb{R}^+$ containing 0. For each $\theta \in (0,1)$ we fix an admissible exact interpolation functor and assume that $A \in \mathcal{G}(E)$. Then we select $\mu \in \rho(-A)$ and $m \in \dot{\mathbb{N}}$, and denote by $\bigl[(E_\alpha, B_\alpha)\ ;\ \alpha \in [-m,\infty)\bigr]$ the interpolation-extrapolation scale generated by (E,B) and $(\cdot,\cdot)_\theta$, $0 < \theta < 1$, where $B := \mu + A$. We also denote by A_α the E_α-realization of A.

Suppose that

$$(x,f) \in E \times L_{1,\mathrm{loc}}(J,E)$$

and put

$$u(t) := e^{-tA}x + \int_0^t e^{-(t-\tau)A} f(\tau)\, d\tau , \qquad t \in J . \tag{2.8.2}$$

Then u is the **mild solution** of (2.8.1). It follows from Remark II.2.1.2(a) that every solution of (2.8.1) coincides with the mild solution, provided $f \in C(J,E)$. Thus (2.8.1) has at most one solution in this case. It is well-known that, in general, a mild solution of (2.8.1) is not a solution. A sufficient condition for (2.8.1) to possess a solution is contained in the following well-known theorem. For the reader's convenience we include its easy proof.

2.8.1 Theorem *Suppose that*

$$(x,f) \in E_1 \times \bigl[C(J,E_1) + C^1(J,E_0)\bigr] .$$

Then the Cauchy problem (2.8.1) *possesses a unique strict solution.*

Proof It suffices to show that the mild solution (2.8.2) is a solution. Since $v(t) := e^{-tA}x$, $t \in \mathbb{R}^+$, is a strict solution of $\dot{u} + Au = 0$ in $\dot{\mathbb{R}}^+$ satisfying $u(0) = x$, if $x \in E_1$, we can assume that $x = 0$.

First suppose that $f \in C(J,E_1)$. Since $A_1 \in \mathcal{G}(E_1)$ by Corollary 2.1.4, it follows that $u \in C(J,E_1)$ and

$$Au(t) = \int_0^t Ae^{-(t-\tau)A} f(\tau)\, d\tau , \qquad t \in J .$$

Also $u \in C^1(J,E_0)$ and

$$\dot{u}(t) = f(t) - \int_0^t Ae^{-(t-\tau)A} f(\tau)\, d\tau = f(t) - Au(t) , \qquad t \in J ,$$

which proves the assertion if $f \in C(J,E_1)$.

Next suppose that $f \in C^1(J, E_0)$. Since

$$u(t) = \int_0^t e^{-\tau A} f(t-\tau)\, d\tau \ , \qquad t \in J \ ,$$

it follows that $u \in C^1(J, E_0)$. Given $t \in J$ and $h > 0$ with $t + h \in J$, it is easily verified that

$$\frac{e^{-hA} - 1}{h} u(t) = \frac{u(t+h) - u(t)}{h} - \frac{1}{h} \int_t^{t+h} e^{-(t+h-\tau)A} f(\tau)\, d\tau \ .$$

Thus, letting $h \to 0$, we see that $u(t) \in \operatorname{dom}(A)$ and $-Au(t) = \dot{u}(t) - f(t)$. This proves the theorem. ■

On the basis of the last theorem we can now give an interpretation of the mild solution if f satisfies suitable 'intermediate regularity', measured by the Banach spaces E_α.

2.8.2 Theorem *Suppose that* $0 \le \alpha \le 1$ *and*

$$(x, f) \in E_\alpha \times \big[C(J, E_\alpha) + C^1(J, E_{\alpha-1}) \cap L_{1,\mathrm{loc}}(J, E) \big] \ . \tag{2.8.3}$$

Then the mild solution (2.8.2) *of* (2.8.1) *is the unique strict solution of the extrapolated Cauchy problem*

$$\dot{u} + A_{\alpha-1} u = f(t) \ , \quad t \in \dot{J} \ , \qquad u(0) = x \tag{2.8.4}$$

in $E_{\alpha-1}$.

Proof By Corollary 2.1.4 we know that $A_\alpha \in \mathcal{G}(E_{\alpha-1})$. Thus (2.8.3) and Theorem 2.8.1 imply that (2.8.4) has a unique strict solution u. Since it is the mild solution of (2.8.4),

$$u(t) = e^{-tA_{\alpha-1}} x + \int_0^t e^{-(t-\tau)A_{\alpha-1}} f(\tau)\, d\tau = e^{-tA} x + \int_0^t e^{-(t-\tau)A} f(\tau)\, d\tau$$

for $t \in J$, where we used (2.1.12). ■

If we restrict the class of admissible Banach spaces and interpolation functors appropriately, we can see that the mild solution (2.8.2) is the unique weak solution of (2.8.1). For this we suppose that

$$\left.\begin{aligned} &E \text{ is a reflexive Banach space and} \\ &(\cdot,\cdot)_\theta \in \big\{ [\cdot,\cdot]_\theta, (\cdot,\cdot)_{\theta,q}, (\cdot,\cdot)^0_{\theta,\infty} \ ; \ 1 < q < \infty \big\}, \quad 0 < \theta < 1. \end{aligned}\right\} \tag{2.8.5}$$

So we can rely on the results of Subsection 2.6 to prove the following:

2.8.3 Theorem *Let assumption* (2.8.5) *be satisfied and suppose that* $0 \le \alpha \le 1$. *Then, given* $p \in [1,\infty]$ *and*

$$(x,f) \in E_\alpha \times L_{p,\mathrm{loc}}(J,E_\alpha)\ , \tag{2.8.6}$$

problem (2.8.1) *has a unique weak* $L_{p,\mathrm{loc}}(J,E_\alpha)$*-solution, namely the mild solution* (2.8.2).

Proof Since $A^\sharp_{-\alpha} \in \mathcal{G}(E^\sharp_{-\alpha})$ by Theorem 2.3.1, it follows from Theorem 2.8.1 that, given $T \in \mathring{J}$ and $f^\sharp \in \mathcal{D}(0,T) \otimes E^\sharp_{1-\alpha}$, the extrapolated Cauchy problem

$$\dot{v}^\sharp + A^\sharp_{-\alpha} v = f^\sharp(t)\ , \quad 0 \le t < T\ , \qquad v^\sharp(0) = 0$$

has a solution, hence a $W^1_{p'}$-solution. Now the asserted uniqueness follows from Lemma 2.5.1 and Proposition 2.6.3.

Given $p \in [1,\infty]$ and $(x,f) \in E_\alpha \times L_{p,\mathrm{loc}}(J,E_\alpha)$, define u by (2.8.2). It remains to show that u is a weak $L_{p,\mathrm{loc}}$-solution of (2.8.1).

First we assume that $f = 0$. Then $u(t) = e^{-tA_\alpha}x$ for $t \in J$ so that

$$u \in C(J,E_\alpha) \subset L_{p,\mathrm{loc}}(J,E_\alpha)\ .$$

Let

$$u^\sharp \in L_{p'}(J,E^\sharp_{1-\alpha}) \cap W^1_{p'}(J,E^\sharp_{-\alpha}) \quad \text{with} \quad \mathrm{supp}(u^\sharp) \subset\subset J\backslash\{\sup(J)\} \tag{2.8.7}$$

be given. From Theorems 2.3.1 and 2.3.2 we easily deduce that

$$\int_J \langle (-\partial + A^\sharp_{-\alpha})u^\sharp(t), u(t)\rangle\, dt = \int_J \langle e^{-tA^\sharp_{-\alpha}}(-\partial + A^\sharp_{-\alpha})u^\sharp(t), x\rangle\, dt\ .$$

Note that

$$e^{-tA^\sharp_{-\alpha}}(-\partial + A^\sharp_{-\alpha})u^\sharp(t) = -\partial\big[e^{-tA^\sharp_{-\alpha}}u^\sharp(t)\big]\ , \qquad t \in J\ .$$

Hence the last integral equals $\langle u^\sharp(0), x\rangle$. This shows that u is a weak $L_{p,\mathrm{loc}}(J,E_\alpha)$-solution if $f = 0$.

Next suppose that $x = 0$ so that

$$u(t) = \int_0^t e^{-(t-\tau)A_\alpha} f(\tau)\, d\tau\ , \qquad t \in J\ .$$

Since there exist $M \ge 1$ and $\omega \in \mathbb{R}$ such that $A_\alpha \in \mathcal{G}(E_\alpha, M, \omega)$,

$$\|u(t)\|_\alpha \le \int_0^t Me^{\omega(t-\tau)}\, \|f(\tau)\|_\alpha\, d\tau\ , \qquad t \in \mathbb{R}\ .$$

Thus we easily deduce from (III.4.2.20) that $u \in L_{p,\mathrm{loc}}(J, E_\alpha)$. Hence, letting $u^\sharp$ satisfy (2.8.7),

$$\begin{aligned}\int_J \Big\langle (-\partial + A_{-\alpha}^\sharp)u^\sharp(t), \int_0^t e^{-(t-\tau)A_\alpha} f(\tau)\, d\tau \Big\rangle\, dt \\ &= \int_J \int_0^t \langle (-\partial + A_{-\alpha}^\sharp)u^\sharp(t), e^{-(t-\tau)A_\alpha} f(\tau)\rangle\, d\tau\, dt \\ &= \int_0^T \int_\tau^T \langle (-\partial + A_{-\alpha}^\sharp)u^\sharp(t), e^{-(t-\tau)A_\alpha} f(\tau)\rangle\, dt\, d\tau\ ,\end{aligned}$$

where $T \in \mathring{J}$ is chosen such that $\mathrm{supp}(u^\sharp) \subset [0,T)$. Similarly as above, we infer that the inner integral in the last expression equals

$$\int_\tau^T -\big\langle \partial_t \big[e^{-(t-\tau)A_{-\alpha}^\sharp} u^\sharp(t)\big], f(\tau)\big\rangle\, dt = \big\langle u^\sharp(\tau), f(\tau)\big\rangle$$

for $0 \leq \tau \leq T$. This shows that

$$\int_J \langle (-\partial + A_{-\alpha}^\sharp)u^\sharp(t), u(t)\rangle\, dt = \int_J \langle u^\sharp(t), f(t)\rangle\, dt\ ,$$

so that u is a weak $L_{p,\mathrm{loc}}(J, E_\alpha)$-solution in this case as well. ■

2.8.4 Corollary *Let E be a Banach space and $A \in \mathcal{G}(E)$. Fix $\mu \in \rho(-A)$, denote by E_{-1} the extrapolation space of E generated by $\mu + A$, and let A_{-1} be the closure of A in E_{-1}. Then $A_{-1} \in \mathcal{G}(E_{-1})$. Suppose that*

$$(x, f) \in E \times C(J, E)\ .$$

Then the mild solution u, defined by

$$u(t) := e^{-tA}x + \int_0^t e^{-(t-\tau)A} f(\tau)\, d\tau\ , \qquad t \in J\ ,$$

of the Cauchy problem

$$\dot{v} + Av = f(t)\ , \quad t \in \dot{J}\ , \qquad v(0) = x \tag{2.8.8}$$

in E is the unique strict solution of the extrapolated Cauchy problem

$$w + A_{-1}w = f(t)\ , \quad t \in \dot{J}\ , \qquad v(0) = x$$

in E_{-1}. If E is reflexive, it is also the unique weak $L_{p,\mathrm{loc}}(J, E)$-solution of (2.8.8) *for each $p \in [1, \infty]$.*

The assumption that E be reflexive is not indispensable. It has been imposed to guarantee that $A^\sharp$ is densely defined. If E is not reflexive, $E^\sharp$ can be replaced by $E^{\odot} := \mathrm{cl}_{E'}\big[\mathrm{dom}(A')\big]$ and $A^{\sharp}$ by the $E^{\odot}$-realization $A^{\odot}$ of $A^{\sharp}$. Then a result of Phillips [Phi55] (also cf. [HiP57, Chapter XIV]) shows that $A^{\odot} \in \mathcal{G}(E^{\odot})$. During the last few years this theory of adjoint semigroups has attracted a certain amount of interest, mainly since it has found some applications in population dynamics (e.g., [ClH$^+$87], [vN92], and the references therein).

In [vN92, Chapter 3] a duality theory for the extrapolation spaces $E_{-\alpha}$ has been developed, where $E_0 := E$ and $E_{-\alpha} := (E_{-1}, E_0)_{1-\alpha,\infty}$, $0 < \alpha < 1$. It is assumed that $A \in \mathcal{G}(E)$, but E need not be reflexive. In fact, it is not even assumed that A be densely defined.

However, experience shows that the duality theory for semigroups in nonreflexive Banach spaces — although rather attractive from the abstract point of view — is quite restricted in its applicability. This stems from the fact that, in concrete situations, it is only rarely possible to identify $E^{\odot}$. (Cf. [Ama83] for a noteworthy exception; also see [AmE94] for another approach to (dual) semigroups on nonreflexive Banach spaces avoiding the need of characterizing $E^{\odot}$.) Thus it is not really meaningful to develop an abstract duality theory for time-dependent problems in nonreflexive spaces that could be a basis for the study of quasilinear problems, since it would be close to impossible to give interesting applications. For these reasons we refrained from doing so.

The fact that mild solutions and weak $L_1(J,E)$-solutions of (2.8.8) coincide is known, even without a reflexivity assumption for E (cf. [Fat83, Theorem 2.4.6]). Related results on weak solutions are given in [Bal77].

Bibliography

[Acq88] P. Acquistapace. Evolution operators and strong solutions of abstract linear parabolic equations. *Diff. Int. Equ.*, **1** (1988), 433–457.

[AcT85] P. Acquistapace, B. Terreni. On the abstract non-autonomous Cauchy problem in the case of constant domains. *Ann. Mat. Pura Appl.* (4), **140** (1985), 1–55.

[AcT86a] P. Acquistapace, B. Terreni. Linear parabolic equations in Banach spaces with variable domain but constant interpolation spaces. *Ann. Scuola Norm. Sup. Pisa, Ser. IV*, **13** (1986), 75–107.

[AcT86b] P. Acquistapace, B. Terreni. On fundamental solutions for abstract parabolic equations. In A. Favini, E. Obrecht, editors, *Differential equations in Banach spaces, Proceedings, Bologna 1985*, pages 1–11. Lecture Notes in Math. #1223, Springer Verlag, 1986.

[AcT87] P. Acquistapace, B. Terreni. A unified approach to abstract linear non-autonomous parabolic equations. *Rend. Sem. Mat. Univ. Padova*, **78** (1987), 47–107.

[Ama78] H. Amann. Invariant sets and existence theorems for semi-linear parabolic and elliptic systems. *J. Math. Anal. Appl.*, **65** (1978), 432–467.

[Ama83] H. Amann. Dual semigroups and second order linear elliptic boundary value problems. *Israel J. Math.*, **45** (1983), 225–254.

[Ama86] H. Amann. Semigroups and nonlinear evolution equations. *Lin. Alg. & Appl.*, **84** (1986), 3–32.

[Ama87] H. Amann. On abstract parabolic fundamental solutions. *J. Math. Soc. Japan*, **39** (1987), 93–116.

[Ama88a] H. Amann. Dynamic theory of quasilinear parabolic equations – I. Abstract evolution equations. *Nonlin. Anal., T.M.&A.*, **12** (1988), 895–919.

[Ama88b] H. Amann. Parabolic evolution equations and nonlinear boundary conditions. *J. Diff. Equ.*, **72** (1988), 201–269.

[Ama88c] H. Amann. Parabolic evolution equations in interpolation and extrapolation spaces. *J. Funct. Anal.*, **78** (1988), 233–270.

[Ama90a] H. Amann. Dynamic theory of quasilinear parabolic equations – II. Reaction-diffusion systems. *Diff. Int. Equ.*, **3** (1990), 13–75.

[Ama90b] H. Amann. *Ordinary Differential Equations. An Introduction to Nonlinear Analysis.* W. de Gruyter, Berlin, 1990.

[Ama91] H. Amann. Highly degenerate quasilinear parabolic systems. *Ann. Scuola Norm. Sup. Pisa, Ser. IV*, **18** (1991), 135–166.

[AmE94] H. Amann, J. Escher. Strongly continuous dual semigroups. *Ann. Mat. Pura Appl.*, (1994). To appear.

[AmHS94] H. Amann, M. Hieber, G. Simonett. Bounded H_∞-calculus for elliptic operators. *Diff. Int. Equ.*, **7** (1994), 613–653.

[Ang90] S.B. Angenent. Nonlinear analytic semiflows. *Proc. Royal Soc. Edinburgh*, **115A** (1990), 91–107.

[Are94] W. Arendt. Gaussian estimates and interpolation of the spectrum in L^p. *Diff. Int. Equ.*, **7** (1994), 1153–1168.

[ArEK94] W. Arendt, O. El-Mennaoui, V. Keyantuo. Local integrated semigroups: evolution with jumps of regularity. *J. Math. Anal. Appl.*, (1994).

[Bai80] J.B. Baillon. Caractère borné de certains genérateurs de semigroupes linéaires dans les espaces de Banach. *C.R. Acad. Sc. Paris*, **290** (1980), 757–760.

[BaiC91] J.B. Baillon, Ph. Clément. Examples of unbounded imaginary powers of operators. *J. Funct. Anal.*, **100** (1991), 419–434.

[Bal77] J.M. Ball. Strongly continuous semigroups, weak solutions and the variation of constants formula. *Proc. Amer. Math. Soc.*, **63** (1977), 370–373.

[Bar76] V. Barbu. *Nonlinear Semigroups and Differential Equations in Banach Spaces.* Noordhoff, Leyden, 1976.

[BCP62] A. Benedek, A.P. Calderón, R. Panzone. Convolution operators on Banach space valued functions. *Proc. Nat. Acad. Sci. USA*, **48** (1962), 356–365.

[BenS88] C. Bennett, R. Sharpley. *Interpolation of Operators.* Academic Press, Boston, 1988.

[Ber74] S.K. Berberian. *Lectures in Functional Analysis and Operator Theory.* Springer Verlag, New York, 1974.

[BerL76] J. Bergh, J. Löfström. *Interpolation Spaces. An Introduction.* Springer Verlag, Berlin, 1976.

[Bou53] N. Bourbaki. *Topologie Général.* Hermann, Paris, 1953.

[Bour83] J. Bourgain. Some remarks on Banach spaces in which martingale differences are unconditional. *Arkiv Mat.*, **21** (1983), 163–168.

[Bur81] D.L. Burkholder. A geometrical characterization of Banach spaces in which martingale difference sequences are unconditional. *Ann. Probability*, **9** (1981), 997–1011.

[Bur83] D.L. Burkholder. A geometrical condition that implies the existence of certain singular integrals of Banach-space-valued functions. In W. Beckner, A.P. Calderón, R. Fefferman, P.W. Jones, editors, *Conference on Harmonic Analysis in Honour of Antoni Zygmund, Chicago 1981*, pages 270–286, Belmont, Cal., 1983. Wadsworth.

[ButB67] P.L. Butzer, H. Berens. *Semi-Groups of Operators and Approximation.* Springer Verlag, Berlin, 1967.

[ClH+87] Ph. Clément, H.J.A.M. Heijmans et al. *One-Parameter Semigroups.* North-Holland CWI Monograph, Amsterdam, 1987.

[ClP90] Ph. Clément, J. Prüss. Completely positive measures and Feller semigroups. *Math. Ann.*, **287** (1990), 73–105.

[CoiW77] R.R. Coifman, G. Weiss. *Transference Methods in Analysis.* Conf. Board Math. Sci., Reg. Conf. Series Math., **31**. Amer. Math. Soc., Providence, R.I., 1977.

[CoL87] T. Coulhon, D. Lamberton. Régularité L^p pour les équations d'évolution. *Séminaire d'analyse fonctionelle 1984–85, Publications mathématiques de l'Université Paris VII*, **26** (1987), 141–153.

[DaK92] D. Daners, P. Koch Medina. *Abstract Evolution Equations, Periodic Problems and Applications.* Pitman Research Notes in Math., #279, Longman Sci. & Tech., Essex, 1992.

[DaP84] G. Da Prato. Abstract differential equations and extrapolation spaces. In *Infinite-Dimensional Systems, Retzhof 1983*, pages 53–61, Berlin, 1984. Lecture Notes in Math. #1076, Springer Verlag.

[DaPG75] G. Da Prato, P. Grisvard. Sommes d'opérateurs linéaires et équations différentielles opérationelles. *J. Math. Pures Appl.*, **54** (1975), 305–387.

[DaPG79] G. Da Prato, P. Grisvard. Equations d'évolutions abstraites non linéaires de type parabolique. *Ann. Mat. Pura Appl.* (4), **120** (1979), 329–396.

[DaPG84] G. Da Prato, P. Grisvard. Maximal regularity for evolution equations by interpolation and extrapolation. *J. Funct. Anal.*, **58** (1984), 107–124.

[Dav80] E.B. Davies. *One-Parameter Semigroups.* Academic Press, London, 1980.

[deGM84] S.R. de Groot, P. Mazur. *Non-Equilibrium Thermodynamics.* Dover Publications Inc., New York, 1984.

[deL87] R. deLaubenfels. A holomorphic functional calculus for unbounded operators. *Houston J. Math.*, **13** (1987), 545–548.

[deL94] R. deLaubenfels. *Existence Families, Functional Calculus and Evolution Equations.* Lecture Notes in Math. #1570, Springer Verlag, Berlin, 1994.

[DeS88] W. Desch, W. Schappacher. Some perturbation results for analytic semigroups. *Math. Ann.*, **281** (1988), 157–162.

[DeSi64] L. De Simon. Un'applicazione della teoria degli integrali singolari allo studio delle equazioni differenziali astratte del primo ordine. *Rend. Sem. Mat. Univ. Padova*, **34** (1964), 205–232.

[Dor93a] G. Dore. H^∞ functional calculus on real interpolation spaces. 1993, Preprint.

[Dor93b] G. Dore. L^p regularity for abstract differential equations. In H. Komatsu, editor, *Functional Analysis and Related Topics, 1991*, pages 25–38, Berlin, 1993. Proceedings Kyoto 1991, Lecture Notes in Math. #1540, Springer Verlag.

[DorF87] G. Dore, A. Favini. On the equivalence of certain interpolation methods. *Boll. UMI* (7), **1-B** (1987), 1227–1238.

[DorV87] G. Dore, A. Venni. On the closedness of the sum of two closed operators. *Math. Z.*, **196** (1987), 189–201.

[DorV90] G. Dore, A. Venni. Some results about complex powers of closed operators. *J. Funct. Anal.*, **149** (1990), 124–136.

[DuS57] N. Dunford, J.T. Schwartz. *Linear Operators. Part* I: *General Theory.* Interscience, New York, 1957.

[EbG91] B. Eberhardt, G. Greiner. Baillon's theorem on maximal regularity. *Tübinger Berichte zur Funkt. Anal.*, **1** (1991,92), 11–16.

[EdE87] D.E. Edmunds, W.D. Evans. *Spectral Theory and Differential Operators.* Oxford Univ. Press, Oxford, 1987.

[Edw65] R.E. Edwards. *Functional Analysis.* Holt, Rinehart & Winston, New York, 1965.

[Est84] J. Esterle. Mittag-Leffler methods in the theory of Banach algebras and a new approach to Michael's theorem. *Contemporary Math., Amer. Math. Soc.*, **32** (1984), 107–129.

[Fat83] H.O. Fattorini. *The Cauchy Problem.* Addison-Wesley, Reading, Mass., 1983.

[Fol84] G.B. Folland. *Real Analysis.* Wiley, New York, 1984.

[Fra79] L.E. Fraenkel. On regularity of the boundary in the theory of Sobolev spaces. *Proc. London Math. Soc.*, **(3) 39** (1979), 385–427.

[Fri69] A. Friedman. *Partial Differential Equations.* Holt, Rinehart & Winston, New York, 1969.

[Gia83] M. Giaquinta. *Multiple Integrals in the Calculus of Variations and Nonlinear Elliptic Systems.* Princeton Univ. Press, Princeton, N.J., 1983.

[GiGS91] M. Giga, Y. Giga, H. Sohr. L^p estimates for abstract linear parabolic equations. *Proc. Japan Acad., Ser. A*, **67** (1991), 197–202.

[GiS91] Y. Giga, H. Sohr. Abstract L^p estimates for the Cauchy problem with applications to the Navier-Stokes equations in exterior domains. *J. Funct. Anal.*, **102** (1991), 72–94.

[Golb66] S. Goldberg. *Unbounded Linear Operators.* McGraw-Hill, New York, 1966.

[Gols85] J.A. Goldstein. *Semigroups of Linear Operators and Applications.* Oxford Univ. Press, Oxford, 1985.

[Gri66] P. Grisvard. Commutativité de deux foncteurs d'interpolation et applications. *J. Math. Pures Appl.*, **45** (1966), 143–290.

[Gui93] D. Guidetti. On elliptic systems in L^1. *Osaka Math. J.*, **30** (1993), 397–429.

[Hen81] D. Henry. *Geometric Theory of Semilinear Parabolic Equations.* Lecture Notes in Math. #840, Springer Verlag, Berlin, 1981.

[HiP57] E. Hille, R.S. Phillips. *Functional Analysis and Semi-Groups.* Amer. Math. Soc., Providence, R.I., 1957.

[Hor66] J. Horváth. *Topological Vector Spaces and Distributions*, volume I. Addison-Wesley, Reading, Mass., 1966.

[Hör83] L. Hörmander. *The Analysis of Linear Partial Differential Operators,* I–IV. Springer Verlag, Berlin, 1983, 1985.

[Jar81] H. Jarchow. *Locally Convex Spaces.* Teubner, Stuttgart, 1981.

[JM91] B. Jawerth, M. Milman. *Extrapolation Theory with Applications.* Memoirs AMS, Vol. 89, #440, 1991.

[Kat59] T. Kato. Remarks on pseudo-resolvents and infinitesimal generators of semi-groups. *Proc. Japan Acad.*, **35** (1959), 467–468.

[Kat61] T. Kato. Abstract evolution equations of parabolic type in Banach and Hilbert spaces. *Nagoya Math. J.*, **19** (1961), 93–125.

[Kat62] T. Kato. Fractional powers of dissipative operators II. *J. Math. Soc. Japan*, **14** (1962), 242–248.

[Kat66] T. Kato. *Perturbation Theory for Linear Operators.* Springer Verlag, New York, 1966.

[KaT62] T. Kato, H. Tanabe. On the abstract evolution equation. *Osaka Math. J.*, **14** (1962), 107–133.

[Kom66] H. Komatsu. Fractional powers of operators (I–VI). *Pacific J. Math.*, (I) **19** (1966), 285–346; (II) **21** (1967), 89–111. *J. Math. Soc. Japan*, (III) **21** (1969), 205–220; (IV) **21** (1969), 221–228. *J. Fac. Sci. Univ. Tokyo, Sect. IA Math.*, (V) **17** (1970), 373–396; (VI) **19** (1972), 1–63.

[Kōm67] Y. Kōmura. Nonlinear semigroups in Hilbert spaces. *J. Math. Soc. Japan*, **19** (1967), 493–507.

[KraZ76] M.A. Krasnosel'skii, P.P. Zabreiko, E.I. Pustylnik, P.E. Sobolevskii. *Integral Operators in Spaces of Summable Functions.* Noordhoff, Leyden, 1976.

[Kre60] S.G. Kreĭn. On the concept of a normal scale of spaces. *Soviet Math. (Doklady)*, **1** (1960), 586–589.

[Kre72] S.G. Kreĭn. *Linear Differential Equations in Banach Space.* Amer. Math. Soc., Providence, R.I., 1972.

[KrP66] S.G. Kreĭn, J.U. Petunin. Scales of Banach spaces. *Russian Math. Surveys*, **21(2)** (1966), 85–159.

[KrPS82] S.G. Kreĭn, J.U. Petunin, E.M. Semenov. *Interpolation of Linear Operators.* Amer. Math. Soc., Transl. Math. Monographs, Providence, R.I., 1982.

[Lam87] D. Lamberton. Equations d'évolution linéaires associeés à des semi-groupes de contraction dans les espaces L^p. *J. Funct. Anal.*, **71** (1987), 252–262.

[Liu89] Gui-Zhang Liu. *Evolution Equations and Scales of Banach Spaces.* PhD thesis, Univ. Eindhoven, Holland, 1989.

[Lun85] A. Lunardi. Interpolation spaces between domains of elliptic operators and spaces of continuous functions with applications to nonlinear parabolic equations. *Math. Nachr.*, **121** (1985), 295–318.

[Lun87] A. Lunardi. On the local dynamical system associated to a fully nonlinear abstract parabolic equation. In V. Lakshmikantam, editor, *Nonl. Analysis and Appl.*, pages 319–326, New York, 1987. M. Dekker.

[Lun95] A. Lunardi. *Analytic Semigroups and Optimal Regularity in Parabolic Equations.* Birkhäuser, Basel, 1995.

[McC84] T. McConnell. On Fourier multiplier transformations of Banach-valued functions. *Trans. Amer. Math. Soc*, **285** (1984), 739–757.

[McI86] A. McIntosh. Operators which have an H_∞-calculus. In B. Jefferies, A. McIntosh, W. Ricker, editors, *miniconference on operator theory and partial differential equations. Proc. Center Math. Anal. A.N.U.*, **14** (1986), 210–231.

[Mie87] A. Mielke. Über maximale L^p-Regularität für Differentialgleichungen in Banach- und Hilbert-Räumen. *Math. Ann.*, **277** (1987), 121–133.

[Mil94] M. Milman. *Extrapolation and Optimal Decompositions with Applications to Analysis.* Lecture Notes in Math. #1580, Springer Verlag, Heidelberg, 1994.

[MS90] M. Milman, T. Schonbek. A note on commutation properties for interpolation spaces. In C. Sadovsky, editor, *Analysis and Partial Differential Equations*, pages 239–248, New York, 1990. M. Dekker.

[Nag83] R. Nagel. Sobolev spaces and semigroups. *Semesterbericht Funktionalanalysis Tübingen*, (1983), 1–20.

[Nag89] R. Nagel. Towards a matrix theory for unbounded operator matrices. *Math. Z.*, **201** (1989), 57–68.

[Nar85] R. Narasimhan. *Complex Analysis in One Variable.* Birkhäuser, Basel, 1985.

[Paz83] A. Pazy. *Semigroups of Linear Operators and Applications to Partial Differential Equations.* Springer Verlag, New York, 1983.

[Pet83] B.E. Petersen. *Introduction to the Fourier Transform and Pseudo-Differential Operators.* Pitman, Boston, 1983.

[Phi55] R.S. Phillips. The adjoint semigroup. *Pacific J. Math.*, **5** (1955), 269–283.

[Prü88] J. Prüss. The Hilbert transform. Lecture Notes, Univ. Delft, 1988.

[PrüS90] J. Prüss, H. Sohr. On operators with bounded imaginary powers in Banach spaces. *Math. Z.*, **203** (1990), 429–452.

[PrüS93] J. Prüss, H. Sohr. Imaginary powers of elliptic second order differential operators in L^p-spaces. *Hiroshima Math. J.*, **23** (1993), 161–192.

[Rie27] M. Riesz. Sur les fonctions conjugées. *Math. Z.*, **27** (1927), 218–244.

[Rub86] J.L. Rubio de Francia. Martingale and integral transforms of Banach space valued functions. In J. Bastero, M. San Miguel, editors, *Probability and Banach Spaces*, pages 195–222. Lecture Notes in Math. #1221, Springer Verlag, Berlin, 1986.

[Sch71] H.H. Schaefer. *Topological Vector Spaces.* Springer Verlag, New York, 1971.

[Schw57a] L. Schwartz. *Lectures on Mixed Problems in Partial Differential Equations.* Tata Institute, Bombay, 1957.

[Schw57b] L. Schwartz. Théorie des distributions à valeurs vectorielles. *Ann. Inst. Fourier*, (chap. I) **7** (1957), 1–141; (chap. II) **8** (1958), 1–207.

[Schw65] L. Schwartz. *Méthodes Mathématiques pour les Sciences Physiques.* Hermann, Paris, 1965.

[Schw66] L. Schwartz. *Théorie des Distributions.* Hermann, Paris, 1966.

[See71] R.T. Seeley. Norms and domains of the complex powers A_B^z. *Amer. J. Math.*, **93** (1971), 299–309.

[Sim92] G. Simonett. *Zentrumsmannigfaltigkeiten für quasilineare parabolische Gleichungen.* PhD thesis, Univ. Zürich, Switzerland, 1992, and Institut für Angewandte Analysis und Stochastik, Report Nr. 2, Berlin, 1992.

[Sob61] P.E. Sobolevskii. Parabolic equations in a Banach space with an unbounded variable operator, a fractional power of which has a constant domain of definition. *Soviet Math. (Doklady)*, **2** (1961), 545–548.

[Sob64] P.E. Sobolevskii. Coerciveness inequalities for abstract parabolic equations. *Soviet Math. (Doklady)*, **5** (1964), 894–897.

[Sob66] P.E. Sobolevskii. Equations of parabolic type in a Banach space. *Amer. Math. Soc. Transl., Ser. 2*, **49** (1966), 1–62.

[Sob75] P.E. Sobolevskii. Fractional powers of coercively positive sums of operators. *Dokl. Akad. Nauk.*, **225** (1975), 1638–1641.

[Sob89] P.E. Sobolevskii. Elliptic and parabolic problems in the space C. *J. Math. Anal. Appl.*, **142** (1989), 317–324.

[Ste70a] E.M. Stein. *Singular Integrals and Differentiability Properties of Functions.* Princeton Univ. Press, Princeton, N.J., 1970.

[Ste70b] E.M. Stein. *Topics in Harmonic Analysis.* Ann. Math. Studies, **63**. Princeton Univ. Press, Princeton, N.J., 1970.

[Tan60] H. Tanabe. On the equation of evolution in a Banach space. *Osaka Math. J.*, **12** (1960), 363–376.

[Tan79] H. Tanabe. *Equations of Evolution.* Pitman, London, 1979.

[Tit48] E.C. Titchmarsh. *Introduction to the Theory of Fourier Integrals.* Clarendon Press, Oxford, 1948.

[Tre67] F. Treves. *Topologic Vector Spaces, Distributions and Kernels.* Academic Press, New York, 1967.

[Tri78] H. Triebel. *Interpolation Theory, Function Spaces, Differential Operators.* North Holland, Amsterdam, 1978.

[Ven93] A. Venni. A counterexample concerning imaginary powers of linear operators. In H. Komatsu, editor, *Functional Analysis and Related Topics, 1991*, pages 381–387, Berlin, 1993. Proceedings Kyoto 1991, Lecture Notes in Math. #1540, Springer Verlag.

[vN92] J. van Neerven. *The Adjoint of a Semigroup of Linear Operators.* Lecture Notes in Math. #1529, Springer Verlag, Heidelberg, 1992.

[vW82] W. von Wahl. The equation $u'+A(t)u = f$ in a Hilbert space and L^p-estimates for parabolic equations. *J. London Math. Soc. (2)*, **25** (1982), 483–497.

[vW85] W. von Wahl. *The Equations of Navier-Stokes and Abstract Parabolic Equations.* Vieweg & Sohn, Braunschweig, 1985.

[Wal86] Th. Walther. *Abstrakte Sobolev-Räume und ihre Anwendungen auf die Störungstheorie für Generatoren von C_0-Halbgruppen.* PhD thesis, Univ. Tübingen, Germany, 1986.

[Yag76] A. Yagi. On the abstract evolution equation in Banach spaces. *J. Math. Soc. Japan*, **28** (1976), 290–303.

[Yag77] A. Yagi. On the abstract evolution equation of parabolic type. *Osaka Math. J.*, **14** (1977), 557–568.

[Yag88] A. Yagi. Fractional powers of operators and evolution equations of parabolic type. *Proc. Japan Acad., Ser. A*, **64** (1988), 227–230.

[Yag89] A. Yagi. Parabolic evolution equations in which the coefficients are the generators of infinitely differentiable semigroups. *Funkcial. Ekvac.*, **32** (1989), 107–124.

[Yag90] A. Yagi. Parabolic evolution equations in which the coefficients are the generators of infinitely differentiable semigroups. ii. *Funkcial. Ekvac.*, **33** (1990), 139–150.

[Yos65] K. Yosida. *Functional Analysis.* Springer Verlag, Berlin, 1965.

[Zim89] F. Zimmermann. On vector valued Fourier multiplier theorems. *Stud. Math.*, **93** (1989), 201–222.

List of Symbols

Abbreviations

TVS 2
LCS 2
OBS 84
OLCS 84
OTVS 84
BVP 237
IBVP 237

Special Symbols

$x \vee y$ 1
$x \wedge y$ 1
x^+ 1
x^- 1
$\dot{M}$ 2
1- 40
2- 41
$[s]$ 143

Sets

$\mathbb{N}$ 1
$\bar{\mathbb{N}}$ 1
$\bar{\mathbb{R}}$ 1
$\mathbb{K}$ 2
$[\ldots]$ 1
$X \backslash A$ 1
A^c 1
Y^X 1
Σ_ϑ 15
S_ϑ 166
J_Δ 45
J_Δ^* 45

Maps

id_X 1
1_X 1
$i \colon X \hookrightarrow Y$ 1
χ_A 1
$\mathbf{1}$ 1
$\mathbf{1}_X$ 1
dom 6
im 6
$\psi\varphi(t,s) := \psi(t)\varphi(t,s)$ 45
$\varphi\psi(t,s) := \varphi(t,s)\psi(s)$ 45
supp 129
$\otimes$ 300

Topological Concepts

$\mathbb{B}(M,\varepsilon)$ 2
$\mathbb{B}(x,\varepsilon)$ 2
$\mathbb{B}$ 4
$\mathbb{B}^n$ 4
$\mathbb{B}_X$ 4
$\mathring{B}$ 2
int 2
int_X 2
$\overline{A}$ 2
cl 2
cl_X 2
$\hookrightarrow$ 1
$\overset{d}{\hookrightarrow}$ 1
$\overset{d}{\subset}$ 1
$\subset\subset$ 2
$\lhook\joinrel\hookrightarrow$ 3

Topological Vector Spaces

$\langle \cdot, \cdot \rangle$ 4
$\langle \xi, x \rangle$ 130
$X_{\mathbb{C}}$ 4
X' 4
$\sigma(X, X')$ 4
X_w 4
$\rightharpoonup$ 4
$E^\sharp$ 268
$\sigma(X', X)$ 4
X'_{w^*} 4

$\overset{w^*}{\rightharpoonup}$ 4
$\kappa\colon X \to X''$ 271
ϑ_E 285

Linear Operators
$T_{\mathbb{C}}$ 4
$D(A)$ 6
$\rho(A)$ 7
$\sigma(A)$ 7
$s(A)$ 17
$A \geq 0$ 33
$A \geq \alpha$ 33
$A^\sharp$ 268
$[A, B]$ 173
r^c 26
type 34
$h \star k$ 49
τ_a 131
λ_s 93
σ_α 132
A_α 275

Spaces of Linear Operators
$\mathrm{Hom}(X, Y)$ 2
$\mathrm{End}(X)$ 2
$\mathcal{L}(X, Y)$ 2
$\mathcal{L}(X)$ 2
$\mathcal{L}_s(X, Y)$ 3
$\mathcal{L}^+(E, F)$ 85
$\mathcal{L}\mathrm{is}(X, Y)$ 2
$\mathcal{L}\mathrm{aut}(X)$ 2
$\mathcal{GL}(X)$ 2
$\mathcal{K}(X, Y)$ 3
$\mathcal{K}(X)$ 3
$\mathcal{L}(E_1, \ldots, E_n; F)$ 283
$\mathcal{L}^n(E, F)$ 283
$\mathcal{C}(X, Y)$ 6
$\mathcal{C}(X)$ 6
$\mathcal{G}(E)$ 34
$\mathcal{G}(E, M, \sigma)$ 34
$\mathcal{H}(E_1, E_0)$ 11
$\mathcal{H}(E_1, E_0, \kappa, \omega)$ 11
$\mathcal{P}(E)$ 147
$\mathcal{P}(K, \vartheta)$ 166
$\mathcal{P}(\vartheta)$ 166
$\mathcal{P}_K$ 147
$\mathcal{BIP}$ 164
$\mathfrak{A}_M$ 211
$\mathfrak{B}(E_1, E_0, F)$ 234
$\mathfrak{B}(E_1, E_0, F, \kappa, \omega)$ 233
$\mathfrak{B}(E_1, E_0, F, \kappa, \omega, \beta)$ 233
$\mathfrak{B}(E_1, E_0, F; \beta)$ 233
$\mathfrak{K}(E, F, \alpha)$ 48
$\mathfrak{K}(E, \alpha)$ 48
$\mathfrak{K}_\infty(E, F, \alpha)$ 49

Pointwise Defined Functions
C 1
C_0 130
C^ρ 40
C^k 129
B 40
BC 40
BUC 40
BUC^ρ 40
$BC^\rho_{\rho,\mu}$ 99
$BUC^\rho_{\rho,\mu}$ 99
$S^{1+\rho}_{\rho,\mu}$ 101
$\mathcal{D}$ 88
$\mathcal{D}(\overline{X})$ 303
$\mathcal{E}$ 129
$\mathcal{S}$ 129
$\mathcal{O}_M$ 130

Integrable Functions
$\mathcal{L}_p$ 89
$\mathcal{L}_{p,\mathrm{loc}}$ 89
L_p 90
$L_{p,\mathrm{loc}}$ 90
W^k_p 89
$W^k_{p,\mathrm{loc}}$ 89

Distributions
$\langle \cdot, \cdot \rangle := \langle \cdot, \cdot \rangle_{\mathcal{D}}$ 89
$\mathcal{D}'$ 88
$\mathcal{E}'$ 129
$\mathcal{S}'$ 129
e_X 302
r_X 302
$\mathrm{PV}(1/t)$ 135

Differential Operators
∂ 41
∂^α 88
D_j 131
∂_+ 194

d_+ 194
$A_{\mathcal{B}}$ 235

Fourier Transforms, Multipliers, and Convolutions

$\widehat{u}$ 130
$u * \varphi$ 132
φ_ε 133
$\mathcal{F}$ 130
M_p 143
$\mathcal{M}_M$ 143
$m(D)$ 142

Norms and Seminorms

$\|\cdot\|_A$ 6
$\|\cdot\|_p$ 88
$\|\cdot\|_{p,Y}$ 88
$\|\cdot\|_{k,p}$ 89
$\|\cdot\|_{C^\rho}$ 40
$\|\cdot\|_{C_\mu}$ 98
$\|\cdot\|_{C^\rho_{\rho,\mu}}$ 99
$\|\cdot\|_{(\alpha),T}$ 48
$[\cdot]_\rho$ 40
$[\cdot]^*_\rho$ 99
$[\![\cdot]\!]_{\rho,\mu}$ 99
$p_{m,K}$ 129
$q_{k,m}$ 129

Interpolation Functors

$[\cdot,\cdot]_\theta$ 29
$(\cdot,\cdot)_{\theta,q}$ 28
$(\cdot,\cdot)^0_{\theta,\infty}$ 29
$(\cdot,\cdot)^\sharp_\theta$ 282

Index

approximate identity, 133
automorphism
    toplinear, 2

ball, 2
Banach couple
    densely injected, 11, 29
Banach scale, 251
    – satisfying interpolation inequalities, 255
    compactly injected, 254
    continuous, 251
    densely injected, 251
    discrete, 251
    dual, 268
    dual interpolation-extrapolation, 282
    equivalent, 255
    extrapolated, 263
    extrapolated fractional power, 266
    fractional power, 258
    interpolation-extrapolation, 275
    isomorphic, 255
    one-sided, 251
    reflexive, 255
    two-sided, 251
BVP, 237
    linear, 237
    regularly elliptic, 238

commutator, 173
compact
    collectively, 20
compactly contained, 2
completion, 261
complexification, 4
cone, 83
    dual, 85
    positive, 84
    proper, 83
    total, 85
continuous
    locally absolutely, 89
convex, 142
convolution, 132
coretraction, 26

derivative
    Dini, 194
dilation, 132
distribution
    – with compact support, 129
    regular-valued, 88
distributions
    temperate, 129
domain
    special Lipschitz, 303
    with minimally smooth boundary, 303
dual inner product, 285
duality
    – pairing, 4
    Riesz – mapping, 285

equation
    linear evolution, 43
    Volterra, 51
extension
    regular, 92
extrapolated power scale, 263

formula
    generalized Green's, 307
    variation of constants, 46
    variation-of-constants – for a final value problem, 306
Fourier multiplier, 142
fractional power, 32
fractional power scale, 258
    extrapolated, 266
function
    – vanishing at infinity, 130

absolutely continuous, 90
characteristic, 1
Heaviside, 168
slowly increasing, 130
functor
complex interpolation, 28
dual interpolation , 282
real interpolation, 28

graph norm, 6
Green's formula
generalized, 307
group
– of dilations, 132
– of translations, 131
general linear, 2
growth bound, 34

Hilbert scale, 255
Hilbert transform, 135
truncated, 135

IBVP, 237
inequality
generalized Gronwall, 52
interpolation, 26
moment, 261
Young – for convolutions, 133
Young's, 26
injection
canonical, 271
compact, 3
continuous, 1
natural, 1
interpolation
K-functional, 28
– couple, 24
– functor, 25
– inequality, 26
admissible – functor, 35
complex – functor, 28
continuous – functor, 29
dual – functor, 282
real – functor, 28
interpolation-extrapolation scale, 275
dual, 282
invariant
continuously translation, 93
translation, 93
isomorphism
toplinear, 2

kernel
resolvent, 51

lemma
Riemann-Lebesgues, 130

map
n-linear, 283
compact, 3
continuous, 1
continuous linear, 2
real linear, 5
trace, 92
martingale, 141
maximal regularity
pair of, 94
property of, 94
mollifier, 133
moment inequality, 261
multiplication, 42
continuous – algebra, 143
point-wise, 42
multiplier
Fourier, 142

neighborhood, 2
norm
euclidean, 4
graph, 6
uniform operator, 3

operator
– of type (K, ϑ), 166
evolution – for a final value problem, 305
extension, 302
form, 284
parabolic evolution, 45
pseudodifferential, 142
restriction, 302
order, 84
natural, 84
natural dual, 313

parabolic, 43
positive, 85
positive definite, 33
power
fractional, 149
power scale, 256
two-sided, 272

preorder, 84
linear, 84
natural, 85, 312
principal value, 135
problem
boundary value, 237
dual final value, 309
extrapolated Cauchy, 295
extrapolated dual final value, 309
final value, 305
initial-boundary value, 237
linear Cauchy, 43
property
almost reiteration, 36, 277
extension, 302
reiteration, 278

realization
E_α-, 275, 288
X-, 7
reflection, 131
reflexive, 271
resolvent commuting, 174
resolvent positive, 85
retract, 26
retraction, 26

semigroup, 85
– of left translations, 93
analytic, 11, 46
seminorm, 3
equivalent, 3
stronger, 3
weaker, 3
semireflexive, 271
solution
generalized, 91
mild, 47, 315
mild – of a final value problem, 306
strict, 43
weak $L_p(J, E_\beta)$-, 308
space
– of E-valued distributions, 88
– of test functions, 88
canonical vector – structure, 2
dual, 4
extrapolation – generated by A, 262
intermediate, 24
interpolation, 25
interpolation – of exponent θ, 25
locally convex, 2
Montel, 88
ordered Banach, 84
ordered Fréchet, 84
ordered locally convex, 84
ordered topological vector, 84
ordered vector, 84
real subspace, 4
Schwartz, 129
singular Hölder, 98
Sobolev, 89
topological vector, 2
trace, 92
UMD, 141
vector, 2
spectral bound, 17
subspace
regularity, 45
support, 129
symbol, 142

tensor product, 300
theorem
Calderón-Zygmund decomposition, 138
convolution, 134
Dore-Venni, 177
Fourier inversion, 131
Fourier-Schwartz, 131
Mikhlin multiplier, 143
Plancherel, 134
reiteration, 31
topology
w^* – of X', 4
bounded convergence, 3
simple convergence, 3
strong – of X', 4
strong operator, 3
uniform operator, 3
weak, 4
translation
left, 93
right, 131
type, 34
operator of positive, 147

Yosida approximation, 75

Monographs in Mathematics

The foundations of this outstanding book series were laid in 1944. Until the end of the 1970s, a total of 77 volumes appeared, including works of such distinguished mathematicians as Carathéodory, Nevanlinna and Shafarevich, to name a few. The series came to its name and present appearance in the 1980s. According to its well-established tradition, only monographs of excellent quality will be published in this collection. Comprehensive, in-depth treatments of areas of current interest are presented to a readership ranging from graduate students to professional mathematicians. Concrete examples and applications both within and beyond the immediate domain of mathematics illustrate the import and consequences of the theory under discussion.

Published in the series since 1983

Volume 78 **H. Triebel, Theory of Function Spaces I**
1983, 284 pages, hardcover, ISBN 3-7643-1381-1.

Volume 79 **G.M. Henkin /J. Leiterer, Theory of Functions on Complex Manifolds**
1984, 228 pages, hardcover, ISBN 3-7643-1477-X.

Volume 80 **E. Giusti, Minimal Surfaces and Functions of Bounded Variation**
1984, 240 pages, hardcover, ISBN 3-7643-3153-4.

Volume 81 **R.J. Zimmer, Ergodic Theory and Semisimple Groups**
1984, 210 pages, hardcover, ISBN 3-7643-3184-4.

Volume 82 **V.I. Arnold / S.M. Gusein-Zade / A.N. Varchenko, Singularities of Differentiable Maps – Vol. I**
1985, 392 pages, hardcover, ISBN 3-7643-3187-9.

Volume 83 **V.I. Arnold / S.M. Gusein-Zade / A.N. Varchenko, Singularities of Differentiable Maps – Vol. II**
1988, 500 pages, hardcover, ISBN 3-7643-3185-2.

Volume 84 **H. Triebel, Theory of Function Spaces II**
1992, 380 pages, hardcover, ISBN 3-7643-2639-5.

Volume 85 **K.R. Parthasarathy, An Introduction to Quantum Stochastic Calculus**
1992, 300 pages, hardcover, ISBN 3-7643-2697-2.

Volume 86 **M. Nagasawa, Schrödinger Equations and Diffusion Theory**
1993, 332 pages, hardcover, ISBN 3-7643-2875-4.

Volume 87 **J. Prüss, Evolutionary Integral Equations and Applications**
1993, 392 pages, hardcover, ISBN 3-7643-2876-2.

Volume 88 **R.W. Bruggeman, Families of Automorphic Forms**
1994, 328 pages, hardcover, ISBN 3-7643-5046-6.

Volume 89 **H. Amann, Linear and Quasilinear Parabolic Problems, Volume I, Abstract Linear Theory**
1995, Approx. 372 pages, hardcover, ISBN 3-7643-5114-4

Birkhäuser Advanced Texts / Basler Lehrbücher

Managing Editors:
H. Amann (Universität Zürich)
H. Kraft (Universität Basel)

This series presents, at an advanced level, introductions to some of the fields of current interest in mathematics. Starting with basic concepts, fundamental results and techniques are covered, and important applications and new developments discussed. The textbooks are suitable as an introduction for students and non-specialists, and they can also be used as background material for advanced courses and seminars.

We encourage preparation of manuscripts in TeX for delivery in camera-ready copy which leads to rapid publication, or in electronic form for interfacing with laser printers or typesetters. Proposals should be sent directly to the editors or to:
Birkhäuser Verlag, P. O. Box 133, CH–4010 Basel, Switzerland

M. Brodmann, **Algebraische Geometrie**
1989. 296 Seiten. Gebunden. ISBN 3-7643-1779-5

E.B. Vinberg, **Linear Representations of Groups**
1989. 152 pages. Hardcover. ISBN 3-7643-2288-8

K. Jacobs, **Discrete Stochastics**
1991. 296 pages. Hardcover. ISBN 3-7643-2591-7

S.G. Krantz, **H.R. Parks**, **A Primer of Real Analytic Functions**
1992. 194 pages. Hardcover. ISBN 3-7643-2768-5

L. Conlon, **Differentiable Manifolds: A First Course**
1992. 369 pages. Hardcover. *First edition, 2nd revised printing.*
ISBN 3-7643-3626-9

M. Artin, **Algebra**
1993. 723 Seiten. Gebunden. ISBN 3-7643-2927-0

H. Hofer / E. Zehnder, **Symplectic Invariants and Hamiltonian Dynamics**
1994. 342 pages. Hardcover. ISBN 3-7643-5066-0

M. Rosenblum, J. Rovnyak, **Topics in Hardy Classes and Univalent Functions**
1994. 264 pages. Hardcover. ISBN 3-7643-5111-X